Quantum Field Theory

Book 5 of Physics from Maximal Information Emanation, a seven-book physics series.

ISBN 979-8-89487-010-6

Quantum Field Theory

by

Stephen Winters-Hilt

ISBN 979-8-89487-010-6

Golden Tao Publishing
Angel Fire, NM
USA

Dedication

This book is dedicated to my family that helped on this lengthy road of discovery: Cindy, Nathaniel, Zachary, Sybil, Eric, Joshua, Teresa, Steffen, Hannah, Anders, Angelo, John and Susan.

Contents

Preface Series – Maximal Information Emergence (Books1-7) xi
Preface Book – Quantum Field Theory (#5) xxxi

1. Introduction 1

2. Quantum Field Theory with Second Quantization 5
   2.1 Background.. 5
   2.2 Electromagnetism and Second Quantization.............................. 7
   2.3 Review of Harmonic Oscillator First Quantization 9
      2.3.1 The one–dimensional harmonic oscillator 9
      2.3.2 Eigenvalues of the Hamiltonian.................................... 10
      2.3.3 Determination of the spectrum...................................... 11
      2.3.4 Eigenstates of the Hamiltonian 13
   2.4 Radiative transitions in atoms 17
   2.5 Electromagnetism with moving charges present 19
   2.6 Thomson Scattering 21

3. Canonical Quantization.. 23
   3.1 Classical Lagrangian Field Theory 23
   3.2 Quantized Lagrangian Field Theory 26
   3.3 Preview of Pion Scattering 27
   3.4 The LSZ Formalism 29
   3.5 Covariant Commutation 30
   3.6 The Reduction formula... 34
   3.7 The Dirac Field and Second Quantization............................... 37
   3.8 The Electromagnetic Field.. 40
   3.9 Scattering... 44
      3.9.1 The S-matrix expansion.. 44
      3.9.2 Wick`s theorem 44
      3.9.3 Pion Scattering 45

4. Path Integral Quantization 51
   4.1 Introduction to Path Integrals 51
   4.2 Path Integral Quantization.. 67
   4.3 Renormalization.. 86

4.3.1 $V = \varphi^2$ and $V = \varphi^3$ renormalization Pion Scattering 95
4.3.2 $V = \varphi^4$ renormalization Pion Scattering 99

4.4 Renormalization Group 117
4.4.1 Renormalization group for φ^4 119
4.5 Grassman variables 132
4.6 Electromagnetism 135
4.7 Quantum electrodynamics Electromagnetism 139
4.7.1 External Field – Purely Electric 150
4.7.2 External Field – Purely Magnetic..................................... 151
4.7.3 The Ward Identities 152
4.7.4 Alpha 158

5. Quantum Field Theory Emergence and Renormalizability 161
5.1 Quantum Field Theory Structure Emergence 161
5.2 Spontaneous Symmetry Breaking.. 163
5.3 Higgs Theory 179
5.4 Yang-Mills Quantization 188
5.5 Quantum Chromodynamics Overview.................................... 210

6. The Standard Model 215

7. Thermal Quantum Field Theory 223
7.1 Thermal Quantum Field Formulation 223
7.1.1 The Propagator and the Partition Function 223
7.1.2 The Propagator with Complexified Time 225
7.1.2.1 Direct Substitution 225
7.1.2.2 Full analyticity – Wick rotation................................ 226
7.1.3 Green's Function – Feynman Propagator 226
7.1.4 Thermal Green's Function 228
7.2 Complexification of time.. 229
7.3 Laws of Thermodynamics are recovered 230
7.3.1 Definition of Heat Bath 230
7.3.2 Entropy 230
7.3.3 First Law of Thermodynamics 231
7.3.4 Second Law of Thermodynamics..................................... 231
7.4 Wightman and Osterwalder-Schrader theorems 232
7.5 The Accelerated Observer → Thermal Quantum Field 233
7.5.1 Particle detector in 4D Minkowski.................................. 233
7.5.2 Adiabatic Vacuum.. 237
7.5.3 Kasner space-time with Kasner Vacuum.......................... 238

7.5.4 Killing Vectors 242
7.5.5 K(100) observer has 10 KV fields 243
7.5.6 Milne 246
7.5.7 Field theory in the K(100) and Milne spacetimes 248
7.6 The Bogoliubov Transform 250
7.7 Physical Distinction among Alternative Vacuum States 256
7.7.1 Flat space-time Fock Spaces 257
7.7.2 Thermal Rindler Particle Number 260

8. Quantum Field Theory on Curved Spacetime 271
8.1 Overview 271
8.2 Quantum Mechanics and QFT in Covariant Form 271
8.3 Curved space-time transforms and diagrams 286
8.4 Covariant quantization 290
8.5 QFT in CST with minimal coupling 292
8.6 Spatially flat isotropically changing metric 294
8.7 Generic Cosmological Particle Production 297
8.8 Spin-statistics 299
8.9 Explicit Fock Space Analysis 300
8.10 High Frequency black-body distribution 301
8.11 Adiabatic Vacuum 302
8.12 Adiabatic regularization 304

Appendix 311
A. The Appearances of alpha 311
A.1 Universality constant C_∞ and its relation to alpha 311
A.2 In electromagnetic coupling constant 312
A.3 A "kabbalistic" spectral-fit parameter 312
A.4 Perturbation parameter in Modern Quantum Mechanics 312
A.5 Perturbation parameter in QED 312
A.6 Perturbation parameter in Emanator Theory 313
A.7 Nonappearance of alpha in manifold physics 314

B. Math Review 317

C. Killing Vectors for Kasner (100) 325

D. Emanator Theory Synopsis 329
D.1 Introduction 329
D.1.1 Dirac used Lorentz Invariance 331
D.1.2 The Maximum Information Emanation Hypothesis 333

D.2 Synopsis of Methods and Prior Results 336
D.2.1 The Cayley Algebras 336
D.2.2 Unit-norm propagation 338
D.2.3 Chiral T-emanation 339
D.2.3.1 Exponential Map Properties 342
D.2.3.2 Alternate 137th count using Exponential Map 343
D.2.3.3 The {α, π} relation using Exponential Map 345
D.2.4 Trigintaduonion Emanation: achirality from chirality..... 346
D.2.4.1 Achiral emanation has 29* effective dimensions ... 348
D.2.4.2 Effective Deck 72 consistent with the $\{\alpha,\pi,C_\infty\}$ 349
D.2.4.3 'Edge of chaos' maximal perturbation hypothesis . 350
D.3 Implication of MIE 351
D.3.1 MIE requires a complex Hilbert Space 351
D.3.2 Time is analytic and Matter is meromorphic 351
D.3.3 Emergent Evolution and Emergent Universal Learning.. 352
D.3.4 Objective Reduction, Zero-Divisors, and Planck 352
D.3.5 Where's the geometry? 353
D.4 Results 353
D.4.1 Two-card Hand, Neuromanifold Two-step Learning 353
D.4.2 The Chiral-Extension Cayley-Family relation 354
D.4.2.1 Sedenion Assoc. extension, Axiom of Choice 355
D.4.2.2 Octonion Assoc. ext., Complex Periodic Time 355
D.4.2.3 Octonion Commutative ext., Analytic Time 356
D.4.3 Extended Standard Model from Emanator Theory 356
D.5 Conclusion 363

References 365

Index 379

Preface to Physics Series on:

Physics from Maximal Information Emanation

The Road goes ever on and on
Down from the door where it began.
Now far ahead the Road has gone,
And I must follow, if I can,
Pursuing it with eager feet,
Until it joins some larger way
Where many paths and errands meet.
And whither then? I cannot say
– **J.R.R. Tolkien, The Fellowship of the Ring**

Variation, Propagation, and Emanation

This is a seven book Physics Series that starts with Classical Mechanics (Book 1 [1]), then Classical Field Theory, such as electromagnetism (Book 2 [2]), then Manifold Dynamics, such a General Relativity (Book 3 [3]). The switch to a quantum mechanics description is given in Book 4 [4], and to a quantum field theory, quantum electrodynamics in particular, in Book 5 [5]. A 'quantum manifold theory' would be the obvious next step except it cannot be done (there is not a renormalizable Field theory for Gravitation). Instead a thermal quantum manifold theory is considered, as well as Black Hole thermodynamics in general, in Book 6 [6]. Book 7 [7] describes a new theory, Emanator Theory, that provides a deeper mathematical construct that undergirds quantum theory, much like quantum theory can be shown to provide a deeper (complexified) mathematical construct based on the classical theory.

This is a modern exposition where subtleties of chaos theory are described in Book 1, of Lorentz Invariance in Book 2, of Covariant Derivatives (General Relativity) and Gauge Covariant Derivatives (Yang-Mills Field Theory) in Book 3. Book 4 on Quantum Mechanics provides an extensive review of quantum mechanics, then considers a full self-adjoint analysis on the full general relativistic solution to the spherical shell in-fall system (a result carried over from Book 3). Book 5 considers quantum field theory basics in detail, along with alternate vacua in specific scenarios. Book 6 considers thermodynamics from the basics to the Hamiltonian thermodynamics of some Black Hole systems. Throughout, the odd recurrence of the alpha parameter is noted. In Book 7 we look to a deeper mathematical formulation from which the Quantum Path Integral formulation would result, as well as explaining the odd parameters and structures that have been discovered (such as alpha and Lorentz Invariance).

The physical description starts with the classic formulations of point particle motion. The first approach to doing this is using differential equations (Newton's 1st and 2nd Law); the second is using a variational function formulation to select the differential equation (Lagrangian variation); the third is using a variational functional formulation (Action formulation) to select the variational function formulation. Historically, it wasn't realized until much later that there are two domains for motion in many systems: non-chaotic; and chaotic.

In a description of particle motion, assuming not in a parameter domain with chaotic motion, several important limits are found to exist. Examples include: the universal constants from the aforementioned chaos phenomenon, that are still encountered in non-chaos regimes if driven "to the edge of chaos". Limits are found where scattering is defined in the asymptotic limit and perturbation theory is well-defined in the sense that it is convergent. Overall, if the evolution is described as a 'process' it is often a Martingale process, which has well-defined limits. So, we have descriptions for motion, typically reducible to an ordinary differential equation (ODE), and for which solutions (requiring limit-definitions) are typically found to exist.

The physical description then contends with field dynamics in 2D, 3D, and 4D (in Book 3 [3]). Two-dimensional ("2D") field dynamics can be described as a complex function (that maps complex numbers to complex numbers). A novelty of the 2D complex function is it also shows how to handle many types of singularities (the residue theorem), thus provides important information about fundamental structures in physics as well as fundamental mathematical techniques for solving many integrals. For the 3D field dynamics we do an analysis of the electromagnetic field in 3D. The level of coverage begins at an overview of electrostatics at the level of the graduate text Jackson [155]. Some problems from Jackson Ch's 1-3 are examined closely in developing the theory itself. For some this material (in Book 2 [2]) might provide a useful accompaniment to Jackson's text in a full course on electromagnetism (based from Jackson's text). A quick review of electrodynamics and electromagnetic wave phenomena is then given. In essence, we see many more examples of ODE problems with solutions, such as for the 3D Laplacian, usually involving separation of variables. We then review the famous transform, discovered by Lorentz in 1899 [156], that relates the electromagnetic field as seen by two observers differing by a relative velocity. With the existence of this transform, that brings in the time dimension along with the relative velocity, we effectively have a 4D theory.

From Lorentz Invariance we have, as a point transformation, rotational invariance under SO(3) or SU(2). If Lorentz Invariance is fundamental, then we should see both forms of rotation invariance, one of vector/tensor type from SO(3), and one of spinorial type from SU(2). This is the case, as gauge fields are vectorial and matter fields are spinorial. From Lorenz Invariance as a local invariance we have the Minkowski (flat) spacetime metric, which then generalizes to the Riemannian metric (in General Relativity).

As with the point particle dynamics, for the field dynamics we have three ways to formulate the behavior: (1) differential equation; (2) function variation (on Lagrangian); and (3) functional variation (on the Action). We will see similar limit phenomena as before, but also new phenomena, including (i) inevitable black holes singularity formation (the Penrose singularity theorem); (ii) FRW Universe formation (from homogeneity and isotropy); (iii) the black holes collapse singularity; (iv) the atomic collapse radiative 'singularity'.

Classical dynamics, thus, has two field-like formulations to describe the world: field and manifold. Such formulations can be interrelated mathematically, so what is happening is more a matter of physics emphasis and convenience. The emphasis on this difference, that appears to be no difference (mathematically), is that different physical phenomenologies are at play. Field descriptions appear to work for 'matter', where the fundamental elements are spinorial. Manifold descriptions appear to work best for geometrodynamics (General Relativity), where the fundamental elements are vectorial (or tensorial, such as the metric). Matter fields are renormalizable, thus quantizable in the standard quantum field theory formulation (to be described in Book 5 [5]), while gravitational manifolds are not renormalizable, and have constraints (weak energy condition and positive energy condition given the existence of spinor fields on the manifold).

The presentation in Books 1-3 [1-3], on 'classical' physics, is partly done to make the transition to quantum physics simple, obvious, and in some cases, trivial. Consider the functional variation (Action) formulation of the behavior (whether point-particle or field), this can be captured in integral form, as was done by D'Alembert very early [157] (then by Laplace [158]). Note the use of a large constant to effect a 'highly damped' integral for selection purposes (on variational extremum of the action). To transition to the quantum theory we also have the large constant from 1/h, and so the only difference is the introduction of a factor of 'i', to effect a 'highly oscillatory' integral for selection purposes.

After the transition to a quantum theory, for the point-particle descriptions, the classical collapse problem for atomic nuclei is eliminated. The spectral predictions have excellent agreement with theory, but there is still fine-structure in the spectra not fully explained. The theory is not relativistic and some initial corrections for this are possible (without going to a field-theory) and these indicate closer agreement and explain most of the fine-structure constant discrepancy (and reveal alpha in another place in the theory). It is shown in Book 3 [3] and Book 4 [4], that the General Relativity singularity problem, however, remains partly unresolved (for the test case of spherical dust shell collapse, done in a full General Relativity analysis, then quantized in a full self-adjoint quantization analysis [4]).

In Book 5 [5], the transition to quantum theory is continued to the field theory descriptions. A precise description/agreement of atomic nuclei is now possible with quantum electrodynamics, and within the nuclei themselves (quark confinement) with QCD. The field theories have a small set of bothersome infinities, however, which is eventually solved by renormalization. As mentioned, the quantization of manifold theories, such as General Relativity, does not appear to be possible due to non-renormalizability. Not to be deterred, in Book 6 [6] we consider a Hamiltonian description of a General Relativity system whose quantization would involve an energy spectrum based on that Hamiltonian, if we then use analytic continuation to take us to the thermal ensemble theory based on the partition function that results, we can consider the thermal quantum gravity of such systems.

This last example (from Book 6), showing a consistent thermal quantum gravity theory if we use analyticity, is part of a long sequence of successful maneuvers involving analytic continuations in different settings. What is indicated is the presence of an actual complex structure to the stated theory. There is the trivial complex structure extension mentioned above that brought us from the standard classical physics theory to the standard path integral quantum theory. But we also see actual complex structure at the component level with time complexation (that ties to thermal version of the theory by defining the partition function), and we have complex structure as the dimension-level in the form of the successfully applied dimensional regularization procedure used in the renormalization program.

As well as covering the breadth of core physics topics at both undergraduate and graduate level (for courses taken at Caltech and Oxford), including extensive presentation of problems and their solutions, the Series also examines, in specific cases, the boundaries of the physical world "from the inside" (and then later "from the outside"). To this end exploration of spherical dust collapse to form a singularity is examined in a fully general relativistic formalism, and then carried-over to a quantum minisuperspace (quantum gravity) analysis (in Books 3 and 4 [3,4]). Also examined in-depth are the topics of black hole thermodynamics and quantum field theory with alternate vacua (part of Books 5 and 6 [5,6]). The in-depth material comprises the topics covered in my PhD dissertation [9], portions of which are published [36,91,142,143].

In recent work on machine learning, that includes statistical learning on neuromanifolds [6], we find a possible new source for a foundational element for statistical mechanics (entropy) via seeking a minimal learning process/path on a neuromanifold [6]. By the time the Series reaches thermodynamics in Book 6, therefore, the foundational thermodynamics elements have all been established from the physical descriptions discovered in Books 1-5, they just haven't been put together in a comprehensive analysis that gives us the fundamental constructs of thermodynamics and statistical mechanics. That said, it would seem that thermodynamics is, thus, entirely derivative from other, truly fundamental theories. Not so, in the joining of the parts to make thermodynamics we have something greater than the sum of the parts. In the 'system' descriptions we find that emergent phenomena exist. This, at least, is unique to thermodynamics, so it is fundamental in this "sum greater than the parts' aspect.

In Book 7 (the last) of the Series, we consider the standard physical world, described by modern physics, "from the outside." In doing this we've already eliminated part of the mystery of entropy by the geometric 'neuromanifold' description. If we can understand other oddities of the standard theory, and arrive at them naturally, then we might have an even deeper dive into modern physics, testing the limits of what is possible, and see possible future developments and unifications of the theory. This is what is described in papers [126,132,148,159-162], and organized along with current results into the final Book of the series.

Efforts in the last book of the Series involve choices and concepts identified in the prior six books of the Series, and theoretical maneuvers gleaned from the most advanced courses in physics and mathematical physics taken while at Caltech (as an undergraduate and then as a graduate) and the Oxford Mathematics Institute (as a graduate), and the University of Wisconsin at Milwaukee (as a graduate).

The broad range of topics covered in the Series is, initially, similar to the Landau & Lifshitz graduate textbook series (see [22]), with a similar exposition on classical mechanics at the start of Book 1. Even with well-established classical mechanics, however, there are significant, modern, updates, such as (modern) chaos theory. In the final two books of the Series (Books 6 and 7 [6,7]) we arrive at statistical mechanics and thermodynamics, together with modern topics such as black hole thermodynamics, thermal quantum gravity, and emanator theory.

Key constants and structures of physics, their discovery from the experimental data, and their theoretical placement in the "Grand Scheme," are emphasized throughout the Series. The constant alpha, a.k.a. the fine structure constant, appears in numerous settings so special note of the occurrence of alpha will be made in each chapter. This is the case even at the outset with Book 1, due to fundamental numerical constants appearing from chaos theory. In Book 7 we see the origin of alpha, as a maximal perturbation amount, appears naturally in a formalism for maximal information 'emanation'. But maximal perturbation in what space and in what manner? In Book 7 of the series [7] we will see a possible representation of such an information entity, and its space of existence, in terms of chiral trigintaduonions.

Thus, in the end, this is an effort to tell of a journey to a special place "where many paths and errands meet", giving rise to emanator theory and an answer to the mystery of alpha. Part of this journey is equivalent to 'finding the arkenstone' (alpha) in the most unlikely of places, the trigintaduonion emanation mathematics underpinning the emanator formalism (e.g., Smaug's Lair, described in Book 7 [7]). Why I should have wandered into such an odd place (mathematically speaking), and why I should posit a deeper form of quantum propagation using hypercomplex trigintaduonions, here called emanation, is why there is such extensive background on standard topics. This extensive background even impacts the classical mechanics description via its modern chaos theory material (due to a possible relation between C_∞ and alpha). The critical role of emergent phenomena is only understood at the end, including for manifolds in geometry and neuromanifolds in statistical mechanics, and leads to a Book 6 that goes from very basic (initial thermodynamics) to very advanced (emergent phenomena). Much is made clear with emanator theory, including how reality is both fractal and emergent. At this point in the journey, as with Tolkien, this much I can say: "The Road goes ever on and on … And whither then? I cannot say".

The seven books in the Series are as follows:

Book 1. Classical Mechanics and Chaos
Book 2. Classical Field Theory
Book 3. Classical Manifold Theory
Book 4. Quantum Mechanics and the Path Integral Foundation
Book 5. Quantum Field Theory and the Standard Model
Book 6. Thermodynamics, Statistical Mechanics, and BH Entropy
Book 7. Maximum Information Emanation and Emanator Theory

Overview of Book 1

Book 1 is a modern exposition of classical mechanics, including chaos theory, and including ties to later theoretical developments as well. The exposition consists, throughout, of the presentation of interesting problems with many solved, the others left for the reader. The problems are drawn from classical mechanics and mathematics courses taken at Caltech, Oxford, and the University of Wisconsin. The courses range from undergraduate level to advanced graduate level. The courses had a rich and sophisticated selection of textbook and reference material, as you might expect, and those reference texts are, similarly, drawn on here. Those classical mechanics texts, listed by author, include: Landau and Lifshitz [163]; Goldstein [164]; Fetter & Walecka [165]; Percival & Richards [166]; Arnold (ODE) [167]; Arnold (CM) [168]; Woodhouse [169]; and Bender & Orszag [170]. Notice how the first Arnold reference and the Bender and Orszag reference involve textbooks focused on ordinary differential equations (ODEs). Likewise, an analysis of the excellent, and rapid, exposition by Landau and Lifshitz, reveals that it partly progresses through the material by going through ODEs of increasing complexity (corresponding to more complicated pendulum motion, for example, such as by adding a frictional force). This strong alignment with the underlying mathematics of ODEs is continued in this exposition, so much so that an appendix is provided for a quick review of ODEs from the applied mathematics perspective.

Particle dynamics, with and without forces, are described, with all arriving at descriptions with chaotic motion, with chaos described in the latter half of Book 1 [1]. Universally it is found that systems transitioning to chaotic behavior do so with a remarkable period-doubling process and this will be described both mathematically and with computer results. In the analysis of such dynamical systems we will find that periodic physical systems can be described in terms of repeated "mappings", e.g., classic dynamic mappings [171], and when described in this way the transition to chaos is made much more mathematically evident (as will be shown). The familiar Mandelbrot set is generated by such a repeated mapping, where it's "edge of chaos" is defined by the fractal boundary of the classic Mandelbrot image.

Properties of the classic Mandelbrot set will be relevant to the physics discussed in Book 1 and Book 7, including the property that the fractal boundary has a fractal dimension of 2 (the fractal dimension of the boundary can be between 1 and 2, to get equal to 2 is special). With the Mandelbrot set we also recover the well-studied constants associated with the universal Feigenbaum constants [133]. In the Mandelbrot set we can clearly see the fundamental constant for maximum perturbation that is at maximum antiphase (negative) with magnitude C_∞, where the same results hold for a family of basic formulations (for a variety of Lagrangian formulations, for example).

From the Lagrangian variational formulation of 'action' for particle motion we will eventually define the path integral functional variational formulation involving that same Lagrangian to arrive at a quantum description for the non-relativistic quantum particle motion (described in detail in Book 4 [4], and relativistic in Book 5 [5]). From the quantum description we arrive at the propagator formalism for describing dynamics (this exists in the classical formulation too, but typically is not used much in that context). Complex propagators will then be found to have ties to statistical mechanics and thermodynamics properties (Book 6 [6]). The ties to statistical mechanics are further emphasized when at the "edge of chaos" but with the orbit motion still confined. This may be associated with an ergodic regime, thus an equilibrium and martingale regime, the existence of which can then be used at the start of Book 6 [6] statistical mechanics and thermodynamics derivations with the existence of equilibria established at the outset. The existence of the familiar entropy measures are already indicated in the neuromanifold description (Book 3 [3]), thus, together with equilibria, the Book 6 thermodynamics description is able to begin with a well-established foundation that is not claimed by fiat, rather claimed as a direct result of what has already been determined in the theory/experiment described in the previous books of the Series.

Overview of Books 2 & 3

When moving from a theory of point particles to a theory of fields, there's not much discussion in the core physics books on fields in a general sense, it usually just directly jumps to the main field of relevance, electromagnetism. If advanced, it may also cover General Relativity, as with [172]. In what follows we will cover these topics, but we will also cover the more basic fields in 1, 2, and 3D (including fluid dynamics), as well as 4D Lorentzian Field formulations (for Special Relativity), the Gauge Field formulation (thus Yang Mills covered in a classical context), and the General Relativity geometric and gauge formulations. This establishes the foundation for the standard forces, and upon quantization (Books 4 and 5 in the Series), lays the foundation for the standard renormalizable forces (all but gravitation).

The gravitational coupling constant 'G' is a dimensionful coupling (not like with alpha in electromagnetism), and gravitation with manifold construct can be described as a gauge field construct, although not renormalizable. Gravitation, and associated geometry/manifolds, appears to relate to its own emergent structure, as will be discussed in Book 6. From the local Lorentzian geometry and Lorentzian field descriptions we also see the first of many examples where there is system information in the complexification of some parameter, here the time component. If the Lorentzian is shifted to complex time, this shifts it to being a Euclidean field, with formally well-defined convergence properties (as occurs in statistical mechanics). Complex time also shows deep connections between classical motion and associated Brownian motion (where random walk reveals pi). Thus, it should not be surprising that an emergent manifold may have complex structure such that there is also an emergent 'thermal' manifold, possibly the neuromanifold described in Book 3 and the related partition functions examined in Book 6. Just like locally flat space-time is a natural construct in General Relativity, so too are optimization "learning" steps on a neuromanifold such that relative entropy is selected as a preferred measure, and from it Shannon entropy and Boltzmann's statistical entropy. Thus, the manifold construct appearing at Book 3 has far reaching impact into the foundations of the thermodynamic and statistical mechanical theory described in Book 6.

Before we even get to the manifold/geometry complexities of General Relativity, however, we have already established much with the electromagnetic field part of the theory: (i) from 'free' electromagnetism without matter we get the speed of light c, Lorentz invariance, and from that special relativity and locally flat space-time; (ii) from electromagnetism with matter we get the dimensionless coupling constant alpha.

In going over field theories to describe matter, force fields, and radiation we first describe the classical field theories of fluid mechanics, electromagnetism, and General Relativity, with many examples shown. This is then carried over to the quantum field theory description in Book 5. A review of the core mathematical constructs employed in classical field theory and quantum field theory is given in the Appendix. Even as the mathematical physics approach grows in sophistication, we still obtain solutions via variational extrema. Thus, determining the evolution of the system from its variational optimum now becomes the focus of the effort. System 'propagation' from one time to a later time can be described by a propagator. Although a 'propagator' formulation is possible mathematically in classical mechanics and classical field theory, which are shown, this is usually not done, in favor of simpler representations for the experimental application at hand. As we move to descriptions in the quantum realm, however, the use of the propagator formalism becomes typical, and when used in the path integral formulations we arrive at a compact formulation describing both the evolution and stationary-phase solution at once.

In Book 2 the focus is on classical field theory in a fixed geometry, the main physical example is electromagnetism. In this setting alpha appears, for example, in the description of an electron-positron pair: $F = e^2/(4\pi\varepsilon a^2)$ for electron-positron distance 'a' apart, where alpha appears as the coupling constant. Later, in quantum mechanics, both modern and in the early Bohr model, we have that alpha = $[e^2/(4\pi\varepsilon)]/(c\hbar)$. The appearance of alpha in these situations is occurring in bound systems. If we examine electromagnetic interactions that are unbound, on the other hand, such as with the Lorentz Force $F = q(E \times v)$, here there arises no alpha parameter, nor with the early quantum mechanical analysis of such systems such as with Compton scattering. Thus, we see an early role for alpha, but only in bound systems, thus only in systems with (convergent) perturbative expansions in system variables.

In Book 3, classical field theory with *dynamic* geometry, i.e. General Relativity, we don't see alpha at all. Instead we see manifold constructs and the mathematics of differential geometry (and to some extent differential topology and algebraic topology). Manifold constructs are entirely encapsulated in the math background given in Book 3 and the Appendix there. An application in the area of neuromanifolds (see [6]), shows the equivalent of a geodesic path in this setting is evolution involving minimum relative entropy steps. Similar to the description of a locally flat space-time we now have a description of 'entropy' increasing/evolving according to minimum relative entropy.

General Relativity stands apart from the other force fields. All the other force fields are part of an adjoint representation of the standard model vis-à-vis the stability subgroup $U(1)xSU(2)_LxSU(3)$. The form of which is derivable from the chiral T one-sided products described in Book 7. The standard model is uniquely obtained in this process, and with no mention of General Relativity. Keep in mind, however, that the adjoint representation has operation on some space (hyperspinorial in case of simple octonion right-products, for example). The 'force' due to gravity is that due to manifold curvature, where the manifold construct is possibly emergent on the space of operation. Thus, the origin of the General Relativity force is entirely different, and it will not allow quantization like the other forces, nor will its singular solutions be resolvable via quantum physics alone, as with electromagnetism in Books 4&5, but will also need thermal physics (as will be described in Book 6).

The existence of singular General Relativity solutions, outside of specially symmetric cases (the classic Black hole solutions), wasn't firmly established until the Penrose singularity theorem [173] (awarded Nobel prize in Physics for this in 2020). Some of this material is covered in Book 3 to show how the mathematical formalism shifts to differential topology methods to describe the singularities, with examples referencing the Hawking and Ellis classic [174] and using Penrose diagrams. This, in turn, will come in handy when describing the classic FRW cosmologies with radiation and matter dominated phases (using notes from Peebles [175], Peebles won the Nobel in Physics in 2019).

The General Relativity development would be remiss if it didn't briefly delve into cosmological models, the classic FRW cosmologies in particular. With the General Relativity tools developed, cosmological results are examined, starting with the entry of the cosmological constant into the formalism (a candidate for Dark energy). Various observational data on galaxy rotations and universe simulations of galaxy cluster formation both indicate the existence of Dark matter. This, then, means we have new matter, non-interacting except gravitationally, and this is actually consistent with the latest observational data on the muon g-2 value [176], where the discrepancy between theory and experiment has grown to 4.2 standard deviations, where an extension in the Standard Model appears to be in the works. This is convenient as Emanator theory (Book 7 [7]), predicts such an extension.

We can thus arrive at field equations for electromagnetism, General Relativity, and Yang-Mills Gauge Fields (Strong and weak). We can obtain wave and vortex phenomena (as hinted in fluid dynamics). We show the classical instability for atomic matter (classical electromagnetic instability) and classical gravitational instability (leading to black hole formation with singularity). From Lagrangian formulations we can then arrive at a quantum field theory formulation (Book 5). The quantum field theory formulation completes the quantum mechanics (Book 4) cure of "non-relativistic atomic instability" with the cure of the fully relativistic atomic description of the radiative-collapse instability. Introduction of quantum field theory also leads to new instability or infinities, but these can be eliminated by renormalization for the electromagnetic and electroweak formulations, and the Yang-Mills strong formulation, but not the General Relativity (gauge) formulation. The current theoretical formulation in modern physics has one glaring gap, therefore: a quantum theory of gravitation. Perhaps this is not a missing element, however, if geometry/General Relativity is a derivative phenomenon, like the field of statistical mechanics and thermodynamics appeared as derivative phenomenon when the complexified quantum propagator gives rise to a real (quantum) partition function. The hint of a deeper emanator theory suggests emergent structures of geometry and thermodynamics are arrived at in the process of emanation, with the information emanated being that of the renormalizable quantum matter fields. In Book 7 [7] a precise mathematical meaning will be found for describing maximal information emanation.

Overview of Book 4

By 1834, with Hamilton's Principle, there was a strong foundation for what is now called classical mechanics. By 1905, with Einstein's publication on the photoelectric effect [177], the rules of classical mechanics were being superseded by the new rules of quantum mechanics. The earliest appearance of quantum mechanics, however, began with the various observations of quantization of light, starting with the strange occurrence of spectral lines for hydrogen. The hydrogen spectrum was made even stranger by a precise fit to a succinct empirical formula by Balmer in 1885 [178]. This is the beginning of an amazing period of discovery. The developments of quantum mechanics from introductory to advanced roughly follows that history.

The early phase of discovery for quantum mechanics moved into the modern quantum mechanics formalism with the discovery of Heisenberg of the successful application of matrix mechanics and the resultant uncertainty principle (1925) [179]. In 1926, Schrodinger showed that the problem of finding a diagonal Hamiltonian matrix in the Heisenberg's mechanics is equivalent to finding wavefunction solutions to his wave equation [180]. An interpretation of the wavefunction was then clarified in 1927 by Born [181]. Dirac developed a manifestly relativistic formalism for the wavefunction and wave-equation for fermionic matter (1928) [182]. An axiomatic reformulation of quantum mechanics was then given by Dirac (1930) [183], laying the foundation for much of modern quantum notation and for critical issues such as self-adjointness. Dirac then described a formulation of a quantum propagation path, with quantum propagator having the familiar phase factor involving the action, in his paper "The Lagrangian in Quantum Mechanics" in 1933 [184]. In essence, Dirac had obtained a single path, in what would eventually be generalized by Feynman to all paths with the invention of the path integral formalism (1942 & 1948) [185,186]. The equivalence of a quantum mechanical formulation in terms of path integrals and the Schrodinger formalism was shown by Feynman in 1948 [186].

In a path integral description, the quantum mixture state, semiclassical physics, and classical trajectories are all given by the stationary phase dominated component. A stationary phase solution that is dominated by a single path is typical for a classical system. Thus, variational methods are fundamental to analysis of physical systems, whether it be in the form of Lagrangian and Hamiltonian analysis, or in various equivalent integral formulations.

Feynman's discovery of the path integral formalism wasn't solely based on the prior work of Dirac (1933) [184], although by appending that paper to his PhD thesis (1946) its importance was clearly emphasized. Feynman also benefited from work going as far back as Laplace [158] for selection process based on highly oscillatory integral constructions that self-select for their stationary phase component. This branch of mathematics eventually became associated with Laplace's method of steepest descents, then to the work of Stokes and Lord Kelvin, then to the work of Erdelyi (1953) [187,188].

Feynman and others then invented quantum field theory for electromagnetism (QED) during 1946-1949 (more on this later). Extension to electroweak occurred in 1959, and to QCD in 1973, and to the "Standard Model" in 1973-1975. Thus, the impact of the path integral revolution in quantum physics was felt well into the 1970's, but this was only the beginning. At their inception path integrals were examined by Norbert Wiener, with the introduction of the Wiener Integral, for solving problems in statistical mechanics in diffusion and Brownian motion. In the 1970's this led to what is now known as "the grand synthesis" which unified quantum field theory and statistical field theory of a fluctuating field near a second-order phase transition, and where use of renormalization group methods enabled significant advances from quantum field theory to be carried over to statistical field theory.

The grand synthesis is one of many instances to come where we see analytic continuation of a constant or a parameter giving rise to familiar physics in the thermodynamic and statistical mechanics domains, showing a deeper connection (still not fully understood, see Book 7). The Schrödinger equation, for example, can be seen to be a diffusion equation with an imaginary diffusion constant. Likewise, the path integral can be seen to be an analytic continuation of the method for summing up all possible random walks.

In Book 4 we also carefully examine the closest gravitational equivalent to the hydrogenic atom (dust shell collapse). What results is an incomplete formulation due to boundary conditions, where to get the time choice you must input that time choice. No specific choice of time is indicated to avoid infall-collapse. The results, however, can show stability and consistency in a "full" thermal quantum gravity description where analyticity is employed. Success in this way, and not others, suggests possible fundamental role of analyticity and thermality (Books 6&7) and also suggests that thermal quantum gravity may 'exist' or be well-formulate-able, while quantum gravity generally might not 'exist'. These results, shown in Book 6, provide the lead-in to the Book 7 discussion on Emanator theory, where core concepts in Books 1-6 that tie to emanator theory are brought together in a new theoretical synthesis.

Overview of Book 5

In Book 5 we show quantum field theory in the gauge field representation, which clearly relates the choice of field theory to a choice of Lie algebra, which, in turn, can be related to a choice of group theory (such as U(1) and SU(3)). From this we can see that non-classical algebraic constructs are ubiquitous in quantum mechanics and quantum field theory, so a review of Group Theory and Lie Algebras is given in the Appendix, as well as a review of Grassman Algebras, and other special algebras needed in quantum mechanics and quantum field theory. Similarly, as regards choice of approach, we find that the Schrodinger and Heisenberg formulations often provide the only tractable way to get a solution for bound systems. In critical theoretical considerations, however, the path integral approach is best (as will be shown). In seeking a deeper theory, the more unified path integral (PI) approach provides important hints as to a deeper theory (see Book 7).

In Book 5 we get the highest precision result for the value of alpha, in its role as perturbation parameter. If a calculation of the electron magnetic moment parameter g-2 is performed, with all of the Feynman diagrams appropriate to expansions up to 5th order, we get a determination of alpha up to 14 digits, where 1/alpha=137.05999..... . This gives us one of the most precise measurements of alpha known. When a similar analysis is done for the muon g-2, given the much larger muon mass, particle production pairs of other particles have a measurable effect, and we are able to probe the lower masses of the standard model that are present. In doing this, in preliminary experiments, there is a discrepancy indicating more particles, e.g. the Standard Model will need to be extended (possibly with a type of 'sterile' neutrino). These missing particles could be the missing "Dark Matter". The prediction of such in Emanator Theory, and why there should be an imbalance between the left and right neutrinos (hint: maximum information transmission) is described in Book 7.

Part of the description of quantum field theory entails use of analyticity and other complex structures to encapsulate more of the physics in a complex extension to the space (or dimension). This often leads to formulations in terms of complex integration, with the choice of complex contour specified, such as with the Feynman propagator. One of the main renormalization methods, for example, is to use dimensional regularization, which entails analytically continuing expressions with dimensionality to dimensionality as a complex parameter. There is also the aforementioned shift to complex and to "Wick rotate" expressions with real time to expressions with pure complex time. In doing this the statistical mechanical partition function for the system is obtained, with well-defined summation. Thus, a connection between 'thermality' and complex structure, in the time dimension at least, is indicated.

The second part of Book 5 describes quantum field theory on curved space-time, where we arrive at an early analysis of Black Hole thermodynamics. Here we find that space-time curvature gives rise to thermality and particle production effects. Black Hole thermality was revealed in Hawking radiation [189], due to the causal boundary at the horizon. Such thermality is even seen in flat space-time (Book 5) if causal boundaries are induced, such as in the case of an accelerated observer [39].

Quantum field theory on curved space-time has one further gift, critical to the statistical mechanics formalism to follow in Book 6, and that's the spin-statistics relation. This relation is usually assumed, along with other critical notions, such as entropy, and the relation between entropy and density of states. These are all shown, with the presentation path chosen in this Physics Series, to be fundamental or derivative to the formalism already established in Books 1-5 (to prepare for Book 6).

The choice of time is related to choice of vacuum, which is related to choice of field geometry or observer motion (such as constant acceleration or expansion). If you have flat spacetime quantum field theory with a boundary, then you have thermodynamic effects (e.g., the Rindler observer). In this setting we can compare the Hawking derivation of Hawking Radiation using the Euclideanization 'trick' vs the Bogoliubov transformations of the field to the Rindler geometry from the

Minkowski geometry (if chosen as the asymptotic vacuum reference). With quantum field theory on curved space-time we also arrive at spin-statistics as mentioned, and get the final extension of the theory by way of Grassman algebras, to arrive at thermodynamically consistent Bose and Fermi statistical descriptions on quantum matter.

Overview of Book 6

Thermodynamics is the oldest of the physics disciplines (fire), with unapologetic use of phenomenological arguments and mysterious thermodynamic potentials (entropy). Obviously, thermodynamics is still prevalent today, including in its more quantified form via statistical mechanics. How is this not a failure of the mechanistic description of the universe indicated by classical mechanics and even quantum mechanics? Concepts that appeared in quantum mechanics, such as probability, are now occurring again. Other new concepts appear as well, including: approximate statistical laws; equations of state; heat as a form of energy; entropy as a variable of state; existence of equilibria; ensembles/distributions; and existence of the partition function. Many of these concepts appear in the path integral descriptions with the analyticity methods/extensions mentioned previously, so there are hints of a deeper theory that arrives at much of thermodynamics/Statistical mechanics foundation from the existing quantum theory.

Book 6 [6] has been placed after the other books in the Series [1-5] to await identification of entropy as fundamental in that it can be identified as an extrinsic system function even before getting to thermodynamics. We also already have experience with many particle systems, via quantum field theory (especially in curved space-time where particle creation is almost unavoidable), without directly tackling that scenario (due to quantum field theory effectively already being many-particle, with analytic determination of many-particle system functions, such as entropy).

With entropy presented at the outset as an important system variable, the derivation of thermodynamic potentials is then a straightforward process, as will be shown. The standard statistical mechanics connections to thermodynamics can then be given. Thus, in covering Thermodynamics and Statistical Mechanics we start with the foundations of the theory mostly established, such as entropy (also with equipartition equivalent to sum on paths with no weightings, etc.), with no assumptions. Everything follows directly from the theoretical discoveries outlined in the preceding books in the Series [1-5]. We don't see new connections to alpha, but we do see new structures/effects, especially manifold constructs (as with General Relativity, where we also saw no role for alpha).

The close ties between quantum mechanics complexified giving rise to the particle ensemble partition function, and quantum field theory complexified and the field ensemble partition function, is a derivative aspect of time-complexation. This complexation will be described in later chapters , where analytic time will allow an exact thermodynamic stability analysis of black hole geometries to be examined.

From atomic physics, described in Book 4, we also obtain the standard rules on electron shell completion (that is encoded in the periodic table). Similarly, we can also understand the origins of the intermolecular quantum chemistry rules. When taken to the statistical mechanics extreme we have thermodynamic equilibrium emergent from (the Law of Large Numbers and reverse Martingale convergence. With completion of application to chemical processes we have clear phase-transition effects, as well as equilibrium and near-equilibrium effects. The familiar chemistry is then observed, with phases of matter.

From chemical equilibrium and near-equilibrium, with 10^{23} elements that interact or not at all, we have two generalizations. The first is to consider chemical near-equilibrium and directly obtain an emergent process at this level, this is the branch that gives us biology/life at its most primitive level. The second is to consider equilibrium and near-equilibrium in general when the elements interact strongly (with10^{10}elements, say), this is the branch that describes biology/life at its most advanced social level and economics. In classic circuit shot noise analysis, the granularity of low-current flow (due to discreteness of electron charge) leads to a noise effect. Thus, as we consider situations with fewer elements, there are more complications, not less, due to granularity noise effects, and we enter the realm of machine learning with sparse data. Noise effects can be significant in complex systems, especially in biology where it is sometimes part of what is selected (such as in hearing, for background noise cancellation).

The second part of Book 6 explores the role of thermodynamics in efforts to extend to thermal quantum field theory and thermal quantum gravity. This is done by exploring Black Hole settings. The recognition of a role for complex structure on system variables becomes apparent in this process. This is done by examining the Hamiltonian thermodynamics of some black hole geometries with stabilizing boundary conditions. In this foray into directly exploring a thermal quantum gravity solution we assume a path integral form for the General Relativity problem and shift directly to a partition function (by 'Wick rotation'). We see that thermal quantum gravity is possible, when positive heat capacity shows stability.

Overview of Book 7

In Books 4,5, and 6 of the Series, we explored examples of quantum mechanics with imaginary time, quantum field theory in curved space-time, Thermal quantum field theory, minisuperspace quantum gravity, and Thermal quantum gravity. In this effort we find the path integral, and PI propagator, to provide the most general representation. In seeking a deeper theory in Book 7 we build on the sum-on-paths with propagator formulation to arrive at a sum-on-emanations with emanator formulation.

Propagation in a complex Hilbert space, in a standard quantum mechanics or quantum field theory formulation, requires the propagator function to be a complex number (not real or quaternionic, etc., [127]). This prohibits what would otherwise be an obvious generalization to hypercomplex algebras. In order to achieve this generalization, we have to introduce a new layer to the theory, one with universal emanation involving hypercomplex algebras (trigintaduonions) that is hypothesized to project to the familiar complex Hilbert space propagation with associated fixed elements (e.g., the emanator formalism projects out the observed constants and group structure of the standard model). The 'projection' is an induced mathematical construct, like having SU(3) on products of octonions, but here it we be the standard model U(1)xSU(2)xSU(3) on products of emanator trigintaduonions. Thus, in Book 7 a unified variational formulation is posed, one that arrives at alpha as a natural structural element, among other things, uniquely specified by the condition of maximal information emanation.

In Book 7 we also make note of the implications of a fundamental mathematical operation on a space that is repeated or added. The non-General Relativity forces are given by the form of the operation (the sequence forming an associative algebra), the General Relativity forces are given indirectly by the form of the space, this leaves the aspect "repeated or added" to be considered with care. If a purely 'repeated' operation, or mapping, occurs we can return to the dynamical mapping discussion of Book 1, where chaos can occur and is ubiquitous. There, the primal 'phase transition', the transition to chaos, is evident. If an operation with addition is involved (in the statistical sense of multiple elements), along with repeated overall steps, we arrive at the general framework of statistical mechanics with effects from the Law of Large Numbers and reverse Martingale convergence, among other things (Book 6). Most notable, however, is the prevalence of a new effect, that of phase transitions and the emergence of new structure (order from disorder), including the remarkable structures of chemistry and biology.

Why the recurring 'Cabbalistic formula'? was a question even in the time of Sommerfeld [190]. Now, the numerological parallel is more exact than realized at that time, so is too much a coincidence to be by chance. The non-coincidence appears to be due to the maximal nature of information transmission in a variety of circumstances (in physics, biology, and even human communication with sufficient optimization) as well as with the fractal-like repetition of key parameter sets that occurs in these different settings {10,22,78,137≅1/alpha}. We see that 10 expresses the dimensionality of propagation (or nodes of connectivity), while 22 corresponds to the number of fixed parameters in the propagation (in Book 7 we explore propagation in a 10 dimensional subspace of the 32 dimensional trigintaduonion space, leaving 22 dimensions at fixed values that appear as parameters in the theory). We will see the number 78 relates to generators of the motion, and that there are 4 chiralities of motion ('doubly chiral'). We will also see that 137 is simply the number of independent tri-octonionic product terms in the general chiral trigintaduonion 'emanation'.

Synopsis – Frodo Lives

Tolkien wrote of eucatastrophes [191], perhaps he anticipated the constructive role of emergent phenomena in maximum information transmission.

Preface to Physics Series, Book #5, on:

Quantum Field Theory

This is a book on quantum field theory. The quantum mechanics solution for the simple harmonic oscillator is reviewed at the outset. In the context of quantum field theory this will be referred to as the first quantization solution because we are going to re-use this solution in a new way (an infinite number of times) to arrive at "second quantization" that will describe a quantized field. Second quantization provides a very convenient and simple pathway to having a quantized field. The proof of the validity of second quantization following from a more rigorous canonical quantization formulation is then given. Many situations of interest lead to calculations that become impractical in the canonical quantization formalism, however, so a transition is then made to the Path Integral formalism, where Path Integral quantization, renormalization, and renormalization group methods are described.

A large number of problems and solutions are provided for the various descriptions throughout, especially those that are fundamental to the construction of quantum electrodynamics in later sections. Thus, the book provides complete material for introductory and advanced graduate courses in quantum field theory, as well as special applications to the Standard Model and possible extensions (circa 2024), and to quantum field theory on curved space-time.

The second part of the book describes thermal quantum field theory and quantum field theory on curved space-time. For thermal quantum field theory, upon shifting (Wick rotating) from real time to imaginary time the statistical mechanical partition function for the system is obtained. Thus, a connection between 'thermality' and complex structure, in the time dimension, is indicated. For quantum field theory on curved space-time, we arrive at an early analysis of black hole thermodynamics, where space-time curvature gives rise to thermality (a thermal vacuum) and particle production effects. Black hole thermality was revealed in Hawking radiation, due to the causal boundary at the horizon. Such thermality is even seen in flat space-time if causal boundaries are induced, such as in the case of an accelerated observer. These topics, with many worked examples, will be described in detail.

Chapter 1. Introduction

This is a book on quantum field theory [5], fifth in the Maximal Information Physics Series consisting of seven books [1-7], and it builds on the quantum mechanics presentation from Book 4 [4]. The main construct that we will need from (first-year undergraduate) quantum mechanics will be the solution for the simple harmonic oscillator, so a review of this will be given in Section 2.3. In the context of quantum field theory this will be referred to as the first quantization solution because we are going to re-use this solution in a new way (an infinite number of times) to arrive at "second quantization" that will describe a quantized field. Second quantization (Chapter 2) provides a very convenient and simple pathway to having a quantized field, but can it be trusted? This wasn't clear at first, but it can be trusted (if used properly), and the proof of the validity of second quantization following from a more rigorous canonical quantization formulation is given in Chapter 3. In Chapter 3 the main relativistic field representations are also described (Klein-Gordon and Dirac). Many situations of interest lead to calculations that become impractical in the canonical quantization formalism, however, so a transition is made to the Path Integral formalism in Chapter 4, where Path Integral Quantization, Renormalization, and Renormalization Group Methods are described.

Detailed notes on quantum field theories in gauge field representation are given in Chapters 4 and 5, which clearly relates the choice of field theory to a choice of Lie algebra, which, in turn, can be related to a choice of group theory (such as U(1) and SU(3)). From this we can see that non-classical algebraic constructs are ubiquitous in quantum mechanics and quantum field theory. As with quantum mechanics, we find that the Schrodinger and Heisenberg formulations, still provide the only tractable way to get a solution for bound systems. In other theoretical considerations, however, the path integral approach is best, as will be shown, with the renormalization examples.

In Section 4.7.4 we get the highest precision result for the value of alpha, in its role as perturbation parameter (further roles for alpha described in Appendix A). If a calculation of the electron magnetic moment parameter g-2 is performed, with all of the Feynman diagrams appropriate to expansions up to 5th order, we get a determination of alpha up to 14 digits, where 1/alpha=137.05999….. . This gives us one of the most precise measurements of alpha known. When a similar analysis is done for the muon g-2, given the much larger muon mass particle production pairs of other particles have a measurable effect, and we are able to probe the lower masses of the standard model that are present. In doing this, in preliminary experiments, there is a discrepancy indicating more particles, e.g. the Standard Model will need to be extended (discussed in Chapter 6). These missing particles could be the missing "Dark Matter". The prediction of such in Emanator Theory, and why there should be an imbalance between the left and right neutrinos (hint: maximum information transmission) is described in Book 7 [7], with a summary in Appendix D.

Part of the description of quantum field theory entails use of analyticity and other complex structures to encapsulate more of the physics in a complex extension to the space (or dimension). This often leads to formulations in terms of complex integration, with the choice of complex contour specified, such as with the Feynman propagator. One of the main renormalization methods, as another example, is to use dimensional regularization, which entails analytically continuing expressions with dimensionality to dimensionality as a complex parameter. There is also the aforementioned shift to complex and to "Wick rotate" expressions with real time to expressions with pure complex time. In doing this the statistical mechanical partition function for the system is obtained, with well-defined summation. Thus, a connection between 'thermality' and complex structure, in the time dimension at least, is indicated.

The second part of the book describes quantum field theory on curved space-time, where we arrive at an early analysis of Black Hole thermodynamics. Here we find that space-time curvature gives rise to thermality (a thermal vacuum) and particle production effects. Black Hole thermality was revealed in Hawking radiation [8], due to the causal boundary at the horizon. Such thermality is even seen in flat space-time (Chapter 7) if causal boundaries are induced, such as in the case of an accelerated observer [9,10].

Quantum field theory on curved space-time has one further gift, critical to the statistical mechanics formalism to follow in Book 6 [6], and that's the spin-statistics relation. The spin-statistics relation is kinematically determined in flat spacetime quantum field theory, but it is separately determined to also be enforced dynamically in curved spacetime quantum field theory.

The choice of time is related to choice of vacuum, which is related to the field geometry and observer motion (such as constant acceleration or expansion). If you have flat spacetime quantum field theory with a boundary, then you have thermodynamic effects (e.g., the Unruh effect [10]). In this setting we can compare the Hawking derivation of Hawking Radiation using the Euclideanization 'trick' versus the Bogoliubov transformations of the field to the Rindler geometry from the Minkowski geometry (if chosen as the asymptotic vacuum reference). With quantum field theory on curved space-time we also arrive at spin-statistics as mentioned, and get the final extension of the theory by way of Grassman algebras, to arrive at thermodynamically consistent Bose and Fermi statistical descriptions on quantum matter.

A large number of problems and solutions are provided for the various descriptions throughout, especially those that are fundamental to the construction of QED in later sections. Thus, the book provides complete material for introductory and advanced graduate courses in quantum field theory, as well as special applications to the Standard Model and possible extensions (ca. 2024), and to quantum field theory on curved space-time.

Chapter 2. Quantum Field Theory with 2nd Quantization

2.1 Background

Other than the spectral lines phenomenon, the first break from classical electromagnetic theory was indicated in Planck's effort to explain the observed spectrum of a black body, i.e., blackbody radiation. Planck postulated that *the process* of emission and absorption was in terms of discrete quanta, such that his integration on a continuum spectrum became a sum on a discrete spectrum, exactly producing the observed blackbody radiation distribution. In 1905, upon describing the photoelectric effect, Einstein postulated that it is not merely the process that is quantized, but the actual elements/packets of electromagnetism itself, e.g., that we have photons. Furthermore, the photons energy is related to its frequency by a factor of Planck's constant. Later, the Compton effect would demonstrate not only this particle interpretation for electromagnetic radiation (photons) but also provide a result consistent with special relativity.

Clearly a quantum theory of fields was indicated given the growing experimental proof of 'photons', but it was not until Dirac's 1927 paper [11] ("The Quantum Theory for the emission and absorption of radiation") that we got a hint of an overall solution. Dirac not only demonstrated a quantum wave equation for an electron field, his equation was consistent with special relativity, demonstrated local rotational invariants consistent with the electron field being a spinor field, and suggested 'anti' solutions whereby Dirac (correctly) predicted the existence of the positron. Furthermore, being a 'field' theory, it was now possible to explain multiple particles, especially the occurrence of electron-positron pairs, and now also provided a means to explore interactions of the quantized electromagnetic field (e.g., photons) with the (charged) electron-positron matter field. In this context, we can think of interactions between electron-positron field quanta as being between electrically-charged particles via "the electromagnetic field' as being mediated by an exchange of photons on that particle description. Diagrammatically, the latter description will appear later in a complete collection of Feynman diagrams in the full QED description. So, Dirac's early description of interactions diagrammatically, and that of others in considering the φ^2 and φ^4 theories to be described in the following sections, lays the foundation for modern quantum field theory in this aspect as well (where it is adopted by Feynman in a well-defined perturbative expansion – with a finite set of 'terms', or diagram types, at a particular level of perturbation).

Dirac's innovation in arriving at a quantum field theory had one other 'trick', the introduction of Second Quantization. For the non-spinorial electromagnetic field the special-relativistically valid Klein-Gordon wave-equation solution had already been proposed in a 'single-particle' field context. If we now view this, like Dirac did, in multi-particle context (as with his Dirac equation description of the electron), we see that imposing canonical-quantization relations on the electromagnetic field will involve a continuum infinity (uncountable) of such relations. In electromagnetism, and in related signal analysis, it is common to go to a Fourier Transform decomposition of a field into a (countable) collection of the field's modes. A countable decomposition is then amenable to a perturbative analysis, among other things (e.g., renormalization), so is a critical maneuver. If doing a perturbative analysis, the mere existence of such a perturbative solution indicates that the solution is a 'semigroup' type such that it is equivalent to other solutions up to a given order. This equivalence on solutions means that the manner of shifting to a countable number of field terms will not matter if it can be 'removed to infinity', so something simple can be chosen. In the case of the electromagnetic field, lets just assume we are not in an infinite volume but a very large box, and impose periodic boundary conditions on the boundaries (like with many simplified electromagnetism problems given in a homework). If we proceed with this Fourier decomposition we will get a description in terms of field modes that will directly suggest 'Second Quantization". Remarkably, as will be shown in a later section, this form of quantization is generally valid.

In the sections that follow use is made of excellent lecture notes taken while at Caltech, UWM, and Oxford, and from various texts. Most notable are the lecture notes from a quantum field theory course taught by Nicholas Papastamatiou at UWM, and those notes, in turn, drew strongly from the text by Ramond [12]. For an excellent user friendly introduction, with much more detail in some areas than can be provided in what follows, see both Mandl & Shaw [13] and Ryder [14]. This book also encompass a broad range of advanced applications, ranging from gauge fields and renormalization to the standard model and curved spacetime.

Notes from these areas, including work done for my PhD thesis [15], and from numerous excellent sources [9,16-24], will be described.

2.2 Electromagnetism and Second Quantization

First recall the formalism of classical electromagnetism, e.g., the Maxwell relations:

$$\nabla \cdot \vec{E} = 4\pi\rho$$
$$\nabla \times \vec{B} - \frac{1}{c}\frac{\partial \vec{E}}{\partial t} = \frac{4\pi}{c}\vec{J}$$
$$\nabla \times \vec{E} + \frac{1}{c}\frac{\partial \vec{B}}{\partial t} = 0$$
$$\nabla \cdot \vec{B} = 0.$$

where $\vec{E}$ is the electric field, a vector field, which acts on charged particles, likewise for $\vec{B}$, the magnetic field. The electric charge density is given by ρ, and the electric charge current by $\vec{J}$. There is no magnetic charge or magnetic charge current. The U(1) part of the standard model gives the freedom for only one (free) charge, where our convention is that it be the electric charge as shown above.

To arrive at a second-order wave-equation from the Maxwell equations, an ever-present reduction in all of the field equations examined, we will now choose gauge fields and make use of gauge invariance that is thereby exhibited to simplify the mathematics. Such maneuvers, to arrive at a second-order wave equation for a gauge field, will be possible in all of the theories examined, and in perturbative analyses this will permit the second quantization methods that follow to be generally applicable. We will choose auxiliary functions, soon to be called gauge fields, such that the vector potential $\vec{A}$ satisfies:

$$\nabla \cdot \vec{B} = 0 \rightarrow \ \vec{B} = \nabla \times \vec{A},$$

and from this we choose the second auxiliary function such that:

$$\vec{E} = -\nabla\varphi - \frac{1}{c}\frac{\partial \vec{A}}{\partial t}.$$

The auxiliary fields are gauge fields because they have the following gauge transformations that leave the values of the $\{\vec{E},\vec{B}\}$ fields unchanged:

$$\vec{A} \rightarrow \ \vec{A}' = \vec{A} - \nabla f \ \ and \ \ \varphi \rightarrow \varphi' = \varphi + \frac{1}{c}\frac{\partial f}{\partial t}.$$

In what follows we will choose Coulomb a.k.a. radiation gauge:

$$\nabla \cdot \vec{A} = 0 \,.$$

We are thereby breaking the relativistic invariance of the formalism, so results will only be applicable to non-relativistic situations. When this formalism is applied later we will have to verify that the solution is non-relativistic to validate the assumption of this gauge choice. And, once done, it will be shown that very accurate agreement is obtained with experiment.

Let's now restrict to a free field, to get

$$\nabla^2 \varphi = 0,$$

and since the scalar field φ must be zero at infinity, we have $\varphi = 0$, and the only equation to be solved is:

$$\left(\frac{1}{c^2}\frac{\partial^2}{\partial t^2} - \nabla^2\right) \vec{A} = 0, \text{ where } \vec{B} = \nabla \times \vec{A} \text{ and } \vec{E} = -\frac{1}{c}\frac{\partial \vec{A}}{\partial t}.$$

Solutions consist of:

$$\vec{A}(x,t) = \vec{A}_0 e^{ik\cdot x - \omega t} \text{ where } k \cdot \vec{A} = 0 \text{ from gauge choice.}$$

Note that transverse propagation is indicated (the non-zero components of $\vec{A}$ are perpendicular to the direction of motion indicated by k). Also note that this transverse vector property is carried over to the $\{\vec{E}, \vec{B}\}$ fields in this gauge as well.

The energy of the waves in the radiation field is:

$$H = \frac{1}{2}\int (E^2 + B^2) d^3x.$$

Now that we've recovered the simple second-order derivative form of the key underlying gauge field, here denoted $\vec{A}$, let's proceed to solve "in a box" (with volume 'V') with periodic boundary conditions as indicated. In this case we have:

$$\vec{A}(0,y,z,t) = \vec{A}(L,y,z,t), \ldots, etc.$$

And we get the modal solutions:

$$\frac{1}{\sqrt{V}} \varepsilon_r(k) e^{ik\cdot x}, \qquad r = 1,2$$

$$k = \frac{2\pi}{L}(n_1, n_2, n_3), \qquad n_1, n_2, n_3 = 0, \pm 1, \pm 2, \ldots$$

The normalization and $k \cdot \vec{A} = 0$ gauge choice then lead to the relations:

$$\varepsilon_r(k)\varepsilon_s(k) = \delta_{rs} \ and \ \varepsilon_r(k) \cdot k = 0.$$

Using the complete set of modes so indicated we now perform Fourier series expansion of the Vector potential:

$$\vec{A}(\vec{x},t) = \sum_k \sum_r \left(\frac{\hbar c^2}{2V\omega_k}\right)^{1/2} \varepsilon_r(\vec{k})\left[a_r(\vec{k},t)e^{ik\cdot x} + a_r^*(\vec{k},t)e^{-ik\cdot x}\right],$$

$$\omega_k = c|k|.$$

Note that we make use of the fact that the vector potential is real, thus $\vec{A} = \vec{A}^*$, which means the grouping with a_r^* as shown. Also note that this expansion is for each time t, and that use of the DeBroglie relation has introduced Planck's constant into the normalization. Solving the wave equation at a time t we then have for each of the modes:

$$a_r(\vec{k},t) = a_r(\vec{k})e^{-i\omega_k\cdot t}$$

And we can now write the energy of the radiation field as:

$$H = \sum_k \sum_r \hbar\omega_k \, a_r^*(\vec{k})a_r(\vec{k}).$$

In this form, the energy in the radiation field looks like the energy of a collection of harmonic oscillators. The energy of each mode is classically decoupled from the others, so may be quantized separately as well. If each mode is quantized as a harmonic oscillator, then a field quantization is thereby described, known as second quantization. Before proceeding with the second quantization description, it will be helpful to first review the first quantization of the harmonic oscillator system that it will be based on.

2.3 The Harmonic Oscillator

Notes from this review section on the quantum mechanics of an oscillator draws from a lengthier analysis presented in Book 4 [4], and in references cited there.

2.3.1 The one–dimensional harmonic oscillator

For the classical harmonic oscillator we have $E = \frac{p^2}{2m} + \frac{1}{2}mw^2x^2$. For the oscillator in quantum mechanics, the classical quantities x and p are replaced by the operators x and p which satisfy:

$$[X,P] = i\hbar$$

And

$$H = \frac{P^2}{2m} + \frac{1}{2}mw^2X^2$$

Since it is time –independent the quantum mechanics study of the harmonic oscillator reduces to the solution of the eigenvalue equation.

$$H|\varphi >= E|\varphi >$$

In the $\{|+>\}$ representation:

$$\left[-\frac{\hbar^2}{2m}\frac{d^2}{dx^2}+\frac{1}{2}m\omega^2x^2\right]\varphi(x)=E\varphi(x)$$

Properties that can be deduced from the form of the potential:

(i) The eigenvalues are positive: $H = T + V \ so \ < H >= E =< T > + < V >$ but $< T > \geq 0$ and $< V > \geq V_m$, so $E > V_m$ (uncertainty principle eliminates equality), and since we have chosen V_m as the zero we then get $E > 0$.

(ii) Eigenfunctions of H have a definite parity since V(x) is even: $V(-x) = V(x)$

(iii) The energy spectrum is discrete.

2.3.2 Eigenvalues of the Hamiltonian

Define dimensionless operators for convenience.

$$\hat{X}=\sqrt{\frac{m\omega}{\hbar}}X \ \ and \ \ \hat{P}=\frac{1}{\sqrt{m\hbar\omega}}P$$

Now,

$$[\hat{X},\hat{P}]=i$$

And $H = \hbar\omega\hat{H} \quad with\ \hat{H} = \frac{1}{2}(\hat{X}^2+\hat{P}^2)$. Now we want solution to

$$\hat{H}|\varphi_v^i >= \varepsilon_V|\varphi_v^i >$$

where i labels degenerate vectors on v. $\hat{H}$ is dimensionless, as is ε_V. Since $\hat{X}\ and\ \hat{P}$ are operators we can't write $\hat{X}^2+\hat{P}^2 \neq (\hat{X}-i\hat{P})(\hat{X}+i\hat{P})$. So. introducing the operators $\hat{a}\ \ and\ \hat{a}^+$:

$$\left.\begin{aligned} a&=\frac{1}{\sqrt{2}}(\hat{X}+i\hat{P})\\ a^\dagger&=\frac{1}{\sqrt{2}}(\hat{X}-i\hat{P})\end{aligned}\right\}\rightarrow \qquad \begin{aligned}\hat{X}&=\frac{1}{\sqrt{2}}(a^\dagger+a)\\ \hat{P}&=\frac{1}{\sqrt{2}}(a^\dagger-a)\end{aligned}$$

$$[a,a^\dagger]=\frac{1}{2}[\hat{X}+i\hat{P},\hat{X}-i\hat{P}]=\frac{1}{2}[\hat{X},-i\hat{P}]+\frac{1}{2}[i\hat{P},\hat{X},]=1$$

$$[a,a^\dagger]=1$$

Also,

$$a^\dagger a=\frac{1}{2}(\hat{X}-i\hat{P})(\hat{X}+i\hat{P})=\frac{1}{2}(\hat{X}^2+\hat{P}^2-1)$$

So,

$$\hat{H} = aa^{\dagger} + \frac{1}{2}$$

Introduce operation N:

$$N = aa^{\dagger}$$

Thus, eigenvectors of $\hat{H}$ are eigenvectors of N.

$$[N, a] = [a^{\dagger}a, a] = a^{\dagger}aa - aa^{\dagger}a = [a^{\dagger}, a]a = -a$$
$$[N, a^{\dagger}] = [a^{\dagger}a, a^{\dagger}] = a^{\dagger}aa^{\dagger} - a^{\dagger}a^{\dagger}a = a^{\dagger}[a, a^{\dagger}] = a^{\dagger}$$

Consider,

$$N|\varphi_j >= v|\varphi_v^i >$$

Then,

$$H|\varphi_j >= \left(v + \frac{1}{2}\right)\hbar\omega|\varphi_v^i >$$

2.3.3 Determination of the spectrum

I. The eigenvalues v of N are positive or zero

Proof

$\|a|\varphi_j >\| \geq 0$

$\|a|\varphi_j >\| =< \varphi_j|a^{+}a|\varphi_j >=< \varphi_j|N|\varphi_j >= v \geq 0$

II. (i) if $v = 0$ $then\ a|\varphi_{v=0}^i >= 0$

(ii) $if\ v > 0\ then\ a|\varphi_v^i > \neq 0$ and is eigenvector of N with value v-1

Proof (i) $v = 0$ $then\ \|a|\varphi_0^i >\| = 0 \quad \rightarrow a|\varphi_0^i >= 0$

(ii) if $v > 0$ $then\ \|a|\varphi_v^i >\| > 0 \quad \rightarrow a|\varphi_v^i > \neq 0$

$N(a|\varphi_v^i >) = a^{+}aa|\varphi_j >= (-1 + aa^{+})a|\varphi_j{}^i >= (-a + aN)|\varphi_j{}^i$

$= (-1 + v)(a|\varphi_j >)$

So, $N(a|\varphi_v^i >) = (v - 1)(a|\varphi_v^i >)$

III. If $|\varphi_v^i >$ is a non zero eigenvector of N of eigenvalue v, then

(i) $a|\varphi_v^i >$ is always non-zero

(ii) $a^{+}|\varphi_v^i >$ is an eigenvector of N with eigenvalue v+1

Proof

(i) $\|a^{+}|\varphi_v^i >\| =< \varphi_v^i|aa^{+}|\varphi_v^i > = < \varphi_v^i|N + 1|\varphi_v^i >= (v + 1) \geq 1$

(ii) $N(a^{+}|\varphi_v^i >) = a^{+}aa^{+}|\varphi_v^i >= a^{+}(1 + a^{+}a)|\varphi_v^i >= (v + 1)a^{+}|\varphi_v^i > \neq 0$

IV. The spectrum of N is composed of non-negative integers. If the eigenvalue of N on $|\varphi_v^i >$ is v and isn't integer valued, then we can write $n < v < n + 1$ where n is an integer. Since (from II)

$a^n|\varphi_v^i>$ is then an eigenvalue of N with strictly positive eigenvalue: v-n, we find that $a^{n+1}|\varphi_v^i>$ is eigenvalue. v-(n+1) which is strictly negative: thus, if V is non-integral, we can therefore construct a tech-zero eigenvector of N with a strictly negative eigenvalue but this contradicts (I), the hypothesis of non-integral V is thus negated.

So, let n<v<n+1 → n=V, what happens?
Rule II reveals that $a^n|\varphi_n^i>$ is non-zero with eigenvalue 0, thus, $a^{n+1}|\varphi_n^i>= a|\varphi_0>= 0$.
Thus, V can only be a non-zero integer. Since $H|\varphi_v^i>= (v + {}^1/_2)\hbar\omega|\varphi_v^i>$, we get

$$E_n = \left(n + \frac{1}{2}\right)\hbar\omega$$

The energy of the harmonic oscillator is quantized, ground state non-zero: a is the destruction or annihilation operator. $a^\dagger$ is the creation operator.

Degeneracy of eigenvalues? Now to show that the energy eigenvalues (levels) are not degenerate for the harmonic oscillator. Consider the eigenstates of $E_0 = \hbar\omega/2$ (eigenstates of N with n=0). Rule II reveals $a|\varphi_0{}^i>= 0$, how many linearly independent kets satisfy this relationship, that will reveal the degeneracy:

$$a|\varphi_0{}^i> = \frac{1}{\sqrt{2}}\left[\sqrt{\frac{mw}{\hbar}}X + \frac{i}{\sqrt{m\hbar\omega}}P\right]|\varphi_0{}^i>= 0$$

In the $\{|x\}$ representation:

$$\left(\frac{\hbar}{mw}x + \frac{d}{dx}\right)\varphi_0{}^i(x) = 0 \ where\ \varphi_0{}^i(0) =< x|\varphi_0{}^i>$$

The solution of which is a

$$\varphi_0{}^i(x) = Ce^{-\frac{1}{2}\frac{mw}{\hbar}x^2}$$

C is a constant of integration resolved upon normalization. Consequently, there is only one ket φ_0 which satisfies $a|\varphi_0>= 0$, and the ground state is not degenerate.

Using recurrence and the fact that the ground state is nondegenerate it is possible to show that all of the states are non-degenerate: If we can show that $E_n = \left(n + \frac{1}{2}\right)\hbar\omega$ non-degenerate implies $E_{n+1} = (n + 1 + {}^1/_2)\hbar\omega$

is non-degenerate, then we are done. To begin, assume only one $|\varphi_n >$ satisfies $N|\varphi_n >= n|\varphi_n$. Then, consider

$$N|\varphi_{n+1} > = (n+1)|\varphi_{n+1}^i >$$

From lemma II: $a|\varphi_{n+1}^i >= c^i|\varphi_n >$. Now, $a^+a|\varphi_{n+1}^i >= c^i a^+|\varphi_n >$ and

$$|\varphi_{n+1}^i > = \frac{c^i}{n+1} a^+|\varphi_n >$$

thus all kets $|\varphi_{n+1}^i >$ are proportional to $a^+|\varphi_n^i >$. They are therefore proportional to each other: the eigenvalue (n+1) is not degenerate. Thus, since n=0 is not degenerate, then the above recurrence relation shows that all of the states are nondegenerate.

2.3.4 Eigenstates of the Hamiltonian

N and H are observables, i.e. their Eigenvectors constitute a basis in the space ε_x. This could be proved by considering the wave functions associated with the eigenstates of N. Since none of the eigenvalues of N (or H) is degenerate, N alone constitutes an operator in ε_x.

Basis vectors in terms of $|\boldsymbol{\varphi}_0>$

$a|\varphi_0 >= 0$ let $|\varphi_0 >$ be normalised (global $e^{i\theta}$ factor remains)
Now,

$|\varphi_1 >= c_1 a^+|\varphi_0 >$ choose c_1 by requiring $|\varphi_1 >$ to be normalised and such that the phase of $|\varphi_1 >$ relative to $|\varphi_0 >$ such that c_1 is real and positive. Then

$$< \varphi_1|\varphi_1 >= |c_1|^2 < \varphi_0|aa^+|\varphi_0 >= 1 \rightarrow c_1 = 1$$

$$|\varphi_1 > = a^+|\varphi_0 >$$

Using similar conventions for $|\varphi_2 >$:

$$|\varphi_2 >= c_2 a^+|\varphi_1 >$$

$$< \varphi_2|\varphi_2 >= |c_2|^2 < \varphi_1|aa^+|\varphi_1 >= 2|c_2|^2 = 1 \qquad c = \frac{1}{\sqrt{2}}$$

Thus, $|\varphi_2 > = \frac{a^+}{\sqrt{2}}|\varphi_1 >$.

Recursion gives

$$|\varphi_n >= \frac{a^+}{\sqrt{n}} \frac{a^+}{\sqrt{n-1}} \cdots \frac{a^+}{\sqrt{1}} |\varphi_0 > = \frac{(a^+)^n}{\sqrt{n!}} |\varphi_0 >$$

$$\boxed{|\varphi_n >= \frac{(a^+)^n}{\sqrt{n!}} |\varphi_0 >}$$

Orthonormalization and closure relations

Since H is Hermitian $|\varphi_n>$ with different n are orthogonal, and since we have normalized we have $<\varphi_n|\varphi_n>=\delta_{nn'}$. Since H is also an observable, which by definition satisfies a closure relation, we have: $\sum_n |\varphi_n><\varphi_n| = 1$.

Given the phase conventions chosen for the basis vectors the action of the $\hat{a}\ and\ \hat{a}^{+}$ operators on the $\{|\varphi_n>\}$ basis is especially simple (in fact X and P, etc., should be calculated in terms of a and a^{+} since it is much less work):

Since

$$\boxed{|\varphi_n>=\frac{1}{\sqrt{n}}a^{+}|\varphi_{n-1}>}$$

given our phase convention shift the index to get.

$$\boxed{a^{+}|\varphi_n>=\sqrt{n+1}|\varphi_{n+1}>}$$

Also

$$a|\varphi_n>=\frac{1}{\sqrt{n}}aa^{+}|\varphi_{n-1}>=\frac{n}{\sqrt{n}}|\varphi_{n-1}>=\sqrt{n}|\varphi_{n-1}>$$

$$\boxed{a|\varphi_n>=\sqrt{n}|\varphi_{n-1}>}$$

Now calculate X and P:

$$X|\varphi_n>=\sqrt{\frac{\hbar}{m\omega}}\frac{1}{\sqrt{2}}(a^{+}+a)|\varphi_n>$$

$$=\sqrt{\frac{\hbar}{2m\omega}}\left[\sqrt{n+1}|\varphi_{n+1}>+\sqrt{n}|\varphi_{n-1}\right]$$

$$P|\varphi_n>=\sqrt{\hbar m\omega}\frac{1}{\sqrt{2}}(a^{+}-a)|\varphi_n>$$

$$=i\sqrt{\frac{m\hbar\omega}{2}}\left[\sqrt{n+1}|\varphi_{n+1}>-\sqrt{n}|\varphi_{n-1}\right]$$

Matrix elements in $\{|\varphi_n>\}$ representation

$<\varphi_{n'}|a|\varphi_n>=\sqrt{n}\delta_{n',n+1}$

$<\varphi_{n'}|a^{+}|\varphi_n>=\sqrt{n}\delta_{n',n+1}$

Wave functions associated with stationary states:

$$\varphi_n(x) = \ <x|\varphi_n> \ = \frac{1}{\sqrt{n!}} <x|(a^+)^n|\varphi_0>$$

$$= \frac{1}{\sqrt{n!}} \frac{1}{\sqrt{2^n}} \left[\sqrt{\frac{m\omega}{\hbar}} x - \sqrt{\frac{\hbar}{m\omega}} \frac{d}{dx} \right]^n \varphi_0(x)$$

$$\varphi_n(x) = \left[\frac{1}{2^n n!} \left(\frac{\hbar}{m\omega} \right)^n \right]^{1/2} \left(\frac{m\omega}{\pi\hbar} \right)^{1/4} \left[\frac{m\omega}{\hbar} x - \frac{d}{dx} \right]^n e^{-\frac{1}{2}\frac{m\omega}{\hbar} x^2}$$

$\varphi_n(x)$ is the product of $e^{-\frac{1}{2}\frac{m\omega}{\hbar} x^2}$ and a polynomial of degree n and parity $(-1)^n$ called a Hermite polynomial.

Consider Δx *and* Δp for the harmonic oscillator:
$<\varphi_n|X|\varphi_n>= 0$, $<\varphi_n|P|\varphi_n>= 0$
$(\Delta X)^2 = <\varphi_n|X^2|\varphi_n> - <\varphi_n|X|\varphi_n>^2 = <\varphi_n|X^2|\varphi_n>$
$(\Delta P)^2 = <\varphi_n|P^2|\varphi_n> - <\varphi_n|P|\varphi_n>^2 = <\varphi_n|P^2|\varphi_n>$
$X^2 = \frac{\hbar}{2m\omega}(a^+ + a)(a^+ + a) = \frac{\hbar}{2m\omega}\left((a^+) + a^+a + aa^+ + a^2\right)$
$P^2 = \frac{m\hbar\omega}{2}(a^+ + a)(a^+ + a) = -\frac{m\hbar\omega}{2}\left((a^+) + a^+a + aa^+ + a^2\right)$

$$\Delta X \Delta P = \hbar\left(n + \frac{1}{2}\right)$$

And $\Delta X \Delta P = \frac{1}{2}\hbar$ for the ground state (recall that the ground state is a Gaussian).

Thus, we arrive at the standard solution to the quantized harmonic oscillator at a particular instant in time. To consider the full time solution we describe our creation and annihilation operators using the Heisenberg picture:

$$i\hbar \frac{da(t)}{dt} = [a(t), H], with\ a(0) = a,$$

With solution:

$$a(t) = ae^{-i\omega \cdot t}.$$

Returning to our effort to quantize the radiation field, we can now write

$$\vec{A}(\vec{x}, t) = \sum_k \sum_r \left(\frac{\hbar c^2}{2V\omega_k} \right)^{1/2} \varepsilon_r(\vec{k}) \left[a_r(\vec{k}, t) e^{ik\cdot x} + a_r{}^\dagger(\vec{k}, t) e^{-ik\cdot x} \right],$$

$$\omega_k = c|k|.$$

Where the amplitudes $\{a_r, a_r{}^*\}$ are now promoted to being operators $\{a_r, a_r{}^\dagger\}$, and we can write $\vec{A}$ in two parts, one consisting only of creation operators and one consisting only of annihilation operators:

$$\vec{A}^{+}(\vec{x},t) = \sum_{k}\sum_{r}\left(\frac{\hbar c^2}{2V\omega_k}\right)^{1/2} \varepsilon_r(\vec{k})\, a_r(\vec{k},t)e^{ik\cdot x-i\omega t}$$

And

$$\vec{A}^{-}(\vec{x},t) = \sum_{k}\sum_{r}\left(\frac{\hbar c^2}{2V\omega_k}\right)^{1/2} \varepsilon_r(\vec{k})\, a_r{}^{\dagger}(\vec{k},t)e^{-ik\cdot x+i\omega t}$$

Example 1

Coherent State Example

The state $|n\rangle$ defined above suggests a form with convenient mathematical properties, that is known as a coherent state. Let's denote a coherent state by $|c\rangle$ and define it thus:

$$|c\rangle = \exp\left(-\frac{1}{2}|c|^2\right)\sum_{n=0}^{\infty}\frac{c^n}{\sqrt{n!}}\,|n\rangle$$

Let's now determine for an electric field E:

(1) $\langle c|c\rangle$

(2) $a|c\rangle$

(3) $\langle c|N|c\rangle$

(4) $(\Delta N)^2$

(5) $\langle c|E(k)|c\rangle$, using $A(k) = \left(\frac{\hbar c^2}{2V\omega_k}\right)^{1/2} \varepsilon_r(\vec{k})\left[a(\vec{k},t)e^{ik\cdot x} + a^*(\vec{k},t)e^{-ik\cdot x}\right]$,

(6) $(\Delta E(k))^2$

Answers

(1) $\langle c|c\rangle = \exp(-|c|^2)\sum_{n'=0}^{\infty}\frac{c^{n'}}{\sqrt{n'!}}\langle n'|\left(\sum_{n=0}^{\infty}\frac{c^n}{\sqrt{n!}}|n\rangle\right) = \exp(-|c|^2)\sum_{n=0}^{\infty}\frac{c^{2n}}{n!} = 1$

(2) $a|c\rangle = \exp\left(-\frac{1}{2}|c|^2\right)\sum_{n=0}^{\infty}\frac{c^n}{\sqrt{n!}}\sqrt{\text{n}}|n-1\rangle = c\exp\left(-\frac{1}{2}|c|^2\right)\sum_{n'=0}^{\infty}\frac{c^n}{\sqrt{n'!}}|n''\rangle = c|c\rangle$

(3) $\langle c|N|c\rangle = \overline{N} = \langle c|a^{\dagger}a|c\rangle = \langle c|c^*c|c\rangle = |c|^2$

(4) $(\Delta N)^2 = \langle c|N^2|c\rangle - [\langle c|N|c\rangle]^2$
$= \langle c|a^{\dagger}aa^{\dagger}a|c\rangle - |c|^4 = \langle c|c^*(1+a^{\dagger}a)c|c\rangle - |c|^4 = |c|^2$

(5) $\langle c|E|c\rangle = \left\langle c\left|-\frac{1}{c}\frac{\partial A}{\partial t}\right|c\right\rangle =$

$$\left\langle c\left|-\frac{1}{c}\left(\frac{\hbar c^2}{2V\omega_k}\right)^{1/2}\varepsilon_r(\vec{k})\left[-i\omega_k a e^{ik\cdot x}+i\omega_k a^* e^{-ik\cdot x}\right]\right|c\right\rangle$$

Let $c = |c|e^{i\delta}$, then

$$\langle c|E|c\rangle = -\varepsilon_r(\vec{k})\left(\frac{\hbar\omega_k}{2V}\right)^{\frac{1}{2}}|c|2\left[e^{i\delta}e^{i(k\cdot x-\omega t)} - e^{-i\delta}e^{-i(k\cdot x-\omega t)}\right]\left(\frac{1}{2i}\right)$$

Thus,

$$\langle c|E|c\rangle = -\varepsilon_r(\vec{k})2\left(\frac{\hbar\omega_k}{2V}\right)^{\frac{1}{2}}|c|\sin(k\cdot x-\omega t+\delta)$$

(6) $(\Delta E)^2 = \langle c|E^2|c\rangle - [\langle c|E|c\rangle]^2 = \left(\frac{\hbar\omega_k}{2V}\right)$

Note that the relative fluctuation in photon numbers goes as $\frac{\Delta N}{\bar{N}} = \frac{1}{\sqrt{N}}$, which tends to zero as $N \to \infty$. Similarly, the fluctuation ΔE becomes negligible for large field strengths and $|c\rangle$ becomes, in the limit, a classical state in which the field is well-defined as $\bar{N} \to \infty$.

2.4 Radiative transitions in atoms

Let's now consider transitions between two states of an atom with emission or absorption of one photon. We use the interaction:

$$H_I = \sum_i -\frac{e_i}{m_i c}A_i \cdot p_i$$

and start with (single photon) emission and consider transition between atomic wavefunctions $|A\rangle$ and $|B\rangle$:

$$\langle B, n(k)+1|H_I|A, n(k)\rangle$$
$$= \frac{e}{m}\left(\frac{\hbar}{2V\omega_k}\right)^{1/2}[n(k)$$
$$+1]^{1/2}\langle B|\varepsilon_r(\vec{k})\cdot \textstyle\sum_i e^{-ik\cdot r_i}p_i\,|A\rangle e^{-i\omega_k t}$$

Thus,

$$W_r d\Omega = \frac{e^2\omega d\Omega}{8\pi^2 m^2\hbar c^3}[n(k)+1]\left|\varepsilon_r(\vec{k})\cdot\langle B|\textstyle\sum_i e^{-ik\cdot r_i}p_i|A\rangle\right|^2$$

These results can be calculated using the electric dipole approximation, to be described next, if

$$e^{-ik\cdot r_i} \approx 1$$

This is justified if the wavelength of the radiation emitted in the transition is very large compared to the (linear) dimensions R of the system of charges. The atomic wavefunctions restrict the effective values of r_i to

$|r_i| < R$, such that $k \cdot r_i \lesssim kR \ll 1$. This is easily satisfied for optical transitions – compare system $R \sim 1\text{Å}$, while optical wavelengths $\lambda \in [4000 - 7500\text{Å}]$.

From the equation of motion $i\hbar \dot{r}_i = [r_i, H]$ we get:

$$\langle B|p_i|A\rangle = m\langle B|\dot{r}_i|A\rangle = -im\omega\langle B|r_i|A\rangle.$$

Electric Dipole Interaction

Let's now consider the electric diploe interaction beginning with a system of N charges $\{e_1 \ldots e_N\}$ which can be described nonrelativistically. This means we can simultaneously specify position information at time t by: $r_i = r_i(t)$. We are interested in finding the transitions between initial and final states and will proceed with two assumptions: (1) the aforementioned $kR \ll 1$, namely, we may neglect variations in the Electric field that is causing transitions over the region of the system; and (2) we neglect magnetic field interactions (so not too high a frequency). Under these conditions we have a transverse electromagnetic field E_T:

$$E_T = -\frac{1}{c}\frac{\partial A(r,t)}{\partial t}$$

and the interaction giving rise to transitions:

$$H_I = -D \cdot E_T(0,t), \;\; D = \sum_i e_i r_i.$$

Transitions from H_I that are first order in perturbation theory are called electric dipole transitions. Since A is first-order in annihilation and creation operators, so is E_T, so first order is equivalent to transitions with a single photon emitted or absorbed. Starting from a state $|X\rangle$ and transitioning to a state $|Y\rangle$ in this situation is then described by:

$$|X, n_r(k)\rangle = |X\rangle|n_r(k)\rangle$$
$$|Y, n_r(k) \pm 1\rangle = |Y\rangle|n_r(k) \pm 1\rangle$$

Writing the dipole operator $D = -e\sum_i r_i = -e\vec{x}$ and using:

$$E_T(0,t) = -\frac{1}{c}\frac{\partial A(0,t)}{\partial t}$$
$$= i\sum_k\sum_r\left(\frac{\hbar\omega_k}{2V}\right)^{1/2} \varepsilon_r(\vec{k})\left[a_r(\vec{k})e^{-i\omega_k t} - a_r{}^{\dagger}(\vec{k})e^{i\omega_k t}\right]$$

Thus

$$\langle Y, n(k)+1|H_I|X, n(k)\rangle$$
$$= i\left(\frac{\hbar\omega_k}{2V}\right)^{1/2} [n(k)+1]^{1/2}\langle Y|\varepsilon_r(\vec{k}) \cdot D|X\rangle e^{i\omega_k t}$$

Thus

$$W_r d\Omega = \frac{e^2 \omega d\Omega}{8\pi^2 m^2 \hbar c^3}[n(k)+1]\left|\varepsilon_r(\vec{k})\cdot\langle Y|\vec{x}|X\rangle\right|^2.$$

From the $[n(k)+1]$ factor we see the spontaneous emission contribution (no classical counterpart) in the '1' part, while the $n(k)$ part gives the induced emission factor.

If we sum over the two polarization states for a given k (the suppressed r index), we then have:

$$\sum_{r=1}^{2}\left|\varepsilon_r(\vec{k})\cdot\langle Y|\vec{x}|X\rangle\right|^2 = |\langle Y|\vec{x}|X\rangle|^2 - \left(\frac{k\cdot\langle Y|\vec{x}|X\rangle}{|k|}\right)\left(\frac{k\cdot\langle Y|\vec{x}|X\rangle^*}{|k|}\right)$$

Thus

$$\sum_{r=1}^{2}\left|\varepsilon_r(\vec{k})\cdot\langle Y|\vec{x}|X\rangle\right|^2 = |\langle Y|\vec{x}|X\rangle|^2 \sin^2\theta$$

For spontaneous emission, the total transition probability per unit time is given by n=0, and using the relation $\int \sin^2\theta \, d\Omega = 8\pi/3$ we then have:

$$W_{Total}(X \to Y) = \frac{e^2\omega^3}{3\pi^2 \hbar c^3}|\langle Y|\vec{x}|X\rangle|^2$$

For lifetime:

$$\frac{1}{\tau} = \sum_n W_{Total}(X \to Y_n)$$

2.5 The electromagnetic field in the presence of moving charges

The Hamiltonian of a system consisting of moving charges has three parts: the matter part (the source charges); the electromagnetic field part; and an interaction part. Starting with the matter part we have for a system of N point charges:

$$H_m = \sum_i \frac{p_i^{\,2}}{2m_i} + H_{ee}; \quad H_{ee} = \frac{1}{2}\sum_{i,j;i\neq j}\frac{e_i e_j}{4\pi\left|r_i - r_j\right|}; \quad p_i = m_i \dot{r}_i$$

Now let's consider a correspondence with the electromagnetic field Hamiltonian for such a circumstance. Staying with Coulomb gauge ($\nabla \cdot A = 0$) in order to have a field decomposition into transverse (E_T) and longitudinal (E_L) fields, to get:

$$E = E_T + E_L; \quad E_T = -\frac{1}{c}\frac{\partial A}{\partial t}; \quad E_L = -\nabla\varphi\,.$$

Thus, $\nabla \wedge E_L = 0$. Recall the formula for the total energy of an electromagnetic field:

$$\mathcal{E} = \frac{1}{2}\int (E^2 + B^2) d^3x = \frac{1}{2}\int ({E_T}^2 + B^2) d^3x + \frac{1}{2}\int {E_L}^2 d^3x.$$

Using the Poisson equation solution $\nabla^2 \varphi = -\rho$ with the last integral we have:

$$\frac{1}{2}\int {E_L}^2 d^3x = \frac{1}{2}\int \frac{\rho(x,t)\rho(x',t')}{4\pi|x-x'|} d^3x d^3x' = \frac{1}{2}\sum_{i,j;i\neq j} \frac{e_i e_j}{4\pi|r_i - r_j|},$$

where the self-energy terms are dropped in the last term (which makes is equal to H_{ee}). Thus, we see that the correspondence will be consistent if we define the full Hamiltonian by:

$$H_m{}' = \sum_i \frac{1}{2m_i}\left(p_i - \frac{e_i}{c}A_i\right)^2 + H_{ee}$$

where the constants are chosen such that the Lorentz relation for moving charges is satisfied:

$$\frac{dv_i}{dt} = e_i\left[E_i + \frac{v_i}{c} \wedge B_i\right].$$

So we've arrived at a consistent description for field and charge dynamics in Coulomb gauge, let's now return to the issue of splitting the Hamiltonian between 'matter' parts and 'interaction' parts:

$$H_m' = H_m + H_I,$$

and

$$H_I = \sum_i \frac{1}{m_i c}\left(\frac{{e_i}^2}{2c}{A_i}^2 - e_i A_i \cdot p_i\right),$$

Thus,

$$H = H_m' + H_{rad} = H_m + H_I + H_{rad}\ .$$

The interaction part will be treated as a perturbation with the non-interacting Hamiltonian have transitions from state to state Y as described with the dipole interaction, except now we have the quadratic term in the vector potential, which will give rise to two-photon processes (at first order in perturbation theory), while the $A_i \cdot p_i$ term will give rise to magnetic interactions.

Example 2

Consider the Lagrangian

$$L(\vec{x}, \dot{\vec{x}}) = \frac{1}{2}m\dot{\vec{x}}^2 + \frac{q}{c}\vec{A}\cdot\dot{\vec{x}} - q\varphi$$

(1) Obtain the equation of motion and show it is Lorentz relation.

(2) Obtain the Hamiltonian form and show it is like that above.

Answer
(1) We have

$$p = \frac{\partial L}{\partial \vec{\dot{x}}} = m\vec{\dot{x}} + \frac{q}{c}\vec{A}$$

Thus, making use of the Euler-Lagrange relations: $m\vec{\ddot{x}} = q[E + \frac{1}{c}\vec{\dot{x}} \wedge B]$

(2) $H = p\dot{q} - L(p,q) = p\vec{\dot{x}} - \frac{1}{2}m\vec{\dot{x}}^2 - \frac{q}{c}\vec{A}\cdot\vec{\dot{x}} + q\varphi$, and using $p = m\vec{\dot{x}} + \frac{q}{c}\vec{A}$:

$$H = \frac{1}{2m}\left(p - \frac{q}{c}A\right)^2 + q\varphi$$

2.6 Thomson Scattering
Let's now consider scattering of photons by atomic electrons, where the photon energy $\hbar\omega$ is large compared to the binding energies of the electrons so that they can be considered free. Yet, we still require the photon energy $\hbar\omega$ to be very small compared to the electron rest mass so as to avoid relativistic effects. In this circumstance the recoil momenta on an electron when a photon scatters will be negligible, and the photon energy is unchanged.

So, we want to consider scattering from an initial state with one photon (momentum $\hbar\vec{k}$) and polarization $\varepsilon_r(\vec{k})$ (with r = 1 or 2) to a final state with one photon with momentum $\hbar\vec{k}'$ and polarization $\varepsilon_s(\vec{k}')$ (with s = 1 or 2). This means there is no number change in photons which is possible via the ${A_i}^2$ interaction, so focusing on that:

$$\vec{A}^2(0,t) = \sum_{k_1,k_2}\sum_{r,s}\frac{\hbar c^2}{2V\sqrt{\omega_1\omega_2}}[\varepsilon_r(k_1)\cdot\varepsilon_s(k_2)]$$
$$\times\left[a_r(k_1)e^{-i\omega_1 t} - {a_r}^\dagger(k_1)e^{i\omega_1 t}\right]\left[a_s(k_2)e^{-i\omega_2 t} - {a_s}^\dagger(k_2)e^{i\omega_2 t}\right]$$

Thus,

$$\left\langle k', s\left|\frac{e^2}{2mc^2}\vec{A}^2(0,t)\right|k, r\right\rangle = \frac{\hbar e^2}{2mV\sqrt{\omega\omega'}}[\varepsilon_r(k)\cdot\varepsilon_s(k')]e^{i(\omega'-\omega)t}$$

where $\omega = c|k|$ and $\omega' = c|k'|$. Thus

$$W_{r\to s}(k')d\Omega = \frac{\mathrm{c}}{\mathrm{v}}\frac{\mathrm{e}^4}{16\pi^2\mathrm{m}^2\mathrm{c}^4}[\varepsilon_r(k)\cdot\varepsilon_s(k')]^2\mathrm{d}\Omega$$

Dividing by the photon flux c/v, we get the cross-section:

$$\sigma_{r\to s}(k')d\Omega = \frac{e^4}{16\pi^2 m^2 c^4}[\varepsilon_r(k)\cdot\varepsilon_s(k')]^2 d\Omega,\quad \frac{e^4}{16\pi^2 m^2 c^4} = r_0^2,\quad r_0 = 2.818fm.$$

For an unpolarized incident photon beam, we get:

$$\sum_r [\varepsilon_r \cdot \varepsilon_s{}']^2 = 1 - (\frac{k}{|k|}\cdot \varepsilon_s')^2$$

and

$$\sum_r (k\cdot\varepsilon_s')^2 = 1 - (k\cdot k')^2 = \sin^2\theta$$

where θ is the angle between the directions k and k' of the incident and scattered photons. Putting this together:

$$\frac{1}{2}\sum_r\sum_s [\varepsilon_r\cdot\varepsilon_s{}']^2 = \frac{1}{2}(1+\cos^2\theta),$$

hence, the unpolarized scattering cross-section for scattering through an angle θ is:

$$\sigma(\theta)d\Omega = \frac{1}{2}r_0^2(1+\cos^2\theta)d\Omega.$$

Integrating over all angles we get the total cross section for Thomas scattering to be:

$$\sigma = \frac{8\pi}{3}r_0^2 = 6.65\times 10^{-25}cm^2.$$

Chapter 3. Canonical Quantization

We will arrive at the Second Quantization formalism from the more formally grounded Canonical Quantization.

In the sections that follow use is made of excellent lecture notes taken while at Caltech, UWM, and Oxford, and from various texts. Most notable are the lecture notes from a quantum field theory course taught by Nicholas Papastamatiou at UWM, and those notes, in turn, drew strongly from the text by Ramond [12]. For an excellent user friendly introduction, with much more detail in some areas than can be provided in what follows, see both Mandl & Shaw [13] and Ryder [14]. This book also encompass a broad range of advanced applications, ranging from gauge fields and renormalization to the standard model and curved spacetime. Notes from these areas, including work done for my PhD thesis [15], and from numerous excellent sources [9,16-24], will be described.

3.1 Classical Lagrangian Field Theory

Denote the Lagrangian: $\mathcal{L} = \mathcal{L}(\varphi_r\,, \varphi_{r,\alpha})$, where $r = 1, \ldots, N$, for several fields in whatever dimensions. The α index indicates the dimension of differentiation (one time dimension, n-space). Although not the most general Lagrangian starting point, it covers most cases of interest in what follows. The Action, S, is then written in terms of the Lagrangian density integrated over spacetime volume Ω:

$$S(\Omega) = \int_\Omega d^4x\, \mathcal{L}(\varphi_r\,, \varphi_{r,\alpha})$$

We will consider the standard variation (e.g., not Palatini form), where

$$\varphi_r(x) \to \varphi_r(x) + \delta\varphi_r(x),$$

and where the variations vanish, $\delta\varphi_r(x) = 0$, on the boundary of the spacetime volume $\Gamma(\Omega)$.

We want the action to have a stationary value, i.e. $\delta S(\Omega) = 0$, so let's now derive the consequences of such a requirement with our current formulation:

$$\delta S(\Omega) = \int_\pi d^4x \left\{ \frac{\partial \mathcal{L}}{\partial \varphi_r} \delta\varphi_r + \frac{\partial \mathcal{L}}{\partial \varphi_{r,\alpha}} \delta\varphi_{r,\alpha} \right\}$$

Recall $\delta\varphi_{r,\alpha} = \frac{\partial \mathcal{L}}{\partial x^\alpha}\delta\varphi_r$, thus:

$$\delta S(\Omega) = \int_\Omega d^4x \left\{\frac{\partial \mathcal{L}}{\partial \varphi_r} - \frac{\partial}{\partial x^\alpha}\left(\frac{\partial \mathcal{L}}{\partial \varphi_{r,\alpha}}\right)\right\}\delta\varphi_r + \int_\Omega d^4x \frac{\partial}{\partial x^\alpha}\left(\frac{\partial \mathcal{L}}{\partial \varphi_{r\alpha}}\partial\varphi_r\right),$$

where the left integral gives rise to the Euler-Lagrange Equations for the Field, and the right integral, being a perfect divergence, is a surface term, thus zero.

To get a visualization of the quantization in more familiar terms, lets approximate the system by an infinitely countable number of degrees of freedom (and ultimately go to the continuum limit). Consider the system at a fixed instant of time t and decompose the flat spacelike surface t-constant, into small cells of equal volume δx_i, that we label by the index $i = 1,2,\ldots$ and we approximate the values of the fields within each cell by their values at, say, the center of the cell $\vec{x} = \vec{x}_i$. The system is now described by the discrete set of generalized coordinates:

$$q_{ri} \equiv \varphi_r(i,t), \quad r = 1, \ldots N, \quad i = 1,2\ldots$$

and

$$L(t) = \sum_i \delta\vec{x}_i \mathcal{L}_i[\varphi_r(i,t), \dot{\varphi}_r(i,t), \varphi_r(i',t)],$$

where the Lagrangian density in the i-th cell, $\mathcal{L}_i$, depends on the fields at the neighbouring lattice sites i' on account of the approximation of the spatial derivatives:

$$p_{r,i}(t) = \frac{\partial L}{\partial \dot{q}_{ri}} \equiv \frac{\partial L}{\partial \dot{\varphi}_r(i,t)} \equiv \pi_r(i,t)\delta x_i \ , \ where\ \pi_r(i,t) \equiv \frac{\partial \mathcal{L}_i}{\partial \dot{\varphi}_r(i,t)}$$

Thus,

$$H = \sum_i p_{ri}\,\dot{q}_{ri} - L = \sum_i \delta\vec{x}_i \left\{\pi_r(i,t)\dot{\varphi}_r(i,t) - \mathcal{L}_i\right\}$$

Let's now take the limit $\delta x_i \to 0$:

$$\pi_r(x) = \frac{\partial L}{\partial \dot{\varphi}_r}\ , \quad L(t) = \int d^3\vec{x}\mathcal{L}\left(\varphi_r, \varphi_{r,\alpha}\right), \quad H = \int d^3x\,\mathcal{H}(x)$$

and

$$\mathcal{H}(x) = \pi_r(x)\dot{\varphi}_r(x) - \mathcal{L}\left(\varphi_r, \varphi_{r,\alpha}\right)$$

Example 1 (the Klein-Gordon equation)

Consider $\mathcal{L} = \frac{1}{2}\left(\varphi_{,\alpha}\varphi_{,}^{\alpha} - \mu^2\varphi^2\right) \ \rightarrow\ (\Box + \mu^2)\varphi(x) = 0$, then:

$$\pi(x) = \frac{1}{c^2}\dot{\varphi}(x)$$

and

$$\mathcal{H}(x) = \frac{1}{2}[c^2\pi^2(x) + (\nabla\varphi)^2 + \mu^2\varphi^2]$$

Example 2 (the Complex Klein-Gordon equation)

Let $\varphi(x) = \frac{a(x)+ib(x)}{\sqrt{2}}$ be a complex field that has Lagrangian $\mathcal{L} = \frac{1}{2}\left(\varphi_{,\alpha}\left(\varphi_{,}^{\alpha}\right)^* - \mu^2\varphi\varphi^*\right)$,

we find that $(\Box + \mu^2)\varphi(x) = 0$ and $(\Box + \mu^2)\varphi^*(x) = 0$.

The complex Klein-Gordon Lagrangian from Example 2 is invariant under global phase transformation:

$$\varphi(x) \to e^{iq\alpha}\varphi(x), \;\; \varphi^*(x) \to e^{-iq\alpha}\varphi^*(x)$$

with infinitesimal transformation:

$$\delta\varphi = iq(\delta\alpha)\varphi, \qquad \delta\varphi^* = -iq(\delta\alpha)\varphi^*$$

and

$$\delta(\partial_\mu\varphi) = iq(\delta\alpha)(\partial_\mu\varphi), \qquad \delta(\partial_\mu\varphi^*) = -iq(\delta\alpha)(\partial_\mu\varphi^*)$$

Complex Klein-Gordon Field with global phase invariance

Let's now compute the variation in the Lagrangian under global phase:

$$\delta\mathcal{L} = (\delta\alpha)\partial_\mu j^\mu, \;\; j^\mu = iq[\varphi^*\partial_\mu\varphi - (\partial_\mu\varphi^*)\varphi].$$

Thus, zero variation under global phase, $\delta\mathcal{L} = 0$, means there is a conserved Noether current:

$$\partial_\mu j^\mu = 0,$$

where q is identified as electric charge in this context.

Complex Klein-Gordon Field with local phase invariance (gauge invariance)

Recall that gauge transformations describe invariance under local (phase) transformations, so let's consider what that would mean in this same context. Let

$$\varphi(x) \to e^{iq\alpha(x)}\varphi(x), \;\; \varphi^*(x) \to e^{-iq\alpha(x)}\varphi^*(x).$$

We now see that

$$\partial_\mu\varphi \to e^{iq\alpha(x)}\left[\partial_\mu\varphi + iq\left(\partial_\mu\alpha(x)\right)\varphi\right],$$

where there is a new term due to the non-zero differentiation $\partial_\mu\alpha(x)$. The method of differentiation is generalized to a gauge-covariant derivative at this juncture to allow the gauge transformation to be invariant. The new $iq\left(\partial_\mu\alpha(x)\right)\varphi$ is offset if we define the gauge covariant derivative to be:

$$\mathcal{D}_\mu = \partial_\mu + iqA_\mu(x),$$

where a gauge field $A_\mu(x)$ has been introduced such that $A_\mu(x) \to A_\mu(x) - \partial_\mu\alpha$ under transformation, so that now:

$$\mathcal{D}_\mu \varphi \to e^{iq\alpha(x)}\mathcal{D}_\mu \varphi,$$

and we have a complex Klein-Gordon scalar field with local gauge invariance according to a gauge vector field A_μ.

Dirac Field with gauge invariance

Let's now consider the Dirac field with a local phase invariance as described for the complex Klein-Gordon Field above. Since the Dirac field is also complex (with adjoint representation slightly more complex than complex conjugate, e.g., $\psi \to \bar{\psi}$, not $\psi \to \psi^*$). Thus, the free Dirac equation:

$$\mathcal{L}_{free} = \bar{\psi}(i\gamma^\mu \partial_\mu - m)\psi$$

generalizes as before

$$\mathcal{L} = \bar{\psi}(i\gamma^\mu \mathcal{D}_\mu - m)\psi = \bar{\psi}(i\gamma^\mu \partial_\mu - m)\psi - qA_\mu \bar{\psi}\gamma^\mu \psi = \mathcal{L}_{free} - J^\mu A_\mu$$

where $J^\mu = q\bar{\psi}\gamma^\mu \psi$, and it is indeed a conserved current. If we now add a kinetic term to make the vector gauge field A_μ dynamical, we arrive at the QED Lagrangian, and we already know what kinetic term will suffice from electromagnetism:

$$\mathcal{L}_{QED} = \mathcal{L}_{free} - J^\mu A_\mu + \frac{1}{4} F^{\mu\nu}F_{\mu\nu} .$$

3.2 Quantization of Lagrangian Field Theory

Thus far we've considered several important classical field theories. Let's now quantize these theories using canonical quantization. Let's start by considering the real Klein-Gordon equation to keep things simple. With Lagrangian and canonical momentum as before, we arrive at the canonical quantization relations:

$$[\phi(x,t), \Pi(x',t')] = i\hbar\delta(x - x')$$
$$[\phi(x,t), \phi(x',t')] = 0$$
$$[\Pi(x,t), \Pi(x',t')] = 0;$$

Let's shift to the notation for positive and negative 'movers':

$$\varphi(x) = \varphi^+(x) + \varphi^-(x),$$

$$\varphi^+(x) = \sum_k \left(\frac{\hbar c^2}{2V\omega_k}\right)^{1/2} a(k)e^{-ikx}, \quad \varphi^-(x) = \sum_k \left(\frac{\hbar c^2}{2V\omega_k}\right)^{1/2} a^\dagger(k)e^{ikx}$$

with

$$\varphi(x) = \sum_k \left(\frac{\hbar c^2}{2V\omega_k}\right)^{1/2} \left[a(k)e^{-ikx} + a^\dagger(k)e^{ikx}\right]$$

where k is a wave 4-vector. In terms of the canonical commutation relations at field level, we can now find the induced relations at the level of the 'a' operator:

$$[a(k), a^\dagger(k')] = \delta_{kk'}$$
$$[a(k), a(k')] = 0$$
$$[a^\dagger(k), a^\dagger(k')] = 0.$$

The relations are precisely those of the harmonic oscillator, with the same operator algebra and spectrum as seen in that carefully examined case. Thus, we have the number operator:

$$N(k) = a^\dagger(k)a(k) \rightarrow n(k) = 0, 1, 2, \ldots$$

and Hamiltonian operator:

$$H = \sum_k \hbar\omega_k \, [N(k) + \frac{1}{2}]$$

and momentum operator:

$$\vec{P} = \sum_k \hbar\vec{k} \left[N(k) + \frac{1}{2}\right].$$

Note that the vacuum has infinite energy $\frac{1}{2}\sum_k \hbar\omega_k$. One way to eliminate this infinity (other than simply subtracting it) is to say it never arises since your choice of operator ordering when going from the classical theory to the quantum can be whatever you choose (e.g., you choose not to introduce an infinity). The standard convention is known as "Normal Ordering" where all absorption (annihilation) operators to the right of creation operators.

3.3 Preview of Pion Scattering – Klein-Gordon field with Second Quantization

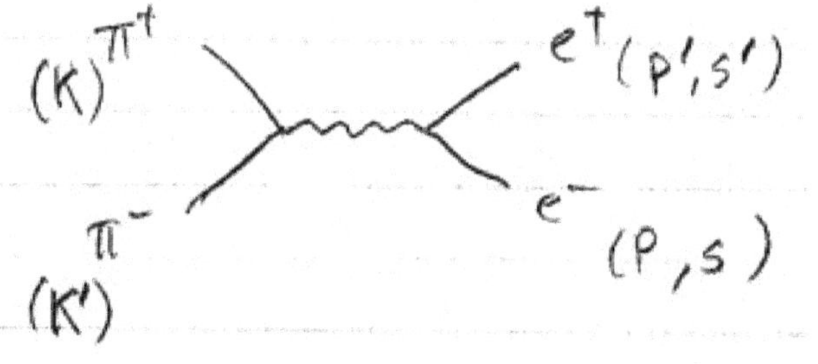

We want

$$ {}_{out}\langle (p.s)(p',s')|k,k'\rangle_{in} $$

For a scalar field $E^2 = p^2 + m^2 \Rightarrow p^2 - E^2 + m = 0$, and we have $p \to i\nabla_x$, $E \to \frac{i\partial}{\partial t}$, so, upon 1st quantization, we have

$$ \left(\frac{\partial^2}{\partial t^2} - \nabla^2 + m\right)\varphi = 0 $$

using

$$ \eta^{\mu\nu} = (+1,-1,-1,-1) \ and\ \Box = \partial^\mu\partial_\mu = \partial_t^2 - \nabla^2 $$

We get:

$$ (\Box + m^2)\varphi = 0 $$

The Klein-Gordon equation where φ is a free field.

Since we will be considering a scattering problem, we will refer to the asymptotically evolved field φ_{in} and φ_{out} as free. The quantum nature of the restrictions of φ come about through the canonical quantization procedure (2nd quantization) and this is evidenced by

$$ [\varphi_{in}(x), \varphi_{in}(y)] \neq 0. $$

Recall

$$ \varphi_{in}(x) = \int \frac{d^3k}{(2\pi)^{3/2}\sqrt{2\omega_k}}\left[a_{\vec{k}}e^{-i\omega_k t+\vec{k}\cdot\vec{x}} + a_{\vec{k}}^\dagger e^{i\omega_k t-\vec{k}\cdot\vec{x}}\right] $$

And, from 2nd Quantization (using Lagrangian and conjugate momentum), we get simple (harmonic oscillator type) commutation relations on the a's. Thus,

$$ [\varphi_{in}(x), \varphi_{in}(y)] $$
$$ = \frac{1}{(2\pi)^3}\int \frac{d^3k}{\sqrt{2\omega_k}}\int \frac{d^3q}{\sqrt{2\omega_q}}\{\delta(\vec{k} - \vec{q})e^{-ikx+iqy} $$
$$ - \delta|(\vec{k} - \vec{q})e^{ikx-iqy}\} $$
$$ = \int \frac{d^3k}{(2\pi)^3 2\omega_k}\left(e^{-ik\cdot(x-y)} - e^{+ik\cdot(x-y)}\right) \equiv i\Delta(x-y). $$

The commutator is Lorentz invariant.

Examples of causality:

$$ [\varphi_{in}(\vec{x},t), \varphi_{in}(\vec{y},t)] = i\Delta(\vec{x} - \vec{y}, 0) = 0 $$

And

$$ [\varphi_{in}(\vec{x},t), \dot{\varphi}_{in}(\vec{y},t)] = i\int \frac{d^3k}{(2\pi)^3 2}\left[e^{ik(x-y)} + e^{-ik(x-y)}\right] = i\delta(\vec{x} - \vec{y}) $$

So far, we've arrived at $\delta(\vec{x}-\vec{y})$, which is a somewhat artificial concept without further definition (distribution theory). Or, we can remove the delta with smearing, but a smeared $[\varphi(x), \varphi(y)]$ is extreme since it would damage causality. Let's focus on causality vis-à-vis identification of a spectrum of a's for the theory. We have that causality can be defined in terms of a spectrum of a operators. This allows the formulation to proceed along two separate tracks: (i) from causality we can define canonical quantization on the field operators, and thereby have the formal canonical quantization of the field theory. On track (ii) we start with the spectrum of a's and define the commutation relations on creation and annihilation operators, and thereby arrive at "2nd quantization" in the theory.

3.4 The LSZ Formalism

Consider a general solution of the Klein-Gordon equation $\Box + m^2 \;\; f_k(x) = 0$, here let

$$f_k = \frac{e^{-ik\cdot x}}{(2\pi)^{3/2}\sqrt{2\omega_k}}$$

although other complete sets exist. Note that the normalization is such that in the Klein-Gordon inner product we have:

$$(f_k, f_l) = i\int d^3x\, f_{\vec{k}}^*(x)\overleftrightarrow{\partial}_0 f_{\vec{k}}^{\square}(x) = \delta_3(\vec{k}-\vec{l})$$

Note $\partial_0(f_k, f_l) = 0 \quad s0 \;\; (f_k, f_l)$ is a conserved quantity in the evolution. Now,

$$a_{\vec{k}} = i\int d^3x\, f_k^*(x)\overleftrightarrow{\partial}_0\varphi_{in}(x)$$

For an interacting field the commutation relations at equal times are identical, but we have no simple relation for a 'Δ' at unequal times short of actually solving a hideous nonlinear equation.

Still, we want to relate φ_{in} with the φ that obeys the interacting Lagrangian theory. As long as our interaction term in the Lagrangian obeys similar symmetries such as Lorentz invariance we can make use of the Noether relation – Noether currents being conserved. Thus,

$$[P^\mu, \varphi(x)] = -i\partial^\mu\varphi(x)$$

(from Heisenberg's equation of motion which follows by Lorentz invariance). The intuitive notion relating fields is that:

$$\varphi \to \varphi^{in} \;\; as\, t \to -\infty$$

In terms of observables, we want to show: $\lim_{t\to\infty} {}_{in}\langle\alpha|\varphi|\beta\rangle_{in} = {}_{in}\langle\alpha|\varphi^{in}|\beta\rangle_{in}$. Thus, consider:

$$ {}_{in}\langle\alpha|\varphi|\beta\rangle_{in} = {}_{in}\langle\alpha|e^{iHt}\varphi(\vec{x},0)e^{-iHt}|\beta\rangle_{in} $$
$$ = e^{i(E_\alpha - E_\beta)t}\,{}_{in}\langle\alpha|\varphi(\vec{x},0)|\beta\rangle_{in} $$

So, we have an oscillating terms that leaves a limit such that $t \to -\infty$ is ill-defined, so perhaps describing an interacting field in a basis consisting of energy eigenstates is ill advised. We get around this with a little smearing. Use as a smearing function a solution to the Klein-Gordon equation: $(\Box + m^2)f(x) = 0$. Thus,

$$ \varphi_f(t) = i\int d^3x\, f^*(x)\overleftrightarrow{\partial_0}\varphi(x) $$

And

$$ \varphi_f^{in}(t) = i\int d^3x\, f^*(x)\overleftrightarrow{\partial_0}\varphi^{in}(x) $$

With this latter form being time independent since φ^{in} satisfies Klein-Gordon equation.

So, now consider the smeared form:

$$ \lim_{t\to-\infty}\langle\alpha|\varphi_f|\beta\rangle = \langle\alpha|\varphi_f^{in}|\beta\rangle $$

this is still too restrictive since we have $[\varphi^{in}(x), \varphi^{in}(y)] = i\delta^3$ and $[\varphi(x), \varphi^{in}(y)] = i\delta^3$ which both enforce a certain scale which may not be in agreement. Since the scale was not the essential feature of the commutation relation we generalize to an adjustable scale such that agreement can be achieved:

$$ \lim_{t\to-\infty}\langle\alpha|\varphi_f|\beta\rangle = Z^{1/2}\langle\alpha|\varphi_f^{in}|\beta\rangle $$

Known as the *LSZ* formalism. In essence, this relation enforces a certain boundary condition of the possible solutions to the nonlinear diff. eq. for φ in the interacting theory. In a more direct sense, we've introduced differing wave function normalization factors so as to achieve agreement.

3.5 Covariant Commutation

Now, study

$$ <0|[\varphi(x), \varphi(y)]|0> = i\Delta'(x-y) $$

Where we start with a less detailed analysis first:

$i\Delta'(x-y) = \int d^4q\left(e^{-iq(x-y)} - e^{iq\cdot(x-y)}\right)\sum_\alpha \delta_4(q - P_x)\ |\langle 0|\varphi(0)|\alpha\rangle|^2$

Let $p(q) = (2\pi)^3\sum_\alpha \delta_4(q - P_x)|\ |\langle 0|\varphi(0)|\alpha\rangle|^2$

Note, $p(\Lambda q) = p(q)$,SO $p(q) = \theta(q^0)\sigma(q^2)$

$$ i\Delta'(x-y) = \int\frac{d^4q}{(2\pi)^3}\sigma(q^2)\epsilon(q^0)\,e^{-q\cdot(x-y)} $$

Recall, $i\Delta(x-y) = \int\frac{d^4q}{(2\pi)^3}\epsilon(k^0)\,\delta(k^2 - m^2)e^{-ik\cdot(x-y)}$ So,

$$i\Delta'(x-y) = \int_0^\infty dk^2\sigma(k^2)\int \frac{d^4q}{(2\pi)^3}\theta(q^0)\,\delta(q^2-k^2)e^{-iq\cdot(x-y)}$$
$$= \int_0^\infty dk^2\,\sigma(k^2)i\Delta(x-y;k^2)$$

Let's now start again, but first find $\langle 0|\varphi_f|\vec{k}\rangle$:

$$\langle 0|\varphi_f|\vec{k}\rangle = i\int d^3x f^*(x)\partial_0\langle 0|\varphi(\vec{x},t)|\vec{k}\rangle$$

Note that: $\langle 0|\varphi(\vec{x},t)|\vec{k}\rangle = < 0|e^{ip\cdot x}\varphi(0)e^{-ip\cdot x}|\vec{k} > = e^{-ik\cdot x} < 0|\varphi(0)|\vec{k} >$, so:

$$\langle 0|\varphi_f|\vec{k}\rangle = i < 0|\varphi(0)|\vec{k} > \int d^3x f^*(x)\vec{\partial}_0 e^{-ik\cdot x}$$

Note that

$$< 0|\varphi_f{}^{in}(x)|\vec{k} > = \frac{e^{-ik\cdot x}}{(2\pi)^{3/2}\sqrt{2\omega_k}}$$

With the limit in LSZ theory such that:

$$\lim_{t\to-\infty} < \alpha|\varphi_f|\beta > = Z^{1/2} < \alpha|\varphi_f^{in}|\beta >$$

So, $< 0|\varphi_f{}^{in}(x)|\vec{k} > = \frac{i}{(2\pi)^{3/2}\sqrt{2\omega_k}}\int d^3x f^*(x)\vec{\partial}_0 e^{-ik\cdot x}$

Thus

$$< 0|\varphi(0)|\vec{k} > = \frac{Z^{1/2}}{(2\pi)^{3/2}\sqrt{2\omega_k}} \quad \textit{from LSZ}$$

Returning to the $\Delta'(x-y)$ calculation:

$i\Delta'(x-y) = < 0|[\varphi(x),\varphi(y)]|0 > = < 0|\varphi(x)\varphi(y) - \varphi(y)\varphi(x)|0 >$
$= \sum_\alpha < 0|\varphi(x)|\alpha >< \alpha|\varphi(y)|0 > - C.C$
$= \sum_\alpha | < 0|\varphi(0)|\alpha > |^2 \left(e^{-p_\alpha(x-y)} - e^{iP_\alpha\cdot(x-y)}\right)$
$= \int d^4q\left(e^{-iq\cdot(x-y)} - e^{iq\cdot(x-y)}\right)\sum_\alpha \delta_4(q-P_\alpha)| < 0|\varphi(0)|\alpha >|^2$
$P(q) \equiv (2\pi)^3\sum_\alpha \delta_4(q-P_\alpha)|< 0|\varphi(0)|\alpha >|^2 = \theta(q^0)\sigma(q^2)$ (since $P(q) = p(\Lambda q)$)

So, $i\Delta'(x-y) = \int \frac{d^4q}{(2\pi)^3}\theta(q^0)\,\sigma(q^2)\left(e^{-iq\cdot(x-y)} - e^{iq\cdot(x-y)}\right) =$
$\int \frac{d^4q}{(2\pi)^3}\sigma(q^2)\,\epsilon(q^0)e^{-iq\cdot(x-y)}$

$i\Delta'(x-y) = \int_0^\infty dk^2\sigma(k^2)\int \frac{d^4q}{(2\pi)^3}\epsilon(q^0)\,\delta(q^2k^2)e^{-iq\cdot(x-y)}$

So,

$$i\Delta'(x-y) = \int_0^\infty dk^2 \sigma(k^2)\, i\Delta(x-y;k^2)$$

Now, $p(q) = Z\theta(q^0)\delta(q^2-m^2) + p'(q)$ (for the single particle and the rest), and similarly, $\sigma(q) = Z\delta(q^2-m^2) + \sigma'(q)$, so we can write:

$$i\Delta'(x-y) = iZ\Delta(x-y;m^2) + \int_{4m^2}^\infty dk^2 \sigma'(k^2)\Delta(x-y;k^2)$$

Where we now take $\frac{\partial}{\partial t_y}|_{t_x=t_y}$: $\frac{\partial}{\partial t_y}\Delta(x-y;m^2)|_{t_x=t_y} = \delta(\vec{x}-\vec{y})$ (which is indep. of m).

So,

$$1 = Z + \int_{4m^2}^\infty dk^2\, \sigma'(k^2)$$

So,

$$0 < Z < 1 \quad and \quad Z = 1 \quad for\ \sigma' = 0\,, i.e. a\ free\ theory.$$

What is $i\Delta_f(x-y)$?

$i\Delta_f(x-y) =< 0|T\varphi^{in}(x)\varphi^{in}(y)|0 >$

We know $(\Box_x + m^2)\varphi^{in}(x) = 0$, so:

$$(\Box_x + m^2)\{\theta(t_x-t_y) < 0|\varphi^{in}(x)\varphi^{in}(y)|0 > +\theta(t_y-t_x) < 0|\varphi^{in}(y)\varphi^{in}(x)|0 >\}$$
$$= [(\partial_{t_x}^2 + \partial_{t_x})\theta(t_x-t_y)] < 0|\varphi^{in}(x)\varphi^{in}(y)|0 >$$
$$+ [(\partial_{t_x}^2 + \partial_{t_x})\theta(t_y-t_x)] < 0|\varphi^{in}(y)\varphi^{in}(x)|0 >$$
$$+\delta(t_x-t_y) < 0\left[\frac{\partial}{\partial t_x}\varphi^{in}(x), \varphi^{in}(y)\right]|0 >$$
$$+< 0\left|T\left((\Box_x + m^2)\varphi^{in}(x)\right)\varphi^{in}(y)\right|0 >$$
$$= \delta(t_x-t_y) < 0|[\dot{\varphi}^{in}(x), \varphi^{in}(y)]|0 >= \delta(t_x-t_y)\{< 0|-i\delta^3(\vec{x}-\vec{y})|0 >\} = -\delta^4(x-y)\,.$$

So,

$$(\Box_x + m^2)i\Delta_f(x-y) = -\delta^4(x-y)$$

If we Fourier transform:

$$\Delta_f(x-y) = \frac{1}{c0nst}\int d^4p e^{-ip\cdot(x-y)}\bar{\Delta}_f(p)\,, where\ (-p^2 + m^2)\bar{\Delta}_f(p) = -1.$$

So,

$$\Delta_f(x-y) = \int \frac{d^4k}{(2\pi)^4} \frac{e^{-ik\cdot(x-y)}}{k^2 - m^2 + i\epsilon'}$$

Where the "$i\epsilon$" term is used to indicate the contour of complex integration.

What is $i\Delta'_f(x-y)$ (with the φ'^s interacting)?

$$i\Delta'_f(x-y) = \int_0^\infty dk^2 \sigma(k^2)\, i\Delta_f(x-y;k^2) = iZ\Delta_f(x-y;m^2)$$

$$+ \int_{4m^2}^\infty dk^2 \sigma'(k^2) i\Delta_f(x-y;k^2)$$

Consider the F.T of Δ'_f:

$\Delta'_f(x) = \int \frac{d^4k}{(2\pi)^4} e^{-ik} \bar{\Delta}'_f(k^2) = iZ\Delta_f(x-y;m^2) + \int_{4m^2}^\infty dk'^2 \sigma'(k^2) i\Delta_f(x-y;k^2)$

Thus,

$$\bar{\Delta}'_f(k^2) = \frac{Z}{k^2 - m^2 + i\epsilon} + \int_{4m^2}^\infty dk^2 \frac{\sigma'(k^2)}{k^2 - k'^2 + i\epsilon}$$

So, the propagator has all of the information pertaining to the spectrum of the theory: a pole at $k^2 = m^2$ and a branch cut (a line of poles) at $k^2 = 4m^2$ out to ∞ along the real axis.

Let's consider a theory with an interaction term, $\frac{\lambda}{4}\varphi^4$, and consider ingoing and outgoing (scattered) momentum states to arrive at and expression known as the Reduction formula.

Consider

$$L = \frac{1}{2}\left(\partial^\mu \varphi \partial_\mu \varphi - m^2\varphi^2\right) - \frac{\lambda}{4}\varphi^4$$

This theory is quantized by demanding that $[\varphi, \pi]_{same\ time} = i\delta_3(\vec{x} - \vec{y})$ and

$$\pi(x) = \frac{\partial L}{\partial(\dot{\varphi})} = \dot{\varphi}(x)$$

Where φ is supposed to do two things:

(1) Satisfy the equation of motion: $(\partial^2 + m^2)\varphi = -\lambda\varphi^3$

(2) Be an operator which can be rep. by a matrix an appropriate Hilbert space.

In Hilbert space $| > in$, a formal relation between φ *and* φ^{in} is:

$$\lim_{t\to-\infty} < \alpha|\varphi_f(t)|\beta > \ = \ < \alpha|\varphi_f^{in}|\beta > Z^{1/2}$$

Where $\varphi_f(t) = \int_{⬚}^{⬚} d^3x\, f^*(x)\overleftrightarrow{\partial}_0\varphi(x)$ (smearing to get rid of oscillations in limit).

The asymptotic field creates only a one particle state whereas the interacting field creates an entire spectrum of particle states.

Now, can also consider an "out" Hilbert space, with $| > out$, etc.

For φ^{out}:

$$\lim_{t\to\infty} < \alpha|\varphi_f(t)|\beta > = < \alpha|\varphi_f^{out}|\beta > Z^{1/2}$$

Of course the in and out states are isomorphic, thus, there is a unitary operator that connects the in and out states known as the S-operator.

$$S|\alpha >_{out} = |\alpha >_{in}$$

Some states are unaffected by the S – operator: the vacuum $|0 >_{in} = |0 >_{out}$ (conservation laws) and the single particle states $|\vec{k} >_{in} = |\vec{k} >_{out}$ (single particle is stable in $\lim_{t\to\infty}$ since $\vec{k} >_{in}$ is defined).

Consider state $|\alpha >_{in}$ at very early times, what is the amplitude that at very late time we arrive in state $|\beta >_{out}$:

$$amplitude\ of\ |\alpha >_{in} \ \rightarrow \ |\beta >_{out} \ = \ {}_{out}< \beta|\alpha >_{in}$$

So, ${}_{out}< \beta|\alpha >_{in} \ = \ {}_{out}< \beta|S|\alpha >_{out}$. Or, using $|\alpha >_{out} = S^{-1}|\alpha >_{in} = S^{\dagger}|\alpha >_{in}$ (since $SS^{\dagger} = 1$ due to unitarity). So,

$${}_{out}< \beta|\alpha >_{in} = {}_{in}< \beta|S|\alpha >_{in}$$

Isomorphism → norm preserved:

$${}_{in}< \alpha|\alpha >_{in} = {}_{out}< \alpha|S^{\dagger}S|\alpha >_{out}$$

Considering field theory in spacetime which isn't geodesically complete (e.g., bad geodetics are of measure zero), we can then express the "Reduction formula".

3.6 The Reduction formula

${}_{out}< k_1 \dots k_n|P_1 \dots P_m >_{in} = {}_{out}< k_2 \dots k_n|a_{k_1}^{out}|P_1 \dots P_m >_{in}$

Recall $a_k^{out} = i\int d^3x\, f^*{}_k(x)\overleftrightarrow{\partial}_0\varphi^{out}(x)$ where $f_k(x) = \frac{e^{-ikx}}{\sqrt{(2\pi)^3 2\omega_k}}$ (this f is not the smearing function…yet). The choice of complete set $f_k(x)$ is the only place where momentum eigenstates are explicitly used, it is possible to simply discuss any $f_k(x)$ which are any complete set of solutions to the wave equation.

From LSZ: $\langle\alpha|\varphi_f^{in}|\beta\rangle = Z^{-1/2} \lim_{t\to-\infty}\langle\alpha|\varphi_f|p\rangle$, and we simply make the smearing function $f_k(x)$

$\langle\alpha|\varphi_f^{in}|\beta\rangle$

$$= iZ^{-1/2} \lim_{t\to-\infty} \int d^3x\, f^*_{k_1}(x_1) \frac{\overleftrightarrow{\partial}}{\partial x_1^0}\ {}_{out}< k_2 \dots k_n|\varphi^{out}|P_1 \dots P_m >_{in}$$

Now ${}_{out}< \beta|\alpha >_{in} = {}_{out}< \beta|S|\alpha >_{out} = {}_{in}< \beta|S|\alpha >_{in}$, $and\ SS^+ = 1$.
The S – matrix can be written:

$$S_{fi} = \delta_{fi} + 2\pi i\delta(\dots)T_{fi}$$

Since we're only interested in the T_{fi} portion of S_{fi} spectator particles should have there amplitudes separated from that of the scattering amplitude in ${}_{out}< \alpha|\beta >_{in}$.
Spectator particles.

$$iZ^{-is} \lim_{t\to-\infty} \int d^3x_1 f^*_{k_1}(x_1)\, \vec{\partial}_0\ {}_{out}< k_2 \dots k_n|\varphi^{out}|P_1 \dots P_m >_{in}$$

$$= {}_{out}< k_2 \dots k_n\big|a^{in}_{k_1}\big|P_1 \dots P_m >_{in}$$

$= zero\ unless\ a\ p\ is = k$ i.e. a spectator.
Subtract off spectators and group under S, for the k_1 at hand call the term S_1:

${}_{out}< k_1 \dots k_n|P_1 \dots P_m >_{in}$

$$= iZ^{-1/2} \left\{ \begin{aligned} &\lim_{t\to+\infty} \int d^3x_1\, f^*_{k_1}(x_1) \frac{\overleftrightarrow{\partial}}{\partial x_1^0}\ {}_{out}< k_2 \dots k_n|\varphi^{out}|P_1 \dots P_m >_{in} \\ &- \lim_{t\to+\infty} \int d^3x\, f^*_{k_1}(x_0) \frac{\overleftrightarrow{\partial}}{\partial x_1^0}\ {}_{out}< k_2 \dots k_n|\varphi^{out}|P_1 \dots P_m >_{in} \end{aligned} \right\}$$

$+ S_1$

$$= iZ^{-1/2} \int d^3x\, f^*_{k_1}(x_0) \frac{\overleftrightarrow{\partial}}{\partial x_1^0}\ {}_{out}< k_2 \dots k_n|\varphi^{out}|P_1 \dots P_m >_{in} \Big\} + S_1.$$

Now, $\partial_0[f^*\partial_0 < in > -(\partial_0 f^*) < in >] = f^*\partial_0^2 < \cdots > -({\partial_0}^2 f^*) < \cdots >$. Integrate twice by parts with $(\nabla^2 f^*) < \cdots |\varphi| \dots >$, surface terms are zero since φ drops off at infinity, so

$$= f^*(\partial_0^2 - \nabla^2 + m^2) < \cdots >.$$

Define the Klein-Gordon operator $K(x) = \partial^\mu \partial_\mu + m^2, then$

${}_{out}< k_2 \dots k_n|P_1 \dots P_m >_{in}$

$$= iZ^{-1/2} \int d^4x_1\, f^*_{k_1}(x_1) k(x_1)\ {}_{out}< k_2 \dots k_n|\varphi(x_1)|P_1 \dots P_m >_{in}$$

Proceeding to "rate out K's" we find commutativity of $\varphi^{'s}$ is an issue:

${}_{out}< k_2 \dots k_n|\varphi(x_1)|P_1 \dots P_m >_{in} = < k_3 \dots k_n\big|a^{out}_{k_2}\varphi(x_1)\big|P_1 \dots P_m >_{in}$

$$= \int d^3x_2\, f^*_{k_2}(x_2) \frac{\overleftrightarrow{\partial}}{\partial x_2^0}\ {}_{out}< k_3 \dots k_n|\varphi^{out}(x_2)\varphi(x_1)|P_1 \dots P_m >_{in}$$

$$= iZ^{-1/2} \lim_{t\to-\infty} \int d^3x_2 f^*_{k_2}(x_2) \frac{\overleftrightarrow{\partial}}{\partial x_2^0}\ {}_{out}< k_3 \dots k_n|\varphi(x_2)\varphi(x_1)|P_1 \dots P_m >_{in}$$

Need to interchange $\varphi(x_2)\varphi(x_1)$ to proceed as before.

Call $S_2 = {}_{out}< k_3 \dots k_n|\varphi(x_1)\varphi_{k_2}^{in}|P_1 \dots P_m >_{in}$ then our integral becomes

=

$iZ^{-1/2}\{\lim_{t\to-\infty} \int d^3x_2 f_{k_2}^*(x_2)\frac{\overleftrightarrow{\partial}}{\partial x_2^0}\,{}_{out}< k_3 \dots k_n|\varphi(x_2)\varphi(x_1)|P_1 \dots P_m >_{in} -$

$\lim_{t_2\to-\infty} \int d^3x_2 f_{k_2}^*(x_2)\frac{\overleftrightarrow{\partial}}{\partial x_2^0} < \varphi(x_2)\varphi(x_1)\} + S_2$

=

$iZ^{-1/2}\{\lim_{t\to-\infty} \int d^3x_2 f_{k_2}^*(x_2)\frac{\overleftrightarrow{\partial}}{\partial x_2^0}\,{}_{out}< k_3 \dots k_n|T\varphi(x_2)\varphi(x_1)|P_1 \dots P_m >_{in} -$

$iZ^{-1/2} \lim_{t\to-\infty} \int d^3x_2 f_{k_2}^*(x_2)\frac{\overleftrightarrow{\partial}}{\partial x_2^0}\,{}_{out}< k_3 \dots k_n|T\varphi(x_2)\varphi(x_1)|P_1 \dots P_m >_{in}\} +$

S_2

$= iZ^{-1/2} \int d^4x_2 \frac{\partial}{\partial x_2^0}\left\{f_{k_1}^*(x_2)\frac{\overleftrightarrow{\partial}}{\partial x_2^0}\,{}_{out}< k_3 \dots k_n|T\varphi(x_1)|P_1 \dots P_m >_{in}\right\} + S_2$

=

$\left(iZ^{-1/2}\right) \int d^4x_2\, f_{k_1}^*(x_2)k(x_2)\,{}_{out}< k_3 \dots k_n|T\varphi(x_2)\varphi(x_1)|P_1 \dots P_m >_{in} +$

S_2

So, ${}_{out}< k_1 \dots k_n|P_1 \dots P_m >_{in} =$

$\left(iZ^{-1/2}\right)\left(iZ^{-1/2}\right) \int d^4x_1\, d^4x_2 f_{k_1}^*(x_1)k(x_1)f_{k_2}^*(x_2)k(x_2)$

$\times\,{}_{out}< k_3 \dots k_n|T[\varphi(x_1)\varphi(x_2)]|P_1 \dots P_m >_{in} + S_1 + S_2$

Etc.

For $|P_1 \dots P_m >= a_{P_1}^{\dagger\, in}|P_2 \dots P_m >=$

$-i \int d^3x\, f_{p_1}(x)\overleftrightarrow{\partial}_0\varphi^{in}(x)|P_2 \dots P_n >_{in}$

${}_{out}< k_1 \dots k_n|a_{P_1}^{\dagger\, in}|P_2 \dots P_m >_{in} =$

$-iZ^{-1/2} \lim_{t\to-\infty} \int d^3x_1 f_{P_1}^{\square}(x_1)\frac{\overleftrightarrow{\partial}}{\partial x_1^0}\,{}_{out}< k_1 \dots k_n|\varphi(x)|P_2 \dots P_m >_{in}$

$= -Z^{-1/2} \int d^4x_1\, f_{P_1}(x_1)k(x_1)\,{}_{out}< \cdots >_{in}$

So, we get the reduction formula:

$${}_{out}< k_1 \dots k_n|P_1 \dots P_m >_{in}$$

$$= \int \prod_{j=1}^{n}\left(iZ^{-1/2}d^4x_j f_j^*(x_j)k(x_j)\right)\prod_{l=1}^{m}\left(iZ^{-1/2}d^4y_l f_{p_l}(y_l)\right)$$

$$< 0|T\varphi(x_1) \dots \varphi(y_m)|0 >$$

The $< 0|T\varphi(x_1)\varphi(x_2) > \cdots \varphi(y_m)|0 >$ is where the unique information lies, the f's and k's are defined by the wave equation and the wavefunctions corresponding to the eigenstates given in the expression $< k|$, $|p >$, etc. So, once we've computed $< 0|T[\dots]|0 >$ we can get the

S-matrix via the reduction formula. We will see that the $< 0|T[...]|0 >$ can be found by path integral methods!

3.7 The Dirac Field and Second Quantization

The Dirac equation

$$i\hbar\frac{\partial\overline{\Psi}(x)}{\partial x^{\mu}}\gamma^{\mu} + mc\,\overline{\Psi}(x) = 0.$$

The Dirac equation can be derived from the Lagrangian density

$$\mathcal{L} - c\,\overline{\Psi}(x)\left[i\hbar\gamma^{\mu}\frac{\partial}{\partial x^{\mu}} - mc\right]\Psi(x)$$

$$\pi_{\alpha}(x) = \frac{\partial\mathcal{L}}{\partial\Psi_{x}} = i\hbar\,\Psi_{\alpha}^{+} + \bar{\pi}_{\alpha}(x) = \frac{\partial\mathcal{L}}{\partial\overline{\Psi}_{\alpha}} \equiv 0$$

$$H = \int d^{3}x\,\Psi(x)\left[-i\hbar c\gamma^{j}\frac{\partial}{\partial x^{j}} + mc^{2}\right]\Psi(x)$$

$$\vec{P} = -i\hbar\int d^{3}\vec{x}\,\Psi^{+}(x)\nabla\Psi(x)$$

The transformation of the field under an infinitesimal Lorentz transformation is in the case of the Dirac field given by

$$\Psi_{\alpha}(x) \rightarrow \Psi_{\alpha}^{A}(x') = \Psi_{\alpha}(x) - \frac{1}{4}\varepsilon_{\mu\nu}\sigma_{\alpha\beta}^{\mu\nu}\,\Psi_{\beta}(x)$$

$\sigma_{\alpha\beta}^{\mu\nu}$ is the (α,β) matrix element of the 4 x 4 matrix : $\sigma^{\mu\nu} \equiv \frac{i}{2}[\gamma^{\mu},\gamma^{\nu}]$

The angular momentum of the Dirac field:

$$\vec{M} = \int d^{3}\vec{x}\,\Psi^{+}(x)[\vec{x}\Lambda(-i\hbar\nabla)]\,\Psi(x) + \int d^{3}x\,\Psi^{+}(x)\left(\frac{\hbar}{2}\vec{\sigma}\right)\Psi(x)$$

Where the 4 x 4 matrices $\vec{\sigma} = (\sigma^{23},\sigma^{31},\sigma^{12})$ are the generalizations for the Dirac theory of the 2 x 2 Pauli spin matrices. The two terms represent the orbital and spin angular momenta of particles of which ½ .

$Q = q\int d^{3}x\,\Psi^{+}(x)\,\Psi(x)$

$S^{\alpha}(x) = \big(c\rho(x),J(x)\big) = cq\,\overline{\Psi}(x)\gamma^{\alpha}\,\Psi(x)$

Consider a cubic enclosure, of volume V, with periodic b.c.'s. A complete set of plane waves can then be defined as follows. Say each momentum $\bar{p}$, allowed by the periodic b.c's, and positive energy: $cp_{o} = E_{p} = +(m^{2}c^{4} + c^{2}\vec{p}^{2})^{1/2}$.

The Dirac equation possesses four independent solutions, these will be written:

$$u_r(\vec{p})\frac{e^{-ipx/\hbar}}{\sqrt{V}}, V_c(\vec{p})\frac{e^{-ipx/\hbar}}{\sqrt{V}} \quad r = 1,2 \quad A \equiv \gamma^\mu A_\mu$$

$$(p - mc)u_r(\vec{p}) = 0, \quad (p + mc)V_r(\vec{p}) = 0 \qquad r = 1,2$$

Interpretation of u and V in the single-particle theory results in difficulties and reinterpretation in terms of hole theory. Second quantization of the theory ($\Psi and\ \Psi^+$ become operators) leads directly to the interpretation in terms of particles and antiparticles without the intellectual contortions of the hole theory. For the Dirac equation, only the longitudinal spin components (i.e. parallel to $\pm\vec{p}$ are constants of the motion, and we shall choose these spin sigenstates for our u and V with $\sigma_p = \frac{\vec{\sigma}\cdot\vec{p}}{|\vec{p}|}$ We choose the spinor such that

$$\sigma_p u_r(p) = (-1)^{r+r} u_r(p), \sigma_p V_r(p) = (-1)^r V_r(p) \ , r = 1,2$$

$u_r^t(\vec{p})u_r(p) = v_r^t(\vec{p})u_r(\vec{p}) = E_p/mc^2$ normalize u and V this way

Thus, we have the orthonormality relations:

$$u_r^t(\vec{p})u_s(\vec{p}) = u_r^t(\vec{p})u_s(\vec{p}) = (E_p/mc^2)\delta_{rs}$$

$$u_r^t(\vec{p})u_s(-\vec{p}) = 0$$

Second Quantization

In order to quantize the Dirac field, we expand it in terms of the complete set of plane

$$\Psi(x) = \Psi^+(x) + \Psi^-(x)$$

$$= \sum_{rp}\left(\frac{mc^2}{VE_p}\right)^{1/2} \left[C_r(\vec{p})u_r(\vec{p})e^{ipx/\hbar} + d_r^t(\vec{p})V_r(\vec{p})e^{ipx/\hbar}\right]$$

$$\bar{\Psi} = \Psi^+\gamma^o$$

$$\bar{\Psi}(x) = \bar{\Psi}^+(x)$$

$$+ \bar{\Psi}^-(x)\sum_{rp}\left(\frac{mc^2}{VE_p}\right)^{1/2} \left[d_r(\vec{p})\bar{V}_r(\vec{p})e^{ipx/\hbar}\right.$$

$$\left. + C_r(\vec{p})u_r(\vec{p})e^{ipx/\hbar}\right]$$

We impose anticommutation relations on the expansion coefficients:

$$[C_r(\vec{p}), C_s^+(\vec{p})]_+ = [d_r(p), d_s^+(\vec{p}')]_+ = \delta_{rs}\delta_{pp'}$$

$$N_r(\vec{p}) = C_s^+(p)C_r(p), \bar{N}_r(\vec{p}) = d_r^t(p)d_r(p)$$

In averaging a product of boson operators in normal order, one treats as through all commutations vanish, for fermion operators, one treats them as through all anticommutations vanish.

$$N(\Psi_\alpha\Psi_\beta) = N[(\Psi_\alpha^+ + \Psi_\alpha^-)(\Psi_\beta^+ + \Psi_\beta^-)]$$

$$= \Psi_\alpha^+ + \Psi_\beta^+ - \Psi_\beta^- + \Psi_\alpha^- + \Psi_\alpha^- + \Psi_\beta^+ + \Psi_\alpha^- + \Psi_\beta^-$$

$$H = \int d^3x\, N\left\{\overline{\Psi}(x)\left[-i\hbar c\gamma^j \frac{\partial}{\partial x_i} + mc^2\right]\Psi(x)\right\}$$

$$H = \sum_{rp} E_p\,[N_r(\vec{p}) + \overline{N}_r(\vec{p})]$$

$$\vec{p} = \sum_{rp} \vec{p}\,[N_r(\vec{p}) + \overline{N}_r(\vec{p})]$$

$$Q = -e\sum_{rp} \vec{p}\,[N_r(\vec{p}) + \overline{N}_r(\vec{p})]$$

We can interpret the particles associated with the C- and d- operators as electrons and positions respectively. Define the longitudinal spin operator by

$$S_p = \frac{\hbar}{2}\int d^3\vec{x} N\left[\Psi^+(x)\sigma_p\Psi(x)\right]$$

$$S_p C_r^+(p)\Big|0> \;= (-1)^{r+1}\frac{\hbar}{2}C_r^+(\vec{p})\Big|0>$$

$$S_p d_r^+(p)\Big|0> \;= (-1)^{r+1}\frac{\hbar}{2}d_r^+(\vec{p})\Big|0>$$

The anticommutation relations for the creation and absorption operators imply anticommutation relations for the Dirac field operators Ψ *and* $\overline{\Psi}$.

$$\left[\Psi_\alpha(x), \Psi_\beta(x)\right]_+ = \left[\Psi_\alpha(x), \Psi_\beta(y)\right]_+ = 0$$

$$\left[\Psi_\alpha(x), \Psi_\beta(y)\right]_+ = i\left(i\gamma^x\frac{\partial}{\partial x^\mu} + \frac{mc}{\hbar}\right)_{\alpha\beta}\Delta^\pm(x-y)$$

Considered as a 4 x 4 matrix equation:

$\left[\Psi^\pm(x), \overline{\Psi}^\pm(y)\right]_+ = iS^\pm(x-y)$

$S^\pm(x) = \left(i\gamma^\mu\frac{\partial}{\partial x^\mu} + \frac{mc}{\hbar}\right)\Delta^\pm(x)$

So, $[\Psi(x), \Psi(y)]_+ = iS(x-y)$

Where,

$S(x) = S^+(x) + S^-(x) = \left(i\gamma^\mu\frac{\partial}{\partial x^\mu} + \frac{mc}{\hbar}\right)\Delta(x)$

The integral representation for $\Delta^\pm(x)$ yields an integral rep. for $S^\pm$:

$$\Delta^\pm(x) = -\frac{1}{(2\pi)^4}\int_{c^+}\frac{d^4k e^{-ikx}}{k^2-\mu^2}$$

$$S^\pm(x) = -\frac{\hbar}{(2\pi\hbar)^4}\int_{c^+} d^4p e^{-ipx/\hbar}\frac{p+mc}{p^2-m^2c^2}$$

$$(p \pm mc)(p \mp mc) = p^2 - m^2c^2$$

$$S^{\pm}(x) = -\frac{\hbar}{(2\pi\hbar)^4}\int_{C^{\pm}} d^4p \frac{e^{-ipx/\hbar}}{p - mc}$$

Connection between spin and statistics: If we demand the existence of a state of lowest energy (i.e. a stable ground state), we must quantize the Dirac equation according to Fermi-Dirac statistics.

The requirement of microcasuality forces us to quantize the Klein-Gordon field according to Bose-Einstein statistics.

The Fermion propagator

We define the Feynman propagator as $\langle 0|T\{\Psi(x)\overline{\Psi}(x')\}|0\rangle$:

$T = \{\Psi(x)\overline{\Psi}(x')\} = \theta(t - t')\Psi(x)\overline{\Psi}(x') - \theta(t' - t)\overline{\Psi}(x')\Psi(x)$

$= \begin{cases} \Psi(x)\overline{\Psi}(x'), if\ t > t' \\ -\overline{\Psi}(x')\Psi(x), \quad if\ t' > t \end{cases}$

$\langle 0|\Psi(x)\overline{\Psi}(x')|0\rangle = \langle 0\big|\Psi^{+}(x)\overline{\Psi}(x')\big|0\rangle = \langle 0\Big|\left[\Psi^{+}(x)\overline{\Psi}(x')\right]_{+}\Big|0\rangle =$
$iS^{+}(x - x')$

$\langle 0|\overline{\Psi}(x')\Psi(x)|0\rangle = iS^{-}(x - x')$

So, $\langle 0|T\{\Psi(x)\overline{\Psi}(x')\}|0\rangle = iS_F(x - x')$

$$S_F(x) = \theta(t)S^{+}(x) - \theta(-t)S^{-}(x) = \left(i\gamma^{\mu}\frac{\partial}{\partial x^{\mu}} + \frac{mc}{\hbar}\right)\Delta_F(x)$$

$$S_F(x) = \frac{\hbar}{(2\pi\hbar)^4}\int d^4p e^{-ipx/t}\frac{p + mc}{p^2 - m^2c^2 + i\varepsilon}$$

3.8 The Electromagnetic Field

We develop a covariant theory starting from an explicitly covariant formulation of classical electrodynamics in which all four components of the four-vector potential $A^{\mu}(x) = (\varphi, \vec{A})$ are treated on an equal fooring. This corresponds to introducing more dynamical degrees of freedom that the system possesses and these will later have to be removed by imposing suitable constants.

$$F^{\mu\nu}(x) = \begin{matrix} \nu\rightarrow \\ \\ \\ \\ \\ \end{matrix}\begin{pmatrix} 0 & 1 & 2 & 3 \\ 0 & E_x & E_y & E_z \\ -E_x & 0 & B_z & -B_y \\ -E_y & -B_x & 0 & B_x \\ -E_z & B_y & -B_x & 0 \end{pmatrix}\begin{matrix} m \\ \downarrow \\ 0 \\ 1 \\ 2 \\ 3 \end{matrix}$$

$S^{\mu}(x) = (cp(x), \vec{j}(x))$

$A^{\mu}(x) = (\varphi, \vec{A})$

$\partial_\nu F^{\mu\nu}(x) = \frac{1}{c} S^\mu(x)$

$\partial^\lambda F^{\mu\nu}(x) + \partial^\mu F^{\nu\lambda}(x) + \partial^\nu F^{\partial x}(x) = 0$ $\qquad$ $\partial_\mu S^\mu(x) = 0$

$F^{\mu\nu}(x)\partial^\nu A^\mu(x) - \partial^\mu A^\nu(x)$

So, $\Box A^\mu(x) - \partial^\mu\left(\partial_\nu A^\nu(x)\right) = \frac{1}{c} S^\mu(x)$

These equations are Lorentz-covariant, and they are invariant under the gauge transformation:

$A^\mu(x) \to A'^\mu(x) = A^\mu(x) + \partial^\mu f(x)$

The field equations can be derived from the following Lagrangian density.

$\mathcal{L} = -\frac{1}{4} F_{\mu\nu}(x) F^{\mu\nu}(x) - \frac{1}{c} S_F(x) A^\mu(x)$

Unfortunately, the Lagrangian density is not suitable for carrying out the canonical quantization.

$\pi^\mu(x) = \frac{\partial \mathcal{L}}{\partial \dot{A}_\mu} = -\frac{1}{c} F^{\mu\nu}(x) \to \pi^o(x) \equiv 0$, which is incomplete with the canonical commutation relations which we must to impose.

A Lagrangian density which is suitable for quantization, first proposed by Fermi, is

$\mathcal{L} = -\frac{1}{2}\left(\partial_\nu A_\mu(x)\right)\left(\partial^\nu \partial^\mu(x)\right) - \frac{1}{c} S_\mu(x) A^\mu(x)$

Whence, $\pi^\mu(x) = \frac{\partial \mathcal{L}}{\partial A_\mu} = -\frac{1}{c^2} \dot{A}^\mu(x)$

The Lagrangian density leads to the field equations:

$\Box A^\mu(x) = \frac{1}{c} S^\mu(x)$

Which is equivalent to maxwells equations if the potential $A^\mu(x)$ satisfies the constant $\partial_\mu A^\mu(x) = 0$. (The lorenty condition)

From gauge invariance $A^\mu(x) \to A'^\mu(x) = A^\mu(x) + \partial^\mu(x)$

$\partial_\mu A^\mu(x) \to \cdots$

$\partial_\mu A^\mu(x) + \partial_\mu \partial^\mu f(x) = \partial_\mu A^\mu(x) + \Box f(x) = 0$ we choose $f(x)$ to be solution for this,

$\partial_\mu A^\mu(x) = 0$ does not specify the potebtials uniquely. If the potentials $A^\mu(x)$ satisfy $\partial_\mu A^\mu(x) = 0$, so with any potentials $A'^\mu(x)$ provided the gauge function $f(x)$ satisfies $\Box f(x) = 0$.

In the free field case ($S^\mu(x) = 0$) we have $\Box A^\mu(x) = 0$, which is the limit of the Klein-Gordon equation for particles with mass zero.

Expand the free electromagnetic field $A^\mu(x)$ in a complete set of solutions of the wave equation.

$A^\mu(x) = A^{\mu+}(x) + A^{\mu-}(x)$

$$A^{\mu+}(x) = \sum_{rk} \left(\frac{\hbar(c)^2}{2V\omega_k}\right)^{1/2} \varepsilon_r^\mu(k) a_r(k) e^{-ikx}$$

$$k^o = \frac{1}{c}\omega_k = |\vec{k}|$$

$$A^{\mu}(x) = \sum_{rk} \left(\frac{\hbar(c)^2}{2V\omega_k}\right)^{1/2} \varepsilon_r^\mu(k) a_r{}^t(k) e^{ikx}$$

For each $\vec{k}$ there are four linearly independent polarization states.
$\varepsilon_r(k)\varepsilon_s(k) = \varepsilon_{r\mu}(k)\varepsilon_s^\mu(k) = -S_r\delta_{rs}$
$r, s = 0, \dots 3$
$\mathcal{J}_f = -1, \mathcal{J}_1 = \mathcal{J}_2 = \mathcal{J}_3 = 1$
$\sum_r \mathfrak{z}_r \varepsilon_r^v(k) = -g^{\mu v}$

A specific choice of polarization vectors in one given frame of reference often facilitates the interpretation. Choose the vectors as

$$\varepsilon_0^\mu(k) = n^\mu \equiv (1,0,0,0) \; ; \; \varepsilon_r^\mu(k) = (0, \vec{\varepsilon}_r(k)) \qquad r = 1,2,3,$$

Where $\varepsilon_1^\mu(k)$ *and* $\varepsilon_2^\mu(k)$ are mutually orthogonal unit vectors which are also orthogonal to $\vec{k}$ and $\varepsilon_r^\mu(k) = \vec{k}/_{im}$

So, $k \cdot \varepsilon_r(k) = 0 \;, r = 1,2$
$\varepsilon_r(k) \cdot \varepsilon_s(k) = \delta_{rs} \; r, s = 1,2,3$
ε_1^μ *and* ε_2^μ are called transverse, ε_3^μ longitudinal polarizations, and ε_o^μ scalar or time-like polarization.

Covariant Commutation

Applying the canonical formation to quantize the free electromagnetic field. The equal-time commutation relations become:
$[A^\mu(\vec{x},t), A^v(\vec{x},t)] = 0, [A^\mu(\vec{x},t), \dot{A}^v(\vec{x},t)] = 0$
$[A^\mu(\vec{x},t), \dot{A}^v(\vec{x},t)] = -i\hbar c^2 g^{\mu v}\delta(\vec{x}', \vec{x})$

$$[A^\mu(x), A^v(\vec{x},t)] = i\hbar c D^{\mu v}(x - x') \quad where \; D^{\mu v}(x) = \lim_{m\to 0}[-g^{\mu v}\Delta(x)]$$

The Feynman propagation:
$\langle 0|A^\mu(x)A^v(x')|0\rangle = i\hbar c D_F^{\mu v}(x - x')$

$$D_F^{\mu v}(x) = \lim_{m\to 0}[-g^{\mu v}\Delta_F(x)] = -\frac{g^{\mu v}}{(2\pi)^2}\int \frac{d^4k e^{-ikx}}{k^2 + i\varepsilon}$$

To gain the photon interpretation of the quantized fields, we substitute the field expansions in the commutation relations, with the result:
$[a_r(k), a_s^t(k')] = \mathcal{J}_r\delta_{rs}\delta_{kk'}$

$[a_r(k), a_s(k')] = (a_t^t(k), a_s^t)(k')$
$\mathcal{J}_r = 1 \ for \, r = 1,2,3 \rightarrow$ standard boson commutation relations $\mathcal{J}_o = -1$ looks as though the usual roles of absorption and creation operations must be interchanged for $a_o(k) \ and \ a_o^t(k)$. However, effecting only this change results in other difficulties, and the standard formation must be modified more radically of the several procedures available, we shall follow that due to Gupta [25] and Blauler [26].

The Photon propagator

Consider $D_F^{\mu v}(x) = \frac{1}{(2\pi)^4} \int d^4k D_F^{\mu v}(k) e^{-ikx}$

Where,

$$D_F^{\mu v}(k) = -\frac{g^{\mu v}}{k^2 + i\varepsilon} = \frac{1}{k^2 + i\varepsilon} \sum_r \mathcal{J}_r \varepsilon_r^\mu(k) \varepsilon_r^v(k)$$

Using the reference frame presented earlier

$$D_F^{\mu v}(k) = \frac{1}{k^2 + i\varepsilon} \left\{ \sum_r \varepsilon_r^\mu(k) \varepsilon_r^v(k) + \frac{[k^\mu - (kn)n^\mu][k^v - (kn)n^v]}{(kn)^2 - k^2} + (-1) n^\mu n^v \right\}$$

$${}_T D_F^{\mu v} \equiv \frac{1}{k^2 + i\varepsilon} \sum_{r=i}^{2} \varepsilon_r^\mu(k) \varepsilon_r^v(k) \ the \ exchange \ of \ transverse \ photons$$

Interpretation of the remaining two terms follows upon regrouping.

$$D_F^{\mu v}(k) = {}_T D_F^{\mu v}(k) + {}_C D_F^{\mu v}(k) + {}_R D_F^{\mu v}(k)$$

Where,

$${}_C D_F^{\mu v}(k) \equiv \frac{n^\mu n^v}{(kn)^2 - k^2}, \ {}_R D_F^{\mu v}(k) \equiv \frac{1}{k^2 + i\varepsilon} \left[\frac{k^\mu k^v - (kn)(k^\mu n^v + k^v k^\mu)}{(kn)^2 - k^2} \right]$$

$${}_C D_F^{\mu v}(x) = \frac{g^{\mu v} g^{vo}}{(2\pi)^4} \int \frac{d^3\vec{k} e^{i\vec{k}\cdot\vec{x}}}{|k|^2} \int dk^o e^{-ik^o x^o} = g^{\mu v} g^{vo} \frac{1}{4\pi|x|} \delta(x^o)$$

Has the time dependence and space dependence characteristics of an instantaneous coulomb potential.

The combination of the remainder term to all observable quantities must vanish. This is indeed the case, the basic reason being that the electromagnetic field only interacts with the conserved change current density $\delta^\mu(x)$. $\quad \partial_\mu S_1^\mu(x) = c \quad \rightarrow k_\mu S_r^\mu(k) = 0$

3.9 Scattering

3.9.1 The S-matrix expansion

In QED, the interacting electron-position and e-m fields are described by the Langrangian density. $\mathcal{L} = \mathcal{L}_o + \mathcal{L}_I$

Free field Lagrangian density: $\mathcal{L}_o = N\left[\overline{\Psi}(x)\left(i\gamma^{\mu}\partial_{\mu} - m\right)\Psi(x) - \frac{1}{2}\left(\partial_i A_{\mu}(x)\right)\left(\partial^{v} A^{\mu}(x)\right)\right]$

Interaction Lagrangian density: $\mathcal{L}_I = N[-s^{\mu}(x)] = N[e\overline{\Psi}(x)A(x)\Psi(x)]$

$\frac{d}{dt}|\phi(t)>= H_I(t)|\phi> \ where \ H_I(t) = e^{it_o(t-t_o)}H_I^{s}e^{iH_o(t-t_o)}$

The formalism we are developing is not approximate for the description of bound states but it is particularly suitable for scattering processes.

In a collision process the state vector $|i>$ will define an initial state, long before the scattering occurs $(t_i = -\infty)$, by specifying a definite number of particles, with definite properties and far apart from each other so that they do not interact.

$|\phi(-\infty)>= |i>$

The s-matrix relates $|\phi(-\infty)>= s|\phi(-\infty)>= S|i>$

$\langle f|S|i\rangle \equiv S_{fi}$

$|\phi(t)>= \Big|i> +(-i)\int_{-\infty}^{t} dt_1 H_I(t_1)\Big|\,\phi(t_1)>$ thus equation can only be solved iteratively.

$$S = \sum_{n=0}^{\infty}(-i)^n \int_{-\infty}^{\infty} dt_1 \int_{-\infty}^{t_1} dt_2 \dots \int_{-\infty}^{t_{n-1}} dt_n i t_a (t_1) H_1(t_2) \dots H_I(t_n)$$

$$= \sum_{n=0}^{\infty}\frac{(-i)^n}{n!}\int_{-\infty}^{\infty} dt_1 \int_{-\infty}^{t_1} dt_2 \dots \int_{-\infty}^{\infty} dt_n T\{H_I(t_1)H_I(t_2)\dots H_I(t_n)\}$$

$$= \sum_{n=0}^{\infty}\frac{(-i)^n}{n!}\int \dots \int d^4x_1\, d^4x_2 \dots d^4x_n\{\mathcal{H}_I(x_1)\dots \mathcal{H}_I(x_n)\}$$

3.9.2 Wick`s theorem

Calculations can be greatly amplified by avoiding the explicit introduction of virtual intermediate particulars. This can be achieved by using writing the s-matrix expansion as a sum of normal products, since in a normal product add absorption operators attend to the right of all creation operators.

The method for exponentiating the s-matrix as a sum of normal product is due to Dyson and Wick.

Let Q,R,,.W be operations linking in any either creation or absorption operators, the

$$N(QR,,,W) = (-1)^p(Q^1R^1,,,W^1)$$

$Q^1,,,W^1$are the operations Q,,,W neordered that all absorption operations attend to the right of all unerthan operations, P is the number of interchanges of neighbouring fewer operations.

$N(RS,,,+VW,,,) = N(RS,,,) + (VW,,,)$

$9b + \{A(x)B(x),,,\}$

Hence we must consider the expansion into a sum of normal products of a ``mixed`` T-product (a T-product whose factors are normal products)

$$AB - N(AB) = \{[A^+, B^-] +\} \text{ Terminus}$$

$$[A^+, B^-] \text{ Boson}$$

$$AB = N(AB) - \langle C|AB|0\rangle$$

Since $N(AB) = \underline{+}N(BA)$, the minus sign applying in the case of terminus!

$T\{A(x_1)B(x_2)\} = N\{A(x_1)B(x_2)\} + \langle 0|T\{A(x_1)B(x_2)\}|0\rangle$

$A(x_1)B(x_2 =)\langle 0|T\{A(x_1)B(x_2)\}|0\rangle$

So

$\varphi(x_1)\varphi(x_2) = 1\Delta_F(x_1 - x_2)$

$\varphi(x_1)\varphi^+(x_2) = \varphi^+(x_2)\varphi(x_1) = 1\Delta_F(x_1 - x_2)$

$\Psi_\propto(x_1)\,\Psi_\propto(x_2) = -\Psi_\beta(X_2)\,\Psi_\alpha(X_1) = iS_{F\alpha\beta}(x_1 - x_2)$

$A^\mu(x_1)A^V(X_2) = iD_F{}^{\mu v}(x_1 - x_2)$

3.9.3 Pion Scattering

Process where a π^+ (pion with momentum p, say) collides with a π^- (momentum p'), where the two particles annihilate and produce and electron-positron pair (electron has momentum k and spin s; positron has momentum k' and spin s'). Express the scattering amplitude for this process in terms of the in-states and out-states. Let's consider scattering (LSZ formalism) for pions (using complex scalar field) into electron-positron (fermion field):

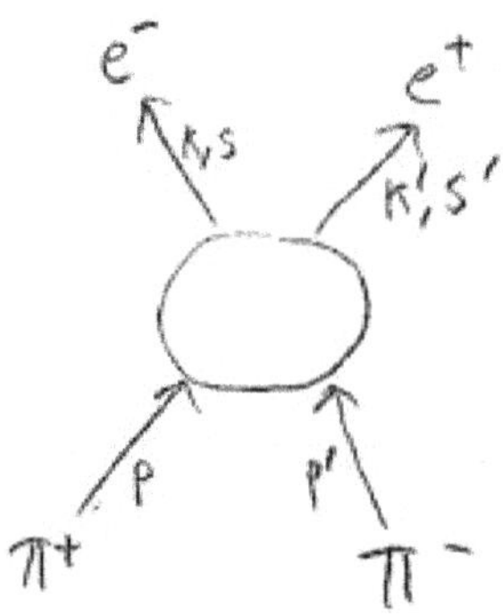

For scalar field:

$$(\Box + m^2)\varphi_{in}(x) = 0 \quad \Rightarrow \varphi_{in}(x)$$
$$= \int \frac{d^3k}{(2\pi)^{3/2}\sqrt{2\omega_k}}\left[a_k e^{-ik\cdot x} + a_k^\dagger e^{ik\cdot x}\right]$$

with $f_k(x) = \frac{e^{-ik\cdot x}}{(2\pi)^{3/2}\sqrt{2\omega_k}} \Rightarrow (\Box + m^2)f_k(x) = 0$, used to simplify expressions. We then add the quantization conditions:

$$[\varphi_{in}(\vec{x},t), \varphi_{in}(\vec{y},t)] = 0$$
$$[\varphi_{in}(\vec{x},t), \pi = \varphi_{in}(\vec{y},t)] = i\delta(\vec{x} - \vec{y})$$

The QFT operator φ that results then gives rise to the spectral operator relations

$$\left[a_{\vec{k}}, a_{\vec{k'}}\right] = 0$$
$$\left[a_{\vec{k}}, a_{\vec{k'}}{}^\dagger\right] = \delta_{kk'}$$

The f's are chosen orthonormal in the sense of the Klein-Gordon inner product, an inner product which is independent of time:

$$i\int d^3x\, f_k^* \overleftrightarrow{\partial}_0 f_l = \delta_3(\vec{k} - \vec{l})$$
$$\varphi_{in}(x) = \int d^3k\left[f_k a_k^{in} + f_k^* a_{\vec{k}}{}^{in\,\dagger}\right]$$

Dropping the 'in' notation in 'a' operators for now:

$$i\int d^3y\, f_l^* \overleftrightarrow{\partial}_0 \varphi_{in} = \int d^3k\left\{\left(\int d^3y\, f_l^* \overleftrightarrow{\partial}_0 f_k\right) a_k + \left(i\int d^3y\, f_l^* \overleftrightarrow{\partial}_0 f_k^*\right) a_k^\dagger\right\}$$
$$= \int d^3k\left\{\delta_3(\vec{l} - \vec{k}) a_{\vec{k}}\right\} = a_{\vec{l}}$$

Thus

$$a_{\vec{k}}^{in} = i\int d^3x\, f_k^* \overleftrightarrow{\partial}_0 \varphi_{in}$$
$$\left(a_{\vec{k}}^{in}\right)^\dagger = -i\int d^3x\, f_{\vec{k}} \overleftrightarrow{\partial}_0 \varphi_{in}^\dagger$$

The Lagrangian density that yields the field equation above (K.G.) is $\mathcal{L} = \frac{1}{2}\left(\partial^\mu\varphi\partial_\mu\varphi - m^2\varphi^2\right)$. If we want to generalize to a field that isn't free adding a $\frac{\lambda}{4}\varphi^4$ interaction term is the next level of complexity(to be discussed later), this is chosen over φ^3 because it retains the inversion symmetry.

Now, let's consider a free *complex* scalar field:

$$\mathcal{L} = \partial_\mu\varphi^\dagger\partial^\mu\varphi - \mu^2\varphi^\dagger\varphi \Rightarrow (\Box + \mu^2)\varphi = 0$$

Which gives:

$$(\Box + \mu^2)\varphi_{in}(x) = 0$$

We can't write $\varphi_{in}(x) = \int d^3k \left[f_k(x)a_k^{in} + f_k^*(x)a_k^{in\,\dagger}\right]$ as before since this form was chosen to be consistent with $\varphi_{in}(x) = \varphi_{in}^\dagger(x)$ which is true for scalar fields but not for the general case of a complex field where there are twice the degrees of freedom. For the general solution we now have:

$$\varphi_{in}(x) = \int d^3k \left[f_k(x)a_k^{in} + f_k^*(x)b_k^{in\,\dagger}\right]$$

Subsequently $\varphi_{in}^\dagger(x) = \int d^3k\left[f_k(x)b_k^{in} + f_k^*(x)a_k^{in\,\dagger}\right]$ and:

$$\Pi_\varphi = \frac{\partial \mathcal{L}}{\partial \dot{\varphi}} = \dot{\varphi}^t \rightarrow [\varphi(x), \dot{\varphi}^\dagger(y)]_{t_x = t_y} = i\delta_3(\vec{x} - \vec{y})$$

So, $\left[a_{k\prime}, a_{k\prime}^\dagger\right] = \delta(k - k')$ then follows. Now,

$$a_k^{in\,\dagger} = -\int d^3x\, f_{\vec{k}} \overleftrightarrow{\partial}_0 \varphi_{in}^\dagger$$

$$b_k^{in\,\dagger} = -i\int d^3x\, f_{\vec{k}} \overleftrightarrow{\partial}_0 \varphi_{in}^{\Box}$$

Notice that there is an invariance of the Lagrangian under $\varphi \rightarrow e^{i\theta}\varphi$, which yields a conserved "charge" (Noether's theorem).

For the electrons we turn to a Fermion Field description and the Dirac equation:

$$\mathcal{L} = \overline{\Psi}(x)(i\gamma \cdot \partial - m)\,\Psi(x) \quad \rightarrow \ (i\gamma \cdot \partial - m)\,\Psi(x) = 0$$

and canonical quantization:

$$\{\Psi, \Pi_\Psi\} = i\delta_3(\vec{x} - \vec{y})$$

(where canonical quantization uses anticommutator for fermions). Thus,

$$\Pi_\Psi = \frac{\partial \mathcal{L}}{\partial \Psi} = i\,\overline{\Psi}\gamma^0 = i\,\Psi^\dagger \Rightarrow \{\Psi, \Psi^\dagger\}_{t_x = t_y} = \delta_3(\vec{x} - \vec{y})$$

Spinor operators, that satisfy the Dirac equation and the commutation relations are now considered:

Let's start with the "C-numbers" (regular real numbers, or complex, not quantum operators often call Q-numbers or spinor, multivalued, variables): $u_{p,r}(x),\ v_{p,r}(x)$ where

$$(i\gamma \cdot \partial - m)u_{p,r}(x) = 0\ , (i\gamma \cdot \partial - m)v_{p,r}(x)$$

Then

$$u_{p,r}(x) = (2\pi)^{-3/2}\sqrt{\frac{m}{E_p}}\ u^{(r)}(p)e^{-ip\cdot x}$$

Where $u^{(r)}(p)$ is a four-component spinor and $r\epsilon 1{,}2$, and also:

$$v_{p,r}(x) = (2\pi)^{-3/2}\sqrt{\frac{m}{E_p}}\; v^{(r)}(p)e^{-ip\cdot x}$$

(the holes).
By construction:

$$\int d^3x\,\bar{u}_{r,p}(x)\gamma^0 u_{r',p'}(x) = \delta_{rr'}\delta(\vec{p}-\vec{p}')$$

Now,

$$\Psi(x) = \sum_r \int d^3p\left[u_{r,p}(x)b_r(p) + v_{r,p}(x)d_r^\dagger(p)\right]$$

$$\overline{\Psi}(x) = \sum_r \int d^3p\left[\bar{v}_{r,p}(x)d_r(p) + \bar{u}_{r,p}(x)b_r^\dagger(p)\right]$$

Dirac field also has invariance under $\Psi \rightarrow e^{i\theta}\,\Psi$, so again there is a conserved charge.

$$b_s^\dagger(p) = \int d^3x\,\overline{\Psi}(x)\gamma^0 u_{s,p}(x) \quad b_s(p) = \int d^3x\bar{u}_{s,p}(x)\gamma^0\,\Psi(x)$$

$$d_s^\dagger(p) = \int d^3x\,\bar{v}_{s,p}(x)\gamma^0\,\Psi(x) \quad d_s(p) = \int d^3x(x)\gamma^0 v_{s,p}(x)$$

LSZ Formalism: $\lim_{t\rightarrow-\infty} < \alpha|\varphi_f|\beta >= Z^{1/2} < \alpha|\varphi_f^{in}|\beta >$ where

$$\varphi_f(t) = i\int d^3x\,f^*(x)\overleftrightarrow{\partial}_0\varphi(x)$$

$$\varphi_f{}^{in}(t) = i\int d^3x\,f^*(x)\overleftrightarrow{\partial}_0\varphi^{in}(x)$$

Here 'f' is an arbitrary solution to the KG equation. Note that $\varphi_f{}^{in}(t)$ is time independent.

Complex field reduction formula

$${}_{out}< k_1 \dots k_n|k_1' \dots k_n' >_{in} = {}_{out}< k_2 \dots k_n|a_{k_1}^{out}|k_1' \dots k_n' >_{in}$$

$$= i\int d^3x_1 f_{k_1}^*(x_1)\frac{\overleftrightarrow{\partial}}{\partial x_1^0}\,{}_{out}< k_2 \dots k_n|a_{k_1}^{out}|k_1' \dots k_n' >_{in}$$

Using LSZ with smearing function f now the momentum eigenstate f^* :

$${}_{out}< k_1 \dots k_n|k_1' \dots k_n' >_{in}$$

$$= iZ^{-1/2}\lim_{t\rightarrow+\infty}\int d^3x_1 f_{k_1}^*(x_1)\frac{\overleftrightarrow{\partial}}{\partial x_1^0}\,{}_{out}< k_2 \dots k_n|\varphi(x)|k_1' \dots k_n' >_{in}$$

If we now separate spectator particles:

$$= iZ^{-1/2}\int d^4x_1\frac{\overleftrightarrow{\partial}}{\partial x_1^0}\left\{f_{k_1}^*(x_1)\frac{\overleftrightarrow{\partial}}{\partial x_1^0}\,{}_{out}< k_2 \dots k_n|\varphi(x)|k_1' \dots k_n' >_{in}\right\} + S_1$$

$$= iZ^{-1/2}\int d^4x_1\, f_{k_1}^*(x_1)(\partial_0^2 - \nabla^2 + m^2) < \cdots > + S_1$$

Where $k(x_1) = (\partial_0^2 - \nabla^2 + m^2)$, thus:
$iZ^{-1/2} \int d^4x_1 \, f_{k_1}^*(x_1) k(x_1) \; {}_{out}\!< k_2 \dots k_n | \varphi(x_1) | k_1' \dots k_n' >_{in} + S_1$
Let's now repeat the reduction analysis for the fermionic field.

Fermion field reduction formula

$${}_{out}\!< (P_1, S_1) \dots (P_n, S_n) | (P_1', S_1') \dots (P_m' S_m') >_{in} =$$
$${}_{out}\!< (P_n, S_n) \dots (P_n, S_n) | b_s(p) | \dots >_{in}$$
$$= \int d^3x \bar{u}_{s,p}(x) \, \gamma^0 \; {}_{out}\!< \cdots | \Psi^{out}(x_1) | \dots >_{in}$$

Using LSZ:

$${}_{out}\!< (P_1, S_1) \dots (P_n, S_n) | (P_1', S_1') \dots (P_m' S_m') >_{in}$$
$$= Z^{-1/2} \lim_{t \to +\infty} \int d^3x_1 \bar{u}_{s,p} \gamma^{\circ} \; {}_{out}\!< \cdots | \Psi(x_1) | \dots >_{in}$$
$$= Z^{-1/2} \left\{ \lim_{t \to +\infty} \int d^3x_1 \bar{u}_{s,p} \gamma^{\circ} < \cdots > - \lim_{t \to -\infty} \int (\dots) \right\} + S$$

Since $\bar{u}(i\gamma \cdot \overleftarrow{\partial} + m) = 0 \rightarrow i \frac{\partial}{\partial x_1^0} \, \gamma^0 = (i\gamma^n \partial_n - m)\vec{u}$ we have:

$$= Z^{-1/2} \int d^4x_1 \frac{\partial}{\partial x_1^0} \{ \bar{u}_{s,p} \gamma^0 < \cdots | \Psi | \dots > \}$$

Let's unpack the notation, recall that:

$$(i\gamma \cdot \partial - m) u_{r,p} = 0 \quad \rightarrow \quad (+i\gamma^0 \cdot \partial_0 - +i\gamma^n \partial_n - m) u_{r,p} = 0$$
$$\bar{u}_{r,p}(-i\gamma \cdot \overleftarrow{\partial} - m) = 0 \quad \rightarrow \quad -i\partial_0 \bar{u}_{r,p} \gamma^0 + i\partial_n \bar{u}_{r,p} \gamma^n - m\bar{u}_{r,p} = 0$$

So,

$$i\partial_o \bar{u} \gamma^o < \cdots | \Psi | \dots >) = i\partial_o \bar{u} \gamma^0 < \cdots | \Psi | \dots > + \, i\bar{u} \gamma^0 \partial_0 < \cdots | \Psi | \dots >$$
$$= \bar{u}(-i\gamma^n \partial_n - m + i\gamma^0 \partial_0) < \cdots | \Psi | \dots > = \bar{u}(i\gamma \cdot \partial + m) < \cdots | \Psi | \dots >$$
$$= \bar{u} D(x) < \cdots | \Psi | \dots >$$

So,

$$(-i) i Z^{1/2} \int d^4x_1 \frac{\partial}{\partial x_1^0} \{ \bar{u}_{s,p} \gamma^0 < \cdots | \Psi | \dots > \} + S_1$$
$$= -iZ^{1/2} \int d^4x_1 \, \bar{u}_{s,p} D(x) < \cdots | \Psi | \dots >$$

Thus, for the final state particle case:

$${}_{out}\!< (P_1, S_1), (P_2, S_2), \dots (P_n, S_n) \mid (P_1', S_1') \dots (P_m', P_m') >_{in}$$
$$= -iZ_{\Psi}^{-1/2} \int d^4x_1 \, \bar{u}_{s,p} D(x_1)$$
$${}_{out}\!< (P_2, S_2) \dots (P_n, S_n) | \Psi(x_1) | (P_1', S_1'), (P', S_1') \dots (P_n', P_n') >_{in}$$

Final state

Particle $\Rightarrow -Z_{\Psi}^{-1/2} \int d^4x \, \bar{u} D(x) < \Psi >$

Antiparticle: $(-i)i\frac{\partial}{\partial x_1^0}\{< \cdots |\overline{\Psi}| \ldots \gamma^0 v_{s,p}(x)\} = (i\partial_o < \Psi > \gamma^o v + i <$ $\overline{\Psi} > \gamma^0 \partial_0 v)(-i)$

$= [i\gamma^0\partial_0 < \Psi > v + < \overline{\Psi} > (i\gamma^n\partial_n + m)v](-i)$

$= -i(i\gamma^0\partial_0 - i\gamma^0\partial_n + m) < \overline{\Psi} > v = (-D(x) < \overline{\Psi} >)v(-i)$

Antiparticle: $iZ_\Psi^{-1/2} \int d^4x < \overline{\Psi} > \overleftarrow{D}(x)V$

Initial state

Initial state has same analysis aside from sign flip due to flip in integration limits:

Particle $\Rightarrow -iZ_\Psi^{-1/2} \int d^4x < \overline{\Psi} > \overleftarrow{D}(x)u$

Antiparticle $\Rightarrow iZ_\Psi^{-1/2} \int d^4x\bar{v}D_x < \Psi >$

For multiple particles we then time-order:

$$\langle (p,s),(p',s')|k,k'\rangle$$

$$\left(-iZ_\Psi^{-1/2} d^4x\bar{u}_{p,s}(x)D(x)\right)\left(-iZ_\varphi^{-1/2} d^4y f_K(y)K(y)\right)$$

$$= \int \times \left(-iZ_\varphi^{-1/2} d^4y'^{f_{K'}(y')}K(y')\right)\langle 0|T\,\Psi(x)\overline{\Psi}(x)\varphi^\dagger(y)\varphi(y')|0\rangle$$

$$\times \left(-iZ_\Psi^{-1/2} d^4x'\overleftarrow{D}(x')\bar{v}_{p',s'}(x')\right)$$

Chapter 4. Path Integral Quantization

Some of the material in this chapter draws from Path Integral First Quantization description from Book 4 of the Series [4]. Some of the material draws from lecture notes taken while at Caltech, UWM, and Oxford, and from various texts. Most notable are the lecture notes from a quantum field theory course taught by Nicholas Papastamatiou at UWM, and those notes, in turn, drew strongly from the text by Ramond [12]. For an excellent user friendly introduction, with much more detail in some areas than can be provided in what follows, see both Mandl & Shaw [13] and Ryder [14]. This book also encompass a broad range of advanced applications, ranging from gauge fields and renormalization to the standard model and curved spacetime. Notes from these areas, including work done for my PhD thesis [15], and from numerous excellent sources [9,16-24], will be described.

4.1 Introduction to Path Integrals and Green's Functions

In this chapter the path integral method and path integral quantization will be described. As the name suggests, a path integral is an integral over spaces of paths (in quantum mechanics) or over spaces of fields (in quantum field theory). If the notion of path integral still sounds mathematically vague, not well-defined in fact, then you would be correct. The process of making the path integral well-defined, in a specific application, or in a more general formulation, is a lengthy history, but the end result will be that it is well-defined, and its utility will be critical in many ways (including the only known way to perform renormalization for the electroweak part of the theory in the standard model).

Path integrals are infinite dimensional integrals, also known as functional integrals or field integrals, and can be defined in terms of extensions of finite-dimensional integrals. How this definition or extension is accomplished can vary greatly. At a high level we must decide if our path integral implementation shall be deterministic (Feynman Path Integral [27]) or probabilistic (Weiner Path Integral). Intermediate between the two are the Gaussian Path Integrals, where the solution space of Gaussian Integrals is the fundamental representation of the propagator, explicitly solvable, providing the basis for a generator for the theory of N-point functions (in quantum field theory), or of a partition function (in statistical mechanical theory), where the latter is undertaken according to a deterministic or probabilistic interpretation of the theory.

In what follows we start with an overview that begins with Dirac's 1933 paper [28] and Feynman's 1948 paper [29], and describes the journey to well-definedness.. The underlying notion of highly oscillatory integrals, however, goes back to the inception of the calculus and classical mechanics (1700's Laplace [30], for more details see Book1 [1]). We will then work with Feynman's path integral defined in terms of classical paths, whose action defines a phase associated with that path, and where a sum on all paths (integral) then defines the path integral. Intuitively, paths will typically add out of phase so the path integral will effectively select for stationary phase solutions. At zeroth order (semiclassical analysis) this will give the classical solution, at higher order it will give the Schrodinger's equation, thus we recover standard quantum mechanics from this seemingly odd formulation (just as can be done in classical mechanics for the Euler-Lagrange equations). Next we will review the Green's function solution provided by the path integral formulation for quantum mechanics. This will then be generalized to its quantum field theory version that will be shown to be well-defined. In what follows the Green's function will provide the basis for a perturbative and renormalization analysis that will give the modern field theory results, e.g., quantum electrodynamics, that have provided the experiments confirmed with the highest precision of any tests of theory known (with 16 decimal places of agreement).

In analysis of bound states, path integrals are weak, it took decades, for example, for a solution to the Hydrogen atom to be accomplished using path integrals [31]. And, even then, the solution required generalizing to path integrals for a curved spacetime and use of analytic time (many of the same methods deployed to make the path integral formulation well-defined). For scattering and perturbative analysis, however, path integrals are strong (already being summation based and explicitly semigroup compatible). Still, in quantum mechanics, the methods, often in the Heisenberg or Interaction Picture with canonical quantization, provide a clear analysis for such problems. So why bother with path integrals? The answer is you don't need to in quantum mechanics. But, in quantum field theory, the only tractable way to proceed in many situations will be with path integral formulations, and in some cases, such as electroweak renormalization, we are only able to complete the theory in terms of path integral derivations. It all comes down to the Green's function. In quantum field theory many problems reduce to knowing the vacuum expectation of the time-ordered product of Heisenberg field operators, i.e., what is $\langle 0|T\hat{\varphi}(x_1)\hat{\varphi}(x_2)\ldots\hat{\varphi}(x_n)|0\rangle$? By use of Wick's theorem (Section 3.9.2), we will see that such n-point functions reduces to a sum on products of 2-point functions. Thus, the core problem reduces to solving a Green's function analysis for the 2-point function. so this derivation is shown in both quantum mechanics and quantum field theory settings.

Although the Overview Section, Feynman Path Integral derivation, and Green's function analysis that follows is mainly geared to prepare for those interested in quantum field theory, the special nature of the path integral formulation will also be revealed in two ways: (1) the formalism naturally provides a generating functional of Green's functions that is analytically related to a partition function, suggesting that time is fundamentally analytic (possibly with periodic boundary conditions related to inverse temperature, see Book 6 [6] and Appendix D for details); (2) the formalism is obtained in Emanator Theory (Book 7 [7], also a synopsis in Appendix D) when emanator projection is to a maximally analytic domain.

Overview

In 1933 Dirac proposed that the propagator of quantum mechanics could be argued to correspond with $\exp\left(\frac{iS}{\hbar}\right)$ [28]. In 1948 Feynman developed this further, simply appending the Dirac paper to his PhD thesis, and continuing with what would be his foundational paper on path integrals [29], where the fundamental object became:

$$\int e^{\left(\frac{i}{\hbar}\right)S[b,a]}.$$

Note that the integral expression is written with no measure, as this is part of what we must decide how to implement at the outset. Making the notion of path integral well-defined can reduced to the following categories [32]:

(1) The sequential approach, makes use of the Trotter product formula (see Appendix B for Math Review) and is what is used in Feynman's 1948 description. If we write the free Hamiltonian operator as: $H = -\frac{\hbar^2}{2m}\Delta$ and the potential operator as V (simply the multiplication operator

for factor V acting on $L^2(\mathbb{R}^d)$), we then get wavefunction solutions in the form:

$$\psi(t) = \lim_{n\to\infty} \left(e^{-\frac{i\,t}{\hbar n}V} e^{-\frac{i\,t}{\hbar n}H} \right)^n \psi(0)$$

In the Green's function analysis that follows we will see that the free Hamiltonian operator satisfies:

$$e^{-\frac{i\,t}{\hbar n}H} = \left(\frac{2\pi i\hbar t}{mn}\right)^{-d/2} e^{-\frac{i}{\hbar}\frac{m}{2t/n}|x-y|^2}$$

which allows us to write the wavefunction solution as:

$$\psi(t,x) = \lim_{n\to\infty} \int_{\mathbb{R}^{nd}} \left(\frac{2\pi i\hbar t}{mn}\right)^{-\frac{dn}{2}} \times \exp\left(-\frac{i}{\hbar}\sum_{j=1}^{n} \frac{t}{n}\left[\frac{m}{2}\frac{\left(x_j - x_{j-1}\right)^2}{\left(\frac{t}{n}\right)^2}\right]\right) \psi(x_0)dx_0 \ldots dx_{n-1}$$

From this form, we can arrive at a rigorous definition of Feynman integration in two ways:

(i) Approximate the paths by piecewise linear paths as done by Feynman and since formalized [33,34] (derivation of this will be shown shortly).

(ii) Generalize he Trotter formula to semigroups (limits now 'strong'):

$$\lim_{n\to\infty} (F(t/n))^n = \exp\,(tF'(0))$$

from which a rigorous definition is possible [35]

(2) An alternate formulation to have a rigorous definition involves analytic continuation in the physical time parameter (Euclideanization if rotated to pure imaginary). Mathematically this not only results in something well defined, it also provides a versatile formalism to address many problems. Physically, however, the meaning of shifting to imaginary time is unclear. Consider the Schrodinger equation and shift to pure imaginary time, the equation then becomes the heat equation:

$$-\frac{\partial}{\partial t}u(t,x) = -\frac{1}{2}\Delta_x u(t,x) + V(x)u(t,x)$$

The solutions are given in terms of Weiner integrals, the specific form of a solution is known as the Feynman-Kac formula:

$$u(t,x) = \int \exp\left(-\int_0^t V(\omega(s)+x)ds\right)u(0,\,\omega(t)+x)\,dW(\omega)$$

which is well-defined in general.

(3) An alternate approach involving Wiener integrals has also been developed by Daubechies and Klauder [37], which gives a well-defined formalism for Hamiltonians with polynomial position and momentum terms that can be carried over to systems with spin.

(4) The white noise approach is possible if the path integral functional is written as the T-transform of a unique Hida distribution [37]. This construction process is widely applicable (requires analyticity), most especially to the time-dependent harmonic oscillator, which is foundational for the path-integral basis of field theory.

(5) The Parseval Identity approach to arrive at a (well-defined) Fresnel Integral [38,39]. Provides a detailed method for stationary phase analysis when working in infinite dimensions. Regarding the latter, we could just proceed directly with a generalization of classic oscillatory integrals to the imaginary form (the Fresnel Integrals).

(6) Infinite dimensional oscillatory integrals, a generalization to the Parseval approach, grounded in work starting with Laplace, provides general applicability, most notably to phase functions up to degree 4, the latter covering the important cases of quantum field theory to be considered [40,41].

A quick review of the sequential approach to obtain the Feynman configuration space path integral will now be given.

Feynman Path Integral derivation
Feynman and Hibbs, pg. 28 [27]:

> The probability $P(a,b)$ to go from a point x_a at the time t_a to the point x_b at t_b is the absolute square $P(a,b) = |K(b,a)|^2$ of an amplitude $K(b,a)$ to go from a to b.

The Feynman Path Integral Hypothesis
The amplitude $K(b,a)$, to go from a to b, shall be the sum over the phase contributions accrued for every path that goes from a to b, where the phase contribution on a given path is taken to be proportional to the action $S[b,a]$ to traverse that path. The functional integral on paths $x(t)$ involves the functional differential '$\mathcal{D}x(t)$':

$$K(b,a) = \frac{1}{A}\int e^{\left(\frac{i}{\hbar}\right)S[b,a]}\,\mathcal{D}x(t)$$

Where the Action for a quantum system with a classical correspondence is given, it shall satisfy the usual definition in terms of the classical Lagrangian of the system:

$$S[b,a] = \int_{t_a}^{t_b} L(\dot{x},x,t)dt.$$

In an attempt to formalize the definition of a path integral, Feynman considers an explicit sum over paths for the free particle [27]:

> The sum over paths is defined as a limit, in which at first the path is specified by giving only its coordinate x at a large number of specified times separated by very small intervals ε. The path sum is then an integral over all these specific coordinates. Then to achieve the correct measure, the limit is taken as ε approaches 0.

Thus,

$$K(b,a) = \lim_{\varepsilon \to 0} \int e^{\left(\frac{i}{\hbar}\right) S[b,a]} \frac{dx_1}{A} \frac{dx_2}{A} \dots \dots \frac{dx_{N-1}}{A}.$$

Notice that this direct approach involves an explicit time-slicing, where all paths have a monotonically increasing t parameter.

A lengthy derivation involving Gaussian integrals then gives the answer:

$$K(b,a) = \left[\frac{2\pi i\hbar(t_b - t_a)}{m}\right]^{-1/2} exp\left\{\frac{im(x_b - x_a)^2}{2\hbar(t_b - t_a)}\right\}.$$

The analysis leading to the above result is only well-founded (based on a Quantum Mechanical formulation based on a Hilbert space, etc.) for a restricted class of Lagrangians (of which the free particle Lagrangian is one [14]). In what follows an explicit derivation will be given without explicit use of time-slicing and as such will be more amenable to generalization.

Starting with the functional integral description:

$$K(b,a) = \frac{1}{A} \int e^{\left(\frac{i}{\hbar}\right) \int_{ta}^{t_b} L(\dot{x},x,t)dt} \mathcal{D}x(t)$$

Let's start by considering classical path $\overline{x}(t)$, where the functional path considered is:

$$x = \overline{x}(t) + y,$$

and we have:

$$S[x(t)] = S[\overline{x}(t) + y(t)]\,.$$

Note that in the phase integral $S[\overline{x}(t) + y(t)] \cong S_{Classical}[\overline{x}(t)] + S[y(t)]$ since elements first order terms in $y(t)$ will have their phases cancelled (by terms with $-y(t)$). We, thus, have:

$$K(b,a) = e^{\left(\frac{i}{\hbar}\right)S_{Cl}[b,a]} \int_0^0 exp\left\{\frac{i}{\hbar}\int_{t_a}^{t_b} \frac{1}{2} m\dot{y}^2 dt\right\} \mathcal{D}y(t),$$

where all paths $y(t)$ begin and end at $y = 0$, thus there is no spatial dependance in the integral, only a functional dependence on t_a and t_b.

For the classical action we have:

$$S_{Cl}[b,a] = \int_{t_a}^{t_b} \frac{1}{2} m \left(\frac{dx}{dt}\right)^2 dt.$$

Since the classical path simply satisfies:

$$\frac{dx}{dt} = \frac{x_b - x_a}{t_b - t_a}$$

So,

$$S_{Cl}[b,a] = \frac{1}{2} m \frac{(x_b - x_a)^2}{(t_b - t_a)}$$

Furthermore, since the classical path has a $\frac{dx}{dt}$ dependence that only depends on Δx and Δt then such will be the case for S_{Cl} and, correspondingly, $\mathrm{F}(t_a, t_b) = \mathrm{F}(t_b - t_a)$. Thus,

$$K(b,a) = F(t_a - t_b)\, exp\left(\frac{im(x_b - x_a)^2}{2\hbar(t_b - t_a)}\right)$$

Also, we know (from the Chapman-Kolmogorov relation):

$$K(b,a) = \int_{x_c}^{\square} K(b,c)K(c,a)dx_c$$

So,

$$F(t_a - t_b)exp\left(\frac{im(x_y - x_a)^2}{2\hbar(t_a - t_b)}\right) = F(t_a - t_c)\mathrm{F}(t_c - t) \cdot I$$

where

$$I = \int_{-\infty}^{\infty} exp\left\{\frac{im}{2\hbar}\left[\frac{(x_b - x_c)^2}{(t_b - t_c)} + \frac{(x_c - x_a)^2}{(t_c - t_a)}\right]\right\} dx_c$$

After regrouping as a Gaussian integral, we get:

$$I = \sqrt{\frac{2\pi\hbar(t_b - t_c)(t_c - t_a)}{(-im)(t_b - t_a)}}\ exp\left(\frac{im}{2\hbar}\left[\frac{(x_b - x_a)^2}{(t_b - t_a)}\right]\right)$$

So,

$$F(t_b - t_a) = F(t_b - t_c)F(t_c - t_a)\sqrt{\frac{2\pi\hbar(t_b - t_c)(t_c - t_a)}{(-im)(t_b - t_a)}}$$

Substituting $\mathrm{F}(t_b - t_a) = \sqrt{\frac{m}{2\pi i\hbar(t_b - t_a)}}\mathrm{f}(t_b - t_a)$ we get

$$\mathrm{f}(t + s) = f(t)f(s) \rightarrow \quad f(t) = e^{at}$$

If we consider a very small increment of time we find that $f(\varepsilon) = \left(\frac{2\pi i\hbar\varepsilon}{m}\right)^{-1/2}$:

Consider

$$\Psi(x_2, t_2) = \int_{-\infty}^{\infty} K(x_2, t_2; x_1, t_1)\Psi(x_1, t_1)dx_1$$

For a short time integral,

$$\Psi(x, t + \varepsilon) = \int_{-\infty}^{\infty} \frac{1}{A} exp\left[\varepsilon\frac{i}{\hbar}\left[\frac{m}{2}\right]\left[\frac{x - y}{\varepsilon}\right]^2\right]\Psi(y, t)dy$$

$$= \int_{-\infty}^{\infty} \frac{1}{A} e^{im\eta^2/2\hbar\varepsilon}\ \Psi(x + \eta, t)d\eta$$

Expanding in ε and equating both sides:

$$\Psi(x, t) + \varepsilon\frac{\partial\Psi}{\partial t} = \int_{-\infty}^{\infty} \frac{1}{A} e^{im\eta^2/2\hbar\varepsilon}\left[\Psi(x, t) + \eta\frac{\partial\Psi}{\partial x} + \frac{1}{2}\eta^2\frac{\partial^2\Psi}{\partial x^2}\right] d\eta$$

Taking the leading terms from both sides we get

$$1 = \int_{-\infty}^{\infty} \frac{1}{A} e^{im\eta^2/2\hbar\varepsilon} d\eta \rightarrow \ A = \left(\frac{2\pi i\hbar\varepsilon}{m}\right)^{1/2}$$

So, the nonrelativistic propagator is:

$$K(b,a) = \sqrt{\frac{m}{2\pi i\hbar(t_b - t_a)}}\, exp\left(\frac{im(x_b - x_a)^2}{2\hbar(t_b - t_a)}\right)$$

So far we've shown how the path integral formalism can explicitly give rise to known quantum solutions for the free particle case. Feynman was able to extend this result to the 4-D Klein Gordon equation by extending to a 5-D formalism where the Klein-Gordon equations takes the form of a free particle Lagrangian (Feynman: PR vol 80 pg 440 (1950) [42]).

Starting with the Klein-Gordon equation:

$$\left(i\frac{\partial}{\partial x_\mu}\right)^2 \Psi = m^2\Psi$$

Define

$$\Psi = \int_{-\infty}^{\infty} exp\left(-\frac{1}{2}im^2u\right)\varphi(x,u)du$$

Then,

$$\left[\left(i\frac{\partial}{\partial x_\mu}\right)^2 - m^2\right]\Psi = \int_{-\infty}^{\infty}\left[\left(i\frac{\partial}{\partial x_\mu}\right)^2 - m^2\right] exp\left(-\frac{1}{2}im^2u\right)\varphi(x,u)du$$

$$= \int_{-\infty}^{\infty}\left\{\left(i\frac{\partial}{\partial x_\mu}\right)^2\varphi - 2i\varphi\frac{\partial}{\partial u}\right\} exp\left(-\frac{1}{2}im^2u\right)du$$

Since $\frac{\partial\Psi}{\partial u} = \left[exp\left(-\frac{1}{2}im^2u\right)\varphi(x,u)\right]_{-\infty}^{\infty} = 0$ (if $\varphi(x,u)$ *bounded*)
we then have:

$$\int_{-\infty}^{\infty} exp\left(-\frac{1}{2}im^2u\right)\frac{\partial\varphi}{\partial u}du = -\int_{-\infty}^{\infty}\left(-\frac{1}{2}im^2\right) exp\left(-\frac{1}{2}im^2u\right)\varphi du$$

Or

$$\int_{-\infty}^{\infty} exp\left(-\frac{1}{2}im^2u\right)\left[\left(i\frac{\partial}{\partial x_\mu}\right)^2 + 2i\left(\frac{\partial}{\partial u}\right)\right]\varphi(x,u)\,du = 0$$

Thus,

$$i\frac{\partial\varphi}{\partial u} = -\frac{1}{2}\left(i\frac{\partial}{\partial x_\mu}\right)^2\varphi$$

Thus, any parabolic partial differential equation can be approached by means of Feynman's method and if it is hyperbolic then the same can be done by considering an extra parameter,

Feynman presents an argument in Rev. Mod. Phys. Vol 20 pg267 (1948) [29] which shows how

$$\Psi(x_{k+1}, t + \varepsilon) = \int exp\left[\frac{i}{\hbar} S(x_{k+1,} x_k)\right] \Psi(x_k, t) \frac{dx_k}{A},$$

along the lines discussed, and how this generalizes and agrees with the Schrödinger equation. Further use of the $exp\left(-\frac{1}{2} im^2 u\right)$ 'regularizer' is made in Feynman Phys. Rev 76, pg. 749 (1949) [43]. These and other methods to formally extend and undergird the path integral formulation will run into further complication, especially when considering the nonrelativistic/noncovariant nature of a particular time slicing. What becomes evident is that a path integral formulation exists that is *generative* such that it generates the correct equations for the quantum system. Much like how the statistical mechanics partition function of a system is generative of a systems thermodynamic potentials. The source of the quantum mechanical path integral formulation as a generative theory is explained in Book 7 of the Series [7], where emanator theory is proposed as giving rise to the quantum theory with its generative expression, along with predictions of the value of alpha and the structure of the Standard Model.

The Path Integral method will be of profound importance when considering quantum field theory in Book 5, where it will provide a pathway to a solution for QED (that will produce the best agreement with experimental results observed in physics).

Propagator for the Schrodinger equation:
In electromagnetism we have Maxwell's equations for the differential point of view, and we have Huygen's principle for the global point of view. The Quantum analogue of Huygen's principle is:

$$\Psi(\vec{r}_2, t_2) = \int d^3r_1\, K(\vec{r}_2, t_2; \vec{r}_1, t_1)\, \Psi(\vec{r}_1, t_1) \qquad (t_2 > t_1)$$

$$(K = 0\ for\ t_2 < t_1),$$

where K is called the propagator.

Proof of the existence of the propagator:
Recall the evolution operator

$$|\Psi(t_2)> = U(t_2, t_1)|\Psi(t_1)\ , \qquad where\ \Psi(\vec{t}_2, t_2) = <\vec{r}_2|\Psi(t_2)>.$$

Thus,

$$\Psi(\vec{r}_2, t_2) = \int d^3r_1 < \vec{r}_2|U(t_2, t_1)|\vec{r}_1 >< \vec{r}_1|\Psi(t_1) >$$
$$= \int d^3r_1 < \vec{r}_2|U(t_2, t_1)|\vec{r}_1 > \Psi(\vec{r}_1, t_1)$$

Thus,

$$K(2,1) = K(\vec{r}_2, t_2; \vec{r}_1, t_1) = < \vec{r}_2|U(t_2, t_1)|\vec{r}_1 > \theta(t_2 - t_1).$$

Known as the retarded propagator, it specifies a unique Green's function. The physical interpretation of $K(2,1)$ is that it represents the probability amplitude that the particle, starting from the point $\vec{r}_1$ at t_1, will arrive at $\vec{r}_2$ at t_2. If H doesn't depend explicitly on time and we consider its eigenstate:

$$H|\varphi_n >= E_n|\varphi_n >,$$

then,

$$U(t_2, t_1) = e^{-iH(t_2,t_1)/\hbar} = e^{-iH(t_2,t_1)/\hbar} \sum_n |\varphi_n >< \varphi_n|$$
$$= \sum_n e^{-iE_n(t_2,t_1)/\hbar} |\varphi_n >< \varphi_n|$$

Thus,

$$K(2,1) = \sum_n e^{-iE_n(t_2,t_1)/\hbar} < \vec{r}_2|\varphi_n >< \varphi_n|\vec{r}_1 > \theta(t_2 - t_1),$$

or

$$K(2,1) = \theta(t_2 - t_1) \sum_n \varphi_n(\vec{r}_2)\varphi_n^*(\vec{r}_1)\, e^{-iE_n(t_2-t_1)/\hbar}$$

Now, if $\varphi_n(\vec{r}_2)e^{-iE_n t_2/\hbar}$ is a solution of the Schrodinger equation :

$$\left\{i\hbar\frac{\partial}{\partial t_2} - H\left(\vec{r}_2, \frac{\hbar}{i}\nabla_2\right)\right\} \varphi_n(\vec{r}_2)e^{-iE_2 t_2/\hbar} = 0$$

So,

$$\left\{i\hbar\frac{\partial}{\partial t_2} - H\left(\vec{r}_2, \frac{\hbar}{i}\vec{V}_2\right)\right\} K(\vec{r}_2, t_2; \vec{r}_1, t_1) = i\hbar\delta(t_2 - t_1) \sum_n \varphi_n(\vec{r}_2)\varphi_n^*(\vec{r}_1)$$

$$\left[i\hbar\frac{\partial}{\partial t_2} - H\left(\vec{r}_2, \frac{\hbar}{i}\nabla_2\right)\right] K(2,1) = i\hbar\delta(t_2 - t_1)\delta(\vec{r}_2 - \vec{r}_1),$$

with boundary condition. $K(2,1) = 0 \quad if\ t_2 < t_1.$

Lagrangian formulation of quantum mechanics:

Consider two pts $(\vec{r}_1, t_1), (\vec{r}_2, t_2)$. Choose N intermediate times t_α, $(i = 1, 2, \ldots N)$:

$$t_1 < t_{\alpha 1} < t_{\alpha 2} < \cdots < t_{\alpha n} < t_2$$

For each $t_{\alpha i}$ choose a $r_{\alpha i}$. As $N \to \infty$ we get a function $\vec{r}(t)$ (which we shall assume to be continuous)such that: $\vec{r}(t_1) = \vec{r}_1$ and $\vec{r}(t_2) = \vec{r}_2$. Now,

$$U(t_2, t_1) = U(t_2, t_{\alpha N})U(t_{\alpha N}, t_{\alpha N-1}) \dots U(t_{\alpha 2}, t_{\alpha 1})U(t_{\alpha 1}, t_1).$$

Now use $\{|r>\}$ rep and use closure relations α_N times:

$$K(2,1) = \int d^3 r_{\alpha N} \int d^3 r_{\alpha N-1} \dots \int d^3 r_{\alpha 2} \int d^3 r_{\alpha 1}\, K(2, \alpha_N)K(\alpha_N, a_{N-1}) \dots K(\alpha_2, \alpha_1$$

As $N \to \infty$: $K(2, \alpha_N)K(\alpha_N, a_{N-1})x \dots K(\alpha_2, 1)K(\alpha_1, 1)$ becomes the probability amplitude of the particle following a given path between 1 and 2. K(2,1) is then the integral which corresponds to the coherent superposition of the amplitudes associated with all possible space-time paths starting from 1 and ending at 2.

Feynman's Postulates:

Thus far we've seen the propagator formulation of Schrodinger's equation and how this is equivalent to a "sum over paths". The propagator formalism has delineated a spacetime formulation of the postulates of quantum mechanics, and thus is more general (e.g., we have a direct relativistic generalization since already in a spacetime formulation). In the Feynman approach we take the propagator object $K(2,1)$ as fundamental and define it directly as the probability amplitude for a particle to go from $(\vec{r}_1, t_1)$ to $(\vec{r}_2, t_2)$ according to the following rules:

(i) $K(2,1)$ is the sum of an infinity of partial amplitudes, one for each of the spacetime paths connecting $(\vec{r}_1, t_1)$ with $(\vec{r}_2, t_2)$.

(ii) The partial amplitude $K_\Gamma(2,1)$ associated with one of these paths (Γ) is given by the classical Action (this is the scenario where such exists), in terms of the classical Lagrangian, according to:

$$S_\Gamma = \int_\Gamma \mathcal{L}(\vec{r}, \vec{p}, t)\, dt$$

where

$$K_\Gamma(2,1) = N e^{\frac{i}{\hbar} S_\Gamma},$$

and N is a normalization constant.

Schrodinger's equation follows from the two postulates above. The Canonical commutation relations for $\vec{R}$ *and* $\vec{P}$ are the same (static). Thus, the above postulates permit a formulation of Quantum Mechanics which is different from that of Schrodinger where the time evolution (the 6th Postulate) was given by explicit time reference: $i\hbar\frac{d}{dt}|\Psi(t)>=H(t)|\Psi(t)>$, it is here given in the Action formulation and bundled into the classical Action. This will afford the Feynman approach greater flexibility via clearer separation of the classical/apparatus parts and the quantum parts of the system. Feynman's postulates, in the classical limit, give Hamilton's principle of least action. Feynman's approach generalizes to system with quantum Actions for which there is no classical counterpart (spin angular momentum) and for any classical system which that has a variational formulation even if there is (seemingly) no mechanical aspect (field descriptions).

The disadvantage of the Feynman approach is often attributed to its complexity -- a summation over an infinite number of paths. This disadvantage will be worsened in the case of generalization to quantum field theory, to the point of being ill-defined. Any 'repair' to the theory to make well-defined and connect with a canonical formulation, or experiment, is in effect defining the Feynman theory as providing a 'generative' formulation, where the representation of the experimental system then governs how that formulation will then generate the local quantum theory appropriate to the experiment. This will actually be seen to be in agreement with the thermal quantum theory that results when generalizing to complex time in the propagator (Book6), where the partition function will result, with known generative properties to define the entire thermodynamics of the system. So, if the entire theory, other than specific of representation and coordinate system, etc., can be defined in the generative Feynman formulation, is there an even deeper layer of the theory that is generative of the Feynman formulation with the known standard models of both particle physics and cosmology? The answer, is apparently yes, and the theory is referred to as emanator theory (emanator instead of propagator) in Book 7, where it is described in detail.

Green's function analysis

In this section we consider how to calculate the quantum field theory n-point function denoted by:

$$\langle 0|T\hat{\varphi}(x_1)\hat{\varphi}(x_2)\dots\hat{\varphi}(x_n)|0\rangle.$$

We begin by consideration of the quantum mechanical equivalent in terms of position operators at time t (not field operators at spacetime coordinate x_1):

$$G^{(n)}(t_1,t_2\dots t_n)=\langle 0|T\hat{q}(t_1)\hat{q}(t_2)\dots\hat{q}(t_n)|0\rangle.$$

We will also see that the n-point Green's function can be written as a sum over 2-point Green's functions. Furthermore, since we have the position operators it is convenient to work in the Heisenberg position representation. Thus, our key calculation reduces to evaluating the 2-point Green's function:

$$G^{(2)}(t_1, t_2) = \langle q', t|T\hat{q}(t_1)\hat{q}(t_2)|q, 0\rangle.$$

Note that if we remove the "two points" we recover the Feynman propagator

$$K = \langle q', t|q, 0\rangle = \int_{q,0}^{q',t} \mathcal{D}q e^{iS}.$$

So, the focus of the Green's function analysis is on evaluating $\langle q', t|T\hat{q}(t_1)\hat{q}(t_2)|q, 0\rangle$, let's start with the case $t_1 > t_2$:

$$\langle q', t|T\hat{q}(t_1)\hat{q}(t_2)|q, 0\rangle = \langle q', t|\hat{q}(t_1)\hat{q}(t_2)|q, 0\rangle$$

and

$$\begin{aligned}
&\langle q', t|\hat{q}(t_1)\hat{q}(t_2)|q, 0\rangle \\
&= \int dq_1 dq_2 \langle q', t|q_1, t_1\rangle\langle q_1, t_1|\hat{q}(t_1)\ \hat{q}(t_2)|q_2, t_2\rangle \langle q_2, t_2|q, 0\rangle \\
&= \int dq_1 dq_2 q_1 q_2 \langle q', t|q_1, t_1\rangle\langle q_1, t_1|q_2, t_2\rangle \langle q_2, t_2|q, 0\rangle
\end{aligned}$$

Let's shift the Feynman propagators to integral form:

$$\begin{aligned}
&\langle q', t|\hat{q}(t_1)\hat{q}(t_2)|q, 0\rangle \\
&= \int dq_1 dq_2 q_1 q_2 \int_{q_1,t_1}^{q',t} \mathcal{D}q e^{iS} \int_{q_2,t_2}^{q_1,t_1} \mathcal{D}q e^{iS} \int_{q,0}^{q_2,t_2} \mathcal{D}q e^{iS}
\end{aligned}$$

from which it then manifest that (for $t_1 > t_2$):

$$\langle q', t|\hat{q}(t_1)\hat{q}(t_2)|q, 0\rangle = \int_{q,0}^{q',t} \mathcal{D}q q_1(t_1) q_2(t_2) e^{iS}.$$

If we repeat the above analysis for $t_1 < t_2$ we get the same result, so the path integral conveniently has one expression for the time-ordered product:

$$\langle q', t|T\hat{q}(t_1)\hat{q}(t_2)|q, 0\rangle = \int_{q,0}^{q',t} \mathcal{D}q q_1(t_1) q_2(t_2) e^{iS}.$$

Let's now shift back to an evaluation in terms of vacuum-to-vacuum elements. For this we shift to the time parameter having a small imaginary phase. If we then evolve to large negative imaginary time, all Hamiltonian eigenstates of $\langle q', t|$ and $|q, 0\rangle$ will be dominated by the ground state. Thus the following proportionality result must exist:

$$\langle q', t|q, -t\rangle \propto \langle \emptyset, t|\emptyset, -t\rangle$$

where an important notational shift has occurred: '$\emptyset$' references the vacuum state and the time parameter is now in a symmetric form that is standard. We can thus, write:

$$\langle \emptyset, t|\emptyset, -t\rangle \propto \langle q', t|q, -t\rangle = \int_{q,-t}^{q',t} \mathcal{D}q e^{iS}.$$

A similar shift can be done with factors of $\hat{q}(t_1)\hat{q}(t_2)\dots\hat{q}(t_n)$ present in the integrand, so we have:

$$\langle \emptyset, t|T\hat{q}(t_1)\hat{q}(t_2)\dots\hat{q}(t_n)|\emptyset, -t\rangle \propto \int_{q,-t}^{q',t} \mathcal{D}q q_1(t_1)q_2(t_2)\dots q_n(t_n)e^{iS}.$$

We now eliminate the proportionality, and cancel the unknown phase term associated with the large negative imaginary time relation, by dividing it out (as manifest in the Feynman propagator), where the vacuum is now written $\langle\emptyset|$:

$$G^{(n)}(t_1, t_2 \dots t_n) = \langle \emptyset|T\hat{q}(t_1)\hat{q}(t_2)\dots\hat{q}(t_n)|\emptyset\rangle$$
$$= \frac{\langle \emptyset, t|T\hat{q}(t_1)\hat{q}(t_2)\dots\hat{q}(t_n)|\emptyset, -t\rangle}{\langle \emptyset, t|\emptyset, -t\rangle}$$

Thus,

$$G^{(n)}(t_1, t_2 \dots t_n) = \frac{\int_{q,-t}^{q',t} \mathcal{D}q q_1(t_1)q_2(t_2)\dots q_n(t_n)e^{iS}}{\int_{q,-t}^{q',t} \mathcal{D}q e^{iS}}$$

The method for solving this is well known in quantum field theory and statistical field theory. We begin by defining a generating functional for Green's functions:

$$Z[J] = \frac{\int \mathcal{D}q e^{i(S+\int J(t)q(t)dt)}}{\int \mathcal{D}q e^{iS}} = \frac{\langle\emptyset|\emptyset\rangle_J}{\langle\emptyset|\emptyset\rangle_{J=0}}$$

from which we see that:

$$\left(\frac{1}{i}\frac{\delta}{\delta J(t_1)}\cdots\frac{1}{i}\frac{\delta}{\delta J(t_n)}Z[J]\right)\Bigg|_{J=0} = \frac{\int \mathcal{D}q q_1(t_1)q_2(t_2)\dots q_n(t_n)e^{iS}}{\int \mathcal{D}q e^{iS}}$$
$$= G^{(n)}(t_1, t_2 \dots t_n)$$

The case of where the Action S is that of a harmonic oscillator is of special interest for quantum field theory generalization, so let's examine that case and evaluate $Z[J]$. The numerator of $Z[J]$ is:

$$N = \int \mathcal{D}q e^{i\int\left[\frac{1}{2}m\dot{q}^2 - \frac{1}{2}m\omega^2 q^2 + J(t)q(t)\right]dt}.$$

For the Action $S = \frac{1}{2}m\dot{q}^2 - \frac{1}{2}m\omega^2 q^2 + J(t)q(t)$ let's denote the classical solutions by $q_c(t)$, we then have:

$$N \propto e^{iS[q_c]}.$$

Making use of the fact that q_c satisfies the equations of motion, we can simply to:

$$S[q_c] = \frac{1}{2}\int J(t)q_c(t)dt.$$

The classical solution can be written in terms of the standard Green's function as:

$$\left(\frac{d^2}{dt^2} + \omega^2\right)G(t,t') = -i\delta(t-t'),$$

where

$$q_c(t) = -i\int J(t')G(t,t')dt'$$

Thus, we can show for the harmonic oscillator:

$$Z[J] = exp\left(\frac{1}{2}\int J(t')G(t,t')J(t)dt'dt\right)$$

For the Green's function solution in momentum space we see that:

$$G(t,t') = \int \frac{dk}{2\pi}\frac{i}{k^2-\omega^2}e^{-ik(t-t')}dt$$

As is the Green's function is ambiguous since it has poles on the axis of integration and we must decide how to do this. Fortunately our prescription from the outset made use of an asymptotic analyticity relation that involved a small negative imaginary part in the time definition. similarly here, to have G go to zero as $t \to \infty$ we need the (precursor to Feynman propagator) prescription:

$$G(t,t') = \int \frac{dk}{2\pi}\frac{i}{k^2-\omega^2+i\epsilon}e^{-ik(t-t')}dt.$$

Note: There is care taken in understanding and specifying the pole structure. Later, the pole structure defines the physical part of propagator in renormalization scheme.

Let's now consider the free Klein-Gordon scalar field. What results is an exact parallel of the preceding analysis. We now have a scalar field φ which has a classical solution that satisfies the Klein Gordon equation:

$$(\partial^2 + m^2)\varphi_c(x) = J(x)$$

and the Klein-Gordon Green's function is defined by:

$$(\partial^2 + m^2)\Delta_F(x,x') = -i\delta^4(x-x').$$

Thus,

$$\varphi_c(x) = i\int d^4xJ(x')\Delta_F(x,x')$$

so,

$$Z = exp\left(\frac{1}{2}\int d^4xd^4x'J(x')\Delta_F(x,x')J(x)\right).$$

Solving the Green's function in 4-momentum space, adopting the same pole prescription as previously, we then have the Feynman propagator:

$$\Delta_F(x,x') = \int \frac{d^4k}{(2\pi)^4}\frac{i}{k^2 - m^2 + i\epsilon}e^{-ik(x-x')}.$$

4.2 Path Integral Quantization

Let's start the derivation of the path integral formulation of field theory by considering a scalar field with Lagrangian:

$$\mathcal{L} = \frac{1}{2}\partial_\mu\varphi\partial^\mu\varphi - V(\varphi)$$

Where we consider the system to be in 1+1 dimension (x,t) for simplicity. In quantum mechanics systems we had $q(t)$ and with the path integral formalism we took $q_n = q(t_i + n\epsilon)$, with each q_n treated as an independent variable for which we had the equation $\int dq_n dp_n$, etc., see Book4 for details. For the field variable we have two continuous variables x and t and now, $\varphi_{nm} = \varphi(x_i + n\delta, t_i + m\epsilon)$, e.g., replacing the continuous spacetime field by a lattice:

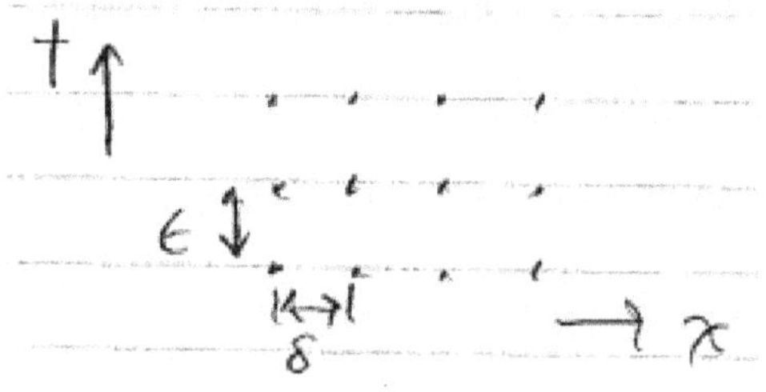

In this setting we have:

$$\int d\varphi_{nm}d\pi_{nm} \to \int \mathcal{D}\varphi(x,t)\mathcal{D}\pi(x,t)e^{i\int dxdt[\pi\dot{\varphi}-H]}$$

Consider

$$\int \mathcal{D}\varphi(x)e^{i\int d^4x\mathcal{L}(\varphi)}$$

(where 'x' now denotes all spacetime coordinates for convenience) and impose

$$Q(x,t \to -\alpha) = Q_i(x)$$
$$Q(x,t \to +\alpha) = Q_f(x)$$

(this is the appropriate path integral for working in the Schrodinger picture). Let's consider the 'vacuum persistence amplitude":

$$W[J] = \frac{1}{N}\int \mathcal{D}\varphi(x)\, exp\left(i\int d^4x\left[\mathcal{L}(\varphi(x)) + \frac{i\epsilon\varphi^2(x)}{2} + J(x)\varphi(x)\right]\right)$$
$$= < 0, out \mid 0, in >_J$$
$$W[J = 0] = 1 \rightarrow fixes\, N,\;\; N = \int \mathcal{D}\varphi(x) exp\left(i\int d^4x\left[\mathcal{L}(\varphi(x))\right]\right)$$

$W[J]$ is a generator function in that:

$$\frac{\delta}{i\delta J(x_{1)}}\cdots\frac{\delta}{i\delta J(x_{n)}}W[J]|_{J=0} = < 0\, |T\varphi(x_1)\ldots\varphi(x_n)|0 >|$$

Knowing $W[J]$, in other words, is largely solving the theory. Usually we can't solve for $W[J]$ exactly, however, many relations can be found from the integral form that are much more difficult to obtain in canonized theory.

$W[J]$ is calculated when $\mathcal{L}$ is quadratic: $\mathcal{L}_0 = \frac{1}{2}\partial_\mu\varphi\partial^\mu\varphi - \frac{1}{2}m^2\varphi^2$:

$$W_0[J] = \frac{1}{N}\int \mathcal{D}\varphi(x)\, exp\left(i\int d^4x\left[\frac{1}{2}\partial_\mu\varphi\partial^\mu\varphi - \frac{1}{2}(m^2 - i\epsilon)\varphi^2 + J(x)\varphi(x)\right]\right)$$

Since $(\Box + m^2 - i\epsilon)\bar{\varphi} = J(x)$ is a stationary point of the exponent, let:

$$\varphi(x) = \varphi'^{(x)} + \bar{\varphi}(x),$$

then

$$\int d^4x\left[\frac{1}{2}\partial_\mu\varphi\partial^\mu\varphi - \frac{1}{2}(m^2 - i\epsilon)\varphi^2 + J(x)\varphi(x)\right] = S[\bar{\varphi}] + S[\varphi']$$

And

$$W[J] = \frac{1}{N}e^{iS[\bar{\varphi}]}\int \mathcal{D}\varphi' e^{iS[\varphi']} = e^{iS[\bar{\varphi}]}$$

Where

$$S[\bar{\varphi}] = \int d^4x\left[-\frac{1}{2}\bar{\varphi}(\Box + m^2 - i\epsilon)\bar{\varphi} + J(x)\varphi(x)\right] = \frac{1}{2}\int d^4xJ(x)\bar{\varphi}(x)$$

Now

$$(\Box + m^2 - i\epsilon)\Delta_f(x) = -\delta_4(x)$$

Let

$$\overline{\varphi}(x) = \int d^4x\, \Delta_f(x - x')J(x')$$

To get:

$$W_0[J] = exp\left\{-\frac{i}{2}\int d^4xd^4x'J(x)\Delta_f(x - x')J(x')\right\}$$

This is the one path integral we know has to do, consider:

$$\Delta_f(x) = \int \frac{d^4k}{(2\pi)^4}e^{-ik\cdot x}\,\hat{\Delta}_f(k) \quad\rightarrow\quad \Delta_f(x) = \int \frac{d^4k}{(2\pi)^4}\frac{e^{-ik\cdot x}}{k^2 - m^2 + i\epsilon}$$

Known as the Feynman propagator. Thus,

$$\Delta_f(x) = -i\int \frac{d^3k}{(2\pi)^3 2\omega_k}\left[\theta(x^0)e^{-i\omega_k x^0} + \theta(-x^0)e^{i\omega_k x^0}\right]e^{i\vec{k}\cdot\vec{x}}$$

Since

$$\frac{\delta}{i\delta J(x)}W_0[J] = -\int d^4x'\Delta_f(x-x')J(x')\cdot W_0[J]$$

and

$$\frac{\delta^2}{i\delta J(x) i\delta J(x')}W_0 = i\Delta_f(x-x')W_0[J] + \int d^4 z\Delta_f(x-z)J(z)\int d^4\omega\Delta_f(x'-\omega)J(\omega)\cdot W_0[J]$$

Note that $\langle 0|T\varphi(x)\varphi(x')|0\rangle = i\Delta_f(x-x')$ is how Δ_f is defined from canonical arguments (not so simple in PI case but two pt function define to be $i\Delta'_f$ in that instance, a function that usually can't be calculated). Continuing:

$$\frac{\delta^3}{i\delta J(y) i\delta J(x) i\delta J(x')}W_0 = i\Delta_f(x-x')\left(-\int d^4\omega\Delta_f(y-\omega)J(\omega)\right)\cdot W_0[J]$$

$$-i\int d^4\omega\,\Delta_f(x'-\omega)J(\omega)\Delta_f(x-y)W_0$$

$$-i\int d^4z\Delta_f(x-z)J(z)\Delta_f(x'-y)W_0 + O(J^3)W_0$$

And

$$\frac{\delta^3}{i\delta J(y) i\delta J(y') i\delta J(x) i\delta J(x')}W_0 = \langle 0|T\varphi(x)\varphi(x')\varphi(y)\varphi(y')|0\rangle$$
$$= -\Delta_f(x-x')\Delta_f(y-y') - \Delta_f(x-y')\Delta_f(x'-y) - i\Delta_f(x-y)\Delta_f(x'-y')$$

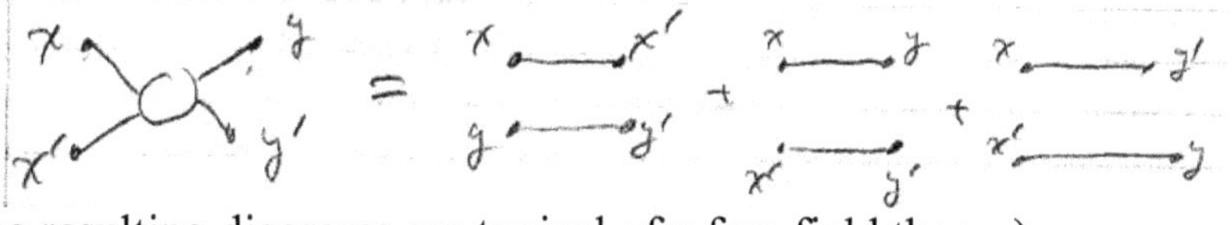

(the resulting diagrams are typical of a free field theory).
Using the reduction formula we get zero with our unconnected diagrams:

$$\langle kk'|pp'\rangle_{in} = -2^{-2}\int dxdx'dydyf_k^*(x)f_{k'}^*(x')f_p(y)f_{p'}(y') \times k(x)k(x')k(y)k(y')\langle 0|T\varphi(x)\varphi(x')\varphi(y)\varphi(y')|0\rangle = 0$$

Since

$k(x)k(x')k(y)k(y')\Delta_f(x-y)\Delta_f(x'-y')$ has $k(x)\Delta_f(x-y) = \delta(x-y)$, thus

$$\int dx dx' k(x)k(x')\left[f_k^*(x)f_{k'}^*(x')f_p(x)f_{p'}(x')\right] = 0$$

To switch to a generating functional that only gives the connected diagrams use:

$$W[J] = e^{-Z[J]} \quad \rightarrow \quad Z(J) = -i\,\ell nW$$

Z produces only the connected parts and for the free field (without connected parts), there is only the two-point function:

$$Z = \frac{1}{2}\int J(x)\Delta_f(x-x')J(x') \; and \; \frac{\delta^n}{\delta Jn}Z\,\Big|_{J=0} = 0, except\; n = 2.$$

For n=2:

$$\frac{\delta}{i\delta J(x)}\frac{\delta}{i\delta J(x')}Z\,\Big|_{J=0} = \Delta_f(x-x') = \langle 0|T\varphi(x)\varphi(x')|0\rangle$$

So, Z generates the connected diagrams:

$$\frac{\delta Z}{i\delta J(x)} = -\frac{1}{W}\frac{\delta W}{\delta J(x)} \; and\; with\; J = 0 \rightarrow \; \langle 0|\varphi(x)|0\rangle$$

And

$$\frac{\delta^2 Z}{i\delta J(x) i\delta J(x')} = +\frac{i}{W}\frac{\delta^2 W}{\delta J(x)\delta J(x')} - i\frac{i}{W^2}\frac{\delta W}{\delta J(x)}\frac{\delta W}{\delta J(x')},$$

and, as $J \rightarrow 0$, this goes to $i\langle 0|T\varphi(x)\varphi(x')|0\rangle - i\langle 0|\varphi(x)|0\rangle\,\langle 0|\varphi(x')|0\rangle$.

Consider

$$W_0[J] = \frac{1}{N}\int \mathcal{D}\varphi e^{i\int\left[\frac{1}{2}\partial_\mu\varphi\partial^\mu\varphi - \frac{1}{2}(m^2 - i\epsilon)\varphi^2 + J\varphi\right]}$$

$$= \frac{1}{N}\int \mathcal{D}\varphi e^{i\int\left\{-\frac{1}{2}\varphi k\varphi + J\varphi\right\}}, where\; k = \Box + m^2$$

Then,

$$\langle 0|\varphi(x)|0\rangle = \frac{1}{N}\int \mathcal{D}\varphi\; \varphi(x) e^{-\frac{i}{2}\int\varphi k\varphi} \rightarrow \; k(x)\langle 0|\varphi(x)|0\rangle$$

$$= \frac{1}{N}\int \mathcal{D}\varphi\; k(x)\varphi(x) e^{-\frac{i}{2}\int\varphi k\varphi}$$

Thus,

$$k(x)\langle 0|\varphi(x)|0\rangle = \int \mathcal{D}\varphi\; \frac{i\delta}{\delta\varphi(x)}\left[e^{-\frac{i}{2}\int\varphi k\varphi}\right] = 0$$

$$\rightarrow \; since\; integral\; of\; total\; derivative.$$

Thus, the vacuum expectation value of $k(x)\langle 0|\varphi(x)|0\rangle = 0$. For interacting fields this turns into:

$$k(x)\langle 0|\varphi(x)|0\rangle = 0\,\langle 0\left|\frac{\delta V}{\delta\varphi(x)}\right|0\rangle$$

the classical Euler equation Now, suppose we want $k(x)\langle 0|T\varphi(x)\varphi(x')|0\rangle$?

$k(x)\langle 0|T\varphi(x)\varphi(x')|0\rangle = k\int \mathcal{D}\varphi\ \varphi(x)\varphi(x')e^{-\frac{i}{2}\int \varphi k\varphi} =$

$i\int \mathcal{D}\varphi\ \varphi(x')\frac{\partial}{\partial\varphi(x)}e^{-\frac{i}{2}\int \varphi k\varphi}$, and with functional integration by parts we have:

$$k(x)\langle 0|T\varphi(x)\varphi(x')|0\rangle = -i\int \mathcal{D}\varphi\ \frac{\delta\varphi(x')}{\delta\varphi(x)}e^{-\frac{i}{2}\int \varphi n} = -i\delta_4(x-x')$$

So, to recap before moving on:

The generating functional for free field theory:

$$W_0[J] = \frac{1}{N}\int \mathcal{D}\varphi e^{i\int\left[\frac{1}{2}\partial_\mu\varphi\partial^\mu\varphi - \frac{1}{2}(m^2-i\epsilon)\varphi^2 + J\varphi\right]}$$

$$W_0[0] = 1$$

$$W_0[J] = e^{-\frac{i}{2}\int d^4x d^4x' J(x)\Delta_f(x-x')J(x')}, where\ i\Delta_f = (x-x')$$

$$= \langle 0|T\varphi(x)\varphi(x')|0\rangle$$

And

$$\langle 0|T\varphi(x_1)\dots\varphi(x_n)|0\rangle = \frac{\delta^n}{i\delta J(x_1),,,i\delta J(x_n)}W_0[J]|_{J=0}$$

Where the result for the last expression is zero unless n is even, then it is a sum over all possible permutations. This result follows from Wick's theorem (where ":….:" denotes normal ordering):

$$T\varphi(x_1)\dots\varphi(\varphi_n) =$$

$$:\varphi(x_1)\dots\varphi(x_n): + \sum_{perm} :\varphi(x_{p_3})\dots\varphi(x_{p_n}): i\Delta_f(x_{p_1} - x_{p_2})$$

$$+\sum_{perm} :\varphi(x_{p_5})\dots\varphi(x_{p_n}): i\Delta_f(x_{p_1} - x_{p_2})\ i\Delta_f(x_{p_3} - x_{p_4})\dots$$

$$+\sum_{p} i\Delta_f(x_{p_1} - x_{p_2})\dots i\Delta_f(x_{p_{n-1}} - x_{p_n})$$

Consider that $\varphi(x) = \varphi^{(+)}(x) + \varphi^{(-)}(x)$ means we can write

$$\varphi(x)\varphi(y) = \varphi^{(+)}(x)\varphi^{(+)}(y) + \varphi^{-}(x)\varphi^{-}(y) + \varphi^{(+)}(x)\varphi^{(-)}(y) + \varphi^{(-)}(x)\varphi^{(+)}(y)$$

where $\varphi^{(+)}(x)\varphi^{(-)}(y)$ is not normal ordered, so rewriting:

$$\varphi^{(+)}(x)\varphi^{(-)}(y) = \varphi^{(-)}(y)\varphi^{(+)}(x) + \left[\varphi^{(+)}(x), \varphi^{(-)}(y)\right]$$

and $\left[\varphi^{(+)}(x), \varphi^{(-)}(y)\right] = i\Delta_f$.
For the vacuum expectation value all the normal ordering terms fall out and we get the result found from the path integral analysis:

$$\langle 0|T\varphi(x_i) \dots \varphi(x_n)|0\rangle = \sum_{perm} i\Delta_f\left(x_{p_i} - x_{p_2}\right) \dots i\Delta_f\left(x_{p_{n-1}} - x_{p_n}\right)$$

Now consider the interacting case, $\boldsymbol{\varphi^4}$ theory:

$$\mathcal{L} = \frac{1}{2}\partial_\mu\varphi\partial^\mu\varphi - \frac{1}{2}m^2\varphi^2 - V(\varphi) \quad where \quad V(\varphi) = \frac{\lambda}{4!}\varphi^4$$

and

$$W(J) = \frac{1}{n}\int \mathcal{D}\varphi e^{i\int d^4x[\mathcal{L}+J\varphi]}$$

Let's now examine the Feynman rules for this theory. Note that most of the headache in Feynman rules is in the combinable factors. Consider the identity:

$$\int dx e^{i(f(x)+g(x)+kx)} = e^{g\left(-i\frac{d}{dk}\right)}\int dx e^{i(f(x)+kx)}$$

This allows us to write:

$$W[J] = \frac{1}{N}e^{-i\int d^4xV\left(\frac{\delta}{i\delta J(x)}\right)}\int \mathcal{D}\varphi e^{i\int d^4x\left[\frac{1}{2}\partial_\mu\varphi\partial^\mu\varphi - \frac{1}{2}m^2\varphi^2 + J\varphi\right]}$$

or

$$W[J] = \frac{1}{N'}e^{-i\int d^4xV\left(\frac{\delta}{i\delta T(x)}\right)}exp\left\{-\frac{i}{2}\int d^4Zd^4Z'J(Z)i\Delta_f(Z - Z')J(2')\right\}$$

We want to first order in λ, but first let's streamline the notation:

$$\varphi(x) \to \varphi_x \quad J(x) = J_x$$

$$i\Delta_f(x - x') \to i\Delta_{fxx'}$$

$$i\Delta_{fxx'}J_{x'} = \int d^4x'i\Delta_f(x - x')J(x')$$

$$(J, \varphi) = \int d^4xJ(x)\varphi(x)$$

$$(J, \Delta_f J) = \int d^4xd^4x'J(x)\Delta_f(x - x')J(x')$$

$$\delta_x = \frac{\delta}{\delta J(x)}$$

To get:

$$W[J] = \frac{1}{N'}e^{-i\int dx\, V(-i\delta_x)}e^{-\frac{1}{2}(J,,\Delta_f J)}$$

And $V = \frac{\lambda}{4!}\varphi^4$, so:

$$W[J] = \frac{1}{N'}\left\{1 - \frac{i\lambda}{4!}\int d^4x(-\delta_x)^4\right\}e^{-\frac{i}{2}(J,,\Delta_f J)}$$

where

$$\delta_x\left(e^{(\square)}\right) = -i\Delta_{fx2}J_2 e^{-\frac{i}{2}(J,,\Delta_F J)}$$
$$\delta_x^2(\square) = \left[-i\Delta_{fxx} + \left(i\Delta_{fx2}J_2\right)^2\right]e^{-\frac{i}{2}(J,,\Delta_F J)}$$
$$\delta_x^3(\square) =$$
$$\delta_x^4(\square) = \left[3\left(i\Delta_{fxx}\right)^2 - 6\left(i\Delta_{fxx}\right)\left(i\Delta_{fxz}J_z\right)^2 + \left(i\Delta_{fx2}J_2\right)^4\right]e^{-\frac{i}{2}(J,,\Delta_F J)}$$

Some diagrams:

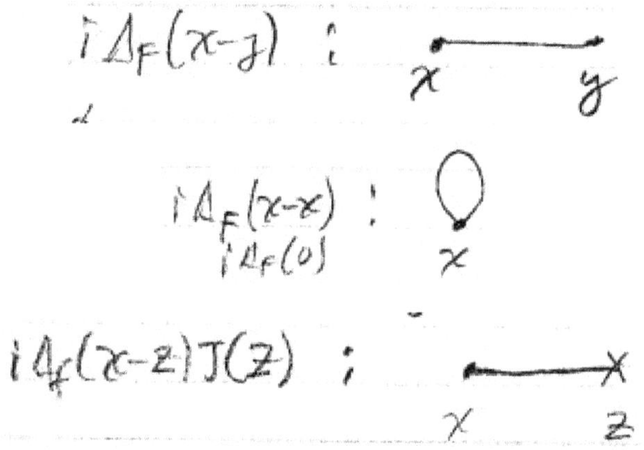

Putting together:

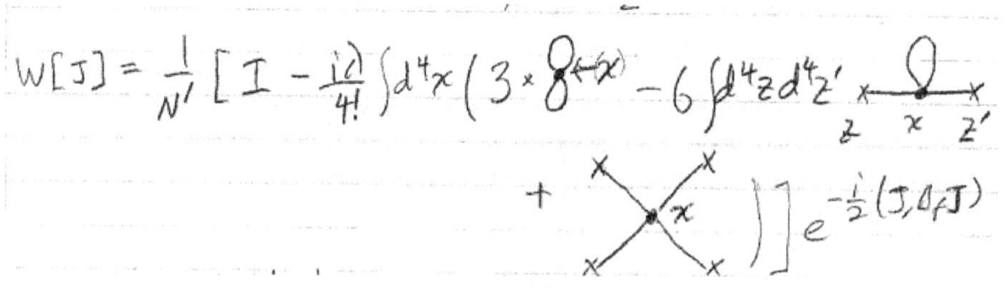

(Note that in finite temp. field theory W is the partition function.)
The "figure-eight diagram" is the only figure that doesn't have a source. Thus, other terms vanish when J→ 0, leaving:

$$N' = 1 - \frac{i\lambda}{4!}\int d^4x\, 3(figure - eight)$$
$$\frac{1}{N'} \simeq 1 + \frac{i\lambda}{4!}\int d^4x\, 3(\text{eight})$$

Where the above normalization amounts to cancellation on such diagrams in he perturbation expansion. The four-point function is:

$$\langle 0 | T\phi(x_1)\phi_{in}\cdots\phi(x_4)|0\rangle = \begin{matrix} x_1 \!-\!\!- x_2 \\ x_3 \!-\!\!- x_4 \end{matrix} \;+\;]\,[\;+\; \times$$

$$\text{from } \tfrac{\partial}{\partial J}(J,\tfrac{\partial}{\partial J}J)$$

$$+\,i\lambda \int dz \left(\begin{matrix} x_1 \!\bullet\!\!-\!\!\overset{Q}{\overset{}{\bigcirc}}\!\!\bullet x_3 \\ x_3\!\bullet \!\!-\!\!-\!\!\bullet x_4 \end{matrix} \;+\text{ perm}\right)$$

$$+\;-i\lambda \int d^4 z \;\begin{matrix} x_1 & x_2 \\ & \overset{\bullet z}{\times} \\ x_3 & x_4 \end{matrix} \;+\text{ order }\lambda^2\cdots$$

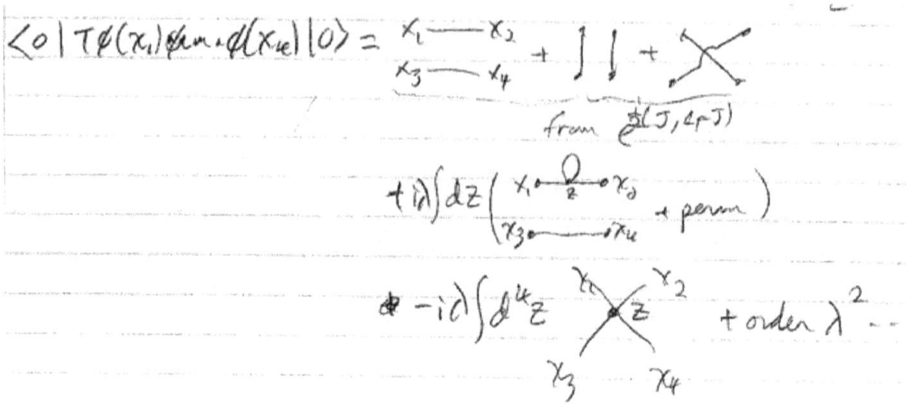

Not the most efficient way to do perturbation ..... and the loop-diagram terms are very divergent. We will manage them with our first renormalization maneuver (on mass).

Recall the simple loop had $i\Delta_F(0) = i(2\pi)^{-4} \int \dfrac{d^4 k}{k^2 - m^2 + i\epsilon}$, while the loop with two arms diagram is as follows:

$$\overset{\overset{Q}{\overset{}{\bigcirc}}}{\underset{z}{\rule{0pt}{0pt}}\;\underset{x}{\rule{0pt}{0pt}}\;\underset{z'}{\rule{0pt}{0pt}}} \;=\; i\Delta_F(0)\int dx\; i\Delta_F(z-x)\, i\Delta_F(x-z')$$

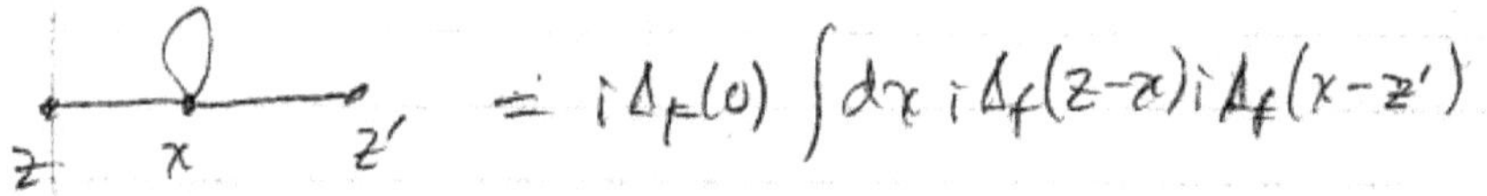

Normal ordering takes the Lagrangian operator to $\mathcal{L} \to\, :\mathcal{L}:$ thus, notably φ^4 is related to $:\varphi^2:$ by the previous Wick's theorem:

$$\varphi^4 \;=\; :\varphi^4: + 6 : \varphi^2: i\Delta_F(x_{1x}) + 3\big[i\Delta_F(x,x)\big]^2$$

In the path integral $:\varphi^4:$ and φ^4 are not distinguished, so we must implement $:\varphi^4: = f(\varphi)$ in the path integral explicitly. So
$$V = \varphi^4(x) - 6i\Delta_F(0)\varphi^2(x) + constant$$
Since we are now considering path integrals the constant can be pulled out to give:

$$\mathcal{L} = \frac{1}{2}\partial_\mu\varphi\partial^\mu\varphi - \frac{1}{2}m^2\varphi^2 - \frac{\lambda}{4!}\varphi^4 + \frac{1}{4}i\lambda\Delta_F(0)\varphi^2$$

If we now repeat the previous analysis, to 1st order in λ, we are able to cancel the loop with two arms diagram term in $W[J]$. Thus, normal ordering gives $m^2 \to m^2 - \frac{1}{2}\lambda i\Delta_F(0)$, so, redefine mass equivalent to normal ordering mass (the "observed mass"), effectively performing mass renormalization to absorb the highly divergent simple loop ($\Delta_F(0)$) infinity using:

$$m_{Obs}^2 = m^2 - \frac{1}{2}\lambda i\Delta_f(0)$$

Let's now consider some combinatorics issues:

Consider that $\frac{1}{2}\Delta_{fxy}\delta_x\delta_y(\ldots\ldots) \to$ usually gives Δ_{fxy} expect when both δ'^s act on the same V, then $\frac{1}{2}$ not cancelled. Consider a V^2 term:

$$V[\varphi]^2 \quad \to \quad \int d\mathfrak{z}_1\varphi^4(\mathfrak{z}_1)\int d\mathfrak{z}_2\varphi^4(\mathfrak{z}_2)$$

and $(\Delta_{fxy}\delta_x\delta_y)$ acting on V^n for different V'^s leads to combinatorics under relabeling with n! relabelings of $V(\mathfrak{z}_n)$ to cancel the n!, similarly, can act on the four φ'^s in $\frac{1}{4!}\int d\mathfrak{z}\varphi^4(\mathfrak{z})$ to eliminate the 4! in $\frac{\lambda}{4!}\int d\mathfrak{z}\varphi^4(\mathfrak{z})$. If all Δ'^s act on the V'^s we get rid of $\left(2n+\frac{k}{2}\right)!$

The main exception is if both δ'^sact on the same V:

$$\left(\frac{i}{2}\Delta_{fxy}\delta_x\delta_y \;\; on\, V(\varphi)\right)$$

$$= \frac{i}{2}\int dxdy\frac{\delta}{\delta\varphi(x)}\frac{\delta}{\delta\varphi(y)}\Delta_F(x-y)\int d^4\mathfrak{z}\varphi^4(\mathfrak{z})$$

$$= \frac{i}{2}\int dxdy\Delta_f(x-y)\frac{\delta}{\delta\varphi(x)}4\varphi^3(y) = \frac{i}{2}\Delta_f(0)4\cdot 3\varphi^2(x)$$

The other exception is if 2 or more $\Delta's$ connect the same pair of V's:

$$\frac{1}{2!}\left(\frac{1}{2}i\Delta_{fxy}\delta_x\delta_y\right)\left(\frac{1}{2}i\Delta_{frs}\delta_r\delta_s\right)\int\varphi^4(\mathfrak{z})d\mathfrak{z}\int\varphi^4(\mathfrak{z}')d\mathfrak{z}'$$

$$= \frac{1}{2!}\left(\frac{i}{2}i\Delta_{fxy}\delta_x\delta_y\right)i\Delta_{frs}\int 4\varphi^3(\mathfrak{z})\delta(r-\mathfrak{z})d\mathfrak{z}\int 4\varphi^3(\mathfrak{z}')\delta(s-\mathfrak{z}')d\mathfrak{z}'$$

Recall abbreviated notation:

$\Delta_F(x-y) \Rightarrow \Delta_{Fxy}$,

$$now\;\frac{\delta}{\delta\varphi(x)} \Rightarrow \delta_{x,} \quad and \quad V[\varphi] = \int d^4x\, V(\varphi(x))$$

And

$$\Delta_{fxy}\delta_x\delta_y = \int dxdy\Delta_F(x-y)\frac{\delta}{\delta\varphi(x)}\frac{\delta}{\delta\varphi(y)}$$

From perturbation theory and the method by Coleman (shown below), we have:

$$\langle 0|T\varphi(x_1)\ldots\varphi(x_k)|0\rangle = e^{\frac{i}{2}\Delta_{fxy}\delta_x\delta_y}\varphi_1\ldots\varphi_k e^{-iV\{\varphi\}}\Big|_{\varphi=0}$$

which we already know is nonzero for n even, so assuming n is even from here let's consider nth order perturbation theory where

$$e^{-iV[\varphi]} \to \frac{(-i)^n}{n!}V[\varphi]^n$$

For a φ^4 theory there are $4n\varphi'^s$ from the potentials, thus $4n + k\ \varphi'^s$ overall. This means that only the $\left(2n + \frac{k}{2}\right)$ term in $e^{\frac{i}{2}\Delta_{fxy}\delta_x\delta_y}$ will contribute at nth order with $\big|_{\varphi=0}$. So, at nth – order

$\langle 0|T\varphi(x_1)\ldots\varphi(x_k)|0\rangle\ at\ nth$

$$= \frac{1}{\left(2n + \frac{k}{2}\right)!}\left[\frac{i}{2}\left(\Delta_{fxy}\delta_x\delta_y\right)\right]^{2n+\frac{k}{2}}\varphi_1\ldots\varphi_k\frac{(-i)^n}{n!}V[\![\phi]\!]^n\big|_{\varphi=0}$$

Method by Coleman

Let's start with the perturbation theory given by:

$$W[J] = \frac{1}{N}e^{-i\int V\left[\frac{\delta}{i\delta J}\right]dx}e^{-\frac{i}{2}\int J(x)\Delta_f(x-y)J(y)dxdy}$$

And

$$\langle 0|T\varphi(x_1)\ldots\varphi(x_n)|0\rangle = \frac{\delta}{i\delta J(x_1)}\cdots\frac{\delta}{i\delta J(x_n)}W[J]\big|_{J=0}$$

Next, let's derive an identity for two functions that are Fourier transformable. To start, we have:

$$G\left(-i\frac{d}{dx}\right)F(x) = F\left(-i\frac{d}{dy}\right)\{G(y)e^{ixy}\}\big|_{y=0}$$

And with Fourier transforms:

$$F(x) = \int dke^{ikx}\tilde{F}(k),\quad G(x) = \int dke^{ikx}\tilde{G}(k)$$

On the left-hand side:

$$G\left(-i\frac{d}{d\lambda}\right)F(x) = \int dke^{ikx}G(k)\tilde{F}(k) = \int dkdk'e^{ik(x+k')}\tilde{F}(k)\tilde{G}(k')$$

On the right-hand side:

$$F\left(-i\frac{d}{dy}\right)\left(G(y)e^{iky}\right) = F\left(-i\frac{d}{dy}\right)\int dk'\tilde{G}(k')e^{i(k'+x)y}$$

$$= \int dk'F(k'+x)\tilde{G}(k')e^{i(k'+x)y}$$

$$= \int dk'\,dk\tilde{F}(k)e^{ik(k'+x)}\tilde{G}(k')e^{i(k'+x)y}$$

So,

$$F\left(-i\frac{d}{dy}\right)\left(G(y)e^{iky}\right)\big|_{y=0} = G\left(-i\frac{d}{d\lambda}\right)F(x)$$

Using this relation, we can rewrite (the Coleman trick):

$$W[J] = \left(\exp\frac{i}{2}\int \Delta_F(x - y)\frac{\delta}{i\delta J(x)}\frac{\delta}{i\delta J(x)}\right)e^{-i\int V(\varphi)dx + i\int J(x)\varphi(x)dx}\Big|_{\varphi=0}$$

Thus

$$\langle 0|T\varphi(x_i),,,\varphi(x_n)|0\rangle = e^{\frac{i}{2}\int \Delta_F(x-y)\frac{\delta}{i\delta J(x) i\delta J(y)}dxdy}\varphi(x_1)\ldots\varphi(x_n)e^{-i\int d^4x V(\varphi(x))}\Big|_{\varphi=0}$$

As before, but shown in a more direct way.

Momentum Space Representation

Let's now turn to the momentum space evaluation to have a simpler notation for many of the results:

$$\langle 0|T\varphi(x_i)\ldots\varphi(x_n)|0\rangle \equiv G(x_1\ldots x_n)$$

$$\tilde{G}(P_1\ldots P_n) = \int d^4x_1\ldots d^4x_n e^{-i\sum_{j=i}^{n} P_i x_j} G(x_1,,,x_n)$$

By definition, $\tilde{G}$ cannot be a function of all $P_1 \ldots P_n$ since:

$$G(x_1\ldots x_n) = \langle 0|Te^{-ip\cdot a}\varphi(x_i)e^{ip\cdot a}e^{-ip\cdot a}\varphi(x_2),,,\varphi(x_n)e^{+ip\cdot a}|0\rangle = \langle 0|T\varphi(x_1 + a)\ldots\varphi(x_n + a)|0\rangle$$

So, translation invariance of the vacuum gives:

$$G(x_1,\ldots,x_n) = G(x_1 + a,\ldots,x_n + a)$$

So, we can pick one coordinate, x_n say, and redefine by $z_j = x_j - x_n$, to get: $G(z_1,\ldots,z_{n-1})$,

$$\tilde{G}(P_1\ldots P_n) = \int d^4z_1\ldots d^4z_{n-1}d^4x_n e^{-i\sum_{j=1}^{n-1} P_j x_j} e^{-i\sum_{j=1}^{n} P_j x_n} G(z_1\ldots z_{n-1})$$

$$= (2\pi)^4\delta_4\left(\sum_{j=1}^{n} P_j\right)G(P_1\ldots P_n)$$

Where the $G(P_1\ldots P_n)$ is now a function of all $P_1\ldots P_n$.

Recall the conventions chosen:

$$\varphi(p) = (2\pi)^{-4}\int dx e^{-ip\cdot x}\varphi(x)$$

$$i\Delta_F(x) = (2\pi)^{-4}\int dp e^{-ip\cdot x} i\Delta_F(p)$$

Where

$$i\Delta_F(p) = \frac{i}{p^2 - m^2 + i\epsilon}$$

$$V[\varphi] = \frac{\lambda}{4!}\int dx\varphi^4(x) = \frac{\lambda}{4!}\int d^4x \int dp_1 \dots dp_4 e^{-(\Sigma(p)\cdot x)}\varphi(p_1)\dots\varphi(p_4)$$
$$= \frac{\lambda}{4!}\int dp_1 \dots dp_4 (2\pi)^4 \delta_4\left(\sum p\right)\varphi(p_1)\dots\varphi(p_4)$$

Recall that

$$i\Delta_{Fxy}\delta_x\delta_\rho = \int d^4x d^4y i\Delta_F(x-y)\frac{\delta}{\delta\varphi(x)}\frac{\delta}{\delta\varphi(y)}$$

Where

$$\frac{\delta}{\delta\varphi(x)} = \int dp \frac{\delta\varphi(p)}{\delta\varphi(x)}\frac{\delta}{\delta\varphi(p)} = \int \frac{dp}{(2\pi)^4}e^{-ip\cdot x}\frac{\delta}{\delta\varphi(p)}$$

So,

$$i\Delta_{Fxy}\delta_x\delta_y = \int \frac{d^4k}{(2\pi)^4} i\Delta_F(k)\frac{\delta}{\delta\varphi(k)}\frac{\delta}{\delta\varphi(-k)}$$

Since

$$G(x_1 \dots x_n) = e^{\frac{i}{2}(\Delta_{Fxy}\delta_x\delta_y)}\varphi(x_1)\dots\varphi(x_n)e^{-iV(\varphi)}\Big|_{\varphi=0}$$

We have:

$$(2\pi)^4\delta_4\left(\sum p\right)G(p_1 \dots p_n)$$
$$= (2\pi)^{4n}e^{\frac{i}{2}\Delta_F\delta_k\delta_{-k}}\varphi(p_1),,,\varphi(p_n)e^{-iV[\varphi]}\Big|_{\varphi=0}$$

This gives us internal lines labeled ‘p’ with

$$i\Delta_F(p) = \frac{i}{p^2 - m^2 + i\epsilon}$$

And external lines labeled ‘k’ with

$$\int \frac{d^4k}{(2\pi)^4}\frac{i}{k^2 - m^2 + i\epsilon}$$

And the four-way intersection diagrams (due to 4th order potential) have:

$$-\lambda(2\pi)^4\delta(k_1 + k_2 + k_3 + k_4)$$

Consider the following Feynman diagram, which has two vertices so is second order in λ:

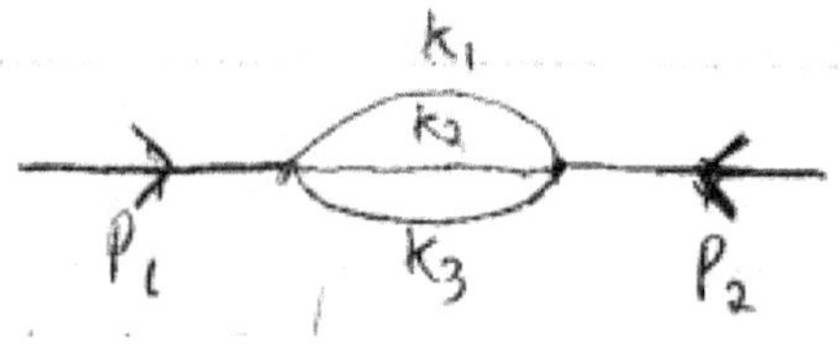

The external lines give $i\Delta_F(p_1)$ and $i\Delta_F(p_2)$ terms and the triple-connect has:

$$\frac{1}{3!}\left(\int \frac{d^4k_1}{(2\pi)^4}i\Delta_F(k_1)\right)^3 [-i\lambda(2\pi)^4\delta(k_1 \dots)]^2$$

So,

$$G^2_{\square}(p_1,p_2)$$

$$G^2_{\square}(p_1,p_2) = \frac{1}{3!} i\Delta_F(p_1) i\Delta_F(p_2)(-i\lambda)^2 \int \frac{dk_1}{(2\pi)^4}\frac{dk_2}{(2\pi)^4} i\Delta_F(k_1) i\Delta_F(k_2) i\Delta_F(p_1 - k_1 - k_2)$$

Let $(p_1 + p_2) = 0$, the two integrations can be spotted in the two independent loops in the diagram. If we drop $\Delta_{F's}$ for every external line we get what is known as the truncated Green's function:

$$G^2_{trunc.}(p_1,p_2) = \frac{1}{3!}(-i\lambda)^2 \int \dots$$

Thus, the four-way intersection in truncated form is just: $G_{trunc} = -i\lambda(\sum p = 0)$, and tree diagrams are simple in general:

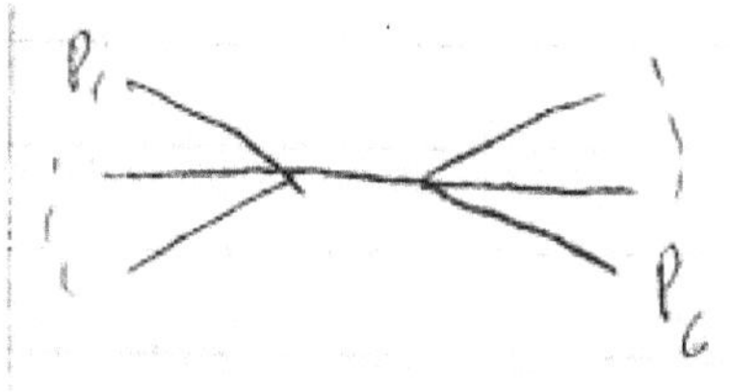

Becomes just $G_{trunc} = (-i\lambda)^2 i\Delta_F(p_1 + p_2 + p_3)$.
Loop diagrams cause the problems. Consider the two loop diagram

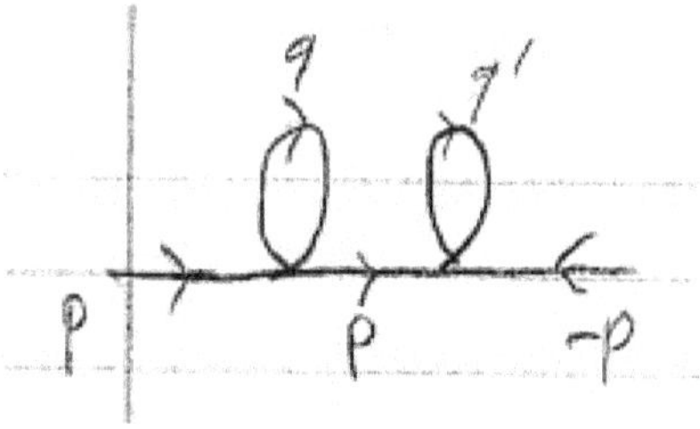

The truncated Green's function is simply:

$$G_{trunc} = \frac{(-i\lambda)^2}{2 \cdot 2} i\Delta_F(p)\left[\int \frac{d^4q}{(2\pi)^4} i\Delta_F(q)\right]^2$$

Which is simply the single loop diagram squared. In essence, a reduction is possible for the diagram to simpler diagrams (when separable by a single cut on an internal line). The two loop diagram is one-particle reducible (cutting one particle internal line reduces to two separate diagrams). The single loop diagrams, or the second order diagram considered at the outset, are examples of 1-particle irreducible diagrams (1-P-I). If we know the 1-P-I diagrams then we have a solution for all, since by definition all other graphs can be constructed from the irreducible ones.

So, we are interested in studying

$$\langle q_1 \dots q_n | p_1 \dots p_m \rangle_{in} = \left(i2^{-1/2}\right)^{n+m} \int \prod_{j=1}^{n} d^4x_j f_{q_j}^*(x_j) k(x_j) \prod_{j=1}^{m} d^4y_j f_{p_j}(x_j) k(y_j) \langle 0 | T\varphi(x_i),,,\varphi(x_n) | 0 \rangle$$

Where $f_p(x) = \frac{e^{-ip\cdot x}}{\sqrt{(2\pi)^3(2\omega_p)}}$. If we write $\langle 0|T\varphi(x_i),,,\varphi(x_n)|0\rangle$ in momentum space:

$$\int \frac{d^4k_1}{(2\pi)^4} \cdots \frac{d^4k_{n+m}}{(2\pi)^4} e^{-i\sum_{i=1}^{n} k_i \cdot x_i - i\sum k_i \cdot y_i} \tilde{G}(k_1 \dots k_{n+m})$$

Then it is apparent that we can do the integrations over the x's and y's. So

$$x_j: \frac{i2^{-1/2}}{\sqrt{(2\pi)^3(2\omega_p)}} \int dx_j e^{iq_j.x_j} k(x_j) e^{-ik_j.x_j}$$

$$= \frac{i2^{-1/2}}{\sqrt{(2\pi)^3(2\omega_p)}} (-q_j^2 + m^2)\delta(q_j - k_j)(2\pi)^4$$

Where $k(x_j) = (-k_j^2 + m^2)$. $\tilde{G}$ has a propagator for each ext. line so it will have factors of $\frac{i}{k_j^2 - m^2}$ (thus killed, thus only need G_{trunc}):

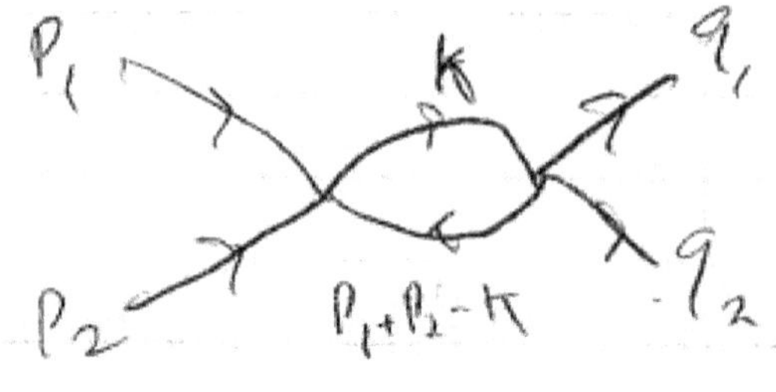

The external lines have: $\frac{1}{\sqrt{Z(2\pi)^3 2\omega_p}}$ and the overall diagram is:

$$\left(\prod_j \frac{1}{\sqrt{Z(2\pi)^3 2\omega_{p_j}}}\right) \int \frac{d^4k,}{(2\pi)^4} i\Delta_F(k)\, i\Delta_F(p_1 + p_2 - k) \left(\frac{(-i\lambda)^2}{2}\right)$$

Thus,

$$\longrightarrow \qquad \frac{1}{p^2 - m^2 + i\epsilon}$$

$$\times\!\!\!\times \quad -i\lambda \quad \Sigma p_i = 0$$

$$(2\pi)^4 \delta_4 (\Sigma p_{ext})$$

$$P_1 + P_2 - k$$

$$P_1 + P_2 = -(P_3 + P_4)$$

$$G(p_1 \cdots p_4) = \left(\prod_{j=1}^{4} \frac{i}{p_j^2 - m^2 + i\epsilon} \right) \frac{(-i\lambda)^2}{2} \times$$

$$\int \frac{d^4 k}{(2\pi)^4} \frac{i}{k^2 - m^2 + i\epsilon} \frac{i}{(p_1 + p_2 - k)^2 - m^2 + i\epsilon}$$

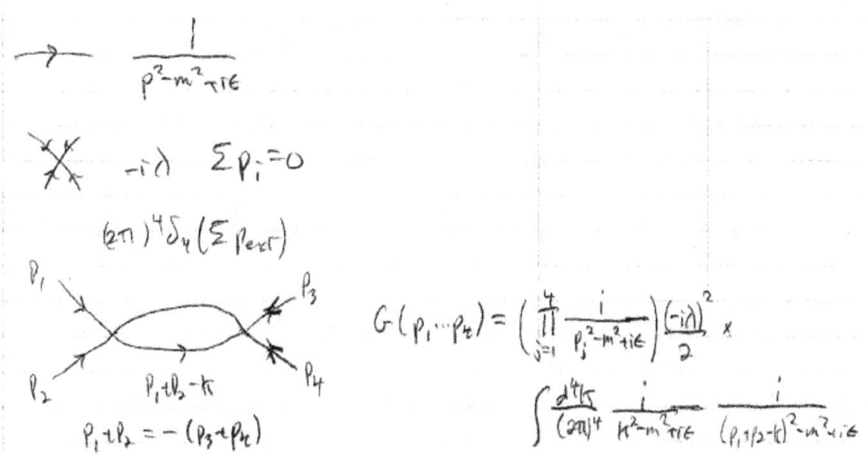

The integration to be done can be broken into an integration equivalent to that of one-loop, so let's focus on that derivation in what follows, where we have:

$$G = \frac{-i\lambda}{2} \int \frac{i d^4 k}{k^2 - m^2 + i\epsilon}$$

Where $G = \frac{-i\lambda}{2} \int \frac{i d^4 k}{k^2 - m^2 + i\epsilon}$. So, we need to solve the integral:

$$I = \int \frac{i d^4 k}{k^2 - m^2 + i\epsilon}$$

Consider $k^2 = k_j^2 - \vec{k}^2 = -K_4^2 - \vec{k}^2 = -k_E^2$, and let's solve the integral by performing a Wick rotation:

$$\omega_k = \sqrt{k^2 + m^2}$$

rotation: $k_0 = i k_4$

Wick rotation

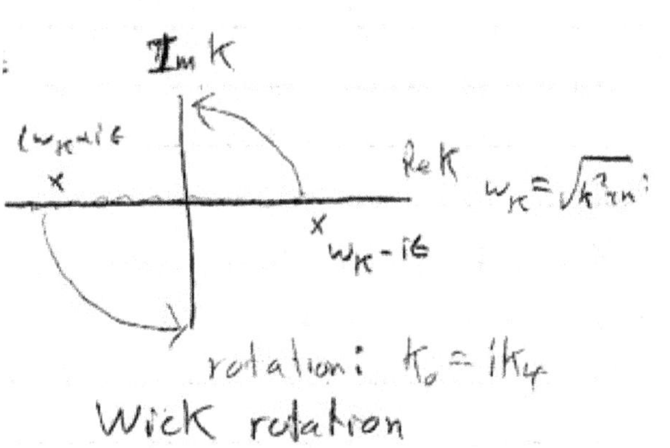

Thus,

$$I = \int \frac{d^4k_{\in}}{k_{\in}^2 + m^2} = 2\pi^2 \int_0^{\infty} \frac{k_{\in}^3 dk_{\in}}{k_{\in}^2 + m^2}$$

Denote

$$I_{\wedge} = 2\pi^2 \int_0^{\wedge} \frac{k_{\in}^3 dk_{\in}}{k_{\in}^2 + m^2} = \pi^2 \left[\wedge^2 - m^2 \ell n \left(\frac{\wedge^2 + m^2}{m^2} \right) \right],$$

Which indicates a quadratic divergence.

Let's now explore the nature of the divergences that arise in the various diagrams. It turns out that the divergence of a graph can be deduced quite easily by looking at the Feynman graph where divergence goes as $D = 4L - 2I$, for L loops, I internal lines, E external lines, and V vertices. Let's apply the rule:

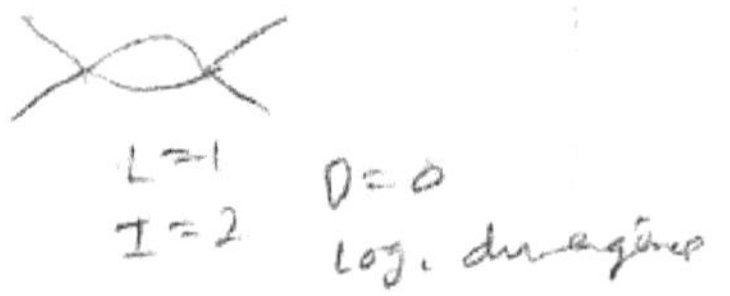

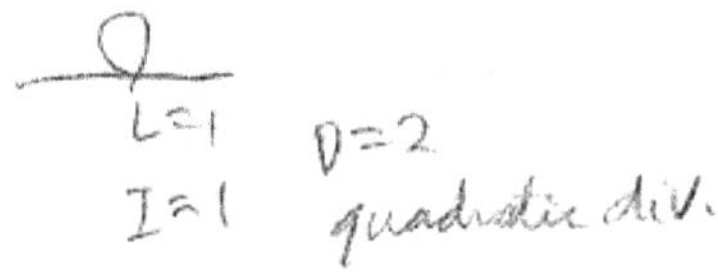

We have the Euler graph identity between Loops and internal lines and vertices: $L = I - V + 1$. For the phi-4 theory we have the relation $E + 2I = 4V$ (Kirckoff circuit type constraint – vertices take 4 connections always). So, $D = 4 - E$. Note that generally would expect

$$D = D(E, V) \; but \; here \; D = D(E)$$

So we have that D is independent of order of perturbation theory! This means that $E = 2$ and $E = 4$ are the only divergent Green's functions. So, the renormalization program to be described shortly will only need to be done on the 2pt and 4pt diagrams.

Consider a general interaction (N-th order): $V_N = \varphi^N$ and also generalize to d space time dimensions, now have:

$$\left. \begin{aligned} D &= dL - 2I \\ L &= I - V_N + 1 \\ E + 2I &= NV_N \end{aligned} \right\} D = d - \left(\frac{d-2}{2} \right) E + \left[\frac{d}{2}(N-2) - N \right] V_N,$$

Where the last term will introduce a dependency on the order of perturbation theory.

For $d = 4$: $D = 4 - E + (N - 4)V_N$ and φ^4 is the only case where D is indep. of the order of perturbation theory. If $N = 3$ instead, have $D = 4 - E - V_3$.$If\ N > 4\ then\ D \geq 4 - E$ and strictly nonrenormalizable.

Dimensional regularization

The dividing line between nonrenormalizable and renormalizable theories is the sign of the $\left[\frac{d}{2}(N-2) - N\right]$ factor preceding V_N. Thus we have renormalizability if $N < \frac{2d}{d-2}$. In $d = 2$ all theories $(N = anything)$ are renormalizable since $D = 2 - 2V_N$, leaving the only divergent graph one with $V_N = 1$. For the $N = 4$ theory discussed this is the one-loop propagator (with quadratic divergence) indicated earlier. (In a 1+1 dimensional theory, we can get renormalizability simply from normal ordering since this eliminates the one-loop lines.) Let's jump ahead to an evaluation of the renormalizability of QED by examination of its graph elements to make Feynman diagrams (which will be the starting point), since from the graph elements we can apply the aforementioned graph rules to determine renormalizability. For QED we have (Feynman gauge):

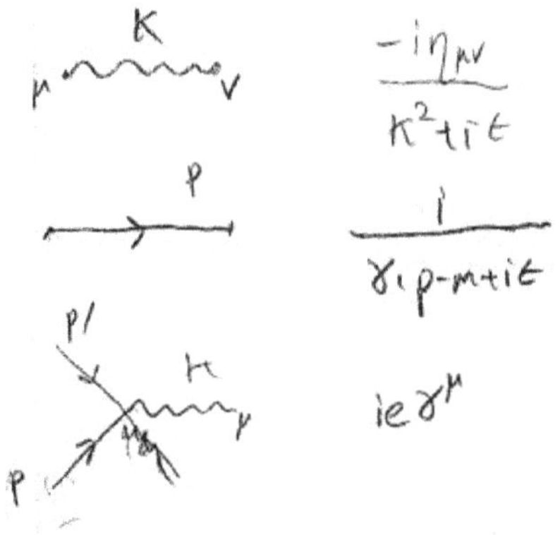

For the electron lines we have: E_e (external) and I_e (internal). For the photon lines have E_γ and I_γ. Thus

$D = 4L - 2I_\gamma - I_e$

$L = I_e + I_\gamma - V + 1$

$E_e + 2I_e = 2V$

$E_\gamma + 2I_\gamma = V$

Remarkably these combine to give

$$D = 4 - \frac{3}{2}E_e - E_\gamma$$

Which is independent of V, which means it is Renormalizable.

Example 1

For the one dimensional harmonic oscillator we have the generator expression shown in Section X:

$$Z[J] = exp\left[\frac{1}{2}\int J(t)G(t-t')J(t')\right]$$

where

$$G(t) = \frac{1}{2m\omega}\left[\theta(t)e^{-i\omega t} + \theta(-t)e^{+i\omega t}\right]$$

Show:

a. $< \Psi_0, t_f|Q(t)|\Psi_0, t_i >= \frac{\delta}{\delta J(t)} Z[J]|_{J=0} = 0.$

b. $< \Psi_0, t_f|T(Q(t)Q(t')|\Psi_0, t_i >= -iG(t-t').$

c. For $t > t'$:

$$< \Psi_0, t_f|T(Q(t)Q(t')|\Psi_0, t_i > \; = \; \frac{1}{2m\omega}e^{-t\omega(t_1-t_f)/2}e^{i\omega(i-t)}$$

Therefore the canonical method gives the same result, except for the factor $e^{-t\omega(t_1-t_f)/2}$. This factor was ignored in the path-integral derivation of the ground-state expectation value when the ground state energy was set equal to 0.

Solution

(a) Recall $\frac{1}{i}\frac{\delta}{\delta J(t)} < x_f, t_f|x_i, t_i >_J |_{J=0} =< x_f, t_f|Q(t)|x_i, t_i >$, thus

$$< \Psi_0, t_f|Q(t)|\Psi_0, t_1 > = \int dx_1 dx_f\, \Psi_0^* < t_{f1} t_f|Q(t)|x_i, t_i > \Psi_0^*(x_i)$$

$$= \frac{1}{i}\frac{\delta}{\delta J(t)} \quad Z[J]|_{J=0}$$

Thus,

$$Z[J] = const. \exp\left\{\underset{t_1}{\Sigma}\int_{t_i}^{t_f} dt \int_{t_i}^{t_f} dt' J(t)\left[\frac{i}{2m\omega}e^{i\omega(t-t')}\right]J(t')\right\},$$

Gives:

$$< \Psi_0, t_f|Q(t)|\Psi_0, t_i > = \frac{2}{i}\left\{\frac{i}{2}\int dt dt'\delta(t-t'')\, G(t-t')J(t')\right\}Z|_{J=0}$$

$$= 0$$

(b)

$$< \Psi_0, t_f|Q(t')Q(t'')|\Psi_0, t_i >$$

$$= \left\{\frac{1}{i\delta J(t'')}\left[\left\{\int dt dt'\delta(t-t'')\, G(t'-t)J(t')\right\}\right]\right\}Z \; |_{J=0}$$

$$= -i\int dt dt'\delta(t-t')G(t-t')\delta(t'-t''')$$

$$= -iG(t'' - t''')$$

(c)

$$TQ(t_1)Q(t_2) = \theta(t_1 - t_2)Q(t_1)Q(t_2) + \theta(t_2 - t_1)Q(t_2)Q(t_1)$$

And using $Q = \sqrt{\frac{\hbar}{2m\omega}}(a + a^{+})$, and choosing units where $\hbar = 1$, we have:

$$< \Psi_0, t_f|TQ(t_1)Q(t_2)|\Psi_0, t_i > = \frac{1}{2m\omega} < \Psi_0, t_f\left|a(t_1)a^{\dagger}(t_2)\right|\Psi_0, t_i > \quad for\ t_1 > t_2$$

$$= \frac{1}{2m\omega} < \Psi_0, t_f\left|e^{+iH(t_f - t_1)}a(t_2)a^{\dagger}(t_2)e^{+iH(t_2 - t_i)}\right|\Psi_0, t_i >$$

$$= \frac{1}{2m\omega} e^{-i\frac{\omega}{2}(t_i - t_f)} e^{-i\omega(t_1 - t_2)}$$

Discrepancy factor arises from setting $E_0 = 0$ in the path integral calculation.

Example 2

Suppose $\boldsymbol{H = \frac{1}{2}m(x^2 + w^2x^2) + \lambda x^3}$

a. Show that to first order in $\boldsymbol{\lambda}$**:**

$$Z[J] = \left[1 - i\lambda \int dt \frac{\partial^3}{(i\delta J(t))^3}\right] Z_0[J]$$

$$= \left[1 + 3\lambda G(0) \int dt G(t - t')J(t')dt' - i\lambda \int dt \left[\int G(t - t')J(t')dt'\right]^3\right] Z_0[J]$$

b. Show,

$$< \Psi_0, t_f|Q(t)|\Psi_0, t_i > = -3i\lambda G(0) \int d\tau G(t - \tau).$$

Solution

(a)

$$Z[J] = \frac{1}{N} \int Dx(t)\, e^{i\int\left[\frac{1}{2}m\dot{x}^2 - \frac{1}{2}m\omega^2x^2 - V(x) + xJ\right]}$$

$$= \frac{1}{N} e^{-i\int V\left(\frac{\delta}{i\delta J(t)}\right)dt} \int Dx(t)\, e^{i\int\left[\frac{1}{2}m\dot{x}^2 - \frac{1}{2}m\omega^2x^2 + xJ\right]}$$

To first order in λ:

$$Z[J]_v = \left(1 - i\int \lambda \left(\frac{\delta}{i\delta J(t)}\right)^3 dt\right) Z_o[J]$$

Let $\delta_t = \frac{\delta}{i\delta J(t)}$, $G_{tt'} = G(t-t')$ and $G_{tt'}J_{t'} = \int dt'\, G(t-t')J(t')$ and

$$(J, GJ) = \int dtdt'\, J(t)G(t-t')J(t')$$

Then

$$-i\lambda \int dt_1 \left(\frac{1}{i}\delta_{t_1}\right)^3 e^{\frac{i}{2}(J,GJ)}$$

$$= -\lambda \int dt_1 \left(\frac{1}{i}\delta_{t_1}\right)^2 \left[\frac{2}{i}\frac{i}{2} G_{t,t'}J_t\right] e^{\frac{i}{2}(J,GJ)}$$

$$= -i\lambda \int dt_1 \left(\frac{1}{i}\delta_{t_1}\right)^2 \left\{\frac{i}{l} G(0) e^{\frac{i}{2}(J,GJ)} + \left(G_{t_1 t'}J_{t'}\right)^2 e^{\frac{i}{2}(J,GJ)}\right\}$$

$$= -i\lambda \int dt_1 \left\{\frac{1}{i} G(0)\left(G_{t,t'}J_{t'}\right) e^{\frac{1}{2}(J,GJ)} + \frac{2}{i} G(0)\left(G_{t,t'}J_{t'}\right) e^{\frac{i}{2}(J,GJ)} + \right.$$
$$\left. \left(G_{t,t'}J_{t'}\right)^3 e^{\frac{i}{2}(J,GJ)}\right\}$$

$$= -\lambda \int dt_1 \left\{-i3G(0)\left(G_{t,t'}J_{t'}\right) e^{\frac{1}{2}(J,GJ)} + \left(G_{t,t'}J_{t'}\right)^3 e^{\frac{1}{2}(J,GJ)}\right\}$$

$$Z[J]_V = \left[1 - 3\lambda G(0)\int dt_1 \int dt_2\, G(t_1 - t_2)J(t_2) - i\lambda \int dt_1 \left[\left[\int dt_2 G(t_1 - t_2)J(t_2)^3\right]\right]\right] Z_0[J]$$

(b) $\langle \Psi_0, t_f | Q(t) | \Psi_0, t_1 \rangle = \frac{\delta}{i\delta J(t)} Z[J]_v |_{J=0} = 3i\lambda G(0) \int dt_1 G(t_1 - t) =$ $-3i\lambda G(0) \int d\tau G(t-\tau)$

4.3 Renormalization

Let's work with Lagrangian Density in the following form

$$\mathcal{L} = \frac{1}{2}\partial_\mu \varphi \partial^\mu \varphi - \frac{1}{2}m^2\varphi^2 - \frac{\lambda}{4!}\mu^{2(2-\nu)}\varphi^4 + \frac{1}{2}(Z_3 - 1)\partial_\mu\varphi\partial^\mu\varphi$$
$$- \frac{1}{2}(Z_m - 1)m^2\varphi^2 - \frac{\lambda}{4!}\mu^{2(2-\omega)}(Z_1 - 1)\lambda\varphi^4$$

where Z_3 is the same as Z from LSZ formalism, and each Z is understood to be a power series expansion in λ:

$$Z_\alpha = 1 + \sum_{n=1}^{\infty} \lambda^n Z_\alpha^n.$$

For which an expansion to second order in λ for four-point functions involves the diagrams:

$$\frac{i\lambda^2}{16\pi^2}\mu^{2(2-\omega)}[\frac{3}{\epsilon} + 3\Psi(1) + 6 - 3\ell n\frac{m^2}{4\pi\mu^2} - A(s) - A(t) - A(u)]$$

and

$$-i\lambda^2\mu^{2(2-\omega)}Z_1^{(1)}$$

Thus, for the minimal subtraction scheme:

$$Z_1^{(1)} = \frac{1}{16\pi^2}\frac{3}{\epsilon}.$$

since $\frac{1}{\epsilon} + \Psi(1) + \ell n 4\pi\mu^2$ always appears as a group, it could be subtracted as a reasonable alternative known as modified minimal subtraction. Having done this it then becomes apparent that a entire families of finite theories can be separated from infinite subtractions. Let's consider this further in the specific case of the mass renormalization. We have for the bare mass:

$$-\frac{1}{2}m_B^2\varphi^2 = -\frac{1}{2}m^2\varphi^2 - \frac{1}{2}(Z_1 - 1)m^2\varphi^2.$$

Consider a free theory:

$$\mathcal{L} = \frac{1}{2}\partial_\mu\varphi\partial^\mu\varphi - \frac{1}{2}m^2\varphi^2 = \frac{1}{2}\partial_\mu\varphi\partial^\mu\varphi - \frac{1}{2}m_1{}^2\varphi^2 - \frac{1}{2}m_2{}^2\varphi^2$$

And let m_1 be in the free theory and m_2 be part of the potential evaluated in perturbation theory. The Feynman rules:

$$i\Delta_F = \frac{i}{p^2 - m_1^2 + i\epsilon}$$

$$-im_2{}^2$$

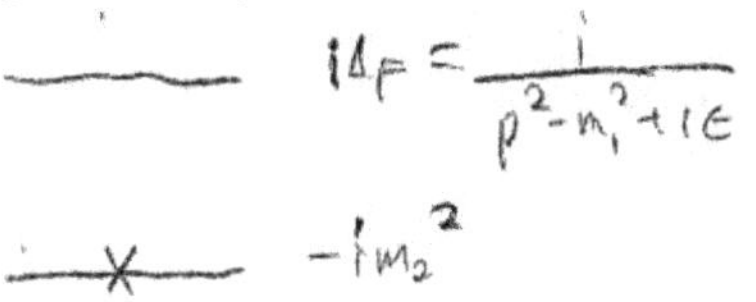

Thus, for $i\Delta_F'(\rho)$:

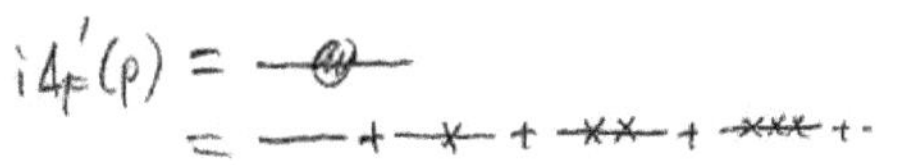

We have:

$$i\Delta_F'(\rho) = i\Delta_F + i\Delta_F(-im_2^2)i\Delta_F + \cdots = i\Delta_F + i\Delta_F(-im_2^2)[i\Delta_F + i\Delta_F(\ldots)i\Delta_F + \cdots]$$

Thus

$$i\Delta_F'(\rho) = i\Delta_F + i\Delta_F(-im_2^2)[i\Delta_F'(\rho)]$$

Or,

$$i\Delta_F'(\rho) = \frac{i\Delta_F(\rho)}{1 - m_2^2\Delta_F(\rho)} = [\Delta_F^{-l} - m_2^2]^{-l} = (\rho^2 - m_1^2 - m_2^2 + i\epsilon)^{-1}$$

So, splitting into m_1^2 and m_2^2 doesn't matter as long at we sum the perturbation series to all orders. Unfortunately in practical applications we can only go to finite order, so the separation of values is important and some give a better behaved convergence. This is where renormalization group arguments will come into play. To summarize: we are free to split the Lagrangian between:

$$\begin{array}{rl} m_{\square}^2, & (Z_m - 1)m^2 \\ 1, & (Z_3 - 1) \\ \lambda\mu^{2(2-\omega)}, & \lambda\mu^{2(2-\omega)}(Z_1 - 1) \end{array}$$

however we like. The choice of the three splittings yields three normalization conditions. These normalization conditions will retain certain physical attributes regardless of normalization. For mass, this will be revealed as a pole in the renormalized 2-point function at the physical mass with residue 1. Consider

$$i\Delta_F'(x) = Z_3 i\Delta_F(x;m^2) + \int_{4m^2}^{\infty} dk^2 i\Delta_F(x;m^2)\,\sigma(k^2)$$

for $\rho^2 \sim m^2 : i\Delta_F'(\rho) \sim \frac{Z_3}{\rho^2 - m^2}$, and since the renormalized $\langle 0|T\varphi(x)\varphi(x')|0\rangle$ will have two factors of $Z_3^{-1/2}\varphi_{real}$, thus:

$$iG_{ren}^{(2)}(p) = \frac{1}{p^2 - m^2} \; as\; \rho \to m^2$$

So, the renormalized 2. pt. function should have a pole at the physical mass with residue = l. Let's now explore this idea in the context of the connected Green's functions, where if we renormalize one we renormalize all. Note that in this process we are not interested in all connected graphs either. Consider the family of 1-particle reducible diagrams, where the diagram can be broken by a single 'cut' (the naïve notion of cut suffices for now, but in Example X we will see some subtleties), such diagrams can simply be eliminated, then reintroduced as perturbation elements. This separation affords an easy diagrammatic summation. The reason for this manipulation becomes apparent if we consider diagrams at second and fourth order:

where the last diagram can be cut (so is reducible), and

etc., where the second and third diagrams (and others not shown) are reducible.

The 1-particle-irriducible Green's function

The 1-particle-irriducible (1PI) Green's function is denoted by $\Gamma^{(n)}$ but for the 2-point Green's function specifically notation is typically not $\Gamma^{(2)}$ but $-\Sigma(p)$. Thus, we have:

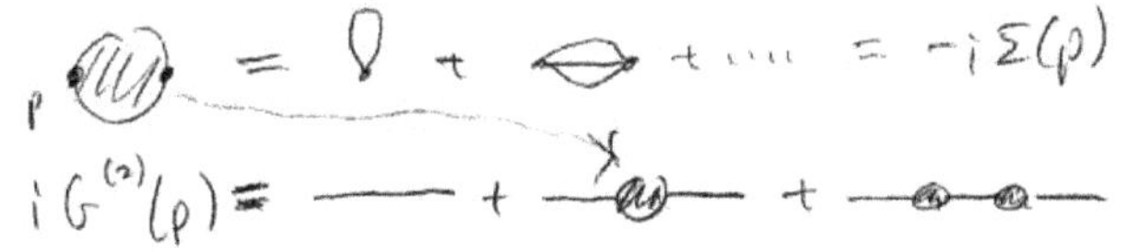

$$iG^{(2)}(p) = i\Delta_F(\rho) + i\Delta_F(p)(-i\Sigma)i\Delta_F(p) + \cdots = i\left[p^2 - m_p^2 - \Sigma(p)\right]^{-1}$$

where the latter expression has $+i\epsilon$ conventions and now incorporates the effect of self-energy. Since we know from previously that

$$G_{ren}^{(2)}(p) \sim \frac{1}{p^2 - m_p^2} \quad (p^2 \to m^2)$$

we must have

$$\Sigma(p^2 = m_p^2) = 0$$

and

$$\left.\frac{\partial\Sigma}{\partial p^2}\right|_{p^2 = m_p^2} = 0$$

To not alter the pole structure. Thus, we can write

$$\Sigma(p^2) = \left(p^2 - m_p^2\right)^2 \Sigma'(p^2).$$

Thus,

$$G^{(2)} = \frac{1}{\left(p^2 - m_p^2\right)\left(1 + \left(p^2 - m_p^2\right)\Sigma'(p^2)\right)}$$

In the coupling constant renormalization for the 4-point irreducible diagrams (details later) we will adopt the convention

$$\Gamma^{(4)}\left(s = t = u = \frac{4}{3}m_p^2\right) = -i\lambda.$$

In effect, we see that the choice of m in the theory can be whatever is convenient, but from the results above, we know for any such choice of mass, there is the following relation to the physical mass:

$$m_p^2 - m^2 - \left(m_p^2 - m^2\right)^2 \Sigma'\left(m_p^2\right) = 0$$

Properties of $W_0[J]$

(1) (a) The Field Theory Path-Integral (PI) Generator (quantum mechanics harmonic oscillator PI basis in Section X), has general form:

$$W_0[J] = \frac{1}{N}\int D\varphi\,(x)e^{i\mathcal{F}} \qquad \mathcal{F}$$

$$= \int d^4x[\mathcal{L}(\varphi(x)) + J(x)\varphi(x) + i\epsilon\varphi^2(x)/2]$$

Now consider a Klein-Gordon scalar field with an interaction term $V(\varphi)$:

$$\mathcal{F} = \int d^4x\left[-\frac{1}{2}\varphi(\partial^2 + m^2)\varphi - V(\varphi) + J\varphi + i\epsilon\varphi^2(x)/2\right],$$

$$k(x) = (\partial^2 + m^2)$$

This also gives rise to the Gaussian form:

$$W_0[J] = \exp\left[-\frac{i}{2}\int d^4xd^4x'\,J(x)\Delta_f(x - x')J(x')\right]$$

where

$$(\Box + m^2 - i\epsilon)\Delta_f(x) = -\delta_4(x).$$

Action of a functional derivative w.r.t. $J(x)$ on $W_0[J]$ will effectively bring a $\varphi(x)$ into the integral, thus we will want to know (dropping epsilon factors that were selecting the proper contour in the path integral, e.g., the Feynman conventions):

$$< 0|\varphi(x)|0 >_J = \frac{1}{N}\int D\varphi(x)\left\{\varphi(x)\exp\left(i\int -\frac{1}{2}\varphi k\varphi - V(\varphi) + J\varphi\right)\right\}$$

Let's solve for above with a $k(x)$ factor to have a trivial solution:

$$k(x) < 0|\varphi(x)|0 >_J = \frac{1}{N}\int D\varphi k(x)\varphi(x)e^{i\int -\frac{1}{2}\varphi k\varphi - V(\varphi) + J\varphi}$$

$$= \frac{1}{N}\int D\varphi\left(i\frac{\delta}{\delta\varphi(x)} - \frac{\partial V(\varphi)}{\partial\varphi(x)} + J(x)\right)e^{i\int -\frac{1}{2}\varphi k\varphi - V(\varphi) + J\varphi}$$

Where we use $i\frac{\delta}{\delta\varphi(x)} \to \left(k(x)\varphi(x) + \frac{\partial V(\varphi)}{\partial\varphi(x)} - J\right)$, however the $\frac{\delta}{\delta\varphi(x)}$ term presents an integral of a total derivative term (whose variation is fixed at the boundary thus zero contribution in what follows) thus:

$$k(x) < 0|\varphi(x)|0 >_J = \frac{1}{N}\int D\varphi\left(-\frac{\partial V(\varphi)}{\partial\varphi(x)} + J(x)\right)e^{i\int -\frac{1}{2}\varphi k\varphi - V(\varphi) + J\varphi}$$

or

$$k(x) < 0|\varphi(x)|0 >_J = -< 0|\frac{\partial V(\varphi)}{\partial\varphi(x)} - J(x)|0 >_J$$

(b) We know from (Section X) that:

$$k(x_1) < 0|T(\varphi(x_1)..\varphi(x_n))|0 >= \left(\frac{\delta}{\delta\varphi(x_n)}\cdots\frac{\delta}{\delta\varphi(x_2)}\right)k(x_1)$$

$$< 0|\varphi(x_1)|0 >_J$$

Given the result of (a), this indicates:

$$k(x_1) < 0|T(\varphi(x_1)..\varphi(x_n))|0> = -\left(\frac{\delta}{\delta\varphi(x_n)} \cdots \frac{\delta}{\delta\varphi(x_2)}\right)$$

$$< 0|\frac{\partial V(x_1)}{\partial\varphi(x)} - J(x_1)|0>_J$$

$$= \langle 0|T(-\partial V(\varphi)/\partial\varphi(x)\ \varphi(x_2) \ldots \varphi(x_n))|0\rangle$$

$$-\sum_j i\delta(x_1$$

$$- x_j)\left\langle 0\middle|T\left(\varphi(x_2) \ldots \varphi(x_{j-1})\varphi(x_{j+1}) \ldots \varphi(x_n)\right)\middle|0\right\rangle$$

(c) We can now consider the relation on the time-ordering of field directly since:

$$k(x)T\varphi(x)\varphi(y) = T[k(x)\varphi(x)\varphi(y)] + \delta(x^0 - y^0)[\dot{\varphi}(x), \varphi(y)]$$

$$= T\left[\frac{\partial V(\varphi)}{\partial\varphi(x)}\varphi(y)\right] - i\delta(\vec{x} - \vec{y}) = -i\delta(\vec{x} - \vec{y})$$

Here we have:

$$k(x_1)T\varphi(x_1)\varphi(x_2) \ldots \varphi(x_n)$$

$$= T\left(\frac{\partial V(\varphi)}{\partial\varphi(x)}\varphi(x_2) \ldots \varphi(x_t)\right)$$

$$-\sum_j i\delta(x_1 - x_j)\,T(\varphi(x_2) \ldots \varphi(x_{j-1})\varphi(x_{j+1}) \ldots \varphi(x_n))$$

2a. Let's use this result in an examination of phi^3 theory, with Lagrangian density:

$$\mathcal{L} - \frac{1}{2}\partial_\mu\varphi\partial^\mu\varphi - \frac{1}{2}m^2\varphi^2 - \frac{\lambda}{3!}\varphi^3$$

where

$$W[J] = \frac{1}{N}e^{-i\int d^4x\, V\left(\frac{\delta}{i\delta J(x)}\right)} \int D\varphi e^{i\int d^4x\left[\frac{1}{2}\partial_\mu\varphi\partial^\mu\varphi - \frac{1}{2}m^2\varphi^2 + J\varphi\right]}$$

$$= \frac{1}{N'}e^{-i\int d^4x V\left(\frac{\delta}{i\delta J(x)}\right)} \exp\left\{-\frac{i}{2}\int d^4z d^4z'\, J(z) i\Delta_f(z$$

$$- z')J(z')\right\}$$

Let's now streamline the notation: $\varphi(x) \to \varphi_x$,$J(x) \to J_{x'}, i\Delta_f(x - x') \to i\Delta_{fxx'}, \frac{\delta}{\delta J(x)} \to \delta_x$, and define contraction rules to imply integration or inner product on common index:

$$\int d^4x' i\Delta_f(x - x')J(x') \to i\Delta_{fxx'}J_{x'}$$

$$\int d^4x' J(x)\varphi(x) \to (J, \varphi)$$

$$\int d^4x d^4x' J(x)\Delta_f(x-x')J(x') \rightarrow (J, \Delta_f J)$$

So, we can now write:

$$W[J] = \frac{1}{N'} e^{-i\int d^4x V(-i\delta_x)} e^{-\frac{i}{2}(J,\Delta_f J)}$$

And with $V = \frac{\lambda}{3!}\varphi^3$ we have to a first order in λ:

$$W[J] = \frac{1}{N'}\left\{1 - \frac{\lambda}{3!}\int d^4x\, (\delta_x)^3\right\} e^{-\frac{i}{2}(J,\Delta_f J)}$$

Evaluating:

$$\delta_x e^{-\frac{i}{2}(J,\Delta_f J)} = -i\Delta_{fxx}J_z e^{-\frac{i}{2}(J,\Delta_f J)}$$

$$\delta_x^2 e^{-\frac{i}{2}(J,\Delta_f J)} = \left[-i\Delta_{fxx} + \left(i\Delta_{fxz}J_z\right)^2\right] e^{-\frac{i}{2}(J,\Delta_f J)}$$

$$\delta_x^3 e^{-\frac{i}{2}(J,\Delta_f J)} = \left[3\left(i\Delta_{fxx}\right)\left(i\Delta_{fxz}J_z\right) - \left(i\Delta_{fxz}J_z\right)^3\right] e^{-\frac{i}{2}(J,\Delta_f J)}$$

Diagrammatically:

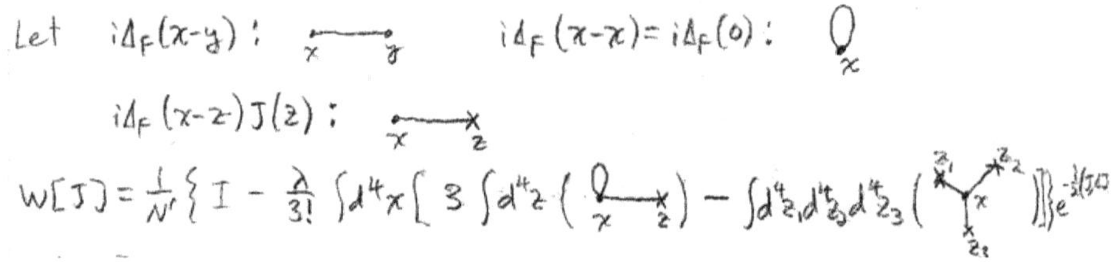

Denote $i\Delta_f(x-y)$ by D1, $i\Delta_f(0)$ by D0, and $i\Delta_f(x-z)J(z)$ by D2. Denote the D0D1 connected on shared x by D3, and denote 3 D2's on shared (internal) x by D4:

$$W[J] = \frac{1}{N}\left\{1 - \frac{\lambda}{3!}\int d^4x \left[3\int d^4x\, (D3) - \int d^4x_1 d^4x_2 d^4x_3\, (D4)\right]\right\} e^{-\frac{i}{2}(J,\Delta_f J)}$$

$W[J=0] = 1$ to 1st order in λ, thus $W[J]$ (at this order) has exponential J dependence via Taylor series on exponential with argument a differential:

$$W[J] = \sum_{n=0}^{\infty} \frac{1}{n!}\int idx \dots idx_n < 0|T\varphi(x_1)\dots\varphi(x_n)|0 > J(x_1)\dots J(x_n)$$

So, at 1st order in λ we have:

$$i\Delta_f'(x-x') = < 0|T\varphi(x)\varphi(x')|0 > = i\Delta_f(x-x')\,.$$

Using the Coleman trick (see Section X)):

$$W[J] = \left(\exp\frac{i}{2}\int \Delta_F(x - y)\frac{\delta}{i\delta J(x)}\frac{\delta}{i\delta J(x)}\right) e^{-i\int V(\varphi)dx + i\int J(x)\varphi(x)dx} \Big|_{\varphi=0}$$

Thus

$$\langle 0|T\varphi(x_i),,,\varphi(x_m)|0\rangle = e^{\frac{i}{2}\int \Delta_F(x-y)\frac{\delta}{i\delta J(x) i\delta J(y)}dxdy}\varphi(x_1)\dots\varphi(x_m)e^{-i\int d^4x V(\varphi(x))}\Big|_{\varphi=0}$$

And

$$\{0|T\varphi(x_i),,,\varphi(x_m)|0\}_{nth\ order\ in\ \lambda} = e^{\frac{i}{2}\int \Delta_F(x-y)\frac{\delta}{i\delta J(x) i\delta J(y)}dxdy}\varphi(x_1)\dots\varphi(x_m)\frac{(-i)^n}{n!}\left[i\int d^4x\, V(\varphi(x))\right]^n\Big|_{\varphi=0}$$

Since each power of V has $3\varphi's$, we need m+3n derivatives.

$$< 0|T\varphi(x_1)\dots\varphi(x_m)|0 >_n = \frac{1}{\left(\frac{m+3n}{2}\right)!}\left\{\frac{i}{2}\Delta_{fxy}\delta_x\delta_y\right\}^{\frac{m+3n}{2}}[\varphi(x_1)\dots\varphi(x_m)]\frac{(-i)^n}{n!}V[n]$$

In what follows, refer to $\left\{\frac{i}{2}\Delta_{fxy}\delta_x\delta_y\right\}$ as Δ, $[\varphi(x_1)\dots\varphi(x_m)]$ as Σ, and $\frac{(-i)^n}{n!}V[n]$ as V. When the derivatives act we get the various integrals over propagators which can be succinctly expressed in a diagrammatic form via Feynman diagrams with their respective rules in this instance. Note that $< 0|T(\dots)|0 > = \frac{\delta^m}{i\delta J\dots}W_0[J]|_{J=0}$ is zero unless m is even. Furthermore, for a specific scattering, the diagram will have no disconnected portions, so we only consider connected diagrams.)

I. Connecting external points in a diagram: $\Delta\ acts\ on\ 2\varphi' s\epsilon\Sigma$, this leads to a disconnected diagram, so omit.

II. Connecting internal point to itself: : $\Delta\ acts\ on\ 2\varphi' s\epsilon V$, this leads to an internal loop.

III. Connecting different internal pts: $\Delta\ acts\ on\ 2\varphi' s\epsilon different\ V's$, which leads to the line-segment diagram.

Combinatorial factors:

(1) $\left(\frac{1}{2}\right)$ $in\ \Delta$ is cancelled by exchange of $2\varphi's$ which δ_x, δ_y act on, except for loop case.

(2) $\left(\frac{1}{3!}\right)$ $in\ V(\varphi(x))$ is cancelled by three derivatives of φ^3.

(3) $\left(\frac{1}{n!}\right)$ $in\ V[n]$ is cancelled by permutation of labels of $z_i's$ of internal pt's. And, $\left(\frac{m+3n}{2}\right)!$ Is cancelled by permutations of the $\frac{m+3n}{2}$ φ^2 pairs

which Δ acts on, except for multi-vertex connections, which retain a factor $\frac{1}{k!}$ multi-connections.

Feynman Rules:

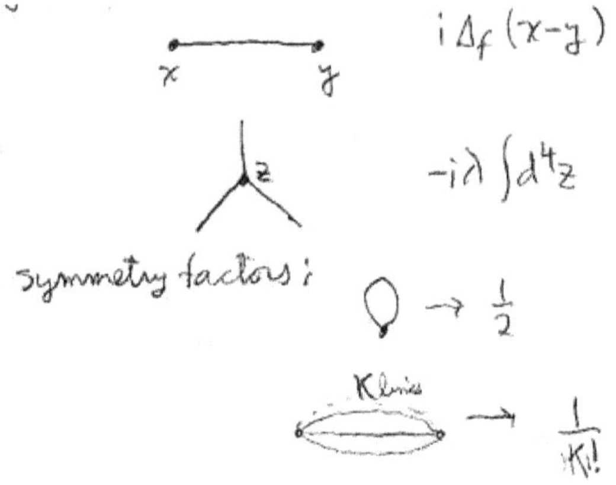

(2b) Let's evaluate $< 0|\varphi(x_1)\varphi(x_2)|0 > = i\Delta_f'(x_1 - x_2)$. We have:

2 external points: m=2

1st order in λ: 3 internal $\varphi's \Rightarrow 5\varphi's$ total, so 0 correction.

2nd order in λ: 6 internal $\varphi's \Rightarrow 8\varphi's$, so nontrivial correction.

So, let's consider 2 internal vertices in the phi^ theory:

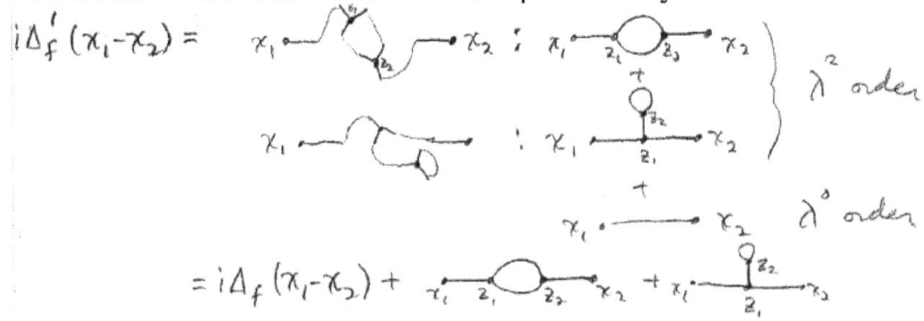

$$i\Delta_f'(x_1 - x_2) = i\Delta_f(x_1 - x_2)$$

$$+(-i\lambda)^2 \int d^4z_1\, d^4z_2 i\Delta_f(x_1 - z_1)\left[i\Delta_f(z_1 - z_2)\right]^2 (\frac{1}{2})\, i\Delta_f(z_2 - x_2)$$

$$+(-i\lambda)^2 \int d^4z_1\, d^4z_2 i\Delta_f(x_1 - z_1) i\Delta_f(z_1 - z_2) i\Delta_f(0)(\frac{1}{2})(z_2 - x_2)$$

What of $< 0|\varphi(x_1)\varphi(x_2)\varphi(x_3)\varphi(x_4)|0 >$? At n=0 we have all disconnected diagrams"

$$i\Delta_f'(x_1 - x_2) = i\Delta_f(x_1 - x_3) i\Delta_f(x_2 - x_4)$$
$$+ i\Delta_f(x_1 - x_2) i\Delta_f(x_3 - x_4)$$
$$+ i\Delta_f(x_1 - x_4) i\Delta_f(x_2 - x_3)$$

At n=2, we have only one connected diagram, which provides a correction term:

$$(-i\lambda)^2 \int d^4z_1\, d^4z_2 i\Delta_f(x_1 - z_1)i\Delta_f(x_2 - z_1)i\Delta_f(z_1 - z_2)i\Delta_f(z_2 - x_3)i\Delta_f(z_2 - x_4).$$

4.3.1 $V = \varphi^2$ and $V = \varphi^3$ renormalization

Consider the $V = \varphi^3$Lagrangian in six spacetime dimensions

(1) Show that the theory is renormalizable, and determine which Green's functions contain divergences.

(2) Write the Lagrangian in terms of renormalized quantities and counter-terms

(3) Compute the renormalization constants to lowest non-trivial order in λ. Assume that the Lagrangian is normal ordered, and therefore ignore the tadpole diagrams.

$$\mathcal{L} = \frac{1}{2}\partial_\mu \varphi \partial^\mu - \frac{1}{2}m^2\varphi^2 - \frac{\lambda}{3!}\varphi^3$$

identity: $L = I - V_N + I$

Identity:$\epsilon + 2I = NV_N$

D = d - $\left(\frac{d-2}{2}\right)\epsilon + \left[\frac{d}{2}(N-2) - N\right]V_N$

$D = 6 - 2\epsilon + (3-3)V_N$

$D = 6 - 2\epsilon$

The two pt. function, E=2, is superficially quadratic ally divergent.

The three pt. function, E=3

Since D is independent of V_N the theory is renormalizable

$$\mathcal{L} = \frac{1}{2}\partial_\mu \varphi \partial^\mu - \frac{1}{2}m^2\varphi^2 - \frac{\lambda}{3!}\varphi^3 + \frac{1}{2}(Z_3 - 1)\partial_\mu\varphi\partial^\mu\varphi - \frac{1}{2}(Z_m - 1)m^2\varphi^2 - (Z_1 - 1)$$

The lowest order correction to the two pt. Function is

$$A = \frac{(-i\lambda)^2\mu^{2\epsilon}}{2}\int \frac{d^{2\omega}k}{(2\pi)^{2\omega}} i\Delta_F(k)i\Delta_F(\rho - k)$$

$$= \frac{1}{2}(-i\lambda)^2\mu^{26}\int \frac{d^{2\omega}k}{(2\pi)^{2\omega}} \frac{i^2}{(k^2 - m^2 + it)(\rho - k)^2 - m^2}$$

$$= \frac{1}{2}(-i\lambda)^2\mu^{26}\int \frac{d^{2\omega}k}{(2\pi)^{2\omega}}\int_o^1 dx \frac{i^2}{(k^2 - m^2 + it)(\rho - k)^2 - m^2} + k^2 - m^2 + (1-x)\rho(\rho - k)$$

$$= \frac{1}{2}(-i\lambda)^2\mu^{26}\int_o^1 dx \int \frac{d^{2\omega}q}{(2\pi)^{2\omega}} \frac{i^2}{[q^2+(1-x)\rho^2+it]^2(\rho-k)^2-m^2}$$

$$(A) = \frac{1}{2}i\lambda^2\mu^{2\epsilon}\int_o^1 dx \int \frac{d^{2\omega}K_\epsilon}{(2\pi)^{2\omega}} \frac{i}{[K_\epsilon^2 - x(1-x)\rho^2+m^2]^2}$$

$$= \frac{1}{2}i\lambda^2\mu^{2\epsilon}\int_o^1 dx\left\{\frac{i\pi^\omega}{(2\pi)^{2\omega}}\frac{\Gamma(2-\omega)}{\Gamma(2)}(m^2-x(1-x)\rho^2)^{\omega-2}\right.$$

$$= \frac{\frac{1}{2}i\lambda^2(4\pi\mu i)^{26}}{(4\pi)^3}\Gamma(\epsilon-1)\int_o^1 dx[m^2-x(1-x)\rho^2]^{1-t}$$

$$= \frac{i\lambda^2}{2(4\pi)^3}\frac{1}{\epsilon}(4\pi\mu)^{26}\frac{\Gamma(1+\epsilon)}{(\epsilon-1)}\int_o^1 dx[\ldots]^{1-6}$$

$$= \frac{i\lambda^2}{2(4\pi)^3}\frac{1}{\epsilon}\left\{-\left[m^2-\frac{1}{6}\rho^2\right]\right\}$$

$$= \frac{i\lambda^2}{2(4\pi)^3}\frac{1}{\epsilon}\left[\frac{1}{6}\rho^2-m^2\right] + finite\ part$$

$(B)+(C) = -i\mathcal{Z}_m^{(2)}\lambda^2 m^2 + i\mathcal{Z}_3^{(2)}\lambda^2\rho^2$

So, $\mathcal{Z}_3 = 1 - \frac{\lambda^2}{12(4\pi)^2}\frac{1}{\epsilon} + O(\lambda^3)$

$\mathcal{Z}_m = 1 - \frac{\lambda^2}{2(4\pi)^3}\frac{1}{\epsilon} + O(\lambda^3)$

To get $\mathcal{Z}_1$ to this order:

$(-i\lambda)^3\mu^{36}\int\frac{d^{2\omega}K}{(2\pi)^{2\omega}} i\Delta_F(k)i\Delta_F(\rho_2+k)i\Delta_F(K-\rho_i)=$

$(-i\lambda)^3\mu^{36}2\int_0^1 dx\int_0^{1-x} dy\int\frac{d^{2\omega}k}{(2\pi)^{2\omega}}\frac{i}{[(1-x-y)(k^2-m^2)+x((k-\rho_i)^2-m^2)+y(k+p_2)^2-m^2]}$

$=(---)\int\frac{d^{2\omega}k}{(2\pi)^{2\omega}}\frac{i^3}{(k^2-A)}\frac{i}{[(1-x-y)(k^2-m^2)+x((k-\rho_i)^2-m^2)+y(k+p_2)^2-m^2]} =$

$-\frac{i\lambda^3}{2(4\pi)^3}(4\pi\mu^2)^\epsilon\frac{\Gamma(\epsilon)}{\Gamma(3)}\int\frac{d^{2\omega}k}{(2\pi)^{2\omega}}\frac{i}{[(1-x-y)(k^2-m^2)+x((k-\rho_i)^2-m^2)+y(k+p_2)^2-m^2]}$

$= -\frac{i\lambda^3}{2(4\pi)^3}\mu^\epsilon\frac{1}{\epsilon} + finite\ teras \Rightarrow \mathcal{Z}_1^{(2)} = 1\sim\frac{\lambda^2}{2(4\pi)^3}\frac{1}{\epsilon} + O(\lambda^3)$

Solution

1. In $d = 2\omega$ dimensions have

   $D = 2\omega L - 2I$

   $L = I - V + 1$; $E + 2I = 3V$

   Therefore $D = 2\omega - (\omega-1)E + (\omega-3)V$. In 6 dimensions, $D = 6 - 2E$ and we have divergences for $E = 2$ *and* $E = 3$. Since D is independent of the number of vertices, the theory is renormalizable.

2. $L = \frac{1}{2}(\partial\phi) - \frac{1}{2}m^2\phi^2 - \frac{\lambda}{3!}u^{\epsilon}\phi^3 + \frac{1}{2}(Z_3 - 1)(\partial\phi)^2 - \frac{1}{2}(Z_m - 1)m^2\phi^2 - (Z_1 - 1)\frac{1}{3!}\mu^{\epsilon}\phi^3$
   where $\epsilon = 3 - \omega$.

3. The lowest order correction to the two-point function is given by the following diagrams.

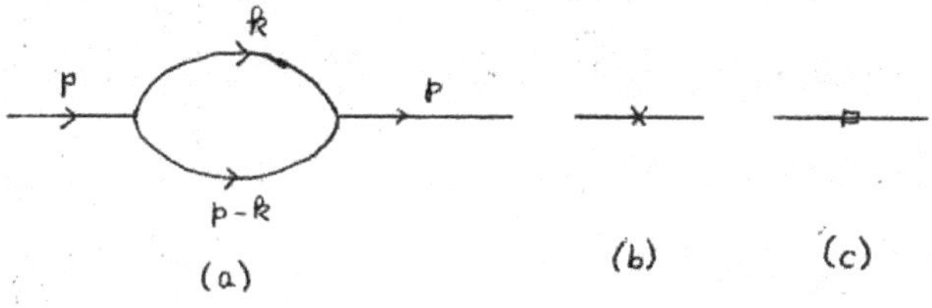

where

$$(a) = \frac{1}{2}(-i\lambda)^2\mu^{2\epsilon}\int\frac{d^{2\omega}k}{(2\pi)^{2\omega}}i\Delta_F(k)i\Delta_F(p-k)$$
$$= \frac{1}{2}(-i\lambda)^2\mu^{2\epsilon}\int_0^1\frac{d^{2\omega}k}{(2\pi)^{2\omega}}\frac{1}{[k^2+x(1-x)p^2-m^2+i\epsilon]^2}$$
$$= \frac{1}{2}i\lambda^2\mu^{2\epsilon}\int_0^1 dx\int\frac{d^{2\omega}k_E}{(2\pi)^{2\omega}}\frac{1}{\left[k_E^2-x(1-x)p^2+m^2\right]^2}$$
$$= \frac{1}{2}\frac{i\lambda^2}{(4\pi)^3}(4\pi\mu^2)^{\epsilon}\Gamma(\varepsilon-1)\int_0^1 dx\,[m^2 - x(1-x)p^2]^{1-\epsilon}$$

For small ϵ this gives

$$(a) = \frac{i\lambda^2}{2(4\pi)^3}\frac{1}{\epsilon}\left[\frac{1}{6}p^2 - m^2\right] + finite\ part$$

The remaining two diagrams give

$$(b) + (c) = -iZ_m^{(2)}\lambda^2 m^2 + iZ_s^{(2)}\lambda^2 p^2$$

Therefore

$$Z_3 = 1 - \frac{\lambda^2}{12(4\pi)^3}\frac{1}{\epsilon} + O(\lambda^3)$$
$$Z_m = 1 - \frac{\lambda^2}{2(4\pi)^3}\frac{1}{\epsilon} + O(\lambda^3)$$

To get Z_1 to this order, look at

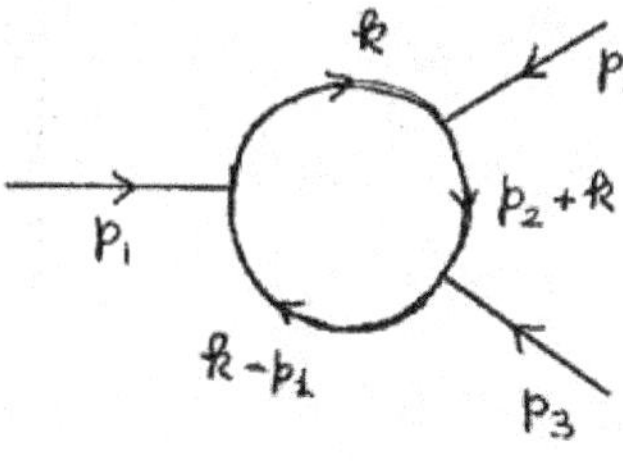

With $\Sigma p_i = 0$ it gives a contribution

$$(-i\lambda)^3\mu^{3\epsilon}\int\frac{d^{2\omega}k}{(2\pi)^{2\omega}}i\Delta_F(k)i\Delta_F(k-p_1)i\Delta_F(k+p_2)$$

$$= (-i\lambda)^3\mu^{3\epsilon}2\int_0^1 dx$$

$$\int_0^{1-x} dy\int\frac{d^{2\omega}k}{(2\pi)^{2\omega}}\frac{i^3}{[(1-x-y)(k^2-m^2)+x((k-p_1)^2-m^2)+y((k+p_2)^2-m^2)]^3}$$

$$= (-i\lambda)^3\mu^{3\epsilon}2\int_0^1 dx\int_0^{1-x} dy\int\frac{d^{2\omega}k}{(2\pi)^{2\omega}}\frac{t^3}{(k^2-A)^3}$$

Where $A = -x(1-x)p_1^2 - y(1-y)p_2^2 - 2xyp_1p_2 + m^2$.

After a Wick rotation the integral becomes

$$-2i\lambda^3\mu^{3\epsilon}\int_0^1 dx\int_0^{1-x} dy\int\frac{d^{2\omega}k}{(2\pi)^{2\omega}}\frac{1}{(k_E^2+A)^3}$$

$$= -\frac{2i\lambda^3}{(4\pi)^3}\mu^{\epsilon}(4\pi\mu^2)^{\epsilon}\frac{\Gamma(\epsilon)}{\Gamma(3)}\int_0^1 dx\int_0^{1-x} dyA^{-\epsilon}$$

$$= -\frac{i\lambda^3}{(4\pi)^3}\mu^{\epsilon}\int_0^1 dx\int_0^{1-x} dy\left[\frac{1}{\epsilon} - \gamma_E - \log\frac{A}{4\pi\mu^2}\right]$$

$$= -\frac{i\lambda^3}{2(4\pi)^3}\mu^{\epsilon}\frac{1}{\epsilon} + finite\ terms$$

The counter-term diagram is:

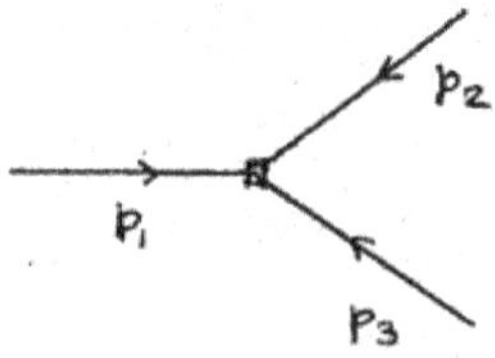

With contribution $-iZ_1^{(2)}\lambda^3\mu^{\epsilon}$, therefore:

$$Z_1 = 1 - \frac{\lambda^2}{2(4\pi)^3}\frac{1}{\epsilon} + O(\lambda^3)$$

Note that use was made of an alternate Feynman form:

$$\frac{1}{ABC} = 2\int_0^1 dx\int_0^1 dy\frac{x}{[xyA+x(1-y)B+(1-x)C]^3} =$$

$$2\int_0^1 dx\int_0^{1-x} dy\frac{1}{[yA+(1-x-y)B+xC]^3}$$

4.3.2 $V = \varphi^4$ **renormalization**

Detailed Renormalization for phi-4 theory. Consider the Lagrangian Density:

$$\mathcal{L} = \frac{1}{2}\partial_\mu\varphi\partial^\mu\varphi - \frac{1}{2}m^2\varphi^2 - \frac{\lambda}{4!}\varphi^4$$

where the mass of the particle 'm' is that carried over from no interactions (e.g., we've so far simply 'added' the phi^4 interaction). For this choice of mass in the 'free' situation we've obtained:

$$i\Delta_F(p) = \frac{i}{p^2 - m^2 + i\epsilon} \quad , \quad p_0 = \sqrt{\vec{p}^2 + m^2}$$

where the restriction to positive infinitesimal 'ϵ' effect a choice of contour integral. We will now consider the renormalized mass to be:

$$(m_{ren})^2 = m^2 + \delta m^2.$$

The renormalized mass will be finite, while the 'bare mass' and the mass correction will both be infinite.

Recall from the LSZ derivation, in scattering calculations we have:

$$\lim_{t\to-\infty}\langle a|\varphi_f(t)|b\rangle = Z^{1/2}\langle a|\varphi_f^{in}|b\rangle, \; Z < 1.$$

We now consider the field itself as renormalized, φ_{ren} measurable (finite), while φ is not measurable. Given the LSZ relation this indicates that there is a revised 'Z' that is not measurable and we have:

$$\varphi_{ren} = Z^{1/2}\varphi.$$

Similarly, compared to the phi^4 analysis that was given previously, we now consider a 'renormalized' coupling:

$$\lambda_{ren} = C\lambda.$$

The revise Lagrangian density becomes:

$$\mathcal{L} = \frac{1}{2}\partial_\mu\varphi_{ren}\partial^\mu\varphi_{ren} - \frac{1}{2}m_{ren}^2\varphi_{ren}^2 - \frac{\lambda_{ren}}{4!}\varphi_{ren}^4$$
$$-\frac{1}{2}(1-Z)\partial_\mu\varphi_{ren}\partial^\mu\varphi_{ren}$$
$$-\frac{1}{2}[m_{ren}^2(Z-1) - \delta m^2 Z]\varphi_{ren}^2 - \frac{1}{4!}\lambda_{ren}\left(\frac{Z^2}{L} - I\right)\varphi_{ren}^4$$

We will now do perturbation theory based on the free Lagrangian given by:

$$\mathcal{L}_{free} = \frac{1}{2}\partial_\mu\varphi_{ren}\partial^\mu\varphi_{ren} - \frac{1}{2}m_{ren}^2\varphi_{ren}^2.$$

The 'counter-terms' from the perturbation part will consist of order $[\varphi_{ren}]^2$:

$$-i[(1-Z)(p^2 - m_{ren}^2) - Z\delta m^2]$$

and order $[\varphi_{ren}]^4$ terms:

$$-i\lambda_{ren}\left(\frac{Z^2}{C}-1\right).$$

These contribution provide corrections for precisely the diagrams which are divergent. For example, we now have

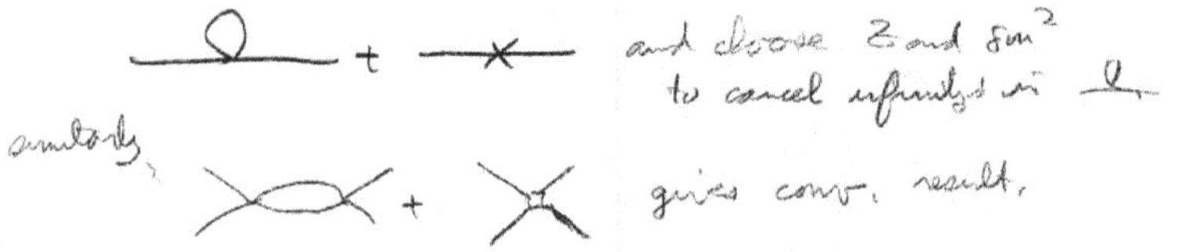

The two main regularization approaches are the cut-off method ($e^{-\Lambda k^2}$ factors are introduced in the integral, or a cut-off is given in the integral bound $\int_0^\Lambda dk$ ...), or use is made of dimensional regularization. Particular attention will be given to dimensional regularization throughout as it is not only the most versatile and usually easiest to work with, it is also precisely the type of regularization indicated by Emanator theory (Book7) where the standard model results from the emanator process. Dimensional regulations originally used because it does not violate gauge invariance (t'Hooft and Veltman). Let's consider one-loop diagrams (t'Hooft and Veltman do up to 2-loops):

$$I(p)=\int d^4k f(k,p),$$

where $f=f(p^2,k^2,k\cdot p)$ is constrained to be a Lorentz scalar. Let's now continue the number of dimensions to be complex:

$$I(\omega,p)=\int d^{2\omega}k\, f(k,p)$$

In this process we want to leave exterior momenta as 4-dimensional (strictly) in order to have compatibility with spinor fields, otherwise there is a mess with higher dimensional Dirac equation representations needed. Define:

$$k_\mu=(\ell_\mu,L_\mu)$$

Where

$$\ell_\mu=(k_1,\dots,k_4,0,\dots,0),\quad L_\mu=(0,0,0,0,\vec{L})$$

Now we can switch to:

$$f(p^2,\ell\cdot p,\ell^2,L^2)$$

And write:

$$I(\omega,p)=\int d^4\ell\int_0^\infty L^{2\omega-5}dL\int d\Omega_L f$$

where $\int d\Omega_L$ describes the surface of a unit-sphere in a $2\omega - 4$, with area: $S_n = \frac{2\pi^{n/2}}{\Gamma\left(\frac{n}{2}\right)}$, thus:

$$I(\omega, p) = \frac{2\pi^{\omega-2}}{\Gamma(\omega-2)} \int d^4\ell \int_0^{\omega} f L^{2\omega-5} dL$$

Besides possible UV divergences for L large we also have a possible IR divergence when $2\omega - 5 \leq -1$. Let's consider the so-called 'tadpole' diagram:

$$I = \int \frac{d^{2\omega}k}{k^2 - m^2 + i\epsilon} = -i \int \frac{d^{2\omega}k}{k^2 + m^2}$$

Let's drop the '-i' factor and work with $f = \frac{1}{k^2+m^2}$:

$$I = \frac{2\pi^{\omega-2}}{\Gamma(\omega-2)} \int d^4\ell \int_0^{\infty} \frac{L^{2\omega-5}}{\ell^2 + L^2 + m^2} dL.$$

For infrared convergence need $2\omega - 5 \leq -1$, or $\omega > 2$. For UV convergence need $\omega < 1$. So, there is no choice of ω that satisfies both. The IR divergence, however, is an artifact of the method, which can be eliminated in the following way:

$$\int_0^{\infty} \frac{L^{2\omega-5}}{\ell^2 + L^2+m^2} = \frac{1}{2(\omega-2)} \int_0^{\infty} \frac{1}{\ell^2 + L^2+m^2} \frac{d}{dL^2}\left(L^{2(\omega-2)}\right) dL^2$$

$$= -\frac{1}{2(\omega-2)} \int_0^{\infty} L^{2(\omega-2)} \frac{d}{dL^2} \frac{1}{\ell^2 + L^2+m^2} dL^2$$

Now we have IR divergence only when $\omega > 1$. Repeating the integration-by-parts maneuver (with zero surface terms) we get:

$$I(\omega) = \frac{\pi^{\omega-2}}{\Gamma(\omega)} \int d^4\ell \int_0^{\infty} L^{2(\omega-1)} \left(\frac{d}{dL^2}\right)^2 \frac{I}{\ell^2 + L^2+m^2} dL^2$$

which is well defined for $0 < \omega < 1$. Let's now use this to extend to higher dimensions using

$$1 = \frac{1}{5}\left(\frac{\partial L}{\partial L} + \frac{\partial \ell_\mu}{\partial \ell_\mu}\right).$$

This identity will be inserted into the integrand and then be 'flipped' using integration by parts on both derivative terms:

$$\frac{\partial \ell_\mu}{\partial \ell_\mu} \to -\ell_\mu \frac{\partial}{\partial \ell_\mu}$$

and the same for $\partial L/\partial L$, to get:

$$I(\omega) = \frac{-3m^2}{\omega - 1} \frac{2\pi^{\omega-2}}{\Gamma(\omega)} \int d^4\ell \int_0^{\infty} L^{2(\omega-1)} \left(\frac{1}{\ell^2 + L^2+m^2}\right)^4 dL^2$$

We now have UV convergence for a wider range:

$$4 + 2\omega - 2 - 8 + 2 < 0 \quad \rightarrow \quad 0 < \omega < 2$$

But note that there is still an explicit pole at $\omega = 1$. Let's use the identity/integration maneuver again:

$$I(\omega) = \frac{2 \cdot 3 \cdot 4m^2}{(\omega-1)(\omega-2)} \frac{\pi^{\omega-2}}{\Gamma(\omega)} \int d^4\ell \int_0^\infty L^{2(\omega-1)} \left(\frac{1}{\ell^2 + L^2 + m^2}\right)^5 dL^2$$

Which is now well-defined on range $0 < \omega < 3$. This allows us to now explicitly consider dimensional regularization from infinitesimally more than 4-dim to the 4-dimensions of spacetime expressed in the standard measure. To do this, note that ω is simply a parameter in $I(\omega)$, so we can now continue analytically in terms of ω (with two poles now indicated at $\omega = 1$ and $\omega = 2$). We are interested in the structure of integral above when $\omega = 2 + \epsilon$, where $\epsilon \ll 1$. Let's focus on the ϵ terms:

$$\epsilon - terms: \quad \propto \frac{\pi^\epsilon}{\epsilon(1+\epsilon)} \frac{1}{\Gamma(\epsilon+2)} L^{2(\epsilon+1)} \times \ldots$$

Where

$$\frac{I}{\Gamma(\epsilon+2)} = \frac{I}{\Gamma(2)} - \epsilon \frac{I}{\Gamma(2)^2} \frac{d\Gamma}{d\epsilon}\Big|_{\epsilon=0}, \quad \Psi(z) = \frac{d}{dZ} ln\Gamma(Z)$$

Thus

$$\epsilon - terms: \quad \propto \frac{1}{\epsilon}(1-\epsilon)(1+\epsilon ln\pi)(1-\epsilon\Psi(2))(1+\epsilon lnL^2)L^2$$

And we want to separate the finite and infinite terms as $\epsilon \to 0$:

$$\epsilon - terms: \ \propto \frac{1}{\epsilon} L^2 + [ln\pi - 1 - \Psi(2) + lnL^2]L^2$$

And putting this back in the integral we have:

$$I(\omega)|_{\epsilon=\omega-2} = \pi^2 m^2 \left(\frac{1}{\epsilon} + \ell n(\pi m^2) - \Psi(2)\right).$$

Thus, using dimensional regularization, we've isolated divergent and convergent parts. Let's now repeat the derivation a quicker way:

$$I_n(\omega) = \int \frac{d^{2\omega}k}{(k^2+m^2)^n} = \frac{\pi^\omega}{\Gamma(\omega)} \int_0^\infty \frac{k^{2\omega-2}dk^2}{(k^2+m^2)^n} = \frac{\pi^\omega}{\Gamma(\omega)} \int_0^\infty \frac{x^{\omega-1}dx}{(x+m^2)^n}$$

Let's now make use of the Beta function:

$$B(x,y) = \frac{\Gamma(x)\Gamma(y)}{\Gamma(x+y)} = \int_0^\infty \frac{t^{x-1}dt}{(1+t)^{x+y}}$$

Thus

$$I_n(\omega) = \frac{\pi^\omega}{\Gamma(\omega)} m^{2(\omega-n)} B(\omega, n-\omega) = \pi^\omega m^{2(\omega-n)} \frac{\Gamma(n-\omega)}{\Gamma(n)}$$

Where we need $n = 1$ case to derive the previous result:

$$I_I = \pi^\omega m^{2(\omega-1)} \Gamma(1-\omega)$$

Since

$z\Gamma(z) = \Gamma(z+1) \rightarrow$

$$\Gamma(1-\omega) = \frac{1}{1-\omega}\Gamma(2-\omega) = \frac{1}{(1-\omega)}\frac{1}{(2-\omega)}\Gamma(3-\omega)$$

We have a form where we can reliably substitute $\omega = 2 + \epsilon$:

$$I = \frac{\pi^\omega m^{2(\omega-1)}\Gamma(3-\omega)}{(1-\omega)(2-\omega)} = \pi^2 m^2\left[\frac{1}{\epsilon} + \ell n\pi m^2 - 1 - \Psi(1)\right]$$

Which is the same as before since:

$$\Psi(x+1) = \Psi(x) + \frac{1}{x}.$$

Let's now repeat the dimensional regularization with the interaction term thrown in:

$$\mathcal{L} = \frac{1}{2}\partial_\mu\varphi\partial^\mu\varphi - \frac{1}{2}m^2\varphi^2 - \frac{\lambda}{4!}\varphi^4$$

First, let's regard the Action as a pure number (technically, it will have an element of its measure that provides dimensions of the Planck length to arrive at the path integral formulation, but ignoring this). So we start with the fact that

$$S = \int d^4x\mathcal{L}$$

is dimensionless. Thus

$$[d^4x] = L^{-4} \rightarrow \ [\mathcal{L}] = L^{-4}.$$

Let $[\varphi] = L^d$, consistency requires that $2d - 2 = -4 \ \rightarrow \ d = -1$. Thus,

$$[\varphi] = L^{-1}.$$

We then have that $[m] = L^{-1}$ and that $[\lambda] = 1$ (is dimensionless).

So far things look like they will generalize easily. But consider what happens if we generalize from 4-Dim. to 2ω (where we work with $4 < 2\omega$ eventually):

$$[d^{2\omega}x] = L^{-2\omega} \rightarrow \ [\mathcal{L}] = L^{-2\omega}$$

Let $[\varphi] = L^d$, consistency requires that $2d - 2 = -2\omega \ \rightarrow \ d = 1 - \omega$. Thus

$$[\varphi] = L^{1-\omega}.$$

We then have that $[m] = L^{-1}$as before, but now:

$$[\lambda] = L^{2\omega-4},$$

i.e., the dimensionless coupling constant takes on dimensions. This will require complication to avoid dimensional inconsistencies, so let's make a minor alteration to the description of our coupling term where an explicit dimensional length quantity μ is introduced

$$\lambda\varphi^4 \rightarrow \lambda(\mu^2)^{2-\omega}\varphi^4$$

We thus keep λ explicitly dimensionless.

Note that even though we've introduced dimension 2ω, for which $[\varphi] = L^{1-\omega}$, we will eventually need to return $2\omega = 4$ dimensions to be consistent with canonical constraints. This is because the simple scalar field (function) relation whereby

$$\varphi(Sx) = S^d \varphi(x)$$

for some d, together with the canonical commutation relations gives:

$$[\varphi(x), \dot{\varphi}(y)]_{x^0=y^0} = i\delta_3(\vec{x} - \vec{y})$$

$$[\varphi(x), \dot{\varphi}(Sy)]_{x^0=y^0} = i\delta_3\big(S(\vec{x} - \vec{y})\big)$$

$$S^{2d-1}[\varphi(x), \dot{\varphi}(y)]_{x^0=y^0} = iS^{-3}\delta(\vec{x} - \vec{y})$$

So, $S^{2d-1} = S^{-3} \Rightarrow d = -1$.

Let's now consider the one-loop interaction diagram:

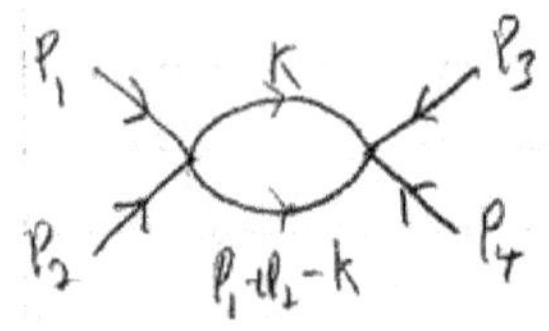

The loop momentum generalizes to: $\int \frac{d^4k}{(2\pi)^4} \to \int \frac{d^{2\omega}k}{(2\pi)^{2\omega}}$ $\quad p = p_1 + p_2$, and recall the modified interaction term, to arrive at:

$$I = \frac{((-i\lambda)^2\mu^{4(2-\omega)}}{2} \int \frac{d^{2\omega}k}{(k^2+m^2)^n} \left[\frac{i}{k^2 - m^2 + i\epsilon}\right]\left[\frac{i}{(p-k)^2 - m^2 + i\epsilon}\right]$$

Evaluate using a math maneuver due to Feynman +Schrodinger. Since

$$\frac{1}{A} = \int_0^\infty dx e^{-xA}$$

We can write

$$\frac{1}{A \cdot B} = \int_0^\infty dx \int_0^\infty dy\, e^{-(xA+yB)}$$

Let $x = \alpha\beta$ and $y = \beta(1-\alpha)$:

$$\frac{1}{A \cdot B} = \int_0^\infty d\alpha \int_0^\infty d\beta e^{-\beta[\alpha A + (1-\alpha)\beta]} = \int_0^1 \frac{d\alpha}{[\alpha A + (1-\alpha)\beta]^2}$$

Using this relation we will replace the product or products of propagators. The generalization for when there is more than two propagators is:

$$\frac{1}{A_1^{n_1} \ldots A_k^{n_k}} = \frac{\Gamma(n_1+n_2 \ldots + n_k)}{\Gamma(n_1)\ldots\Gamma(n_k)} \int_0^1 d\alpha_1 \ldots d\alpha_k \delta(1 - \sum_k \alpha_k)\left(\frac{(\alpha_1)^{n_1-1}\ldots(\alpha_k)^{n_k-1}}{[\alpha_1 A_1 + \alpha_2 A_2 + \cdots \alpha_k A_k]^{n_1+n_2\ldots+n_k}}\right).$$

So, the difficult part is doing the parametric integrations:

$$I = \frac{(-i\lambda)^2 \mu^{4(2-\omega)}}{2}$$

$$\int_0^\infty d\alpha \int \frac{d^{2\omega}k}{(2\pi)^{2\omega}} \frac{1}{\left[\alpha(k^2 - m^2 + i\epsilon) + (1-\alpha)[(p-k)^2 - m^2 + i\epsilon]\right]^2}$$

Let $q = k - (1-\alpha)p$ and $s = (p_1 + p_2)^2$ (Mandelstam variables) in the inner integral to get:

$$\int \frac{d^{2\omega}q}{(2\pi)^{2\omega}} \frac{1}{[q^2 + \alpha(1-\alpha)s - m^2 + i\epsilon]^2}$$

Wick rotating with the freedom indicated by "$i\epsilon$" leads to the Euclideanized q integral:

$$= i \int \frac{d^{2\omega}q_E}{(2\pi)^{2\omega}} \frac{1}{[q_E^2 + m^2 - \alpha(1-\alpha)s]^2}$$

$$= \frac{i\pi^\omega}{(2\pi)^{2\omega}} \frac{\Gamma(2-\omega)}{\Gamma(2)} (m^2 - \alpha(1-\alpha)s)^{\omega-2}$$

Let's now introduce $\varepsilon = 2 - \omega$ and consider the ε dependence in the latter term:

$$\frac{\mu^{2\varepsilon}}{(4\pi)^2 (4\pi)^{-\varepsilon}} \frac{1}{\varepsilon} \Gamma(1+\varepsilon)(m^2 - \alpha(1-\alpha)s)^\varepsilon$$

And consider to lowest order:

$$= \frac{1}{(4\pi)^2} \frac{1}{\varepsilon} [(1 + \varepsilon \ell n\mu^2)(1 + \varepsilon \ell n 4\pi)\left(1 + \varepsilon \Psi(1)\right)(1 + \varepsilon \ell n[m^2 - \alpha(1-\alpha)s])]$$

$$= \frac{1}{(4\pi)^2} \left\{\frac{1}{\varepsilon} + \left[\ell n\mu^2 + \ell n 4\pi + \Psi(1) + \ell n[m^2 - \alpha(1-\alpha)s]\right]\right\}$$

We therefore have:

$$I = G^{(4)}_{truncated} = \frac{i\lambda^2}{16\pi^2} \mu^{2(2-\omega)} \left[\frac{1}{\varepsilon} + \Psi(1) + 2 - \ell n \frac{m^2}{4\pi\mu^2} - A(s)\right]$$

where

$$A(s) = 2\sqrt{\frac{4m^2}{s} - 1}\ tan^{-1}\left(\frac{1}{\sqrt{\frac{4m^2}{s} - 1}}\right)$$

when $s > 4m^2$ $A(s)$ becomes imaginary.

Notice how the regularization schemes appear to be based on some continuation that separates to an infinity and a constant term, and then, with a renormalizable theory, we can eliminate the infinites in a well-defined manner by a finite number of 'renormalizations' (that pair with a finite number of constants or other parameters in the theory).

As indicated by the reference to the Mandelstam variables, there are three such permutations on the external lines, consider the t- variable: $t = (p_1 + p_3)^2$ corresponding to switching p_2 and p_3:

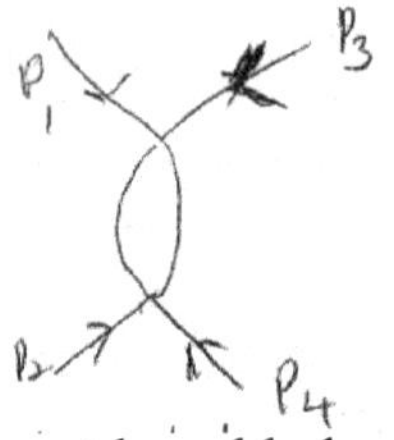

So, there are 3 diagrams that must be added, and the full expression for $G^{(4)}_{trunc}$ is then:

$$G^{(4)}_{truncated} = \frac{i\lambda^2}{16\pi^2}\mu^{2(2-\omega)}$$

$$\times\left[\frac{3}{\varepsilon} + 3\Psi(1) + 6 - 3\ell n\frac{m^2}{4\pi\mu^2} - A(s) - A(t) - A(u)\right]$$

Having regularized the theory, let's now consider renormalization.

The Lagrangian with 'bare' fields specified is:

$$\mathcal{L} = \frac{1}{2}\partial_\mu\varphi_B\partial^\mu\varphi_B - \frac{1}{2}m_B^2\varphi_B^2 - \frac{\lambda_B}{4!}\varphi_B^4$$

Let's introduce three renormalization parameters that relate 'bare' to 'real':

$$\varphi_B = Z_3^{1/2}\varphi_R,\ \lambda_B = \frac{Z}{Z_3^{1/2}}\lambda_R,\ m_B^2 = \frac{Z_m}{Z_3}m_R^2$$

And rewrite the Lagrangian to get:

$$\mathcal{L} = \frac{1}{2}\partial_\mu\varphi_R\partial^\mu\varphi_R - \frac{1}{2}m_R^2\varphi_R^2 - \frac{\lambda_R}{4!}\varphi_R^4$$

$$+\frac{1}{2}(Z_3 - 1)\partial_\mu\varphi_R\partial^\mu\varphi_R - \frac{1}{2}(Z_m - 1)m_R^2\varphi_R^2 - \frac{1}{4!}(Z_1 - 1)\lambda_R\varphi_R^4$$

And we can see new 'renormalization' terms, including one for the wavefunction renormalization: $i(Z_3 - 1)p^2$; one for the mass renormalization: $-i(Z_m - 1)$; and one for renormalization on the coupling: $-i(Z_1 - 1)\lambda_R\mu^{2(2-\omega)}$.

For the coupling renormalization expressed in terms of power series in coupling we have:

$$Z_1 = 1 + \lambda_R Z_1^{(1)} + \lambda_R^2 Z_1^{(2)} + \cdots$$

Since $G_{trunc}^{(4)}$ is second-order in λ_R, we want to consider $(Z_1 - 1)$ from the $-i(Z_1 - 1)\lambda_R\mu^{2(2-\omega)}$ renormalization only at first-order, to have a second-order renormalization contribution. So, need to add $-iZ_1^{(1)}(\lambda_R)^2\mu^{2(2-\omega)}$ to $G_{trunc}^{(4)}$ to renormalize. So let $Z_1^{(1)} = \frac{3}{16\pi^2}\frac{1}{\varepsilon}$ to get the renormalized 4-point function.

Note that renormalization can destroy unitarity if Z's are complex because then the Lagrangian and Hamiltonians will have imaginary terms and will no longer be Hermitian (thus exp(iH) no longer unitary).

The prescription outlined above, to subtract only the $1/\varepsilon$ term, is the minimal subtraction method. Since $\left(\frac{1}{\varepsilon} + \Psi(1) - \ell n\frac{m^2}{4\pi\mu^2}\right)$ always appear together in any diagram, they are often subtracted as well (known as 'modified minimal subtraction scheme').

Doing minimal subtraction for other diagrams:

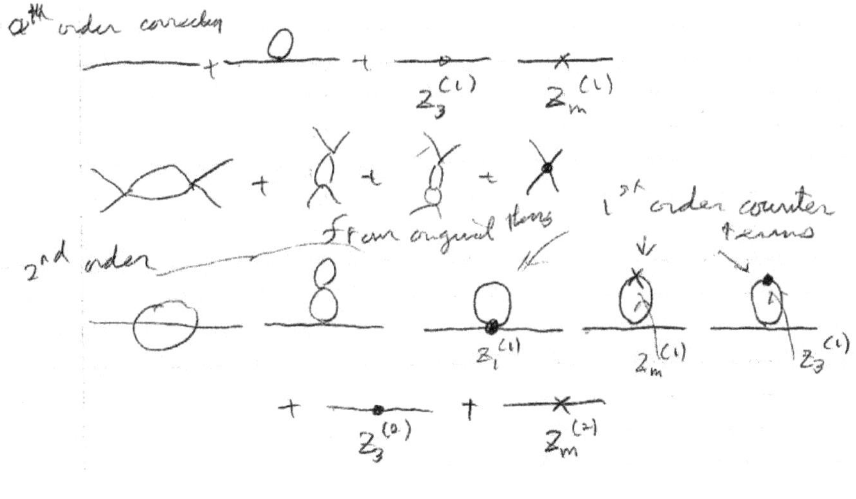

A series of Examples will now be explored.

Example 1
Consider phi^4 theory. What are the 2-point functions of order λ^3 (connected)?

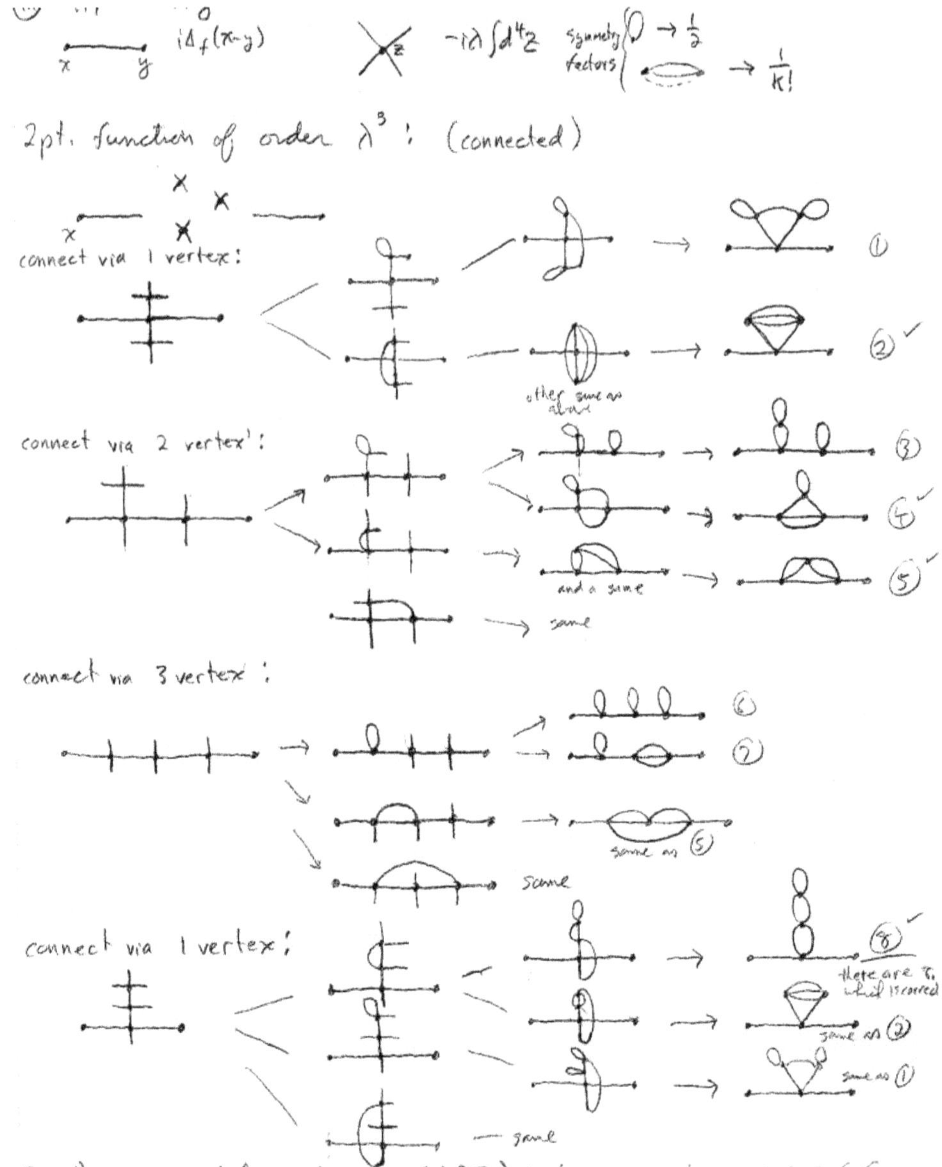

For the one-particle irreducible (1 P.I.) diagrams we have ①,②,④,⑤,⑧

b. With the abbreviation $(dk) \equiv \frac{d'k}{(2\pi)'}$, we have:

$$(a) = \frac{(-i\lambda)^3}{4}\int (dk)(dq)(dt)[i\Delta_\Gamma(k)]^2$$
$$i\Delta p(\neg)i\Delta_F(\neg)i\Delta_F(1)i\Delta_F(p-k-l)$$

$$(b) = \frac{(-i\lambda)^3}{4}\int (dk)(dq)(dt) i\Delta_\Gamma(k)\, i\Delta p(\neg) i\Delta_F(l)$$
$$i\Delta_F(p-k-l) i\Delta_F(p-q-k)$$
$$(c)\ \frac{(-i\lambda)^3}{8}\int (dk)(dq)(dt)[i\Delta_\Gamma(k)]^2[i\Delta_F(q)]^2 i\Delta_F(l)$$
$$(d) = \frac{(-i\lambda)^3}{2.3}\int\int (dk)(dq)(dt)[i\Delta_\Gamma(k)]^2[i\Delta_F(q)]^2 i\Delta_F(l)\, i\Delta_F(k-q-l)$$

Example 2

Write analytic expressions for the diagrams (a) in momentum space.

Review of Feynman Rules in Momentum space

Need momentum space Feynman Rules:

$< 0|T(\varphi(x_1)) \dots (\varphi(x_n)) \equiv G(x_1 \dots x_n)$

Translation invariance of the vacuum means $\tilde{l}$ is a function of only n-1 of the $\varphi's$

$G(x_1 \dots x_n) = < 0|T\left(e^{-ip\cdot a}\varphi(x_1)e^{ip\cdot a}e^{-ip\cdot a}\varphi(x_2) \dots \varphi(x_1)e^{-ip\cdot a}|0>\right)$

$=< 0|T(\varphi(x_1 + a) \dots \varphi(x_n + a)|0)$

$= G(x_1 + a, \dots, x_n + a)$

So we can eliminate an x_n conclude: $Z_j = x_j - x_n \quad j = 1 \dots n-1$

$$\tilde{G}(P_1 \dots P_n) = \int dZ_1 \dots dZ_{n-1} dx_n e^{-i\sum_{j=1}^{n} P_j Z_j} G(Z_1 \dots Z_n)$$
$$= (2\pi)^4 \delta_4\left(\sum_{j=1}^{n} P_j\right) G(P_1 \dots P_n)$$

Now the quantity of internal.

$$\text{Let}\begin{cases}\varphi(p) = (2\pi)^{-4}\int dx e^{-ip\cdot x}\varphi(x)\\ i\Delta_f(x) = (2\pi)^{-4}\int dp e^{-ip\cdot x} i\Delta_f(p)\end{cases}$$

$$V[\varphi] = \frac{\lambda}{4!}\int d^4x\varphi^4(x) = \frac{\lambda}{4!}\int d^4x\int dp_1 \dots dp_4\, e^{i\left(\sum_{j=1}^{n} P_j x_i\right)}\varphi(P_1) \dots \varphi(P_2)$$
$$= \frac{\lambda}{4!}\int dp_1 \dots dp_4 (2\pi)^4 \delta_4\left(\sum_{j=1}^{n} P_j\right)\varphi(P_1) \dots \varphi(P_4)$$

$$i\Delta_{fxf}\delta_x\delta_y = \int d^4x d^4y + \Delta_f(x-y)\frac{\delta}{i\delta\varphi(x)}\frac{\delta}{i\delta\varphi(0)} = \int \frac{dp}{(2\pi)^4} e^{-ik\cdot x}\frac{\delta}{i\delta\varphi(p)},$$

So

$$i\Delta_{fxy}\delta_x\delta_y = \int d^4x d^4y\, (2\pi)^{-4}$$
$$\int d^4k e^{-ik\cdot(xy)} i\Delta_f(k)\int \frac{d^4p_1}{(2\pi)^4} e^{-ip_i\cdot x}\frac{\delta}{i\delta\varphi(x)}\int \frac{d^4p_1}{(2\pi)^4} e^{-iB\cdot y}\frac{\delta}{i\delta\varphi(p)}$$
$$= \int \frac{d^4p_1}{(2\pi)^4} i\Delta_f(k)\frac{\delta}{\delta\varphi(k)}\frac{\delta}{\delta\varphi(-k)}$$

(1b)

ⓑ $G(x_1 \cdots x_n) = e^{\frac{i}{2}\Delta_{fxy}\delta_x\delta_y}\phi(x_1)\cdots\phi(x_n)e^{-iV[\phi]}\Big|_{\phi=0}$

$(2\pi)^4\delta_4(\Sigma p)G(p_1\cdots p_n) = (2\pi)^{4n}e^{\frac{i}{2}\Delta_{fK}\delta_K\delta_{-K}}\phi(p_1)\cdots\phi(p_n)e^{-iV[\phi]}\Big|_{\phi=0}$

external lines:  $i\Delta_F(p) = \frac{i}{p^2-m^2+i\epsilon}$

internal lines:  $\int\frac{d^4K}{(2\pi)^4}\frac{i}{K^2-m^2+i\epsilon}$

vertex; 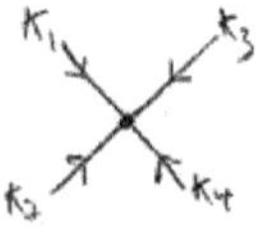 $\underline{(-i\lambda)}(2\pi)^4\delta(K_1+K_2+K_3+K_4)$

$G_{truncated}(p_1\cdots p_n) \Rightarrow$ we drop the "$i\Delta_F(p)$"'s associated with the external lines.

For the 1.P.I's in ⓐ we get:

① 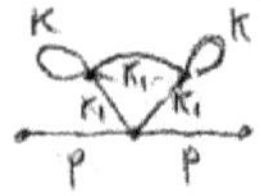 $G_{tr} = \frac{(-i\lambda)^3}{2^3}\int\frac{d^4K_1}{(2\pi)^4}[i\Delta_f(K_1)]^3\left[\int\frac{d^4K}{(2\pi)^4}i\Delta_f(K)\right]^2$

② 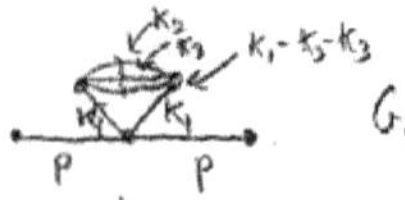 $G_{tr} = \frac{(-i\lambda)^3}{2\cdot 3!}\int\frac{d^4K_1}{(2\pi)^4}\frac{d^4K_2}{(2\pi)^4}\frac{d^4K_3}{(2\pi)^4}[i\Delta_f(K_1)]^2 i\Delta_f(K_2)i\Delta_f(K_3) \times i\Delta_f(K_1-K_2-K_3)$

④ 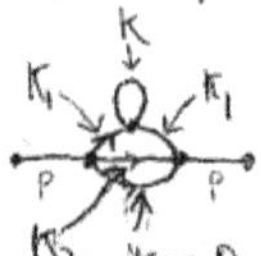 $G_{tr} = \frac{(-i\lambda)^3}{2^2}\int dK_1 dK_2 dK\, i\Delta_f(p-K_1-K_2)i\Delta_f(K_2)[i\Delta_f(K_1)]^2 i\Delta_f(K)$

$K_3 = p-K_1-K_2$

⑤ 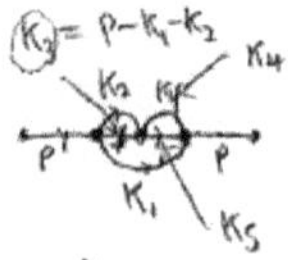 $K_4+K_5 = K_2+K_3$ $\quad K_5 = p-K_1-K_4$ $\quad G_{tr} = \frac{(-i\lambda)^3}{2^2}\int dK_1 dK_2 dK_4\, i\Delta_f(K_1)i\Delta_f(K_2)i\Delta_f(p-K_1-K_2) \times i\Delta_f(K_4)i\Delta_f(p-K_1-K_4)$

⑧ 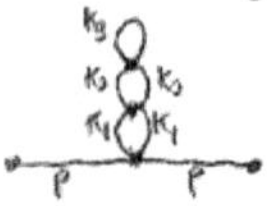 $G_{tr} = \frac{(-i\lambda)^3}{2^3}\int dK_1 dK_2 dK_3[i\Delta_F(K_1)]^2[i\Delta_f(K_2)]^2[i\Delta_f(K_3)]$

(1b) $G(x_1 \dots x_n) = e^{\frac{i}{2}\Delta_{fxy}\delta_x\delta_\sigma}\varphi(x_1 \dots \varphi(x_n))e^{-iv[\varphi]}|_{\varphi=0}$

$(2\pi)^4\delta_4(\varepsilon p)G(p_1 \dots p_n) = (2\pi)^{4n}e^{\frac{i}{2}\Delta_{fxy}\delta_k\delta_{-k}}\varphi(\varphi p_1 \dots \varphi p_n)e^{-iv[\varphi]}|_{\varphi=0}$

External lines: $i\Delta_F(p) = \frac{i}{p^2-m^2+i\epsilon}$

Internal lines: $\int\frac{d^4k}{(2\pi)^4}\ \frac{i}{k^2-m^2+i\epsilon}\ (-i\lambda)(2\pi)^4\delta(k_1+k_2+k_3+k_2)$

$G_{truncated}(p_1 \dots p_n)$ we drop the $i\Delta_F(p)$ is associated with the external lines. For the 1PI's in (1a) we get ($k = k_1+k_2+k_3$):

$$(1)\ \ G_{tr} = \frac{(-i\lambda)^3}{2^3} \int \frac{d^4k}{(2\pi)^4} [i\Delta_F(k_1)]^3 \ \ \left[\int \frac{d^4k}{(2\pi)^4} i\Delta_F(k)\right]^2$$

$$(2)\ G_{tr} = \frac{(-i\lambda)^3}{2.3!} \int \frac{d^4k}{(2\pi)^4}\frac{d^4k}{(2\pi)^4}\frac{d^4k}{(2\pi)^4} [i\Delta_F(k_1)]^2 i\Delta_F(k_2) i\Delta_F(k_3) x i\Delta_F(k)$$

$$(3)\ G_{tr} = \frac{(-i\lambda)^3}{2^2} \int dk, dk, dk\, i\Delta_F(p-k_1-k_2) i\Delta_F(k_1)[i\Delta_F(k_1)]^2 i\Delta_F(k)$$

$$(4)\ G_{tr} = \frac{(-i\lambda)^3}{2^2} \int dk_1 dk_2 dk_3\, [i\Delta_F(k_1)]^2 [i\Delta_F(k_2)]^2 [i\Delta_F(k_3)]$$

Example 3

The order λ^3 , one particle irreducible contributions to the four point function are given by the following diagrams, plus those obtained by permutation of the external momenta.

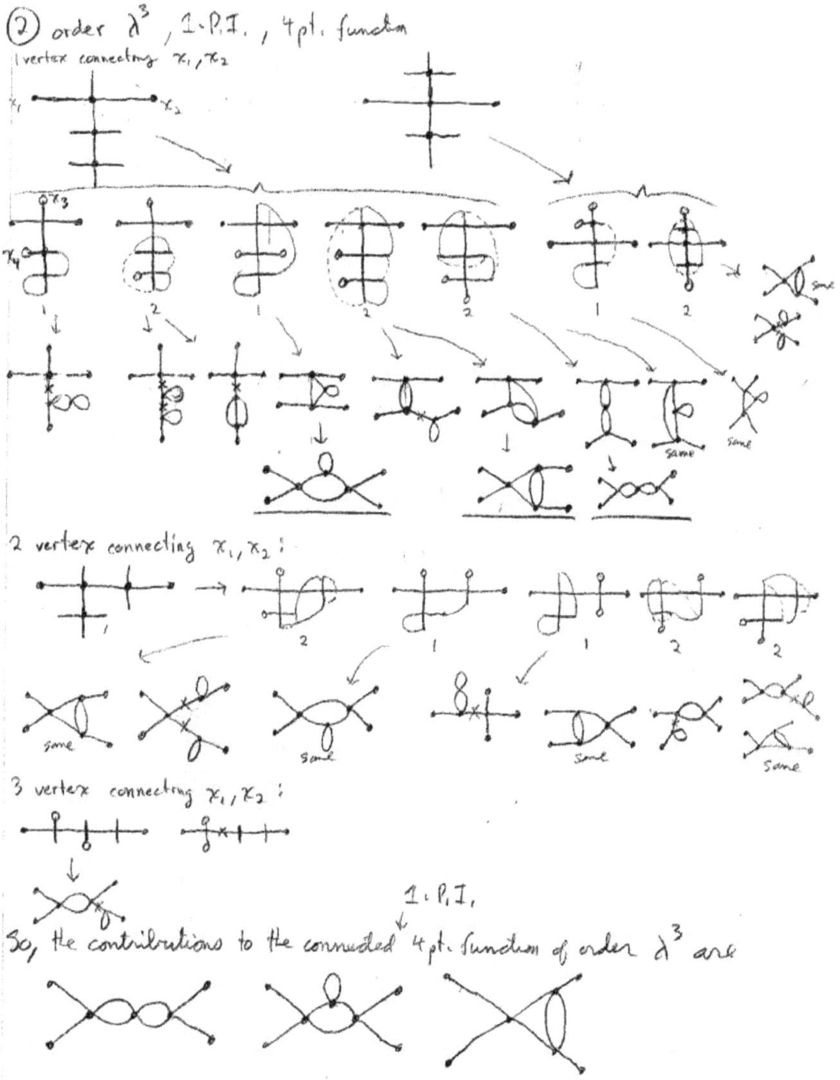

Much easier if directly proceeding to 1.P.I. graphs:

(1a) To not be 1.P.I. must have:

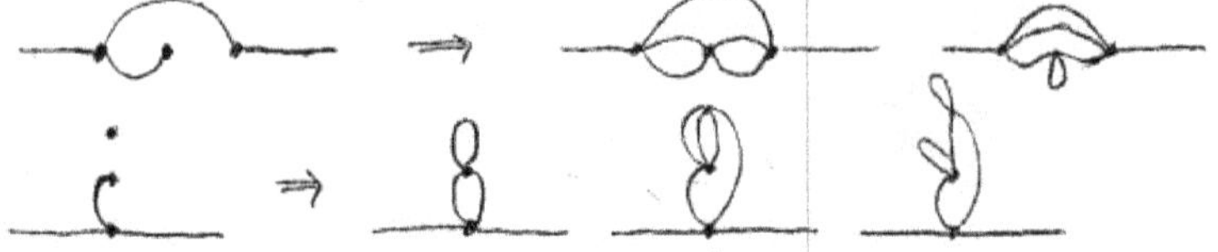

So,

(2) For the 1.P.I. in this part: 4 ext. lines, 3 vertex'

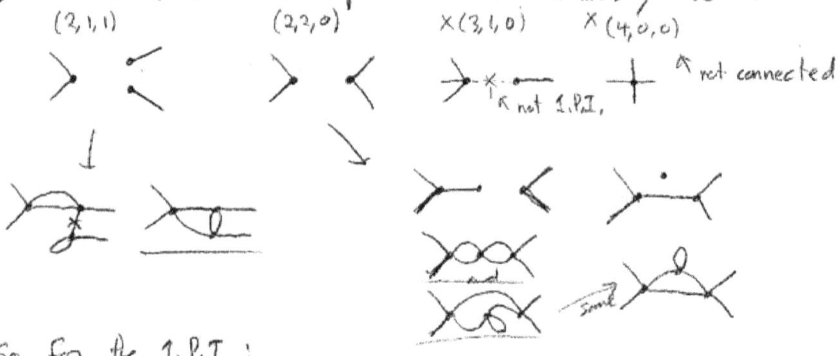

So, for the 1.P.I.:

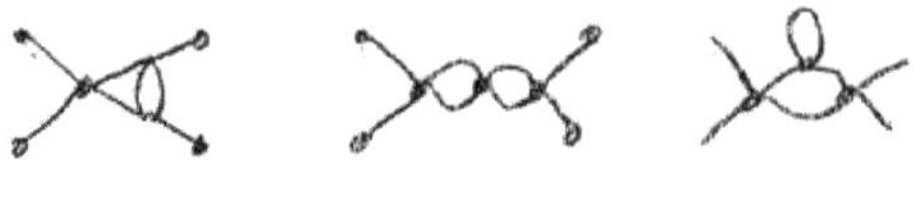

Example 4

Compute the corrections to the propagator in phi^4 theory to second order. Include relevant counterterms. So, need corrections to the propagator in $\lambda\varphi^4$ theory to order λ^2 including counterterms. First consider the calculation of , the "tad pole" diagram, without normalization.

$G_{tr} = \frac{(-i\lambda)}{2}\int d^4k \frac{i}{k^2-m^2+i\epsilon}$

Wick rotation $k_o = ik_4$

$\omega_k = \sqrt{k^2+m^2}$

Solve $I = \int \frac{id^4k}{k^2-m^2+i\epsilon}$

$k^2 = k_o^2 - \vec{k}^2 = \omega_k^2 - \vec{k}^2$

Poles at $\omega_k^2 - \vec{k}^2 - m^2 = 0$

$\omega_k = \sqrt{\vec{k}^2 + m^2}$

$k^2 = -k_4^2 - \vec{k}^2 = -k_E^2$

$I = \int \frac{d^4 k_E}{k_E^2 + m^2}$

$I = 2\pi^2 \int_0^\infty \frac{k_E^2 dk_E}{k_E^2 + m^2}$

$I_\Lambda = 2\pi^2 \int_0^\Lambda \frac{k_E^2 dk_E}{k_E^2 + m^2} = \pi^2 \left[\Lambda^2 - m^2 l_n \left(\frac{\Lambda^2 + m^2}{m}\right)\right]$

Quadratic divergence

In a divisionary regularization the important integral is:

$\int \frac{d^{2\omega} k}{(k^2+m^2)^m} \quad \pi^\omega m^{2(\omega-n)} \frac{\Gamma(1-\omega)}{\Gamma(1)}$

For the tadpole diagram this gives $I = \pi^\omega m^{2(\omega-n)} \frac{\Gamma(1-\omega)}{\Gamma(1)}$

(note! $\Psi(Z) = \frac{d}{dZ} l_n \Gamma(Z)$ will be useful upon expanding $\Gamma(Z)$ also

$\Psi(x+1) = \Psi(x) + \frac{1}{x}$

A useful equation from Feynman and Schrodinger:

Formally $\frac{1}{A_1^{"} \ldots A_k^{nk}} = \frac{\Gamma(n_1 + \cdots + n_k)}{\Gamma(n_1)\Gamma(n_2)\ldots\Gamma(n_k)} \int_0^1 d\alpha_1 \ldots d\alpha_k \delta\left(1 - \frac{\varepsilon}{k}\alpha_k\right) \left(\frac{(\alpha_1)^{n_1-1} \ldots (\alpha_k)^{n_k-1}}{[\alpha_1 A_1 + \alpha_2 A_2 \ldots \alpha_k A_k]^{n_1+n_1+n_k}}\right)$

To actually solve this the difficult part is the parameter integration. The Lagrangian upon which own field theory is based its producing divergent results. So, consider the standard renormalization scheme. Thus for we have countered the bare Lagrangian, in effect, what we want is the renormalized Lagrangian.

$\mathcal{L} = \frac{1}{2}\partial_\mu \varphi_B \partial^\mu \varphi_B - \frac{1}{2} m_B^2 \varphi_B^2 - \frac{\lambda_B}{4!} \varphi_B^4$

$\varphi_B = Z_3^{1/2} \varphi_{Rev}\ , \lambda_B = \frac{Z_1}{Z_3^2} \quad \lambda_{ren}\quad , m_B^2 = \frac{Z_m}{Z_3} \quad m_R^A$

$\mathcal{L} = \frac{1}{2}\partial_\mu \varphi_B \partial^\mu \varphi_B - \frac{1}{2} m_B^2 \varphi_B^2 - \frac{\lambda_B}{4!} \varphi_R^4 + \frac{1}{2}(Z_3 - i)\partial_\mu \varphi_R \partial \varphi_n - \frac{1}{2}(Z_m - 1) m_R^2 \varphi_R^2 - \frac{1}{4!}(Z_1 - 1)\lambda_R \varphi_R$

"Counter terms"

Where each Z is understood to be a power series in λ:

$$Z_\alpha = 1 + \sum_{n=1}^{\infty} \lambda^n Z_\alpha^{(n)}$$

The old Feynman rules:

$i\Delta_f(x-y)$

$-i\lambda \int d^4 Z$

Momentum space:

Ext. lines $\overset{\leftrightarrow}{p} i\Delta_f(p) =$

$\frac{i}{p^2-m^2+i\epsilon}$; $int.\,lines \overset{\leftrightarrow}{k} \int \frac{d^4k}{(2\pi)^4}\frac{i}{k^2-m^2+i\epsilon}$; $vector\ X\ -i\lambda(2\pi)^4\delta(Sx)$

$-\frac{1}{2}(Z-1)m_R^2\varphi_R^2 \Rightarrow$ mass normalization (in momentum space)

$-i(Z_m-1)m^2$ 1st order in λ $0(\lambda)$

Counter terms considered like additional potential terms subsequent Feynman rules.

$\frac{1}{2}(Z_3-1)\partial_\mu\varphi_R\partial^4\varphi_R \Rightarrow$ wave function renormalization (expressed in momentum space)

$i(Z_3-1)p^2 \quad \sigma(\lambda)$

$-\frac{1}{4!}(Z_1-1)\lambda_R\varphi_R^4 \Rightarrow \qquad -i\lambda_R(Z_1-1) \quad \sigma(\lambda^2)$

Now we can consider corrections to $G_{trunc}^{(3)}$ $to\ \vartheta(\lambda^2)$:

Regularize diagrams using Dimensional Regularization method:

$$G_{trunc} = \frac{(-i\lambda)}{2}\int\frac{d^4k}{(2\pi)^4}\frac{i}{k^2-m^2+i\epsilon} = \frac{-i\lambda}{2(\pi)^4}\int i\frac{d^4k}{k^2-m^2+i\epsilon}$$

(Wick rotation)

$$= \frac{-i\lambda}{2(\pi)^4}\int\frac{d^4k_E}{k_E^2+m^2} = -\frac{\lambda}{2(2\pi)^4}\quad \pi^\omega m^{2(\omega-1)}\Gamma(1-\omega)\frac{\Gamma(3-\omega)}{(1-\omega)(2-\omega)}$$

$$= \frac{-i\lambda}{2(\pi)^4}\int\frac{d^4k_E}{k_E^2+m^2} = \frac{-\lambda}{2(2\pi)^4}\pi^\epsilon m^{2\epsilon} \qquad \Gamma(1-\epsilon) = \Gamma(1) - \epsilon\frac{d\Gamma(z)}{dz}$$

$$=$$

$$\left(\frac{-i\lambda}{2(2\pi)^4}\pi^2m^2\right)\left(\frac{\frac{1}{\epsilon}(1+\epsilon l_n\pi)(1+\epsilon l_n m^2)(1-\epsilon\Psi(1))}{x(1-\epsilon)}\right)\Gamma(z) + (z)|_z$$

$$= (\dots)\left(\frac{1+\epsilon(l_n\pi + l_n m^2 - \Psi(1) - 1)}{\epsilon}\right)$$

$$= \frac{-i\lambda}{2(\pi)^4}\pi^2m^2\left[\frac{1}{\epsilon} + l_n\pi m^2 - \Psi(2)\right]$$

$$G_{tranc.} = \frac{(-i\lambda)^2}{2.21}\int\frac{d^4k}{(2\pi)^2}\left(\frac{i}{k^2-m^2+it}\right)^2\int\frac{d^4q}{(2\pi)^4}\left(\frac{i}{q^2-m^2+it}\right)$$

$$= \frac{(-i\lambda)^2}{4}\frac{(i)}{(2\pi)^8}\int d^4k_\epsilon\left(\frac{i}{k^2+m^2}\right)^2\int d^4q_\epsilon\left(\frac{i}{q^2+m^2}\right)$$

$$= \frac{i(-i\lambda)^2}{4(2\pi)^8}\pi^2m^2\left[\frac{1}{\epsilon} + \ell n\pi m^2 - \varphi(2)\right]\pi^2\left[\pi^\epsilon m^{2\epsilon}\frac{\Gamma(3-\omega)}{\pi(2)(2-\omega)}\right]$$

$Z\Gamma(Z) = \Gamma(Z+1)$

$\Gamma(2)$=I

$$\frac{(1+\epsilon\ell nm^2)\big(\Gamma(1)-\epsilon\Gamma(1)\varphi(1)\big)}{\Gamma(2)\epsilon}=\left[\frac{1}{\epsilon}+\ell nm^2-\varphi(1)\right]$$

$$=\frac{-i\lambda^2m^2}{1024\pi^2}\left[\frac{1}{\epsilon^2}+\frac{1}{\epsilon}\big(\ell n\pi^2m^4-\varphi(2)\big)\big(\ell n\pi m^2-\varphi(1)\big)\right]$$

$$G_{tranc}=\frac{1}{(2\pi)^4}\int\frac{d^4ki}{k^2-m^2+it}(-i\lambda)(\mathrm{Z}_1-1)$$

$$=\frac{-i(\lambda)^2\left(\overline{\mathrm{Z}}_{,}^{o}\right)}{(2\pi)^4}\pi^2m^2\left[\frac{1}{\epsilon}+\ell n\pi m^2-\varphi(2)\right]$$

$$G_{tranc}=\frac{(-i\lambda)}{2}\int\frac{d^4k}{(2\pi)^4}\left(\frac{i}{k^2-m^2+it}\right)^2(-i[\mathrm{Z}_m-1])m^2$$

$$=\frac{i(-i\lambda)^2\mathrm{Z}_m^{(1)}m}{2}\int\frac{d^4k}{(2\pi)^4}\frac{i}{(k^2-m^2+it)^2}=\frac{-i\lambda^2\mathrm{Z}_m^{(1)}}{2(2\pi)^4}\int d^4k_\epsilon\left(\frac{i}{k_\epsilon^2+m^2}\right)^2$$

$$=\frac{-i\lambda^2\mathrm{Z}_m^{(1)}m^2}{32\pi^2}\left[\frac{1}{\epsilon}+\ell n\pi m^2-\varphi(1)\right]$$

$$G_{tranc}=\frac{(-i\lambda)}{2}\int\frac{d^4k}{(2\pi)^4}\left(\frac{i}{k^2-m^2+it}\right)^2 i\lambda\mathrm{Z}_3^{(1)}k^2$$

$$=\frac{i\lambda^2\mathrm{Z}_3^{(1)}}{2(2\pi)^4}\int\frac{d^4k}{(2\pi)^4}\left(\frac{i}{k_\epsilon^2-m^2+1T}\right)^2 I\lambda\mathrm{Z}_3^{(i)}K^2$$

$$=\frac{I\lambda^2\mathrm{Z}_3^{(i)}}{2(2\pi)^4}\int d^4k_\epsilon\frac{k_\epsilon^2}{(k_\epsilon^2+m^2)^2}$$

Define $G(t)=\frac{i}{2xn\omega}\left[\theta(t)e^{-i\omega t}+\theta(-t)e^{i\omega t}\right]$

$$\mathrm{Z}[J]=const.\,e^{\frac{i}{2}}\int_{\alpha_i}^{\alpha_\alpha}dt\int_{\alpha_i}^{\alpha}dt'JG(t-t')J(t')$$

$$G(t\,\dagger)=\frac{1}{2m\pi}\int_{-\alpha}^{\alpha}d\epsilon\frac{e^{i\epsilon t}}{\omega^2-\epsilon^2-i\eta}$$

Represents to otter expressions, so

$G(t)$ is a green's function !

$$\mathrm{m}\left(\frac{d^2}{dt}+\omega^2-i\eta\right)G(t\,\dagger)=\delta(t\,\dagger)$$

$$\int\mathscr{L}x(t)e^{i\int(xAx+Jx)dt}=e^{i\int JA^{-1}J}$$

$$\mathrm{Z}[J]=const.\int\mathscr{L}xe^{1/2\int_{\alpha_i}^{\alpha\alpha}\left[mx\left(\frac{d^2}{d\dagger^2}+\omega^2-i\eta\right)x+Jx\right]}$$

$$-m(x^2-\omega^2x^2+i\eta x^2)$$

$$\omega^2\to\omega^2-i\eta$$

$$G_{tranc.} = \frac{(-i\lambda)^2}{2.21}\int \frac{d^4k}{(2\pi)^2}\left(\frac{i}{k^2 - m^2 + it}\right)^2 \int \frac{d^4q}{(2\pi)^4}\left(\frac{i}{q^2 - m^2 + it}\right)$$

$$= \frac{(-i\lambda)^2}{4}\frac{(i)}{(2\pi)^8}\int d^4k_\epsilon \left(\frac{i}{k^2 + m^2}\right)^2 \int d^4q_\epsilon \left(\frac{i}{q^2 + m^2}\right)$$

$$= \frac{i(-i\lambda)^2}{4(2\pi)^8}\pi^2 m^2 \left[\frac{1}{\epsilon} + \ell n\pi m^2 - \varphi(2)\right]\pi^2 \left[\pi^\epsilon m^{2\epsilon}\frac{\Gamma(3-\omega)}{\pi(2)(2-\omega)}\right]$$

$$Z\Gamma(Z) = \Gamma(Z+1)$$

$\Gamma(2)$=I

$$\frac{(1+\epsilon\ell n m^2)\big(\Gamma(1) - \epsilon\Gamma(1)\varphi(1)\big)}{\Gamma(2)\epsilon} = \left[\frac{1}{\epsilon} + \ell n m^2 - \varphi(1)\right]$$

$$= \frac{-i\lambda^2 m^2}{1024\pi^2}\left[\frac{1}{\epsilon^2} + \frac{1}{\epsilon}\big(\ell n\pi^2 m^4 - \varphi(2)\big)\big(\ell n\pi m^2 - \varphi(1)\big)\right]$$

$$G_{tranc} = \frac{1}{(2\pi)^4}\int \frac{d^4ki}{k^2 - m^2 + it}(-i\lambda)(Z_1 - 1)$$

$$= \frac{-i(\lambda)^2 \left(\overline{Z}_{,}^{o}\right)}{(2\pi)^4}\pi^2 m^2 \left[\frac{1}{\epsilon} + \ell n\pi m^2 - \varphi(2)\right]$$

$$G_{tranc} = \frac{(-i\lambda)}{2}\int \frac{d^4k}{(2\pi)^4}\left(\frac{i}{k^2 - m^2 + it}\right)^2 (-i[Z_m - 1])m^2$$

$$= \frac{i(-i\lambda)^2 Z_m^{(1)} m}{2}\int \frac{d^4k}{(2\pi)^4}\frac{i}{(k^2-m^2+it)^2} = \frac{-i\lambda^2 Z_m^{(1)}}{2(2\pi)^4}\int d^4k_\epsilon \left(\frac{i}{k_\epsilon^2+m^2}\right)^2$$

$$= \frac{-i\lambda^2 Z_m^{(1)} m^2}{32\pi^2}\left[\frac{1}{\epsilon} + \ell n\pi m^2 - \varphi(1)\right]$$

$$G_{tranc} = \frac{(-i\lambda)}{2}\int \frac{d^4k}{(2\pi)^4}\left(\frac{i}{k^2 - m^2 + it}\right)^2 i\lambda Z_3^{(1)} k^2$$

$$= \frac{i\lambda^2 Z_3^{(1)}}{2(2\pi)^4}\int \frac{d^4k}{(2\pi)^4}\left(\frac{i}{k_\epsilon^2 - m^2 + 1T}\right)^2 1\lambda Z_3^{(i)} K^2$$

$$= \frac{1\lambda^2 Z_3^{(i)}}{2(2\pi)^4}\int d^4k_\epsilon \frac{k_\epsilon^2}{(k_\epsilon^2 + m^2)^2}$$

$$\int \frac{d^{2\omega}KK^2}{(K^2+m^2)} = \frac{\pi^2}{\Gamma(\omega)}\int_o^\infty \frac{k^{2\omega}dk^2}{(k^2+m^2)^n} = \frac{\pi^\omega}{\Gamma(\omega)}\int_o^\infty \frac{x^\omega dx}{(x+m^2)^n}$$

$$B(x,y) = \frac{\Gamma(x)\Gamma(y)}{\Gamma(x+y)} = \int_o^\infty \frac{t^{x-1}dt}{(1+t)^{x+y}} \quad x = m^2 t$$

$$= \frac{\pi^\omega}{\Gamma(\omega)}\int_o^\infty \frac{m^{2\omega}\,t^\omega\, m^2 d\,t}{m^{2n}(t+1)^n} = \frac{\pi^\omega m^{2(\omega+1)}}{\Gamma(\omega)}\int_0^\infty \frac{t^\omega dt}{(t+1)^n}$$

$$= \frac{\pi^\omega m^{2(\omega+1)}}{\Gamma(\omega)}\frac{\Gamma(\omega+1)\Gamma(n-\omega-1)}{\Gamma(n)}$$

$$= \pi^{\omega} m^{2(\omega+1)} \omega \frac{\Gamma(n-\omega-1)}{\Gamma(n)}$$

$$\Gamma(\omega+1) = \omega\Gamma(\omega)$$

$$\Gamma\big((\epsilon-1)+1\big) = (\epsilon-1)\Gamma(\epsilon-1)$$

$$\Gamma(\epsilon+1) = \epsilon\Gamma(\epsilon)$$

$$= \pi^2 m^6 \left\{\pi^{\epsilon} m^{2\epsilon}(2+\epsilon)\frac{\Gamma(-1+\epsilon)}{\Gamma(2)}\right\}$$

$$= \pi^2 m^6 \left\{(1+\epsilon \ell n\pi)(1+\epsilon \ell n m^2)(2+\epsilon)\frac{\Gamma(1+\epsilon)}{\epsilon(\epsilon-1)}\right\}$$

$$= \pi^2 m^6 \left\{(1+\epsilon \ell n\hbar)(1+\epsilon \ell n m^2)(2+\epsilon)\frac{\Gamma(1+\epsilon)}{\epsilon}\right\}$$

$$= -2\pi^2 m^6 \left\{\frac{1}{\epsilon} + \ell n\pi m^2 + \frac{1}{2} + \varphi(1) + 1\right\}$$

$$= -2\pi^2 m^6 \left\{\frac{1}{\epsilon} + \ell n\pi m^2 + \frac{1}{2} + \varphi(1) + 1\right\}$$

$$G_{tranc} = \frac{-i\lambda^2}{16\pi^2} m^6 z_3^{(1)} \left[\frac{1}{\epsilon} + \ell n\pi m^2 + \frac{1}{2} + \varphi(2)\right]$$

Adding, this guess $G^{(2)}$ tranc $\big(O(\lambda^2)\big)$ aside from the term

4.4 Renormalization Group

Background

There are a variety of ways that renormalization can be accomplished in quantum field theory, but the main idea is that mass and charge are renormalized by introducing a cutoff to the infinity in the momentum scale. In essence, if a theory allows for a form of scale transformation that is self-similar (same resulting theory but with parameter values shifted), then that theory is renormalizable. In 1954 Gell-Mann and Low [44] determined that the variation of the electromagnetic coupling $\alpha(\mu)$ in QED, at energy scale μ, had a simple relation:

$$\alpha(\mu) = W^{-1}\left[\left(\frac{\mu}{M}\right)^d W(\alpha(M))\right],$$

if referenced to energy scale M, where W is a merely a function (known as Wegner's function in this context), and d is constant. In renormalization the (scale) transformation shifts from bare terms to physical terms, according to the above, and in so doing the transformations form a group (this was actually noted in 1953 [45]). What is remarkable, however, is not just that this "renormalization group" exists for quantum field theory as regards renormalization, but that it exists in all situations with the same form of self-similarity under scale transformation (known as "universality"). Restating the scale-relation and Renormalization Group transformation in general form, we have:

$$g(\mu) = W^{-1}\left[\left(\frac{\mu}{M}\right)^{d} W(g(M))\right],$$

and

$$\beta(g) = \frac{\partial g(\mu)}{\partial \ln \mu}$$

where now the system parameter described by the rescaling could be a coupling, a Hamiltonian, a partition function, etc., as long as it captures the whole description of the physics of the system (necessary to have a self-similar relation). Examination of $\beta(g)$, provides an understanding of the variation of the coupling (or whatever variable in the scaling relation) with energy.

Consider a theory with scaling function Z, state variables S, and coupling constants J. If we group some of the state variables in a local way (blocking [46,47]), we now have fewer state variables 's' and changed J'. What β tells us is $\{s\} = \beta(\{S\})$, and for this reason it is sometimes described as inducing renormalization group flow on J-space. The values of J in this process are known as running coupling constants. Obviously structure in this flow space will be highly significant. Most notably, any fixed point of the flow is likely to correspond with a macroscopic state of the system. If the fixed point correspond with a free field theory, then the related quantum theory is more manageable (as with QED). Note that with the above brief description it is clear that the mathematical system is that of a semigroup not a group as the name would suggest, since as we shift to coarser scales we lose individual state variable information. Thus, the process can't be inverted (has no unique inverse), thus it is a semigroup. The entire perturbation expansion analysis is itself a semigroup analysis, so in the context of application it makes no difference.

It might seem odd at first that the coupling to something as basic as charge could change. But keep in mind that the types of scale transformations we speak of, looking at finer scale for example, involve peering past the virtual electron-pair screening around an electron, say, according to the energy scale used, and thus will see different charge accordingly. Thus, it is in quantum field theory, where such contributions can arise, as captured in the necessary rescaling relation, that such phenomena can exist. Consider for example the famous coupling parameter from electromagnetism α. We know at low energies there is the familiar, $\alpha \cong 1/137$, but according to the group flow equation, the predicted coupling at 200GeV energy scales should have $\alpha' \cong 1/127$, and this is precisely what is observed [48].

4.4.1 Renormalization group for φ^4

Let's consider the renormalization group further. To do so we will begin by introducing a generator formalism for the various Green's functions much like what was done with Lagrangians in the past, we here do for Lagrangian densities. Let's start with the following form:

$$\mathcal{L} = \frac{1}{2}\partial_\mu\varphi\partial^\mu\varphi - \frac{1}{2}m^2\varphi^2 - \frac{\lambda}{4!}\varphi^4 + \frac{1}{2}(Z_3 - 1)\partial_\mu\varphi\partial^\mu\varphi - \frac{1}{2}(Z_m - 1)m^2\varphi^2 - \frac{\lambda}{4!}(Z_1 - 1)\lambda\varphi^4$$

Consider the generating functional

$$Z[J] = \mathcal{N}^{-1}\int \mathcal{D}\varphi\, e^{i\int[\mathcal{L}(\varphi)+J\varphi]} = e^{iW[J]}$$

Where choosing $Z[0] = 1$ is a convention with all vacuum diagrams cancelled, and $W[J]$ generates the connected diagrams:

$$W[J] = \sum_n \frac{i^n}{n!}\int dx_1 \dots dx_n \langle 0|T\varphi(x_1)\dots\varphi(x_n)|0\rangle_c J(x_1)\dots J(x_n)$$

As mentioned previously, we don't need all connected diagrams, just the 1PI diagrams. It turns out that a generating functional for 1PI diagrams is possible via a functional version of a Legendre Transform (from Jona Lasinio in the late 1960's). Recall that for the Legendre transformation given a Lagrangian:

$$L(q,\dot{q});\ p = \frac{\partial L}{\partial \dot{q}};\quad H(p,q) = (p\dot{q} - L);\ \frac{\partial H}{\partial p} = \dot{q}$$

In what follows, the role of $\dot{q}$ is played by J, and $W[J]$ is the analog of L:

$$\frac{\delta W[J]}{i\delta J(x)} = \langle 0|\varphi(x)|0\rangle_J \equiv \varphi_c(x)$$

Where no 'C' denotes 'c-number', and when $J \to 0$, $\varphi_c(x) \to 0$ for theories without degenerate vacua.

Now consider

$$i\Gamma[\varphi_c] = W[J] - i\int dx J(x)\varphi_c(x)|_{J=J(\varphi_c)}$$

We have:

$$\frac{i\delta\Gamma[\varphi_c]}{\delta\varphi_c(x)} = \int dy \frac{\delta W}{\delta J(y)}\frac{\delta J(y)}{\delta\varphi_c(x)} - i\int dy \frac{\delta J(y)}{\delta\varphi_c(x)}\varphi_c(y) - iJ(x)$$

Thus,

$$\frac{\delta\Gamma[\varphi_c]}{\delta\varphi_c(x)} = -J(x)$$

Note that when $J \to 0$, $\varphi_c(x)$ extremizes Γ, making Γ the effective action. Thus, Γ should reduce to the classical action to $O(\hbar)$:

$$\Gamma = \sum_n \frac{1}{n!}\int dx_1 \dots dx_n\, \Gamma^n(x_1 \dots x_n)\varphi_c(x_1)\dots\varphi_c(x_n)$$

where the $\Gamma^n(x_1 \dots x_n)$ coefficients are the 1PI Green's functions.

Consider

$$\frac{\delta}{\delta\varphi_c(y)}\left\{\frac{\delta W}{i\delta J(x)} = \varphi_c(x)\right\} \quad \to \quad \delta_4(x-y) = \int dz \frac{\delta^2 W}{i\delta J(x)\delta J(z)}\frac{\delta J(z)}{\delta\varphi_c(y)}$$

When $J \to 0$ we get:

$$\int dz G^{(2)}(x-z)\Gamma^{(2)}(z-y) = i\delta_4(x-y)$$

Thus

$$G^{(2)}(p)\Gamma^{(2)}(p) = i$$

And since

$$G^{(2)}(p) = \frac{1}{p^2 - m^2 - \Sigma(p)}$$

then

$$\Gamma^{(2)}(p) = p^2 - m^2 - \Sigma(p)$$

Repeating the process of derivative and letting $J \to 0$ we generate $\Gamma^{(3)}$:

$$\Gamma^{(3)}(y,x,y') = \int dx'dzdz' G_c^{(3)}(x',z,z')\Gamma^{(2)}(x,z')\Gamma^{(2)}(y,z)\Gamma^{(2)}(z',y)$$

Semiclassical Analysis requires reintroducing h to properly track perturbation order

To proceed with the semiclassical analysis we need to reintroduce ђ:

$$Z[J] = \mathcal{N}^{-1}\int D\varphi e^{\frac{i}{ђ}\int[\mathcal{L}[\varphi]+J\varphi]} = e^{\frac{1}{ђ}W[J]}$$

From this we can see:

$$\Delta_F \sim ђ$$
$$V \sim ђ^{-1}$$

So, a diagram with I internal lines and V vertices has order of $Order(ђ) = I - V$. Recall that loops in diagrams satisfy $Order(ђ) = I - V + 1$. So, a semiclassical expansion in powers of ђ is equivalent to the expansion in the number of (loops-1) – first described by Nambu. So,

$$Order(ђ)\ in\ W[J] = Loop\ number = I - V + 1$$

A standard perturbation analysis splits the Lagrangian and in doing so can break some symmetry or gauge freedom, but when done diagrammatically by loop number (effecting an ħ expansion) these breaks do not occur, and the Ward Identities are preserved, for example.

Consider the Action $S[\varphi, J]$ with $\varphi = \varphi_0 + \hat{\varphi}$:

$$S[\varphi, J] = S[\varphi_0, J] + \int dx \left[\frac{1}{2}\left(\partial_\mu \hat{\varphi}\right)^2 - \frac{1}{2}m^2\hat{\varphi} - \frac{1}{2}V''(\varphi_0)\hat{\varphi} - \sum_{n\geq 3} \frac{\hat{\varphi}^n}{n!} V\,(\varphi_0) \right] = S + S'$$

So,

$$Z[J] = e^{\frac{i}{\hbar}S[\varphi_0, J]} \smallint \mathcal{D}\hat{\varphi} \exp\left(\frac{iS'}{\hbar}\right)$$

where $\varphi_0 = 0$ for $J = 0$. This leads to $\smallint \mathcal{D}\hat{\varphi} \exp\left(\frac{iS'}{\hbar}\right) = 1$ when $J = 0$ from $Z[0] = 1$ normalization. Let's write $\hat{\varphi} = \hbar^{1/2}\varphi$, thereby shifting the ħ dependence into the higher order potential expansion:

$$Z[J] = e^{\frac{i}{\hbar}S[\varphi_0, J]} \smallint \mathcal{D}\varphi \exp\left\{ i \int dx \left\{ \frac{1}{2}(\partial\varphi)^2 - \frac{1}{2}m^2\varphi^2 - \frac{1}{2}V''(\varphi_0)\varphi^2 - \sum_{n\geq 3} \frac{\hbar^{\frac{n}{2}-1}}{n!} \varphi^n V^{(n)}(\varphi_0) \right\}\right.$$

From here we can integrate the $\hbar^{(0)}$ order terms by explicit integration (the classic Gaussian Integral). Let

$$K_v(x, x') = [D_x^2 + m^2 + V''(\varphi_0)]\delta(x - x')$$
$$K_0(x, x') = [\partial_x^2 + m^2]\delta(x - x')$$

And the zeroth order integral is then

$$det[K_0^{-1}K_v]^{-1/2} = det[I - \Delta_F V''(\varphi_0)]$$

since

$$K_0\Delta_F = -I.$$

From the relation

$$det[A] = e^{Tr\ell nA}$$

We then have

$$W = iS[\varphi_0, J] - \frac{\hbar}{2} Tr\ell n\big(I - \Delta_F V''(\varphi_0)\big) + \mathcal{O}(\hbar^2)$$

Which is the semiclassical derivation to order ħ of the generating functional for the connected greens function. To proceed from here we need φ_c and them need the Legendre transformation to get Γ.

Let's now compute φ_c to zeroth order:

$$\varphi_c(x) = \frac{\delta W}{i\delta J(x)}|_{\hbar=0} = \frac{\delta S}{i\delta J(x)} = \int \frac{\delta S}{i\delta\varphi_o(x)}\frac{\delta\varphi_o(x)}{\delta J(x)} + \frac{\delta S}{\delta J(x)}$$

And since $S = \int[\mathcal{L} + J\,\varphi_o] \rightarrow \frac{\delta S}{\delta J(x)} = \varphi_o(x)$, we have:

$$\varphi_c = \varphi_0 + \mathcal{O}(\hbar)$$

and

$$S[\varphi_c] = S[\varphi_0] + \mathcal{O}(\hbar^2)$$

since φ_0 is an extremum.
Thus,

$$\Gamma[\varphi] = S[\varphi] + \frac{i}{2}\hbar Tr\ell n\big(I - \Delta_F V''(\varphi)\big) + \mathcal{O}(\hbar^2)$$

and Γ is classical action in classical limit ($\hbar \rightarrow 0$).

Thus, Γ is the effective action as well as the generating function for 1PI graphs. What is the $(1 - \Delta_F)$ term? Recall that it is considered as a matrix in x – space, and since:

$$ln(1 - x) = \sum_{n=l}\frac{1}{n}x^n$$

we have

$$Tr\ ln(I - \Delta_F V''(\varphi)) = \sum_n \frac{1}{n}\int \ldots$$

$$= \sum_n \frac{1}{n}\int dz_1 \ldots dz_n [i\Delta_F(z_1 - z_2)\left(-iV''(\varphi(z_2))\right) i\Delta_F(z_2 - z_3)\left(-iV''(\varphi(z_3))\right) \ldots i\Delta_F(z_n - z_1)(-iV''(\varphi(z_1)))]$$

Let's consider this in the specific case of phi^4 theory whose bare Lagrangian is:

$$\mathcal{L} = \frac{1}{2}\partial_\mu\varphi_0^2 - \frac{1}{2}m_0^2\varphi_0^2 - \frac{l}{4!}\lambda_0\varphi_0^4$$

In general we expect

$$\varphi_0 = Z_3^{1/2}\varphi_R$$

to relate the bare field to the real field for some renormalization factor Z_3. We also expect, due to self-interactions, that

$$m_0^2 = \frac{Z_m}{Z_3}m_R^2$$

and

$$\lambda_0 = \frac{Z_1}{Z_3^2}\mu^{2\epsilon}\lambda_R$$

where $\epsilon = \frac{1}{2}(4-d)$. In this process the Z'^s turn out to be infinite. This is managed by considering each Z to be an infinite power series in λ. To each order in λ we are able to pick a $Z_\alpha^{(n)}$ to get a finite quantity. Substituting the renormalization relations:

Now,

$$\mathcal{L} = \frac{1}{2}\partial_\mu\varphi\partial^\mu\varphi - \frac{1}{2}m^2\varphi^2 - \frac{\lambda}{4!}\mu^{2\epsilon}\varphi^4 + \frac{1}{2}(Z_3-1)\partial_\mu\varphi\partial^\mu\varphi - \frac{1}{2}(Z_m-1)m^2\varphi^2 - \frac{\lambda}{4!}\mu^{2\epsilon}(Z_1-1)\lambda\varphi^4,$$

where the fields and couplings are now ‘real’, but the ‘R’ subscripting is dropped. To each order in λ we pick the corresponding $Z^{(n)}$ to get a finite quantity, where:

$$Z_\alpha = 1 + \sum_{n=1}^{\infty}\lambda^n Z_\alpha^{(n)}$$

Let’s consider $\mathcal{O}(\lambda)$ contributions:

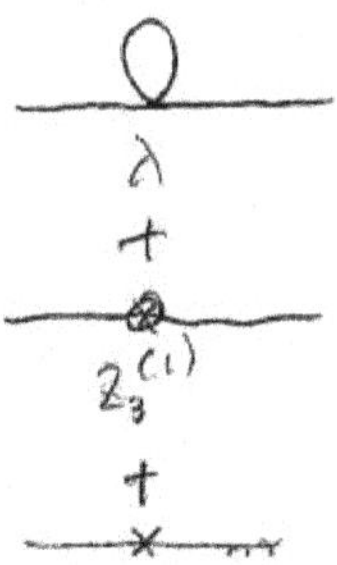

$$\frac{1}{4!}\lambda_R\mu^{2\epsilon}\varphi_R^4$$

$$\frac{1}{2}(Z_3-1)(\partial\varphi_R)^2$$

$$-\frac{1}{2}(Z_m-1)m_R^2\varphi_R^2$$

We get to order λ in the propagator:

$$\frac{i\lambda}{2}\frac{m^2}{(4\pi)^2}\left[\frac{1}{\varepsilon}+1-\gamma_E-\ell n\frac{m^2}{4\pi\mu^2}\right]+iZ_3^{(1)}p^2\lambda - iZ_m^{(1)}m^2\lambda$$

From which we immediately se that $Z_3^{(1)} = 0$ since there is no matching term in p^2. Using minimal subtraction for $Z_m^{(1)}$ we have:

$$Z_m^{(1)} = \frac{1}{2}\frac{1}{(4\pi)^2}\frac{1}{\varepsilon}$$

and we now have a finite self-energy.

Before moving to order λ^2 in the propagator (2-point function), let’s consider the 4-point function. For this we have the diagram studied previously, which had three permutations, from which we obtain $-iZ_1^{(1)}\lambda$ from the $\lambda\mu^{2\epsilon}(Z_1 - 1)\lambda\varphi^4$ term (again, R subscripting dropped). Thus:

$$Z_1 = 1 + \frac{3\lambda}{(4\pi)^2\varepsilon} + \mathcal{O}(\lambda^2)$$

And we have now we have continued all the counter-terms to order λ.

Let's now consider the propagator to $\mathcal{O}(\lambda^2)$:

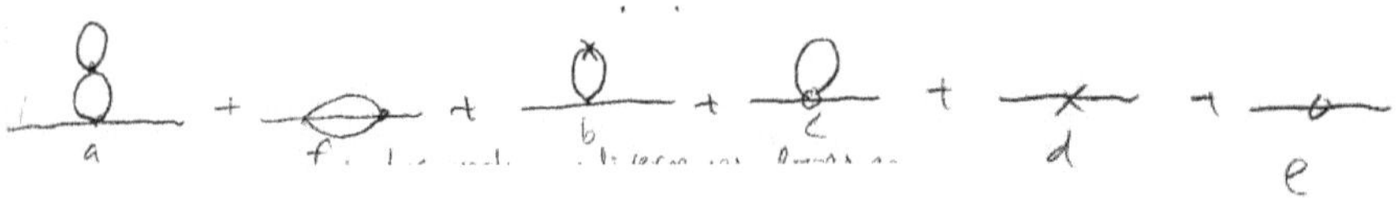

From previous analysis we already have (a)+(b) terms:

$$(a) + (b) = \frac{i\lambda^2 m^2}{4(4\pi)^4}\left\{\frac{1}{\varepsilon}\left(1 - \gamma_E - \ell n\frac{m^2}{4\pi\mu^2}\right) - \left(\gamma_E + \ell n\frac{m^2}{4\pi\mu^2}\right)\left(1 - \gamma_E - \ell n\frac{m^2}{4\pi\mu^2}\right)\right\}$$

For the (c) and (f) terms:

$$(c) = \frac{3i\lambda^2 m^2}{4(4\pi)^4}\frac{1}{\varepsilon}\left(\frac{1}{\varepsilon} + 1 - \gamma_E - \ell n\frac{m^2}{4\pi\mu^2}\right)$$

and

$$(f) = \frac{-i\lambda^2 m^2}{6(4\pi)^4}\left(\frac{3}{2\varepsilon^2} + \frac{3}{\varepsilon}\left[\frac{3}{2} + \Psi(1) - \ell n\frac{m^2}{4\pi\mu^2}\right] + \frac{1}{4\varepsilon}\frac{p^2}{m^2}\right) + finite\ part$$

Note that the renormalization constants can often be calculated in closed form, as done here, but that the physical 'finite part' often can't be computed in closed form. Also note that terms like $\frac{1}{\epsilon}\ell\eta p^2$, which could arise in overlapping loops, must cancel in a renormalizable theory since the Z's are polynomial in p^2 and can't cancel them otherwise.

For the (d) and (e) terms:

$$(d) + (e) = -i\lambda^2 m^2 Z_m^{(2)} + i\lambda^2 p^2 Z_3^{(2)}\mu^{2\epsilon}$$

we have:

$$Z_m^{(2)} = \frac{\lambda^2}{4(2\pi)^4}\left(\frac{2}{\epsilon^2} - \frac{1}{\epsilon}\right) \quad , \quad Z_3^{(2)} = \frac{\lambda^2}{24(2\pi)^4}\frac{1}{\epsilon}.$$

In principle we could continue to higher and higher order. In this process, the general feature of the Z's in the minimal subtraction method is that they only depend on λ. In other words, we generally have:

$$Z_\alpha = Z_\alpha\left(\lambda, \frac{m}{\mu}\right) \to Z_\alpha(\lambda)$$

with minimal subtraction.

Let's reconsider the truncated 2-point function (sum of all such diagrams that are 1PI) using the prior 'sigma' notation, but now showing the mass variable (scale) and the renormalization parameter mu: $-\Sigma(p^2, m, \mu)$ is the p-to-p truncated 1-PI diagram. As before, we construct the 2-pont Greens function from the free propagator and zero, one, and more $-\Sigma$ diagrammatic elements in what amount to a geometric series, whose summation is simply:

$$G^{(2)} = \frac{1}{p^2 - m^2 - \Sigma(p^2, m, \mu)}.$$

Recall that we want the propagator to have a pole at the physical mass of the particle:

$$m_{ph}^2 - m^2 - \Sigma(m_{ph}^2, m, \mu) = 0.$$

Solving this gives mass as a function of mu: $m = m(\mu)$. We can, similarly, examine the residue in the aforementioned pole:

$$\Sigma(p^2, m, \mu) = \Sigma(m_{ph}^2, m, \mu) + (p^2 - m_{ph}^2)\frac{\partial\Sigma}{\partial \mathrm{p}^2}\bigg|_{p^2=m_{ph}^2} + (p^2 - m_{ph}^2)^2(\ldots) + \cdots$$

So,

$$G^{(2)} \approx \frac{i\hat{z}}{p^2 - m^2} \qquad p^2 \to m_{ph}^2$$

where

$$\hat{z} = \left(1 - \frac{\partial\Sigma}{\partial \mathrm{p}^2}\bigg|_{p^2=m_{ph}^2}\right)^{-1}$$

which, again, depends on μ. Note that if we use a physical renormalization scheme, $\hat{z} = 1$, but here we use minimal subtraction, so:

$$\varphi_R \to \hat{z}^{1/2}\varphi_{out}$$

where the $\hat{z}^{1/2}$ is a *finite* renormalization of the field. Similarly, $G^{(2)}(p, \lambda, m, \mu)$ gives $\lambda(\mu)$.

The renormalization group for the phi^4 Lagrangian with standard renormalization constants starts with 'bare' Lagrangian:

$$\mathcal{L} = \frac{1}{2}\partial_\mu \varphi \partial^\mu \varphi - \frac{1}{2}m^2\varphi^2 - \frac{\lambda}{4!}\varphi^4$$

where the bare subscripting is dropped. The standard renormalization relations:

$$\varphi_0 = Z_3^{1/2}\varphi_R$$
$$m_0^2 = \frac{Z_m}{Z_3}m_R^2$$
$$\lambda_0 = \frac{Z_1}{Z_3^2}\mu^{2\epsilon}\lambda_R$$

where $d = 2\omega = 4 - 2\epsilon$. We now see that the different renormalization schemes considered correspond to different values for μ. Therefore, we can write:

$$\Gamma_R^{(n)}(p_1, \dots, p_n, \lambda, m, \mu) = Z_n \Gamma_R^{(n)}(p_1, \dots, p_n, \lambda', m', \mu')$$

Or, relating real to bare and adding the dimensionality as a parameter:

$$\Gamma_R^{(n)}(p_1, \dots, p_n, \lambda_R, m_R, \mu, \epsilon) = Z\Gamma_o^{(n)}(p_1, \dots, p_n, \lambda_o, m_o, \epsilon).$$

Let's solve for Z. Recall that

$$G_o^{(n)} = \langle 0|T\varphi_o(x_1) \dots \varphi_o(x_n)|0\rangle = Z_3^{n/2}G_R^{(n)}$$

and that

$$G_R^{(n)} = \left[G_R^{(2)}\right]^n \Gamma_R^{(n)}$$

so

$$\Gamma_R^{(n)} = Z_3^{n/2}\Gamma_o^{(n)} \ \rightarrow \ Z = Z_3^{n/2}.$$

Let's now consider the operation of $\mathcal{R} \equiv \mu\frac{d}{d\mu}$, where the bare terms have no μ dependance:

$$\mathcal{R}\Gamma_R^{(n)} \equiv \mu\frac{d}{d\mu}\Gamma_R^{(n)} = \frac{n}{2}\frac{d}{d\mu}(\ln Z_3)\Gamma_R^{(n)}.$$

Let's now rewrite the total derivative $d/d\mu$ in terms of partial derivatives involving the dependent variables on μ that have already been established:

$$\left[\mu\frac{\partial}{\partial\mu} + \mu\frac{d\lambda_R}{d\mu}\frac{\partial}{\partial\lambda_R} + \mu\frac{dm_R}{d\mu}\frac{\partial}{\partial m_R}\right]\Gamma_R^{(n)} = \frac{n}{2}\frac{d}{d\mu}(\ln Z_3)\Gamma_R^{(n)}$$

Now, introduce new variables:

$$\beta = \mu\frac{d\lambda_R}{d\mu}, \quad \gamma_m = \frac{\mu}{m_R}\frac{dm_R}{d\mu}, \quad \gamma = \frac{1}{2}\mu\frac{d}{d\mu}(\ln Z_3)$$

to get:

$$\left[\mu\frac{\partial}{\partial\mu} + \beta\frac{\partial}{\partial\lambda_R} + m_R\gamma_m\frac{\partial}{\partial m_R} - n\gamma\right]\Gamma_R^{(n)}(p_1, \dots, p_n, \lambda_R, m_R, \mu, \epsilon) = 0$$

Where $\beta = \beta(\lambda_R, m_R/\mu)$, and same similar dependence in γ_m and γ. If we use a mass-independent renormalization, such as with minimal subtraction, we have $\beta = \beta(\lambda_R)$, etc.

Let's now consider combinations of real parameters $\{\lambda_R, m_R, \mu\}$which are invariant under the $\mathcal{R} \equiv \mu \frac{d}{d\mu}$ operation. In other words, want functions $\bar{\lambda}(\lambda, \mu)$ and $\bar{m}(\lambda, m, \mu)$ that satisfy:

$$\mathcal{R}\bar{\lambda}(\lambda, \mu) = 0$$

and

$$\mathcal{R}\bar{m}(\lambda, m, \mu) = 0$$

because then we have:

$$\mathcal{R}G(p, \bar{\lambda}, \bar{m}) = 0.$$

To solve for the new functions it is convenient to change variables $\mu = \mu_o e^{-t}$ since then:

$$\mu \frac{d}{d\mu} = -\frac{\partial}{\partial t}$$

Also, let $\bar{\lambda}(\lambda, t = 0) = \lambda$. We then have:

$$\mathcal{R}\bar{\lambda}(\lambda, \mu) = 0 \rightarrow \left[-\frac{\partial}{\partial t} + \beta(\lambda)\frac{\partial}{\partial \lambda}\right]\bar{\lambda}(\lambda, \mu) = 0$$

To solve the partial differential equation indicated, lets consider the following integral form of solution and verify that it solves the pde:

$$\int_{\lambda}^{\bar{\lambda}(\lambda,\mu)} \frac{dx}{\beta(x)} = t.$$

If we take $\frac{\partial}{\partial t}$ on the above integral expression we get back $\frac{\partial \bar{\lambda}}{\partial t} = \beta(\bar{\lambda})$. If we take $\frac{\partial}{\partial \lambda}$ we get

$$-\frac{\partial \bar{\lambda}}{\partial t} + \beta(\lambda)\frac{\partial \bar{\lambda}}{\partial \lambda} = 0$$

Thus,

$$\frac{\partial \bar{\lambda}}{\partial t} = \beta(\bar{\lambda})$$

and $\bar{\lambda}(t = 0) = \lambda$.

Similarly,

$$\left[-\frac{\partial}{\partial t} + \beta(\lambda)\frac{\partial}{\partial \lambda} + m\gamma_m \frac{\partial}{\partial m}\right]\bar{m}(\lambda, m, \mu) = 0$$

where $\bar{m}(t=0)=m$, has solution:

$$\bar{m} = m exp\left[\int_0^t \gamma_m(\bar{\lambda}(t'))dt'\right]$$

with

$$\frac{\partial \bar{m}(\lambda, m, t)}{\partial t} = \bar{m}\,\gamma_m\left(\bar{\lambda}(t)\right).$$

We can now write:

$$\Gamma_R^{(n)}(p_1, \dots, p_n, \lambda, m, \mu)$$
$$= \Gamma_R^{(n)}(p_1, \dots, p_n, \bar{\lambda}(t), \bar{m}(t), \mu_o)\, exp\left(-n\int_0^t \gamma(\bar{\lambda}(t'))dt'\right)$$

Let's now consider the 2-point Greens function:

$$G_{\square}^{(2)}(p) = \frac{i\hat{z}}{p^2 - m_{ph}^2} + G'$$

where G' doesn't have a pole term. Using our relation for the n=2 case, we also have:

$$(\mathcal{R} + 2\gamma)G_{\square}^{(2)}(p) = 0.$$

Consistency requires that $\hat{z}$ and $m_{ph}^{\square}$ be renormalization group invariants:

$$(\mathcal{R} + 2\gamma)\hat{z} = 0, \quad \mathcal{R}m_{ph}^{\square} = 0$$

Can show that the S-matrix is $\mathcal{R}$ invariant too:

$$S = \lim_{p^2 \to m_{ph}^2}\left(i\hat{z}/G_{\square}^{(2)}\right)^n G_{\square}^{(n)} \rightarrow \mathcal{R}S = 0$$

Example 1

1. Derive the connection between the renormalization group functions and he poles of the renormalization group constants for the phi^3 theory in six space-time dimensions.
2. Using the renormalization constants for phi^3 in six dimensions:
(a) Determine the Renormalization group functions.
(b) Compute the running coupling constants and the running mass.
(c) Express the one-particle irreducible n-point function with momenta $s \cdot p_\mu$ in terms of the same function with momenta p_μ. What do you conclude about the high-momentum behavior of the theory?

Relations between ren. Group function and poles of ren. Constants

$$\mathcal{L} = \frac{1}{2}(\partial\varphi_o)^2 - \frac{1}{2}m_o^2\varphi_o^2 - \frac{i}{3!}\lambda_o\varphi_o^3 \quad \epsilon = d-3$$

$$\varphi_o = Z_3^{1/2}\varphi R, m_o^2 = \frac{Z_m}{Z_3}m_R^2, \lambda_o = \frac{Z_i}{Z_3^{1/2}}\mu^\epsilon\lambda_R$$

$$\lambda_o = Z_\lambda\lambda\mu^\epsilon \quad Z_\lambda = Z_i/Z_3^{1/2}$$

$$O = \epsilon\mu^\epsilon(\lambda Z\lambda) + \mu^\epsilon\beta(\lambda,\epsilon)\frac{d(\lambda_Z\lambda)}{d\lambda}$$

$$Z_\lambda = 1 + \sum_{k=1}\frac{a_k}{\epsilon^k} \Rightarrow \beta(\lambda,\epsilon) = \frac{-2\epsilon Z_\lambda\lambda}{\frac{d}{d\lambda}(\lambda Z_\lambda)} = \frac{-2\lambda(1+\Sigma)}{1+\sum_{k=1}\frac{a_k+\lambda a_k'}{\epsilon^k}}$$

$$= -2\epsilon\lambda(1+\Sigma)\left(1-\sum_{k=1}\frac{a_k+\lambda a_k'}{\epsilon^k}\right)$$

We get consistency relations:

$$O = \epsilon_\mu^\epsilon\left(\lambda+\sum_{k=1}\frac{\lambda a_k'}{\epsilon^k}\right) + \mu^6(-2\lambda\epsilon + 2\lambda^2 a_,')\left(1+\sum_{k=1}\frac{a_k+\lambda a_k'}{\epsilon^k}\right)$$

$$\Rightarrow \underline{a_{k+1}' = (a_k + \lambda a_k')a_,'}$$

$m_o^2 = \frac{Z_m}{Z_3}m^2$ so, $\beta(\lambda,\epsilon)\frac{d}{d\lambda}\frac{Z_m}{Z_3}m^2 + 2m^2\frac{Z_m}{Z_3}\gamma_m = o$

If $\frac{Z_m}{Z_3} = \sum_{k=1}\frac{bk}{\epsilon^k}$ $\quad then\ \gamma_m = \frac{1}{2}\lambda b_,', \underline{b_{k+1}' = b_,'b_k' + \lambda a_,'b_k'}$

And , $\boldsymbol{\gamma} = \underline{\frac{1}{2}\mu\frac{d}{d\mu}\log Z_3}, if\ \underline{Z_3 = 1+\sum_{k=1}\frac{k}{(\epsilon)^k}} \Rightarrow \underline{\gamma = -\frac{1}{2}\lambda C_,'}$

$$Z_1 = 1 - \frac{\lambda^2}{2(4\pi)^3}\frac{1}{\epsilon} + \mathcal{O}(\lambda^3)$$

$$Z_m = 1 - \frac{\lambda^2}{2(4\pi)^3}\frac{1}{\epsilon} + \mathcal{O}(\lambda^3)$$

$$Z_3 = 1 - \frac{\lambda^2}{2(4\pi)^3}\frac{1}{\epsilon} + \mathcal{O}(\lambda^3)$$

So, $Z_\lambda = \frac{\lambda^2}{2(4\pi)^{3/2}} = 1 - \frac{\lambda^2}{2(4\pi)^3}\frac{1}{\epsilon} + \mathcal{O}(\lambda^3) \Rightarrow a_; = \frac{-3\lambda^2}{8(4\pi)^3\epsilon} + \mathcal{O}(\lambda^3) \Rightarrow a_; = \frac{-3\lambda^2}{8(4\pi)^3}$

$$\frac{Z_3}{Z_3} = 1 - \frac{\lambda^2}{2(4\pi)^3}\frac{\cdot}{\epsilon} + \mathcal{O}(\lambda^3) \Rightarrow b_1 = \frac{-5\lambda^2}{12(4\pi)^3}$$

$$\frac{Z_3}{Z_3} = 1 - \frac{\lambda^2}{2(4\pi)^3}\frac{\cdot}{\epsilon} + \mathcal{O}(\lambda^3) \Rightarrow C_1 = \frac{-3\lambda^2}{12(4\pi)^3}$$

$$\beta(\lambda,\epsilon) = \lambda\epsilon + \lambda^2 a_{\prime}' = \lambda^2\left(\frac{-3\lambda^2}{8(4\pi)^3}\right) = \frac{-3}{4}\frac{\lambda^2}{(4\pi)^3}$$

$$\gamma_m = \frac{1}{2}\lambda b_{\prime}' = -\frac{-5\lambda^2}{12(4\pi)^3}$$

$$\gamma = -\lambda C_{\prime}' = \frac{\lambda^2}{12(4\pi)^3}$$

$$\int_{\lambda}^{\bar{\lambda}(\dagger)} \frac{dx}{(3(x)} = \dagger\ \frac{\partial\bar{\lambda}}{\partial\ \dagger} = \beta(\lambda(\dagger), \bar{\lambda}(\dagger = o)\lambda$$

$$\int_{\lambda}^{\bar{\lambda}(\dagger)} \frac{dx}{\frac{-3}{4}\frac{x^3}{(4\pi)^3}} = \frac{1}{\left(-\frac{3}{4\frac{1}{(4\pi)^3}}\right)}\left[\frac{1}{2}\left(\frac{1}{x^2}\right)\right]^{\bar{\lambda}(\dagger)} = \frac{1}{\left(-\frac{3}{4\frac{1}{(4\pi)^3}}\right)}\left(\frac{V}{\bar{\lambda}(\dagger)^2} -\right)$$

$$\frac{3}{2}\frac{1}{(4\pi)^3}\dagger + \frac{1}{\lambda^2} = \bar{\lambda}(\dagger)^2 \Rightarrow \boxed{\bar{\lambda}(\dagger)^2 = \frac{\lambda^2}{1 + \frac{3}{2}\frac{1}{(4\pi)^3}\dagger}}$$

$$\overline{m}(S) = m exp\int_{o}^{\dagger} \gamma_m\left(\bar{\lambda}(x)\right)dx \ \ from\ \frac{\partial\overline{m}}{\partial\ \dagger} = \overline{m}\gamma_m\left(\bar{\lambda}(x)\right), \overline{m}(\dagger = o)$$
$$= u$$

$$\left\{m exp\int_{o}^{\dagger} \gamma_m\left(\bar{\lambda}(x)\right)dx\right\} = \left\{m exp\int_{o}^{\dagger} \gamma_m\left(\bar{\lambda}(x)\right)dx\right\}$$

$$\left\{m exp\int_{o}^{\dagger} \gamma_m\left(\bar{\lambda}(x)\right)dx\right\} = \left\{m exp\int_{o}^{\dagger} \gamma_m\left(\bar{\lambda}(x)\right)dx\right\} = m\left(\frac{\bar{\lambda}(\dagger)}{\lambda}\right)^{5/9}$$

$$\Gamma^{(n)}(p_1, , , , p_n, \lambda, m, \mu)$$
$$= exp\left[-n\int_{o}^{x} \gamma\left(\bar{\lambda}(\dagger\,')\right)d\ \dagger\,'\right]\Gamma^{(n)}(p_1, , , , p_n, \bar{\lambda}, (\dagger)\overline{m})$$

$$\Gamma^{(n)}(sp, \lambda, sm, s\mu) = S^{dn}\Gamma^{(n)}(p, \lambda, m, \mu)$$
$$d_n = 6 - n$$
$$\Gamma^{(n)}(sp, \lambda, sm, s\mu)$$
$$= exp\left[-n\int_{o}^{\dagger} \gamma(\bar{\lambda}(\dagger\,')d\ \dagger\,'\right]\Gamma^{(n)}(sp, \bar{\lambda}(\ell ns), \overline{m}(\ell ns), \mu$$

$$\int_{o}^{3} \gamma(\bar{\lambda}(x))dx = \int_{o}^{3} \frac{1}{12(4\pi)^3}\left(\frac{1\lambda^2}{1 + \frac{3}{2}\frac{\lambda^2}{(4\pi)^3}}\right)dx = \frac{1}{9}\ell n\, 1 + \frac{3}{2}\frac{\lambda^2}{(4\pi)^3}s$$

$$= \frac{1}{18} log \frac{\lambda^2}{\bar{\lambda}(s)^2} = \frac{-1}{9} log \left(\frac{\bar{\lambda}(s)}{\lambda}\right)$$

$$\Gamma^{(n)}(sp, \lambda, sm, s\mu) = S^{dn} \left(\frac{\bar{\lambda}(s)}{\lambda}\right)^{n/9} \Gamma^{(n)}\left(p, \bar{\lambda}(S), \frac{\overline{m}(S)}{S}, \mu\right)$$

$$G^{(n)}(x_{,}, , , x_n) = \langle o|T\varphi(x_1), , , \varphi(x_n)|o\rangle [G^{(n)}(x_{i}, , , x_n)] = [m]^n$$

$$G^n(\rho)\delta_4\left(\sum \rho\right) = \int dx_{i}, , , dx_n G^{(n)}(x) \quad [G^{(n)}(\rho)] = [m]^{n-6_n+6}$$

$$G^n(\rho) = \frac{G^n(\rho)}{[G^{(2)}(12)]} = [m]^{(n-6_n+6)-n(-4)}$$

$$= [m]^{6-n}$$

At high momenta S→ ∞, $\bar{\lambda}(S) \rightarrow O$, the theory is asymptotically free; furthermore, the masses become negligible at high momenta.

Part 2

1. $\lambda_E = Z_\lambda \mu^2 \lambda$. Differentiate with respect to μ.

   $\Psi(\lambda Z_\lambda) + \mu^\epsilon \beta(\lambda, \epsilon) = 0$

If $Z_\lambda = 1 + \Sigma_{k=1} \frac{a_k}{\epsilon}$, we find by equating powers of c.

$\beta(\lambda, c) = -\lambda_c + \lambda^2 a_1'$

$a_{k+1}^1 = (a_k + \lambda a_k')a_k^1$

$m_B^2 = \frac{Z_m}{Z_3} = 1 + \Sigma_{k+1} \frac{b_\beta}{c^k}$, we find

$\gamma_m = \frac{1}{2}\lambda u_1$

$b_{1+1} = b_1' b_x + \lambda d_1 b_k'$

Finally, $\gamma = \frac{1}{2}\mu \frac{d}{d_\mu} \log Z_s$ *if* $Z_s = 1 + \Sigma_{k=1} \frac{a}{\square}$ *we find*

$\tau = -\frac{1}{2}\lambda c_1'$

$c_{x+1}' = c_1' c_k + \lambda c_1' c_k'$

2. For the $\lambda\phi^3$ theory

   $$Z_\lambda = \frac{Z_1}{Z_3^{1/2}} = 1 - \frac{3\lambda^2}{8(4\pi)^3\epsilon} + O(\lambda^3)$$

   $$\frac{Z_m}{Z_3} = 1 - \frac{5\lambda^2}{12(4\pi)^3\epsilon} + O(\lambda^3)$$

   $$Z_3 = 1 - \frac{\lambda^2}{12(4\pi)^3\epsilon} + O(\lambda^3)$$

   a. Using the above, we find:

   $$\beta(\lambda) - \frac{3\lambda^2}{8(4\pi)^3\epsilon} + O(\lambda^5)$$

   $$\gamma_m = -\frac{5\lambda^2}{12(4\pi)^3\epsilon} + O(\lambda^3)$$

$$\gamma_m = -\frac{\lambda^2}{12(4\pi)^3\epsilon} + O(\lambda^3)$$

b. $\int_{\lambda}^{x(t)} \frac{dx}{\beta(x)} = t \Rightarrow \lambda(s)^2 = \frac{\lambda^3}{1+\frac{3\lambda^3}{2(4x)^3}\log s}$

$$\overline{m}(s) = mexp \int \gamma_m(\lambda(x))\, dx = m\left(\frac{\overline{\lambda(s)}}{\lambda}\right)^{s/g}$$

Finally, $\int_0^0 \gamma\,(\lambda(x))dx = -\frac{1}{9}\,l0g\frac{\overline{\lambda}(s)}{\lambda}$

c. $\Gamma^{(n)}(sp, \lambda, m, \mu) = S^{\sigma n}\left(\frac{\lambda(s)}{\lambda}\right)^{n/g}\Gamma^{(n)}\left(p, \lambda(s), \frac{\overline{m}(s)}{s}, \mu\right)$
Where $d_n = 6 - 2n$. The theory is asymptotically free, and the masses become negligible at high momenta.

4.5 Grassman Variables

Recall

$$G^{(2)}(x_1, x_2) = \langle 0|T\varphi(x_1)\varphi(x_2)|0\rangle = \int \mathcal{D}\varphi\, \varphi(x_1)\varphi(x_2)\, e^{iS}$$

Thus,

$$G^{(2)}(x_1, x_2) = G^{(2)}(x_2, x_1),$$

e.g., there is symmetry on exchange, a characteristic of Bose Theory. The method for generalizing the theory to allow for Fermi Theory is to consider the time-ordered product from Bose theory:

$$T\varphi(x)\varphi(x') = \theta(x^0 - x^{0\prime})\varphi(x)\varphi(x') + \theta(x^{0\prime} - x^0)\varphi(x')\varphi(x)$$

If for Fermi theory we define the T product with anticommutation instead, then a consistent theory results, e.g., for fermi fields ψ we have:

$$T\psi(x)\psi(x') = \theta(x^0 - x^{0\prime})\psi(x)\psi(x') - \theta(x^{0\prime} - x^0)\psi(x')\psi(x)$$

For the proposed Fermionic T product we get:

$$G^{(2)}(x', x) = -G^{(2)}(x, x').$$

Canonical quantization then follows in the familiar way (see Section X). For generalization of the Path Integral formalism, however, significantly more mathematical structure is needed – the Grassman Algebra. Consider that we want to have

$$[\psi(x), \psi(x')]_+ = 0$$

where the '+' symbol indicates an anticommutator. This is known as a Grassman algebra. Note that $[\theta, \theta]_+ = 0$ means that $\theta^2 = 0$. So functions of a Grassman variable θ, are very simple:

$$f(\theta) = a + \theta b$$

with no higher terms.

Let's consider derivative operations with Grassman variables and adopt the standard definition consistent with $\frac{d}{d\theta}\theta = 1$ to get:

$$\frac{df(\theta)}{d\theta} = \frac{d}{d\theta}a + \frac{d}{d\theta}\theta b = b.$$

Let's consider the derivative for the function f' that swaps the grassman variable and grassman constant $\theta b \to b\theta$:

$$\frac{df(\theta)}{d\theta} = \frac{d}{d\theta}a + \frac{d}{d\theta}b\theta = \frac{d}{d\theta}a + \left[\frac{d}{d\theta}, b\right]_{+} - b\frac{d}{d\theta}\theta = \left[\frac{d}{d\theta}, b\right]_{+} - b$$

Consider that Leibnitz distributional property exists:

$$\theta^2 = 0 \to \frac{d}{d\theta}\theta^2 = 0 \;\to\; \left(\frac{d}{d\theta}\theta\right)\theta + \theta\left(\frac{d}{d\theta}\theta\right) = 0 \to \left[\frac{d}{d\theta}, \theta\right]_{+} = 0$$

Since we can consider the neighborhood of $\theta = b$, we must have:

$$\left\{\left[\frac{d}{d\theta}, \theta\right]_{+}\right\}_{\theta = b} = 0 \to \left[\frac{d}{d\theta}, b\right]_{+} = 0$$

So, if

$$f(\theta) = a + b\theta$$

then

$$\frac{df(\theta)}{d\theta} = -b$$

Note that

$$\left[\frac{d}{d\theta}, \frac{d}{d\theta}\right]_{+} = 0 \to \frac{d^2}{d\theta^2} = 0$$

Differentiation is a little odd with Grassman variables, and integration is even more so. Part of the problem is that integration is not the inversion of differentiation in some sense (due to the idempotent property $d^2 = 0$). Consider that we want transitive invariance of the measure, e.g.,

$$\int d\theta f(\theta + c) = \int d\theta f(\theta)$$

for a shift by the Grassman constant 'c'. For this to hold for the classes of functions possible we must have:

$$\int d\theta = 0.$$

Returning to our basic function form:

$$f(\theta) = a + \theta b$$

consider

$$\int d\theta\, f(\theta) = \int d\theta\, a + \int d\theta\, \theta b = \left(\int d\theta\, \theta\right) b.$$

We can adopt a 'normalization' condition on the Grassman variables such that:

$$\left(\int d\theta\, \theta\right) = 1,$$

and with that convention we then have:

$$\int d\theta\, f(\theta) = b.$$

Let's now consider the integral form with multiple Grassman variables:

$$\int d\theta_1\, d\theta_2\ \theta_1\theta_2 = -\int d\theta_1\, d\theta_2\ \theta_2\theta_1 = -1 \quad \rightarrow \quad \int d\theta_1\, d\theta_2\ \theta_2\theta_1 = 1$$

We want to evaluate the general form:

$$\int d\theta_1 \dots\, d\theta_N e^{-\frac{1}{2}\theta^T\cdot A\cdot\theta}$$

where

$$\theta^T \cdot A \cdot \theta = \sum_{i,j} A_{ij}\, \theta_i\theta_j,$$

and since the θ, the resulting A is antisymmetric in all generality. Notice, also, that 'N' must be even to have a nonzero contribution since the exponent will only give even powers of θ in its series expansion (definitional) form. To evaluate, first consider the N=2 case:

$$\int d\theta_1\, d\theta_2 e^{-\frac{1}{2}(\theta_1 A_{12}\theta_2 + \theta_2 A_{21}\theta_1)} = \int d\theta_1\, d\theta_2 e^{(-A_{12}\theta_1\theta_2)}$$

$$= \int d\theta_1\, d\theta_2(-A_{12}\theta_1\theta_2) = A_{12} = (detA)^{1/2}$$

Notice how the exponent on the determinant is the opposite of what is obtained for the Gaussian Integral when using normal variables, where

$$\int dx_1\, dx_2 e^{-\frac{1}{2}x^T\cdot A\cdot x} = C\cdot (detA)^{-1/2}$$

Let's now apply the formalism to Dirac fields ψ and shift to a stationary point of the action to make the integration easier:

$$\psi(x) = \psi'(x) - \int d^4y\, S_F(x-y)\eta(y), \quad (i\gamma\cdot\partial - m)S_F(x) = \delta_4(x)$$

and

$$\bar{\psi}(x) = \bar{\psi}'(x) - \int d^4y\, \bar{\eta}(y)\, {S_F}^{\dagger}(x-y)$$

We have the generator function

$$W_o[\eta, \bar{\eta}] = exp\left[-i\int dxdy\, \bar{\eta}(x) S_F(x-y)\eta(y)\right]$$

From which

$$\langle 0|T\psi(x)\bar{\psi}(y)|0\rangle = \left(\frac{\delta}{i\delta\bar{\eta}(x)} W_o \frac{\overleftarrow{\delta}}{i\delta\bar{\eta}(x)}\right)_{\eta=0=\bar{\eta}} = iS_F(x-y)$$

Exactly as with the canonical method. Using the $+i\epsilon$ prescription we can then write:

$$S_F(x) = \int \frac{d^4p}{(2\pi)^4} \frac{e^{-ip\cdot x}}{(\gamma\cdot p - m + i\epsilon)} = \int \frac{d^4p}{(2\pi)^4} \frac{(\gamma\cdot p + m)}{(p^2 - m^2 + i\epsilon)} e^{-ip\cdot x}$$

4.6 Electromagnetism

For electromagnetic field we have the Lagrangian:

$$\mathcal{L} = -\frac{1}{4} F_{\mu\nu}F^{\mu\nu}, \;\; F_{\mu\nu} = \partial_\mu A_\nu - \partial_\nu A_\mu$$

with gauge invariance under $A_\mu \to A_\mu + \partial_\mu f$, etc. If we immediately proceed with the Path Integral formulation using the Feynman process we get:

$$W_o[J_\mu] = N^{-1} \int \mathcal{D}A_\mu(x)\, exp\left\{i \int d^4x \left[-\frac{1}{4} F_{\mu\nu}F^{\mu\nu} + J^\mu A_\mu\right]\right\}.$$

This is not be rigorously defined, however, since it is not firmly grounded in the phase space implementation. For electromagnetism we can make things work anyway since non-abelian, but eventually a proper theoretical basis is needed and this was provided by Fadeev and Popov, for which the correct path integral formulation is:

$$W_o[J_\mu] = N^{-1} \int \mathcal{D}A_\mu(x)\, exp\left\{i \int d^4x\, d^4x' [A^\mu(x) K_{\mu\nu}(x,x') A^\nu(x') + J^\mu(x)\delta(x-x')A_\mu(x')]\right\}$$

and with use of the Gaussian integral transformation:

$$W_o[J_\mu] = N^{-1}(\det K_{\mu\nu})^{-1} exp\left\{-\frac{1}{2}\int d^4x\, d^4x' \left[J^\mu(x)[K_{\mu\nu}(x,x')]^{-1} J^\nu(x')\right]\right\}$$

The problem is that the gauge freedom permits an infinite number of zero eigenvalues which makes the determinant of $K_{\mu\nu}$ zero, which means $(\det K_{\mu\nu})^{-1}$ and $[K_{\mu\nu}(x,x')]^{-1}$ in the above process are not well-defined. Since not a proper Hamiltonian phase space formulation, the measure $\mathcal{D}A_\mu(x)$ is also not well-defined at the outset either. The solution to both of these problems, by Fadeev and Popov [49], was also shown to be equivalent to a 'proper' Hamiltonian phase space formulation. The direct

Lagrangian analysis can thus be retained in what follows with corrections as indicated by the Fadeev method. The 'fix' by Fadeev actually serves to recognize more clearly the impact of gauge freedom and gauge group volume on the measure and thus explicitly on the form of the path integral used in the variation, and will also generalize to solving similar issues when dealing with non-abelian gauge fields.

Let's start by finding $K_{\mu\nu}$ and then examining gauge choices and their 'delta-selection' with care:

$$\mathcal{L} = -\frac{1}{4}F_{\mu\nu}F^{\mu\nu}, \; F_{\mu\nu} = \partial_\mu A_\nu - \partial_\nu A_\mu \quad \rightarrow \quad \mathcal{L} = -\frac{1}{2}\partial_\mu A_\nu F^{\mu\nu}$$

So,

$$\int d^4x\, \mathcal{L} = \frac{1}{2}\int d^4x\, \left[A_\mu \Box A^\mu - A_\mu \partial^\mu \partial^\nu A_\nu\right]$$

$$= \frac{1}{2}\int d^4x\, A_\mu(x)[\Box \eta^{\mu\nu} - \partial^\mu \partial^\nu]A_\nu(x)$$

Thus,

$$K_{\mu\nu}(x, x') = [\Box \eta^{\mu\nu} - \partial^\mu \partial^\nu]\delta(x - x').$$

Notice that:

$$[\Box \eta^{\mu\nu} - \partial^\mu \partial^\nu]\partial_\nu f(x) = 0$$

and requiring gauge invariance, $A_\nu \rightarrow A_\nu + \partial_\nu f(x)$ then means that we must have:

$$[\Box \eta^{\mu\nu} - \partial^\mu \partial^\nu]A_\nu = 0$$

Thus, an infinite number of zero eigenvalues are indicated.

Also consider that we know our naïve choice of measure makes no accounting for the gauge invariance. What is needed is a path integral with added measure term to explicitly enforce gauge invariance:

$$\int \mathcal{D}A_\mu(x)\, \mathcal{M}(A_\mu(x)) exp\left\{i \int d^4x\, \mathcal{L}\right\}$$

Where the measure factor $\mathcal{M}(A_\mu(x))$ is introduced to make the integrand $\mathcal{M}(A_\mu(x)) exp\{i \int d^4x\, \mathcal{L}\}$ gauge invariant. This is analogous to:

$$\int d^3r\, f(r) = 4\pi \int dr\, r^2 f(r),$$

where 4π is the group volume of the (compact) rotation group. So, in essence, what Fadeev and Popov did was seek to isolate the gauge group volume. To do this we begin by imposing a gauge condition according to $F(A) = 0$ and introducing the identity functional that selects on this gauge choice as:

$$1 = \Delta_F[A_\mu] \int \mathcal{D}f \, \delta(F(A_\mu^f)), \quad A_\mu^f = A_\mu + \partial_\mu f.$$

Let's consider a different gauge choice, A_μ^g, as starting point to test:

$$1 = \Delta_F[A_\mu^g] \int \mathcal{D}f \, \delta(F(A_\mu^{g+f})).$$

Using translation invariance on the functional measure we can rewrite:

$$1 = \Delta_F[A_\mu^g] \int \mathcal{D}f \, \delta(F(A_\mu^f)).$$

Thus,

$$\Delta_F[A_\mu^g] = \Delta_F[A_\mu],$$

And "$\Delta_F[A]$" is, indeed, gauge invariant.

Let's now put this identity into our (formative) path integral:

$$\int \mathcal{D}A_\mu(x) \left[\Delta_F[A] \int \mathcal{D}f \, \delta(F(A_{\square}^f))\right] \mathcal{M}(A_\mu(x)) exp\left\{i \int d^4x \, \mathcal{L}\right\}$$

Or, using translation invariance again:

$$\int \mathcal{D}f \int \mathcal{D}A_\mu(x) \, \Delta_F[A] \mathcal{M}(A) \delta(F(A)) exp\left\{i \int d^4x \, \mathcal{L}\right\}$$

Note, however, that this last form has cleanly separated $\int \mathcal{D}f$ from an integrand with no f dependence. Thus, the $\int \mathcal{D}f$ contributes a group volume factor that is absorbed into the normalization factor N^{-1} in W given previously.

For $F(A) = \partial_\mu A^\mu + \square f = 0$ gauge condition we have:

$$1 = \Delta_F[A] \int \mathcal{D}f \, \delta(\partial_\mu A^\mu + \square f).$$

Using the matrix generalization to the relation $\int dx \, \delta(a + bx) = 1/b$ we then have:

$$1 = \Delta_F[A](\det \square)^{-1} \rightarrow \Delta_F[A] = \det \square.$$

Note, however, that $\Delta_F[A] = \det \square$ means that $\Delta_F[A]$ is independent of A, thus can be moved out of the integral and absorbed into the normalization factor. Suppose we choose $F(A) = \nabla \cdot A = 0$, then get $\Delta_F[A] = \det \nabla$, which is still independent of A.

The insertion of $\Delta_F[A]$ is shown by Fadeev, for both abelian and non-abelian cases, to be equivalent to starting with the actual phase space formalism, thus it is the modification to the naïve application of the Feynman prescription that makes it complete.

Let's now turn to Lorentz gauge, $\partial_\mu A^\mu = 0$, and consider the functional integral:

$$\int \mathcal{D}A_\mu(x)\,\mathcal{M}(A)\delta(\partial_\mu A^\mu)exp\left\{i\int d^4x\,\mathcal{L}\right\}$$

Let's represent the delta function by the following limit:

$$\delta(x) = \lim_{\alpha\to 0} e^{ix^2/2\alpha}$$

to get:

$$\lim_{\alpha\to 0}\int \mathcal{D}A_\mu(x)\,\mathcal{M}(A)e^{i\int d^4x\,[\mathcal{L}+(\partial_\mu A^\mu)^2/2\alpha]}$$

For the more general gauge $\partial_\mu A^\mu = \sigma(x)$ this becomes:

$$\int \mathcal{D}A_\mu(x)\,\mathcal{M}(A)\delta(\partial_\mu A^\mu - \sigma(x))exp\left\{i\int d^4x\,\mathcal{L}\right\}$$

and if the integral is independent of $\sigma(x)$, we can multiply by the constant factor:

$$\int \mathcal{D}\sigma(x)\,e^{\frac{i}{2\lambda}\int \sigma^2 dx}$$

To get

$$\int \mathcal{D}\sigma(x)\,e^{\frac{i}{2\lambda}\int \sigma^2 dx}\int \mathcal{D}A_\mu(x)\,\mathcal{M}(A)\delta(\partial_\mu A^\mu - \sigma(x))e^{iS}$$
$$= \int \mathcal{D}A_\mu(x)\,\mathcal{M}(A)e^{i\int d^4x\,[\mathcal{L}+(\partial_\mu A^\mu)^2/2\lambda]}$$

So, either way, we have a form built for handling gauge choice with the modified action $\int d^4x\,[\mathcal{L}+(\partial_\mu A^\mu)^2/2\lambda]$, so let's now develop this in the standard way:

$$W[J_\mu] = N^{-1}\int \mathcal{D}A_\mu(x)\,exp\left\{i\int d^4x\left[\begin{matrix}-\frac{1}{4}F_{\mu\nu}F^{\mu\nu}+\frac{(\partial_\mu A^\mu)^2}{2\lambda}\\ +J^\mu A_\mu\end{matrix}\right]\right\}.$$

Or,

$$W[J_\mu]$$
$$= N^{-1}\int \mathcal{D}A_\mu(x)\,exp\left\{i\int d^4x\left[\begin{matrix}\frac{1}{2}A^\mu[\Box\eta^{\mu\nu}-(1+\lambda^{-1})\partial^\mu\partial^\nu]A^\nu\\ +J^\mu A_\mu\end{matrix}\right]\right\}$$

Thus, we can write the (photon) propagator as:

$$[\Box\eta^{\mu\nu}-(1+\lambda^{-1})\partial^\mu\partial^\nu]D_{F\,\mu\rho}(x) = \delta^4(x)\delta^\mu{}_\rho$$

Using Fourier Transforms:

$$D_{F\,\mu\rho}(x) = \int \frac{d^4k}{(2\pi)^4}e^{-ik\cdot x}\widetilde{D}_{F\,\mu\rho}(k)$$

We get:

$$[k^2\eta^{\mu\nu} - (1+\lambda^{-1})k^\mu k^\nu]\widetilde{D}_{F\,\mu\rho}(k) = -\delta^\mu{}_\rho$$

This is then solved to get:

$$\widetilde{D}_{F\,\mu\rho}(k) = -\left[\eta^{\mu\nu} - \frac{(1+\lambda)}{k^2}k^\mu k^\nu\right]\frac{1}{k^2+i\epsilon}$$

Which is known as the Feynman propagator for photons. There is a clear separation of transverse and longitudinal parts, with gauge dependence only on the longitudinal term:

$$\widetilde{D}_{F\,\mu\rho}(k) = -\left[\eta^{\mu\nu} - \frac{1}{k^2}k^\mu k^\nu\right]\frac{1}{k^2+i\epsilon} + \frac{\lambda k^\mu k^\nu}{k^2}\frac{1}{k^2+i\epsilon},$$

where the first term is transverse. Since photons couple to sources that are conserved ($\partial_\mu J^\mu = 0$), we have $k_\mu \tilde{J}^\mu(k) = 0$, and the longitudinal (second) term will be eliminated, thus gauge invariance. Common gauge choices: $\lambda = 0$ (Landau) and $\lambda = -1$ (Feynman).

Repeating the Gaussian type integration we now have:

$$W[J_\mu] = exp\left\{-\frac{1}{2}\int d^4x\, d^4x'[J^\mu(x)D_{F\,\mu\nu}(x)J^\nu(x')]\right\}.$$

4.7 QED

Background

Quantum electrodynamics (QED) is an abelian gauge theory with symmetry group U(1). The Lagrangian consists of a spin-1/2 field interacting with the gauge field according to:

$$\mathcal{L} = -\frac{1}{4}F^{\mu\nu}F_{\mu\nu} + \bar{\psi}(i\gamma\cdot D - m))\psi,$$

where D is the gauge covariant derivative, $F^{\mu\nu}$ is the electromagnetic field tensor, and ψ is the spinor field.

Evaluation of radiative corrections to order α

Let's now consider Fermionic fields with the Dirac equation, with its global gauge invariance that must be managed to recover a gauge invariant formulation. Let's start by writing the Fermion field Lagrangian:

$$\mathcal{L}_\psi = \bar{\psi}(i\gamma\cdot\partial - m)\psi,$$

which is invariant under:

$$\psi \to e^{ie\theta}\psi, \quad \bar{\psi} \to \bar{\psi}e^{-ie\theta},$$

e.g., a global gauge transformation exists for θ constant. Let's elevate this to a true gauge freedom with $\theta = \theta(x)$, with whatever modification is needed to maintain gauge invariance. Let's gauge transform with the Lagrangian as is:

$$\mathcal{L}_\psi \to \mathcal{L}_\psi - e\,\bar{\psi}\gamma_\mu\psi\partial^\mu\theta(x).$$

To recover gauge invariance in such a formalism we need an extra gauge-field interaction term, with gauge field A_μ:

$$\mathcal{L}_{interaction} = e\,\bar{\psi}(\gamma \cdot A)\psi,$$

and have the gauge field transform under gauge transformation as:

$$A_\mu \to A_\mu + \partial_\mu\theta\,.$$

Once we've introduced A_μ in this way, it generalizes the gauge freedom as desired, but it is not a truly independent field with its own propagation. To enable this, a kinetic energy term is needed, and we want to maintain all the symmetries, so let's add the term:

$$-\frac{1}{4}F_{\mu\nu}F^{\mu\nu},\;\; F_{\mu\nu} = \partial_\mu A_\nu - \partial_\nu A_\mu.$$

Having done this, we see that the gauge freedom impact on the measure is accounted for in the naïve formulation with Lagrangian having an additional correction term as indicated previously: $\frac{1}{2\lambda}\left(\partial_\mu A^\mu\right)^2$. Thus, the full Lagrangian for gauge invariant fermionic field results in the Lagrangian foundation for QED, the theory of photon-lepton interactions:

$$\mathcal{L} = -\frac{1}{4}F_{\mu\nu}F^{\mu\nu} + \bar{\psi}(i\gamma\cdot\partial - m + e\gamma\cdot A)\psi + \frac{1}{2\lambda}\left(\partial_\mu A^\mu\right)^2.$$

From this we then generate the Feynman diagrams for internal lines, with two kinds of propagators and one type of interaction term:

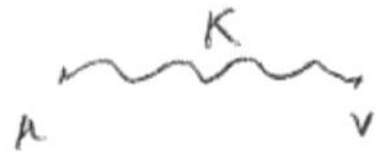

$$iD_{F\mu\nu}(k) = i\left(\eta_{\mu\nu} - \frac{k_\mu k_\nu}{k^2}\right)\frac{1}{k^2+i\epsilon} + \frac{i\lambda k_\mu k_\nu}{k^2(k^2+i\epsilon)}$$

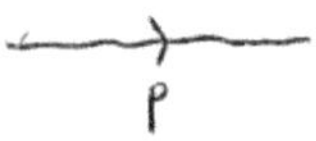

$$iS_F(p) = \frac{i}{\not{p} - m + i\epsilon} = \frac{i(\not{p}+m)}{p^2 - m^2 + i\epsilon}$$

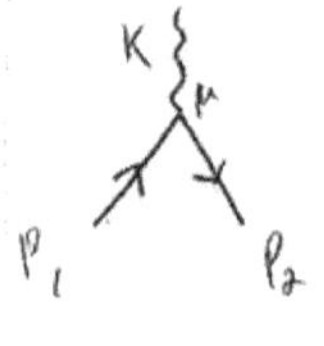

$$ie\gamma^\mu$$

also rules for external lines,

The Grassman nature of the fields is reflected in a (-1) factor for each closed loop.

Let's now consider the order of divergence for a given diagram.

$d = 2\omega$ dimension

I_γ, E_γ internal and external photon lines

I_e, E_e " fermion (electron) lines

photon prop. $\propto \frac{1}{p^2}$

electron prop. $\propto \frac{1}{p}$

L: loop. 2ω momenta for each loop.

Superficial Degree of divergence: $D = 2\omega L - 2I_\gamma - I_e$

$L = I_e + I_\gamma - V + 1$ (# of vertices)

$E_e + 2I_e = 2V$

$E_\gamma + 2I_\gamma = V$

$D = 2\omega - (\omega - \frac{1}{2})E_e - (\omega - 1)E_\gamma + (\omega - 2)V$

renorm when D doesn't depend on $V \rightarrow \omega = 2 \Rightarrow$ 4 dimensions

$\omega = 2$:

$D = 4 - \frac{3}{2}E_e - E_\gamma$

Let's consider some of the divergent diagrams that will arise:

(1) $E_e = 0,\ E_\gamma = 2$ (the photon self-interaction) has $D = 2$, indicating a quadratic divergence. The actual divergence is only $D = 0$ (logarithmic), however, due to the $\eta^{\mu\nu} - \frac{1}{k^2}k^\mu k^\nu$ term. Renormalization is via the wavefunction amplitude constant.

(2) $E_e = 2,\ E_\gamma = 0$ (the electron self-interaction) has $D = 1$, with renormalization via mass constant.

(3) $E_e = 0,\ E_\gamma = 3$, has $D = 1$ and to counter need F^3 in $\mathcal{L}$, but $\dim(F^3)$=6, which is non-renormalizable. So, here we need cancellation identically. It is shown in Furry's Theorem, that diagrams with an odd number of photon lines are identically zero (in QED). One way to see this is to consider the impact of charge conjugation invariance.

(4) $E_e = 2,\ E_\gamma = 1$ (the electron-photon interaction), has $D = 0$ (logarithmic), and is renormalized via the coupling constant. This is the only interaction in the Lagrangian that can be countered.

(5) $E_e = 0,\ E_\gamma = 4$, has $D = 0$ and to counter need F^4 in $\mathcal{L}$, again non-renormalizable. However, term is actually convergent due to gauge invariance, so the matter is moot.

Let's write the renormalized form of the QED Lagrangian, beginning with the renormalized variables:

$$\psi_B = Z_2^{1/2}\psi_R,\ e_B = \frac{Z_1}{Z_2 Z_3^{1/2}} e_R \mu^\varepsilon,\ \ \varepsilon = 2 - \omega,\ \ m_B = \frac{Z_m}{Z_2} m_R$$

$$\lambda_B = Z_3\lambda_R, \quad A_{B\mu} = Z_3^{1/2} A_{R\mu},$$
where the last relations follow from the $\frac{1}{2\lambda}\left(\partial_\mu A^\mu\right)^2$ term.

The Lagrangian is then written:
$$\mathcal{L} = -\frac{1}{4}F_{\mu\nu}F^{\mu\nu} + \bar{\psi}(i\gamma\cdot\partial - m + e\mu^{\varepsilon}\gamma\cdot A)\psi + \frac{1}{2\lambda}\left(\partial_\mu A^\mu\right)^2 + \mathcal{L}_{c.t.}$$
where the counter-term Lagrangian is:
$$\mathcal{L}_{c.t.} = -(Z_3 - 1)\frac{1}{4}F_{\mu\nu}F^{\mu\nu} + (Z_2 - 1)\bar{\psi}(i\gamma\cdot\partial - m)\psi$$
$$-(Z_m - 1)m\bar{\psi}\psi + e\mu^{\varepsilon}(Z_m - 1)\bar{\psi}\gamma\cdot A\psi$$
So, we have the added diagram:

And the added counter-terms:

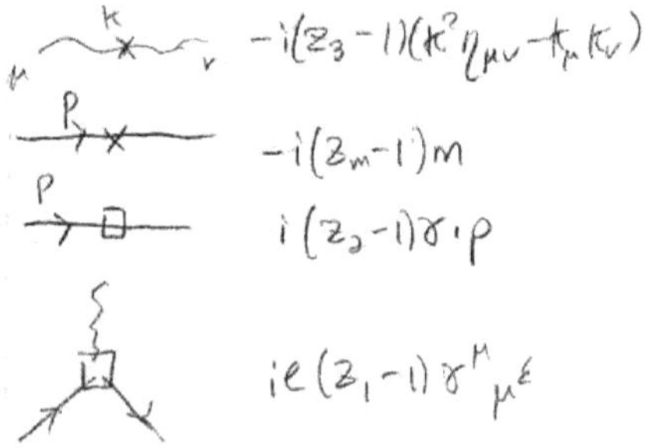

Note that we want the full theory to be gauge invariant, so the specific form
$$\bar{\psi}(i\gamma\cdot\partial + e\gamma\cdot A)\psi$$
Must be invariant, which then indicates that:
$$Z_1 = Z_2,$$
which is the First Ward Identity. That completes the Feynman diagram components, with no symmetry factors like with phi^4 theory, but there is the property that each loop carries an extra (-1) factor. In 2ω-dim we have $\{\gamma_\mu, \gamma_\nu\} = 2\eta_{\mu\nu}\cdot I$, and the γ matrices are $f(2\omega)\times f(2\omega)$ dimensional, where $f(4) = 4$. So,
$$Tr\gamma_\mu\gamma_\nu = \frac{1}{2}Tr\{\gamma_\mu, \gamma_\nu\} = \eta_{\mu\nu}\, Tr\, I = f(2\omega)\eta_{\mu\nu}$$
Let's evaluate the diagram:

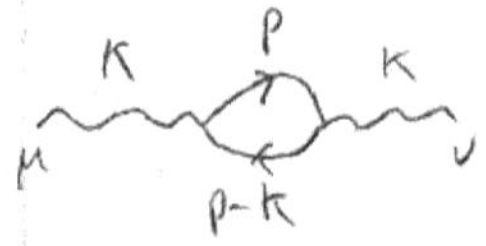

For which we have for the 1.P.I. part:

$$iD(k)_{\mu\nu}\{i\pi^{\lambda\lambda'}(k)]iD(k)_{\lambda'\nu}$$

which is known as the vacuum polarization tensor in electrodynamics. The trace is over Dirac indices. Instead of "slash notation" using "bold notation":

$$i\pi_{\mu\nu}(k) = (-1)(\,ie\mu^{\varepsilon})^2 \int \frac{d^{2\omega}p}{(2\pi)^{2\omega}}\, tr\big[iS_F(p-k)\gamma_\nu iS_F(p)\gamma_\mu\big]$$

$$= -(\,e\mu^{\varepsilon})^2 \int \frac{d^{2\omega}p}{(2\pi)^{2\omega}} \frac{tr\big[(\mathbf{p}-\mathbf{k}+m)\gamma_\nu(\mathbf{p}+m)\gamma_\mu\big]}{[(p-k)^2-m^2+i\varepsilon][p^2-m^2+i\varepsilon]}$$

Notice that:

$$tr\big[(\mathbf{p}-\mathbf{k}+m)\gamma_\nu(\mathbf{p}+m)\gamma_\mu\big] =$$

$$f(2\omega)[2p_\mu p_\nu - (p_\mu k_\nu + k_\mu p_\nu - p\cdot(p-k)\eta_{\mu\nu} + m^2\eta_{\mu\nu}$$

And, using the standard integral maneuver for the denominator:

$$\int_0^1 dx \frac{1}{(x[(p-k)^2-m^2+i\varepsilon]+(1-x)[p^2-m^2+i\varepsilon])}$$

And let $l = p-(1-x)k$:

$$\int_0^1 dx \frac{1}{(l^2+x(1-x)k^2-m^2+i\varepsilon)^2}$$

So, in the $\pi_{\mu\nu}(k)$ integral we have symmetric integration. So, terms with odd $l's$ give zero and $l_\mu l_\nu = \frac{1}{2\omega}\eta_{\mu\nu}l^2$, thus:

$$i\pi_{\mu\nu}(k) = -(\,e\mu^{\varepsilon})^2 \int_0^1 dx \int \frac{d^{2\omega}l}{(2\pi)^{2\omega}} \frac{A_{\mu\nu}}{(l^2+x(1-x)k^2-m^2+i\varepsilon)^2}$$

$$A_{\mu\nu} = f(2\omega)[(\frac{1}{\omega}-1)l^2\eta_{\mu\nu} -$$

$$2x(1-x)k_\mu k_\nu + x(1-x)k^2\eta_{\mu\nu} + m^2\eta_{\mu\nu}]$$

Thus,

$$i\pi_{\mu\nu}(k) = -(\,e\mu^{\varepsilon})^2 f(2\omega) i\frac{\Gamma(2-\omega)}{(4\pi)^\omega}\int_0^1 dx \frac{2x(1-x)\big(k^2\eta_{\mu\nu}-k_\mu k_\nu\big)}{(m^2-x(1-x)k^2)^{2-\omega}}$$

Note the appearance of the projection operator, $\big(k^2\eta_{\mu\nu}-k_\mu k_\nu\big)$, mentioned earlier.

Now, define $\pi_{\mu\nu} = \big(k^2\eta_{\mu\nu}-k_\mu k_\nu\big)\pi(k^2)$, write $\omega = 2-\epsilon$, and expand out using the $\int_0^1 dx$ integral as before, to get:

$$\pi(k^2) = \frac{-e^2}{12\pi^2}\frac{1}{\epsilon} + \pi_c(k^2)$$

$$\pi_c(k^2) = \frac{-e^2}{12\pi^2}\left\{-\gamma_E - \ln\frac{m^2}{4\pi\mu^2} - 6\int_0^1 dx(1-x)x\ln\left[1 - x(1-x)\frac{k^2}{m^2}\right]\right\}$$

Now $f(4+\epsilon) = 4 + A\epsilon$ ...., so $\Gamma(2-\omega) = \Gamma(\epsilon) = \frac{1}{\epsilon}\Gamma(1+\epsilon)$. By setting $f(2\omega) = 4$ we have shifted the constant $A\Gamma(1+\epsilon)$ to the infinite part.

Set $Z_3 = 1 - \frac{\alpha}{3\pi}\frac{1}{\epsilon} + \mathcal{O}(\alpha^2)$ where $\alpha = \frac{e^2}{4\pi}$ to get $-i(Z_3 - 1)\left(k^2\eta_{\mu\nu} - k_\mu k_\nu\right)$ to cancel the infinite part:

$$iD'_{F\mu\nu}(k) = iD_{F\mu\nu} + iD_{F\mu\lambda}i\pi^{\lambda\lambda'}iD_F\lambda'\nu + \cdots = iD_{F\mu\lambda}\left[\delta^\lambda_\nu + i\pi^{\lambda\lambda'}iD'_{F\lambda'\nu}\right]$$

So

$$D_F^{-1\,\lambda\mu}D_F^{-1}{}_{\mu\nu} = \delta^\lambda{}_\nu - \pi^{\lambda\lambda'}D'_{F\lambda'\nu}$$

and

$$D_F^{-1}{}_{\mu\nu} = -\left(k^2\eta_{\mu\nu} - k_\mu k_\nu\right) + \lambda^{-1}k_\mu k_\nu$$

Thus,

$$D'_{\lambda\nu} = A(k^2)\left(-\eta_{\mu\nu} + \frac{k_\mu k_\nu}{k^2}\right) + B(k^2)\frac{k_\mu k_\nu}{k^2}$$

where

$$A(k^2) = \frac{1}{k^2(1-\pi_c)}, \quad B(k^2) = \frac{\lambda}{k^2}$$

Now,

$$D'_{\lambda\nu} = -\left(-\eta_{\mu\nu} + \frac{k_\mu k_\nu}{k^2}\right)\frac{1}{k^2(1-\pi_c) + i\epsilon} + \frac{\lambda}{k^2}\frac{k_\mu k_\nu}{(k^2 + i\epsilon)}$$

Notice that the π_c acts as a dielectric constant. The photon still has a pole at $k = 0$, still has zero mass, and the π_c affects only the wave function renormalization. Also the longitudinal part is not affected by renormalization, thus, λ does not get renormalized, which justifies not introducing another renormalization for λ.

Note that $\frac{e^2}{k^2}$ is the F.T. of $\frac{e^2}{|\vec{x} - \vec{x}'|}$. So, if we have $\frac{e^2}{k^2(1-\pi_c)}$ instead, we've reduced e^2 by $\frac{1}{(1-\pi_c)}$. This means that the vacuum is a polarizable medium which is frequency dependent (since $\pi_c(k^2)$).

Let's consider the following diagram:

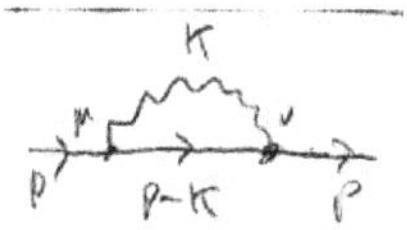

$$Diagram = iS_F(p)i\Sigma(p)iS_p(p)$$

where

$$i\Sigma(p) = (ie\mu^{\epsilon})^2 \int \frac{d^{2\omega}k}{(2\pi)^{2\omega}} \gamma_\nu iS_F(p-k)\gamma_\mu iD_F^{\mu\nu}(k)$$

$$= -e^2\mu^{2\epsilon} \int \frac{d^{2\omega}k}{(2\pi)^{2\omega}} \frac{\gamma_\mu(\mathbf{p} - \mathbf{k} + m)\gamma^\mu}{((p-k)^2 - m^2 + i\epsilon)(k^2 + i\epsilon)}$$

Note $\gamma_\mu(\mathbf{p} - \mathbf{k} + m)\gamma^\mu = (-\mathbf{p} + \mathbf{k} + m)\gamma_\mu\gamma^\mu + 2(p-k)_\mu\gamma^\mu$ and $\gamma_\mu\gamma^\mu = 2\omega$. Also, doing the usual denominator maneuver:

$$\frac{1}{((p-k)^2 - m^2 + i\epsilon)(k^2 + i\epsilon)}$$

$$= \int_0^1 dx \frac{1}{[x((p-k)^2 - m^2) + (1-x)k^2 + i\epsilon]^2}$$

Let $l = k - xp$ and the overall integral becomes:

$$i\Sigma(p) = -2e^2\mu^{2\epsilon} \int_0^1 dx \int \frac{d^{2\omega}l}{(2\pi)^{2\omega}} \frac{(1-\omega)(1-x)\mathbf{p} + \omega m}{[l^2 + x(1-x)p^2 - xm^2]^2}$$

$$= \frac{-2ie^2\mu^{2\epsilon}}{(4\pi)^\omega}\Gamma(2-\omega) \int_0^1 dx \frac{(1-\omega)(1-x)p + \omega m}{D^{2-\omega}}$$

where $D = xm^2 - x(1-x)p^2$. There is a pole at $\omega = 2$ (from the gamma function), so let's set $\omega = 2 - \epsilon$ and let $\epsilon \to 0$:

$$i\Sigma(p) = (\mathbf{p} - m)B(p^2) + A(p^2)$$

where

$$A = -\frac{2e^2}{(4\pi)^2}\Gamma(\epsilon)(4\pi\mu^2)^\epsilon \int_0^1 dx\,(1 + x - \epsilon x)D^{-\epsilon}$$

$$B = -\frac{2e^2}{(4\pi)^2}\Gamma(\epsilon)(4\pi\mu^2)^\epsilon \int_0^1 dx\,(\epsilon - 1)(1-x)D^{-\epsilon}$$

Now let $\epsilon \to 0$:

$$A = \frac{-3e^2m}{(4\pi)^2}\frac{1}{\epsilon} + A_c(p^2)$$

where

$$A_c(p^2) = \frac{e^2 m}{(4\pi)^2}\left[1 + 3\gamma_E + 2\int_0^1 dx(1+x)\, ln\frac{xm^2 - x(1-\gamma)p^2}{4\pi\mu^2}\right]$$

and

$$B = \frac{e^2}{(4\pi)^2}\frac{1}{\epsilon} + B_c(p^2)$$

where

$$B_c(p^2) = \frac{-e^2}{(4\pi)^2}\left[1 + \gamma_E + 2\int_0^1 dx(1-x)\, ln\frac{xm^2 - x(1-\gamma)p^2}{4\pi\mu^2}\right]$$

There are two pole terms, one in A and one in B, and they are cancelled by mass and wavefunction renormalization counter-terms:

$$i(Z_m - 1)m + i(Z_2 - 1)\boldsymbol{p}$$

And matching terms we get:

$$Z_2 = 1 - \frac{\alpha}{4\pi}\frac{1}{\epsilon} + O(\alpha^2), \qquad \alpha = \frac{e^2}{4\pi}$$

$$Z_m = 1 - \frac{\alpha}{\pi}\frac{1}{\epsilon} + \cdots$$

So,

$$iS'_F(p) = iS_F(p) + iS_F i\Sigma iS_F + iS_F i\Sigma iS_F i\Sigma iS_F + \cdots = iS_F + iS_F i\Sigma iS'_F$$

Thus

$$S'_F(p) = S_F(p) - S_F(p)\Sigma(p)S'_F(p)$$
$$S_F^{-1}(p)S'_F(p) = I - \Sigma(p)S'_F(p)$$
$$(S_F^{-1}(p) + \Sigma(p))S'_F(p) = I$$
$$S'_F(p)^{-1} = S_F^{-1}(p) + \Sigma(p) = \boldsymbol{p} - m + \Sigma(p)$$

We have from previously (dropping the c subscripting on A and B):

$$\Sigma = (\boldsymbol{p} - m)B(p^2) + A(p^2)$$
$$S'_F(p)^{-1} = (\boldsymbol{p} - m)(1 + B(p^2) + A(p^2))$$

Suppose we let $p^2 = m_{phys}^2$, then:

$$S'_F(p)^{-1} = (\boldsymbol{p} - m_{phys})\left(1 + B(m_{phys}^2)\right) + (m_{phys} - m)\left(1 + B(m_{phys}^2)\right) + A(m_{phys}^2)$$

Where the physical pole structure is recovered if the latter two terms cancel:

$$(m_{phys} - m)\left(1 + B(m_{phys}^2)\right) + A(m_{phys}^2) = 0$$

This gives $m = m(m_{phys}, \mu)$ a function of m_{phys}, μ. We can use the fact that μ is the arbitrary to adjust μ such that $A(m^2) = 0$ (not most general, but good enough for what follows).

Looking at A_c (just called A now) let $p^2 = m^2$ to get

$$A(m^2) = (stuff) + (constant) \int_0^1 dx(1+x)\left[2lnx + ln\frac{m^2}{4\pi\mu^2}\right]$$

$$= -\frac{5}{2} + \frac{3}{2} ln\left(\frac{m^2}{4\pi\mu^2}\right).$$

So, the $A(m^2) = 0$ imposes the constant:

$$3\left(\gamma_E + ln\left(\frac{m^2}{4\pi\mu^2}\right)\right) = 4.$$

From here on we will assume m to be chosen to be the physical mass. So will have the above relation and μ no longer appears in any result. So, now we see the extra terms in the propagator:

$$S_F'(p)^{-1} = (\mathbf{p} - m)\left(1 + B(p^2)\right) + A(p^2)$$

Let's now consider the diagram describing the lowest order QED radiative correction:

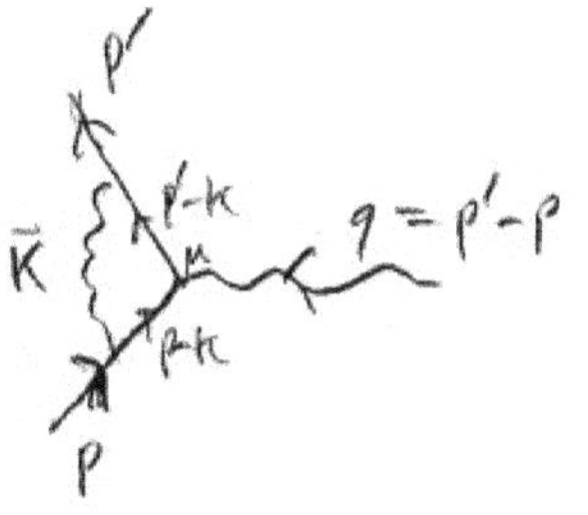

Let's assume the external lines are physical electrons. For scalar particle this means $p^2 = p'^2 = m^2$, and we have $\bar{u}(p')(\dots)u(p)$ where $(\mathbf{p} - m)u(p) = 0$ and $\bar{u}(p')(\mathbf{p}' - m) = 0$. Let's consider the 1PI forms where we have $\gamma_\mu \to \gamma_\mu + \Lambda_\mu$ and Λ_μ is the radiative correction. We then have the lowest order diagram (no loop) with $ie\mu^\varepsilon\gamma_\mu$ and the vertex correction term (diagram above) with $ie\mu^\varepsilon\Lambda_\mu(p, p')$. Let's write the vertex correction explicitly (an integration over the internal loop k):

$$\Lambda_\mu(p, p') = (ie)^2\mu^{2\varepsilon}\int\frac{d^{2\omega}k}{(2\pi)^{2\omega}}\gamma_\rho iS_F(p' - k)\gamma_\mu iS_F(p - k)\gamma_\sigma iD_F^{\rho\sigma}(k)$$

Let's rewrite

$$\Lambda_\mu(p, p') = -e^2\mu^{2\varepsilon}\int\frac{d^{2\omega}k}{(2\pi)^{2\omega}}\frac{\mathcal{N}_\mu}{\mathcal{D}}$$

and deal with the numerator $\mathcal{N}_\mu$ and denominator $\mathcal{D}$ terms separately, where:

$$\mathcal{N}_\mu = \gamma_\rho(\mathbf{p}' - \mathbf{k} + m)\gamma_\mu(\mathbf{p} - \mathbf{k} + m)\gamma^\rho$$
$$\mathcal{D} = ((p' - k)^2 - m^2)((p - k)^2 - m^2)(k^2)$$

As long as sandwiched between $\bar{u}$ and u we have
$\mathcal{N}_\mu \to [4p \cdot p' + 4(p + p') \cdot k - 2(1 - \omega)k^2]\gamma_\mu - 4\mathbf{k}(p + p')_\mu +$
$[4m + 4(1 - \omega)\mathbf{k}]k_u$

And using Feynman reparameterization $\mathcal{D}$:

$$\frac{1}{\mathcal{D}} = 2\int_0^1 dx \int_0^{1-x} dy\, [(1 - x - y)k^2 + x((p - k)^2 - m^2)$$
$$+ y((p' - k)^2 - m^2)]^{-3}$$

If we write $k = l - xp - yp'$ all linear terms in l may be dropped in $\mathcal{N}_\mu$, etc., if sandwiched between $\bar{u}$ and u, and redefining $z = x + y$, $w = x - y$ (note w is not ω) we get:

$$\Lambda_\mu(p, p') = -e^2\mu^{2\varepsilon} \int \frac{d^{2\omega}l}{(2\pi)^{2\omega}} \frac{\mathcal{N}_\mu}{\mathcal{D}}$$

$$\mathcal{N}_\mu = 2\left\{\left[-\left(1 - z - \frac{1 - \omega}{4}(z^2 - w^2)\right)q^2\right.\right.$$
$$\left. + (2(1 - 2z) - (1 - \omega)z^2)m^2 + \frac{(1 - \omega)^2}{\omega}l^2\right]\gamma_\mu$$
$$+ mz\big(1 + z(1 - \omega)(p + p')_\mu$$
$$\left. + mw\big(1 - z(1 - \omega)\big)(p - p')_\mu\big)\right\}$$

where the $\frac{(1-\omega)^2}{\omega}l^2$ term is divergent in 4D, while the $mw\big(1 - z(1 - \omega)\big)(p - p')_\mu$ term is linear in l, thus dropped. We write $\Lambda_\mu = I_{1\,\mu} + I_{2\,\mu}$ in what follows, where the divergent term is placed in $I_{1\,\mu}$. Let's compute $1/\mathcal{D}$ first however:

$$\frac{1}{\mathcal{D}} = \int_0^1 dz \int_{-1}^1 dw\, [l^2 - m^2G^2]^{-3}, \quad G^2 = \left(1 - \frac{q^2}{4m^2}\right)z^2 + \frac{q^2}{4m^2}w^2$$

(Note that for electron scattered by soft photon, $q \to 0$, this becomes $G^2 \sim z^2$, and the integral above can be solved approximately to give: $1/\mathcal{D} = [l^2 - m^2G^2]^{-3}$.)

We can solve for the divergent $I_{1\,\mu}$ term:

$$I_{1\,\mu} = e^2\mu^{2\epsilon}2\frac{(1-\omega)^2}{\omega}\gamma_\mu\int_0^1 dz\int_{-1}^1 dw\int\frac{l_E^2}{(l_E^2+m^2G^2)^2}\frac{d^{2\omega}l_E}{(2\pi)^{2\omega}}$$

Where use of Euclideanization is made on the l integration, thus the shift to "Euclideanized l": l_E. We now have:

$$I_{1\,\mu} = \frac{2e^2}{(4\pi)^2}\left[\frac{1}{\epsilon}-2-\gamma_E-ln\frac{m^2}{4\pi\mu^2}-\frac{1}{2}\int_0^1 dz\int_{-1}^1 d\omega\ lnG^2\right]\gamma_\mu\,.$$

We have the counter-term:

 $i(Z_1-1)e\gamma_\mu\mu^\epsilon$

For minimal subtraction we, thus, have:

$$Z_1 = 1-\frac{e^2}{8\pi^2}\frac{1}{\epsilon}$$

To this order we've thereby verified $Z_1 = Z_2$ (recall that this is required at all orders by gauge invariance, known as the First Ward Identity).

By use of $m = m(m_{ph},\mu)$ it can then be shown that $\gamma_E + ln\frac{m^2}{4\pi u^2} = -\frac{4}{3}$, so the finite part of $I_{1\,\mu}$can be written:

$$I_{\mu R} = -\frac{2e^2}{(4\pi)^2}\left[\frac{4}{3}+\frac{1}{2}\left(-\frac{q^2}{4m^2}\right)+\cdots\right]\gamma_\mu$$

Now to find I_2 (finite, where $\mathcal{N}'_\mu$ is the finite part of $\mathcal{N}_\mu$):

$$I_2 = -\frac{e^2}{(4\pi)^2}\left(\frac{4\pi\mu^2}{m^2}\right)^2\frac{1}{2m^2}\Gamma(1+\epsilon)\int_0^1 dz\int_{-1}^1 dw\,(\mathcal{N}'_\mu)(G^2)^{-(1+\epsilon)}$$

Trying q^2 small we find infra-red divergence (a common sequence in QED):

$$q^2 = 0 \rightarrow G^2 = z^2 \rightarrow I_2 \propto \int_0^1\frac{dz}{z^{2(1+\epsilon)}}$$

Note that a UV divergence would indicate the renormalization was incomplete. For the IR divergence let's evaluate the integral within our dimensional regularization context (now a critical advantage to the traditional method to isolate IR divergence by giving the photon a small mass and contend with the problem as an aspect of the arbitrariness of Bremsstrahlung radiation). Here we simply evaluate:

$$\int_{-1}^{1} d\omega \int_0^1 \frac{dz}{z^{2(1+\epsilon)}} = 2\int_0^1 dx \int_0^{1-x} dy\,(x+y)^{-2(1+\epsilon)}$$

$$= -\frac{1}{1+2\epsilon}\int_0^1 dx\left[1 - x^{-2(1+\epsilon)}\right] = -\frac{1}{\epsilon}$$

This term can be omitted as it is cancelled in any physical measurement. Thus, we obtain the final (renormalized) result (for real vacuum polarization vertex correction):

$$\Lambda_{\mu\,R} = \frac{e^2}{8\pi^2}\left[\frac{4}{3}\gamma_\mu + \frac{i}{2m}\sigma_{\mu\nu}q^\nu - \frac{1}{4}\frac{q^2}{m^2}\gamma_\mu\right] + \mathcal{O}\left(\left(\frac{q^2}{m^2}\right)^2\right)$$

Where use of Gordon decomposition gives:

$$\boldsymbol{a}(\boldsymbol{p}-m) + (\boldsymbol{p}'-m)\boldsymbol{a} = 0$$

with $\boldsymbol{a} = \gamma_\mu$ gives $\gamma_\mu(\boldsymbol{p}-m) + (\boldsymbol{p}'-m)\gamma_\mu = 0$, thus:

$$\gamma_\mu \boldsymbol{p} = \frac{1}{2}\left[\{\gamma_\mu,\gamma_\nu\} - [\gamma_\mu,\gamma_\nu]\right]p^\nu = p_\mu - i\sigma_{\mu\nu}p^\nu$$

where $\sigma_{\mu\nu} = \frac{i}{2}[\gamma_\mu,\gamma_\nu]$, and

$$\boldsymbol{p}\gamma_\mu = p'_\mu + i\sigma_{\mu\nu}p'^\nu,$$

which are combined to give the relation that is used:

$$2m\gamma_\mu = (p+p')_\mu + i\sigma_{\mu\nu}(p'-p)^\nu$$

Physical application

Let's consider vertex contributions to one-loop order:

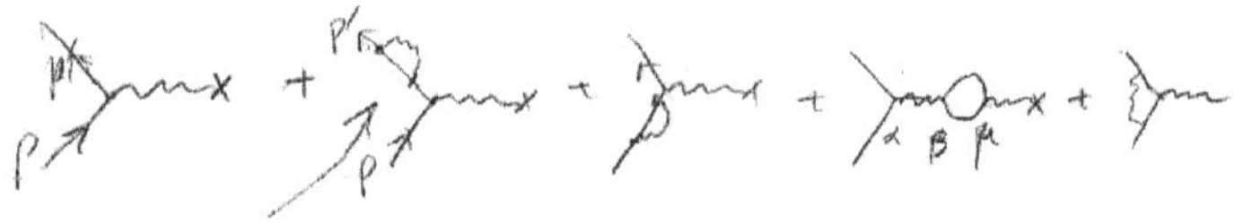

In the second and third terms above we have

$$\bar{u}(p')\Sigma(p')\,... = 0$$

(the advantage of using physical mass renormalization) thus eliminated. This leaves only the second to last diagram to evaluate:

$$= ie\bar{u}(p')\left[\gamma_\mu + i\gamma_\alpha D_F{}^{\alpha\beta}(q)\pi_{\beta\mu}(q) + \Lambda_\mu(p,p')\right]u(p)\bar{A}^\mu_{ext}(q)$$

$$= ie\bar{u}(p')\left[(1+\pi(\pi))\gamma_\mu + \Lambda_\mu(p,p')\right]u(p)\bar{A}^\mu_{ext}(q)$$

4.7.1 External field purely electric

Suppose the external field were a Coulomb field, then the only nonzero component of $\bar{A}^\mu_{ext}(q)$ is $\mu = 0$, then $\bar{u}(p')\gamma^\mu u(p) \to 1$ for p, p' static (this is how change is defined from electrostatics). Thus, the last diagram becomes:

$$= ie\bar{u}(p')\big[\big(1+\pi(0)\big)\gamma_\mu + \Lambda_\mu(0)\big]u(p)$$

where $\pi(0) = \frac{e^2}{12\pi^2} - \frac{4}{3}$ and $\Lambda_\mu(0) = \frac{e^2}{8\pi^2} \cdot \frac{4}{3}\gamma_\mu$, so:

$$e_{phys} = e\left(1 + \frac{5e^2}{18\pi^2}\right) + \cdots$$

Inverting we get:

$$e = e_{phys}\left(1 - \frac{Se^2}{18\pi^2} e_{phys}^2\right) + \cdots$$

We can now write the diagram as:

$$= ie\bar{u}(p')\big[F_1(q^2)\gamma_\mu + iF_2(q^2)\sigma_{\mu\nu}q^\nu\big]u(p)\bar{A}^\mu{}_{ext}(q)$$

where

$$F_1 = e\left(1 + \frac{5e^2}{18\pi^2}\right) - \frac{23e^3}{15 \cdot 32\pi^2}\frac{q^2}{m^2} + \cdots = e_{phys} - \frac{23e_{phys}{}^3}{15 \cdot 32\pi^2}\frac{q^2}{m^2} + \cdots$$

$$F_2 = \frac{e^3}{15\pi^2 m} + \cdots = \frac{e^3{}_{phys}}{16\pi^2 m} + \cdots$$

Consider static coulomb potential $q \to 0$ thus $A^0_{ext} = \frac{Q}{4\pi|\vec{x}|} \rightarrow \bar{A}^0_{ext} = \frac{Q}{q^2}$ and we get:

$$F_1\bar{A}^0_{ext} = \frac{e_{phys}Q}{q^2} - \frac{23e_{phys}{}^3}{15 \cdot 32\pi^2 m^2}$$

Reverting back to configuration space (and all charges physical from here):

$$\frac{eQ}{4\pi}\left[\frac{1}{|\vec{x}|} + \frac{23\alpha}{30m^2}\delta(\vec{x})\right] + \cdots$$

Where the correction term is the Lamb shift (only effects s states).

4.7.2 External field purely magnetic

For an external field that is purely magnetic:

$$\bar{A^i} \neq 0 \ \ , \bar{A}^0 = 0$$

Consider the interaction term from the field with a (Dirac) electron: $\bar{u}\gamma_\mu u\bar{A}^\mu{}_{ext}$ (Dirac's result). Using the use Gordon decomposition again:

$$\gamma_\mu \rightarrow \frac{1}{2m}\big[(p+p')_\mu + i\sigma_{\mu\nu}q^\nu\big]$$

And from the second term:

$$\bar{u}\sigma_{ij}q^j u \sim i\varepsilon_{ijk}\chi^\dagger\sigma_k q^j \bar{A^i}(q)$$

where

$$\sigma_{ij} = \varepsilon_{ijk} \begin{pmatrix} \sigma_k & 0 \\ 0 & \sigma_k \end{pmatrix}.$$

Thus,

$$\bar{u}\sigma_{ij}q^j u \sim \chi^\dagger [\vec{\sigma} \cdot (\vec{q} \; x \; \vec{A})] \chi$$

Focusing on the $\vec{\sigma} \cdot (\vec{q} \; x \; \vec{A})$ term and shifting to configuration space (using an F.T.), we have:

$$\vec{\sigma} \cdot (\vec{q} \; x \; \vec{A}) \;\; \rightarrow \;\; \vec{\sigma} \cdot (\vec{\nabla} \; x \; \vec{A}) = \vec{\sigma} \cdot \vec{B}.$$

Note that, without renormalized QED correction, the link from quantum electron and classical field then has:

$$\frac{e}{2m} \vec{\sigma} \cdot \vec{B} = 2 \frac{e}{2m} \vec{s} \cdot \vec{B},$$

where the classical g-factor, g=2, is recovered directly from the Dirac equation. Let's now consider the QED correction, when to lowest order instead of $\frac{e}{2m}$ we have $\frac{e}{2m} + F_2(0)$ from previously:

$$\left[\frac{e}{2m} + F_2(0)\right] \vec{\sigma} \cdot \vec{B} = \frac{e}{2m}\left[1 + \frac{\alpha}{2\pi}\right] \vec{\sigma} \cdot \vec{B},$$

which predicts a g-factor of:

$$\boldsymbol{g = 2\left[1 + \frac{\alpha}{2\pi}\right] + \mathcal{O}(\alpha^2)}$$

The second most accurate verification of QED:

Let $X = \frac{1}{2} g - 2$, we have for theoretical computation to $\mathcal{O}(\alpha^3)$:

$$10^9 X = 1159652.4 \pm 0.4$$

This includes contributing diagrams from other leptons and quarks (e.g., whatever is relevant from entire Standard Model). This agrees with the experimental results to the current known levels of precision:

$$10^9 X = 1159652.41 \pm 0.2 \; .$$

4.7.3 The Ward identities

Let's restate the basic generator and gauge formalism we've arrived at for the path integral formulation:

$$Z[J_\mu, \eta, \bar{\eta}] = \int D\Psi D\bar{\Psi} DA_\mu e^{i\phi}$$

$$\phi = \int d^4x \left[\mathcal{L} + \mathcal{L}_{g.f} + \bar{\Psi}\eta + \bar{\eta}\Psi + J_\mu A^\mu\right]$$

$$\mathcal{L} = -\frac{1}{4} F_\mu F^{ku} + \bar{\Psi}(i\gamma \cdot \partial - m)\Psi + e\bar{\Psi}\gamma \cdot A\Psi$$

and gauge field Lagrangian:

$$\mathcal{L}_{g\cdot f} = \frac{1}{2\lambda}\left(\partial_\mu A^\mu\right).$$

We have from gauge invariance:

$$\Psi(x) \to e^{ie\theta(x)}\,\Psi(x)$$
$$\Psi(x) \to \Psi e^{ie\theta(x)}$$
$$A_\mu(x) \to A_\mu(x) \to \partial_\mu\theta(x)$$

A very useful property of the path integral formulation is that invariance properties from gauge invariance are easily dealt with. Consider the infinitesimal transformation:

$$\Psi(x) \cong \Psi(x) + ie\theta\,\Psi(x)$$

and want $\frac{\delta Z}{\delta\theta(x)} = 0$. We have:

$$\delta\phi = \int d^4x\left[\frac{1}{\lambda}\partial_\mu A^\mu\Box\theta + ie(\bar{\eta}\Psi - \bar{\Psi}\eta)\theta + J_\mu\partial^\mu\theta\right]$$

at lowest order in θ. In order to have

$$\frac{\delta Z}{\delta\theta(x)} = \int D\Psi D\bar{\Psi} DA_\mu i\frac{\delta\phi}{\delta\theta}e^{i\phi} = 0$$

The master equation for the Ward identities follows a change of notation:

$$\int D\Psi D\bar{\Psi} DA_\mu F[\Psi,\bar{\Psi}]\,e^{i\phi} = < F[\Psi,\bar{\Psi}] >_{J,\eta,\bar{\eta}}$$

So,

$$\frac{\delta Z}{\delta\theta} =< \frac{1}{\lambda}\Box\partial_\mu A^\mu + ie(\bar{\eta}\Psi - \bar{\Psi}\eta) - \partial_\mu J^\mu(x) >_{J,\eta,\bar{\eta}}= 0$$

Let's act on the relation above with the familiar generator operation

$$\frac{\delta}{i\delta J_v(j)}(\ldots)|_{sources\to 0}$$

To get:

$$\frac{1}{\lambda}\Box_x\partial_\mu^{(x)} < 0|TA^\mu(x)A^u(y)|0 > + i\partial_{(x)}^u\delta_4(x-y) = 0$$

In momentum space we have:

$$\langle 0|TA^\mu(x)A^u(y)|0\rangle = i\int\frac{d^4k}{(2\pi)^4}e^{ik\cdot(x-y)}D_F'(k)^{\mu v}$$

(using the complete propagator D_F'). So the first identity:

$$K^2K_\mu D_F'(k)^{\mu v} = \lambda K^v$$

Consider $D_F'(k)^{\mu v}$ consists of $\eta^{\mu v}$ and $\eta^\mu k^v$ terms, so write:

$$D_F'(k)^{\mu v} = \left(\eta^{\mu v} - \frac{k^\mu k^v}{k^2}\right)A(k^2) + \frac{k^\mu k^v}{k^2}\;B(k^2)$$

So, $k^2 k_\nu B = \lambda K_\nu \rightarrow B = \frac{\lambda}{K^2}$. Thus, the B term is completely determined by the Ward identities. The B term becomes exactly the longitudinal part of the free propagator, thus the Ward identity also tells us that the full propagator has longitudinal part the same to all orders.

The original form of the Ward Identity (before Takahashi generalization) is still instructive. Consider:

$$\frac{\delta}{i\delta_\eta(Z)}\frac{\delta}{i\delta_\eta(y)} < \cdots > |_{\substack{sources \\ =0}} = \frac{1}{\lambda}\Box_x \partial_\mu^{(x)} < 0|TA^\mu(x)\,\Psi(y)\,\overline{\Psi}(z)|0 >$$
$$+\, e < 0|T\,\Psi(x)\,\overline{\Psi}(z)|0 > \delta_4(x-y)$$
$$-e < 0|T\,\Psi(y)\,\overline{\Psi}(z)|0 > \delta_4(x-z)$$
$$= 0$$

Going to momentum space:

$$\int dxdydz\, e^{-i(p\cdot y+p'\cdot z+q\cdot x)} < 0|TA^\mu(x)\,\Psi(y)\,\overline{\Psi}(z)|0 >$$
$$= (2\pi)^4\delta_4(p+p'+q)G^\mu(p,p')$$

where $G^\mu(p,p')$ is exact to all orders. So,

$$-\frac{1}{\lambda}(p'+p)^2(p'+p)_\mu G^\mu(p,p') + eS_F'(p') - eS_F'(p) = 0$$

And we now finally use the relation from Section X where all that's needed for the full $G^\mu(p,p')$ is the 1PI part:

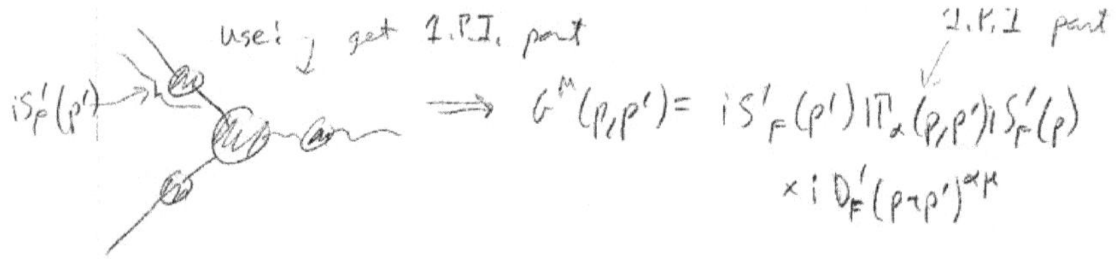

Thus

$$G^\mu(p,p') = iS'(p')i\Gamma_\alpha(p,p')iS_F'(p)iD_F'(p+p')^{\alpha\mu}$$

So,

$$(p'+p)_\mu S_F'(p')\Gamma^\mu(p,p')S_F'(p) = eS_F'(p') - eS_F'(p)$$
$$(p+p')_\mu\Gamma^\mu(p,p') = e[S_F'(p)^{-1} - S_F'(p')^{-1}]$$

Perform an operation:

$$\frac{\partial}{\partial p_\mu}\{\ldots\}\bigg|_{p+p'=0}$$

To get:

$$\Gamma^\mu(p,-p) = e\frac{\partial}{\partial p_\mu}[S_F'(p)]^{-1}$$

Thus last expression allows calculation of the self energy of the electron for all higher orders and bypasses the problem of overlapping divergencies as the problem originally occurred in the 1950's.

Since

$$\Gamma^\mu(p,p') = e\gamma^\mu + e\Lambda^\mu(p,p')$$

and

$$S_F'(p) = (\mathbf{p} - m) + \Sigma(p)$$

we have

$$e\gamma^\mu + e\gamma^\mu(p,-p) = e\gamma^\mu + e\frac{\partial\Sigma(p)}{\partial p_\mu}$$

Thus the Ward Identity in the form:

$$\Lambda^\mu(p,-p) = \frac{\partial\Sigma}{\partial p_\mu}.$$

Recall that the coupling counter-term, $ie\Lambda^\mu_{c.t} = ie(Z_1 - 1)\gamma^\mu$, had its own self-energy counter-term part: $i\Sigma(p)_{c.t} = -i(Z_m - 1)m + i(Z_1 - 1)p$, from which we must have:

$$(Z_1 - 1)\gamma^\mu = (Z_2 - 1)\gamma^\mu \rightarrow Z_1 = Z_2$$

(sometimes this alone is referred to as the Ward Identity).

Example 3

1. a. Show that the free Dirac propagator $S_F(x)$ can be written as
   $S_F(x) = (i\gamma\cdot\theta + m)\Delta_F(x)$
   Where $\Delta_F(x)$ is the propagator for a Klein-Gordon field of the same mass.
   b. Use the above result to show that
   $S_F^T(-x) = CS_F(x)C^{-1}$
   Where the superscript T denotes transposition in Dirac space and C is the charge conjugation matrix for the Dirac equation, defined by the property.
   $C\gamma_\mu C^{-1} = -\gamma_\mu^T$
   c. Prove Furry's theorem: the contributions of the two fermion loops shown below cancel each other if the number of photon lines is odd.

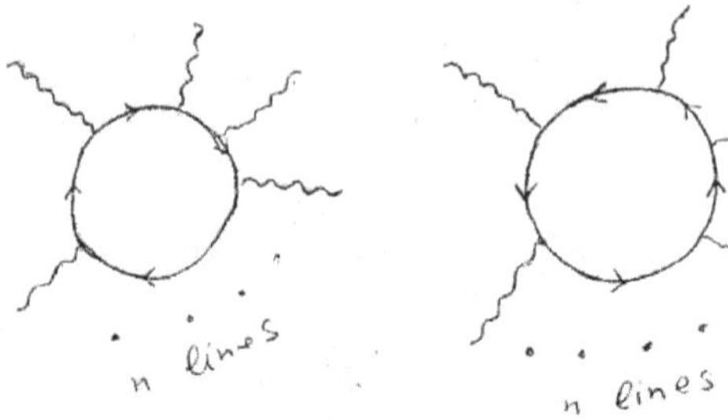

2. Show that one-loop diagram for light scattering by light, which is superficially logarithmically divergent, is actually convergent.

Answer

Since $\Delta_f(x) = \int \frac{d^4 p}{(2\pi)^4} \frac{e^{-ip.x}}{p^2 - m^2 + it}$

We have

$$(i\gamma, \partial + m)\Delta_f(x) = \int \frac{d^4 p}{(2\pi)^4} \frac{\gamma . p + m}{p^2 - m^2 + it} e^{-ip.x} = s_F(x)$$

$$s_F(x)C^{-1} = (-i\gamma^{\dagger} . \partial + m)\Delta_f(x)$$

$$= \int \frac{d^4 p}{(2\pi)^4} \frac{\gamma . p + m}{p^2 - m^2 + it} e^{-ip.x} = s_F(x)$$

$$I_1 = \int tr\left[iS_F(x_2 - x_1)\gamma_{\mu}, iS_F(x_1 - x_n)\gamma_{\mu},, \gamma_{\mu}, iS_F(x_3 - x_2)\gamma_{\mu}\right] dx_1,,, dx_{\mu}$$

$$I_2 = \int tr\left[iS_F(x_2 - x_1)\gamma_{\mu}, iS_F(x_1 - x_n)\gamma_{\mu},, \gamma_{\mu}, iS_F(x_3 - x_2)\gamma_{\mu}\right] dx_1,,, dx_{\mu}$$

The trace is under transportation

$$I_2 = \int tr\left[iS_F(x_2 - x_1)\gamma_{\mu}, iS_F(x_1 - x_n)\gamma_{\mu},, \gamma_{\mu}, iS_F(x_3 - x_2)\gamma_{\mu}\right] dx_1,,, dx_{\mu}$$

$$= \int tr\left[iS_F(x_2 - x_1)\gamma_{\mu}, iS_F(x_1 - x_n)\gamma_{\mu},, \gamma_{\mu}, iS_F(x_3 - x_2)\gamma_{\mu}\right] dx_1,,, dx_{\mu}$$

$$= (-i)^n I,$$

So, for odd n the diagrams cancel.

One – loop diagrams for scattering of light by light

3 independent diagrams to order α^2:

$$-(ie)^4 \mu^{4\epsilon} \int \frac{d^{2\omega} k}{(2\pi)^{2\omega}} tr\left[iS_F(x_2 - x_1)\gamma_{\mu}, iS_F(x_1 - x_n)\gamma_{\mu},, \gamma_{\mu}, iS_F(x_3 - x_2)\gamma_{\mu}\right]$$

… the only potential divergence comes from

$$= \left[(\eta\mu\nu\eta p^{o} - \eta\mu p\eta\omega + \eta\mu o\ell\eta u p)(\ell^2)^2 - 2(\eta\mu\ell p\ell\sigma + \eta\mu\sigma\ell\eta u p + \eta\mu\ell p\ell\sigma + \eta\mu\sigma\ell\eta u p)\ell^2 8\ell_\mu\ell_\nu\ell_\nu\ell_\sigma\right]$$

Use symmetric integration: $\ell_\mu\ell_\nu \to \frac{1}{2\omega}\eta\mu\nu\ell^2$

$$\ell_\mu\ell_\nu\ell_\nu\ell_\sigma \to \frac{1}{4\omega(1+\omega)}\eta\mu\nu\eta p^{o} + \eta\mu p\eta\omega + \eta\mu o\ell\eta u p$$

Becomes: $4I_{\mu\nu\rho\sigma}(\ell^2), will$

$$I_{\mu\nu\rho\sigma} = \frac{\omega-1}{\omega+1}(\eta\mu p\eta\omega + \eta\mu o\ell\eta u p) - \frac{(\omega-1)(\omega+2)\eta\mu p\eta\sigma}{\omega(\omega+2)}$$

With the other two independent Diagrams we get

$$I_{\mu\nu\rho\sigma} + I_{\mu\nu\rho\sigma} + I_{\mu\nu\rho\sigma} + \frac{\omega-2}{\omega}(\eta_{\mu\nu}\eta_{\rho\sigma} + \eta_\mu p\eta_\omega + \eta_\mu\eta_{u\rho})$$

Alt. Solution

1. A. since $\Delta_F(x) = \int \frac{d^4p}{(2\pi)}\frac{e^{-ipt}}{p^2-m^2+ie}$

   We have $(i_\gamma \cdot \partial + m)\Delta_F(x) = \int \frac{d^4p}{(2\pi)}\frac{\gamma\cdot p+m}{p^2-m^2+if}e^{-ipz} = S_F(x)$

   b. since $\Delta_p(x)$ is an even function of x,

   $CS_F(x)C^{-1} - (-i\gamma^T\cdot\partial + m)\Delta_F(x) - S_F^T(-x)$

   The fermion loops in the two diagrams above give:

   $I_1 = \int tr\left[iS_F(x_2-x_1)\gamma_{1,2}iS_F(x_1-x_\eta)\gamma_{1,2}\ldots\gamma_{1,2}S_F(x_3-x_2)\gamma_{1,2}\right]d_{x1}\ldots d_{xn}$

   $I_2 = \int tr\left[iS_F(x_1-x_2)\gamma_{\mu 2}iS_F(x_2-x_3)\gamma_{1,2}\ldots\gamma_{1,2}S_F(x_n-x_1)\gamma_{1,2}\right]d_{x1}\ldots d_{xn}$

Since the trace is invariant under transposition,

$I_2 = \int tr\left[\gamma_1^T i{S_F}^T(x_n-x_1)\ldots\gamma_1^T i{S_F}^T(x_2-x_3)\gamma_1^T{S_F}^T(x_1-x_2)\gamma_{1,2}\right]d_{x1}\ldots d_{xn}$

$= \int tr\left[C^{-1}\gamma_{\mu 1}^T CiS_F(x_1-x_n)C^{-1}\ldots C^{-1}\gamma_{\mu 1}^T CiS_F(x_3-x_2)C^{-1}\gamma_{\mu 1}^T CiS_F(x_2-x_1)\right]dx_1\ldots dx_n$

$= (-1)^n I_1$

Hence even n, the two diagrams give the same contribution; for odd n, they cancel.

2. There are three independent to order C^2

With $\Sigma^k = 0$. The first one gives

$$-(ie)^4\mu^{4c}\int\frac{d^{2\omega}k}{(2\pi)^{2\omega}}tr\left[iS_F(p)\gamma_\mu iS_F(p-k_1)\gamma_\sigma iS_F(p+k_2+k_3)\gamma_\mu iS_F(p-k_2)\gamma_\mu\right]$$

Since the superficial degree of divergent in 4 dimensions is logarithmic, the only potentially dangerous term is the one with the highest power of the integration variable. Introduce Feynman parameters and shift integration variable to allow the use of asymmetric integration. [the new variable is I, we have $t = p + \cdots$ and the only potential divergence comes from.

$$tr\gamma_\mu\gamma_\sigma\gamma_\sigma\gamma_\mu = \begin{bmatrix} (\eta_{uv}\eta_{\sigma\sigma} - \eta_{nv}\eta_{vp})(l^3)^2 \\ -2(\eta_{nv}l_pl_\sigma + \eta_\mu l_vl_p + \eta_{vp}l_pl_\sigma + \eta_{p\sigma}l_\mu l_v)l^2 \\ +8l_\mu l_vl_\sigma l_\sigma \end{bmatrix}$$

Use symmetric integration:

$$l_\mu l_v \rightarrow \frac{1}{2\omega}\eta_{uv}l^{2`}$$

$$l_\mu l_\mu l_pl_a \rightarrow \frac{1}{4\omega(1+\omega)}(\eta_{\mu v}\eta_{p\omega} + \eta_{\mu p}\eta_{va} + \eta_\alpha\eta_{\mu p})(l^2)^2$$

Then (1) reduces to $4l_{\mu vpe}(l^2)^2$, with

$$l_{\mu vpa} = \frac{\omega-1}{\omega+1}(\eta_{aa}\gamma l_{ia}\gamma l_{up}) - \frac{(\omega-1)(\omega+2)}{\omega(\omega+1)}\eta_{\mu p}\eta_{u\sigma}$$

When the other two independent diagrams are included, $I_{\mu vp}$ is replaced by

$$I_{\mu vpc} + I_{\mu p\mu c} + I_{\mu v\sigma p} + \frac{\omega-2}{\omega}(\eta_{\mu v}\eta_{pv} + \eta_{\mu v}\eta_{up} + \eta_{\mu a}\eta_{\mu p})$$

Which vanishes in 4 dimensions.

4.7.4 Alpha and the anomalous magnetic moment

In what follows we will derive $a_e \equiv \frac{g-2}{2} = \frac{\alpha}{2\pi} + O(\alpha^2)$, the anomalous moment to first order in alpha. The magnetic moment of a particle is revealed when it traverses a magnetic field, especially if scattering with a magnetic field present. If we consider electron scattering by a static external magnetic field we have the term:

$$ieu(p')\gamma \cdot A(q = p' - p)u(p),$$

which has Feynman diagram at zeroth order:

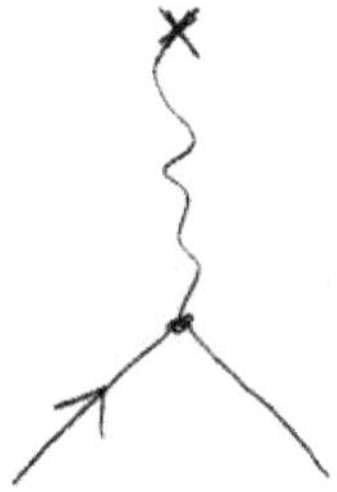

Denoting $a_e = a_e^{(1)} + a_e^{(2)} + a_e^{(3)} + a_e^{(4)} + a_e^{(*)}$, for the corrections at higher orders in α, and lastly with $a_e^{(*)}$, a separate group of corrections that appear in higher order graphs due to pair production with other leptons than the electron (muon and tau) and the hadrons.

For the radiative corrections to order α, these are given by the following Feynman diagrams:

We will soon derive the result that

$$a_e^{(1)} = \frac{\alpha}{2\pi},$$

which was first derived by Schwinger in 1948 [ref] (and experimentally measured Kusch and Foley 1947, 1948a [ref]).

Radiative corrections to order α^2 we have the Feynman diagrams:

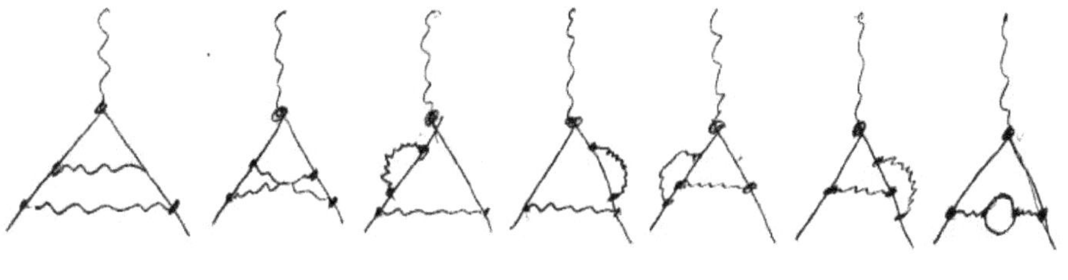

These graphs have been solved analytically to give $a_e^{(2)}$:

$$a_e^{(2)} = \left(\frac{\alpha}{\pi}\right)^2 \left[\frac{197}{144} + \frac{1}{12}\pi^2 + \frac{3}{4}\zeta(3) - \frac{1}{2}\pi^2 \ln 2\right],$$
where $\zeta(3)$ is the Riemann zeta function and $\zeta(3) = 1.20206$.

Radiative corrections to order α^3 involve 72 Feynman graphs [deWit], and is partly solved analytically and partly based on numerical solutions:

$$a_e^{(3)} = 1.1765(13)\left(\frac{\alpha}{\pi}\right)^3$$

Radiative corrections to order α^4 involve 891 Feynman graphs [Kinoshita and Lindquist], and they are solved based on numerical solutions:

$$a_e^{(4)} = (-0.8 \pm 2.5)\left(\frac{\alpha}{\pi}\right)^4.$$

At higher order the loop contributions from other leptons and the hadrons begin to factor in, with the following correction terms of similar magnitude to the $a_e^{(4)}$ correction:
Muon contribution: 2.8×10^{-12}
Tau contribution: 0.1×10^{-12}
Hadron contribution: $1.6(2) \times 10^{-12}$

The theoretical result for a_e to fourth-order in alpha is:

$$a_e = 1159652459(43) \times 10^{-12},$$

and this has been experimentally verified to within the precision indicated. Using the analytic relation to alpha, we can invert to get the theoretical(and experimental) value for alpha to high precision as [21]:

$$\alpha^{-1} = 137.035993(10).$$

Chapter 5. QFT

5.1 Multispecies and Multiparticle: Structure Emergence

So far we've considered field theories with one species of a particular type, be it bosonic or fermionic or a combination of the two (one of each with coupling as with QED). For two species of the same type there was the complex scalar field description, but mainly, the discussion has been for Lagrangian $\mathcal{L}(\varphi, \partial_\mu \varphi)$. Let's now consider multiple species of a particular type such that we have $\mathcal{L}(\varphi^i, \partial_\mu \varphi^i)$, where i indexes the fields in the family. We now consider exact symmetries (generated by Lie groups) of the Lagrangian, i.e., situations where $\mathcal{L}(\varphi^i, \partial_\mu \varphi^i)$ is invariant under transformations of type:

$$\varphi^i \to \varphi'^i = D(g)^i{}_j \varphi^j$$

If group that results is continuous we can look at infinitesimals and apply Noether's Theorem:

$$\varphi^i \to \varphi'^i + \delta\varphi^i$$
$$\delta\varphi^i = i(\theta \cdot t)^i_j \varphi^j$$

Shifting to abstract notation on dot-products where convenient: $\theta \cdot t = \theta^a t^a$, we apply Noether's Theorem to get conserved currents:

$$\theta^a j^{a\mu} = \frac{\partial \mathcal{L}}{\partial(\partial_\mu \varphi^i)} \delta\varphi^i = i\delta_\mu \overline{\varphi}_i (\theta \cdot t)^i_j \varphi^j,$$

Where the latter result assumes a Lagrangian with standard kinetic terms like $\partial_\mu \overline{\varphi}_i \partial_\mu \varphi^i$. So,

$$j^{a\mu} = i(\partial_\mu \overline{\varphi}_i)(t^a)^i_j \varphi^j$$

where

$$\partial_\mu j^{a\mu} = 0.$$

And there is the usual conserved charge relation that can be expressed:

$$Q^a = -\int d^3x j^{a\,0}(x) = -i\int d^3x\, \pi_i (t^a)^i_j \varphi^j.$$

To show the group structure given this invariance we need to have a representation to work with… it so happens that the conserved charges (cardinality the same as the number of species) is itself a representative of the algebra. Let's start by verifying this amazingly convenient result:

$$[Q^a, Q^b] = -(t^a)^i_j (t^b)^k_\ell \int d^3x d^3y [\pi_i(\lambda)\varphi^j(x), \pi_k(y)\varphi^\ell(y)]$$

And since $[\varphi^i(x), \pi_j(y)] = i\delta^i{}_j\delta(x-y)$ we get:

$$[Q^a, Q^b] = -i(t^a)^i_j(t^a)^k_\ell \int d^3x d^3y [\delta^i_k \pi_i(x)\varphi^\ell(y) - \delta^i{}_l \pi_k(y)\varphi^j(x)]\,\delta(x-y)$$

$$= -\int d^3x d^3y [\pi_i(x)(t^a t^b)^i{}_l \varphi^\ell(x), \pi_k(y) - \pi_k(x)(t^b t^a)^k{}_j \varphi^j(x)]$$

$$= -\int d^3x \pi_i(\lambda)[t^a, t^b]^i{}_\ell \varphi^\ell(x) = if^{abc}Q^c$$

Also from Noether's theorem, the Q'^s generate the group transformation itself:

$$[Q^a, \varphi^i(x)] = -(t^a)^i_j \varphi^j(x)$$

So,

$$[\theta\cdot Q, \varphi^i(x)] = -(\theta\cdot t)^i_j \varphi^j(x) = i\delta\varphi^i$$

$$\Rightarrow e^{i\theta\cdot Q}\varphi^i(x)e^{-i\theta\cdot Q} = D(g)^i_j\varphi^j, \quad D(g) = e^{i\theta\cdot t}$$

Let's apply the above to the situation of asymptotic fields where there is, presumably, a simple field decomposition with creation and annihilation operators acting on the different modes, .ie., consider $U\varphi_{asy}U^{-1}$, where $U = e^{i\theta\cdot Q}$, and for asymptotic creation operators, therefore:

$$|i\rangle = a_i^\dagger|0\rangle \rightarrow \ Ua_i^\dagger U^{-1}|0\rangle,$$

But want $|i\rangle \rightarrow \ U|i\rangle$, so need $U|0\rangle = |0\rangle$ (and $Q^a|0\rangle = 0$).

What do these invariance requirements indicate vis-à-vis the Hamiltonian operator? Assuming the (unitary) operation of the group transformation commutes with the Hamiltonian (e.g., is not time dependent) we have: $[H, U] = 0$, thus:

$$H|i\rangle = E_i|i\rangle \rightarrow \ HU|i\rangle = UU^{-1}HU|i\rangle = UH|i\rangle = E_iU|i\rangle$$

Which indicates that the states are degenerate (share the same energy level).

From Reduction Formula analysis:

$$\langle 0|T\varphi_i(x_,)\varphi_{i2}(x_2),,,,\varphi_{in}(x_n)|0\rangle$$

Switch to up indices:

$$\varphi'^i = U\varphi^i U^{-I} = D(g)^i_j\varphi^i$$

Now, above time ordered

$$\langle 0|T\varphi_i(x_,)\varphi_{i2}(x_2),,,,\varphi_{in}(x_n)|0\rangle$$
$$= D(g)^i_j \ldots D(g)^{in}_{in}\langle 0|T\varphi_i(x_,)\varphi_{i2}(x_2),,,,\varphi_{in}(x_n)|0\rangle$$

5.2 Spontaneous Symmetry Breaking

With gauge invariance there is manifest symmetry as indicated, but what if that symmetry is explicitly broken: $U|0\rangle \neq |0\rangle$ and $Q^a|0\rangle \neq 0$ for some a? Strictly speaking, if the symmetry is spontaneously broken as indicated, the Q^a are not well defined:

$$Q = -\int d^3x j_o(\overline{x}, t)$$

since $Q|0\rangle \neq 0$ means there is an infinite norm:

$$\langle 0|QQ|0\rangle = -\int d^3x \langle 0|j_o(\overline{x}, t)Q|0\rangle = -\int d^3x \langle 0|j_0(0)Q|0\rangle = \infty$$

To be really careful define integral in Q to be over a finite volume but then Q is no longer time independent:

$$Q_V^a = -\int_V^{\square} d^3x j_0(\overline{x}, t)$$

Where for integral over volume V, we take the limit:

$$\lim_{V\to\infty} \frac{d}{dt}[Q_V^a, A(z)] = 0.$$

However, note that

$$\frac{d}{dt}[Q_V^a, A(z)] = -\int_V^{\square} d^3x[\partial^0 j_0(\overline{x}, t)A(z)] = -\int_{\partial V}^{\square} d\Sigma \cdot [j(\overline{x}, t), A(z)]$$

and the commutator in the last is only non-zero for time-like separated x and z. Once the volume is taken sufficiently large that x and z become space-like separated, we get zero commutator and the desired limit holds. Note that this is a nontrivial causality relationship that is chosen to be enforced axiomatically in the Wightman description in Section 7.3.

So, we have

$$\frac{d}{dt}[Q_{\square}^a, A(z)] = 0,$$

Thus we have the two cases $[Q_{\square}^a, A(z)] = 0$ and $[Q_{\square}^a, A(z)]$ equal to a constant (non-zero). The former corresponds to the manifest symmetry, the latter to a case where we have spontaneously broken symmetry. Let's consider the consequences of the new ($[Q_{\square}^a, A(z)] \neq 0$) situation:

$$\lim_{V\to\infty} \langle 0|[Q_V^a, A(z)]|0\rangle = \delta A(t) \neq 0$$

$$\delta A(t) = \lim_{V\to\infty} \sum_n [\langle 0|Q_V^a(t)|n\rangle\langle n|A|0\rangle - \langle 0|A|n\rangle\langle n|Q_V^a(t)|0\rangle]$$

$$\delta A(t) = -\lim_{V\to\infty}\sum_n \int d^3x[\langle 0|j_0(\overline{x},t)|n\rangle\langle n|A|0\rangle - \langle 0|A|n\rangle\langle n|j_0(\overline{x},t)|0\rangle]$$

$$\delta A(t) = -\sum_n (2\pi)^3\delta_3(\vec{p}_n)\,[\langle 0|j_0(0)|n\rangle\langle n|A|0\rangle e^{-iE_nt} - \langle 0|A|n\rangle\langle n|j_0(0)|0\rangle e^{iE_nt}]$$

which is generally non-zero, so consistent. Now consider the time-derivative:

$$\lim_{V\to\infty}\frac{d}{dt}\langle 0|[Q_V^a, A(z)]|0\rangle = 0$$

To get:

$$0 = \sum_n (2\pi)^3\delta_3(\vec{p}_n)E_n\,[\langle 0|j_0(0)|n\rangle\langle n|A|0\rangle e^{-iE_nt} + \langle 0|A|n\rangle\langle n|j_0(0)|0\rangle e^{iE_nt}]$$

Or

$$E_n(\vec{p}_n = 0) = 0.$$

These are known as Goldstone modes.

Different ways in which a symmetry can be broken:

Starting with

$$[Q^a, \varphi^i(x)] = -(t^a)^i_j\varphi^j(x)$$

Taking the vacuum expectation value on the left-hand side can give nonzero, so we can have:

$$\langle 0|[Q_V^a, A(z)]|0\rangle \neq 0 \rightarrow\ (t^a)^i_j\langle 0|\varphi^j(x)|0\rangle \neq 0.$$

If this occurs, we have spontaneous symmetry breaking.

As we proceed in what follows we could use a real basis or complex. Often the complex is more compact or elegant, but here use of complex fields would require determination of the adjoints to then compute the invariances, a lot of extra work and notational headache. So, we take the other extreme, that all fields are strictly indexed as reals. Let's begin with a "tree level" description with Lagrangian

$$\mathcal{L} = \frac{1}{2}\partial_\mu\varphi^i\partial^\mu\varphi^i - V(\varphi)$$

and associated Hamiltonian

$$\mathcal{H} = \frac{1}{2}\partial_0\varphi^i\partial^0\varphi^i + \frac{1}{2}\partial_k\varphi^i\partial^k\varphi^i - V(\varphi)$$

The ground state minimizes $\mathcal{H}$, so we expect φ's that are constant or zero, and expect:

$$\frac{\partial V(\varphi)}{\partial \varphi^i} = 0$$

Let's denote

$$v^i = \langle 0|\varphi^i(x)|0\rangle$$

in what follows, and explore this further.

Let's try the theory of $V(\varphi) = \frac{1}{2}m^2|\varphi|^2 + \frac{\lambda}{4!}|\varphi|^4$, so v^i is just v for the one species, for which we get:

$$\frac{\partial V(\varphi)}{\partial \varphi^i} = m^2\varphi + \frac{\lambda}{6}|\varphi|^2\varphi = 0 \rightarrow m^2 v + \frac{\lambda}{6}|v|^2 v = 0.$$

Thus, we need

$$|v| = \sqrt{-6m^2/\lambda}.$$

Solutions only exist for $m^2 < 0$. In other words, spontaneous breakdown only occurs if $m^2 < 0$.

Now lets consider the specific form:

$$\left.\frac{\partial V(\varphi^i)}{\partial \varphi^i}\right|_{\varphi^i = v^i} = 0$$

Let's start by shifting the φ^i's so they match with the reduction formalism:

$$\varphi^i = \varphi'^i + v^i \rightarrow \left.\frac{\partial V(\varphi'^i)}{\partial \varphi'^i}\right|_{\varphi'^i = 0} = 0.$$

Dropping primes, we now write the Taylor expansion for $V(\varphi^i)$:

$$V(\varphi^i) = C + \frac{1}{2}\left.\frac{\partial^2 V(\varphi^i)}{\partial \varphi^i \partial \varphi^j}\right|_{\varphi^i = v^i} \dot{\varphi}^i \dot{\varphi}^j + \cdots$$

The first term of the series is the constant C, to be eliminated in the ∂V expressons to follow, so precise value irrelevant. The linear term is zero as indicated, so the leading relevant term is the second derivative term, whose coefficient is often described in terms of a mass matrix M_{ij}^2 for the φ^i's:

$$V(\varphi^i) = C + \frac{1}{2}M_{ij}^2 \dot{\varphi}^i \dot{\varphi}^j + \cdots$$

(convention is to use mass squared as a reminder that it is positive). But we also have the expression for

$$\frac{\partial V}{\partial \varphi^i}(t^a)^i_j \varphi^j = 0$$

Thus we also have

$$\frac{\partial V}{\partial \varphi^i \partial \varphi^k}(t^a)^i_j \varphi^j + \frac{\partial V}{\partial \varphi^i}(t^a)^i_k = 0$$

If we set $\varphi^i = v^i$, we get:

$$M^2_{ik}(t^a)^i_j v^j = 0.$$

Thus, $(t^a)^i_j v^j$ can be an eigenvector of M^2_{ik} with zero eigenvalues (for a spontaneously broken theory).

Let's break the generators (t^a) into two groups:
(1) Unbroken Symmetry Group: $(t^a)^i_j v^j = 0$, i.e., the t^a's annihilate the v^j's, and not a zero-value of M^2_{ik}.
(2) Broken symmetry Group: $(t^a)^i_j v^j \neq 0$, thus to satisfy $M^2_{ik}(t^a)^i_j v^j = 0$, must have $(t^a)^i_j v^j$ be an eigenvector of M^2_{ik} with zero eigenvalue.

Suppose we denote Y^p for unbroken symmetry generators ($Y^p v = 0$) and X^a for broken. We can write (with use of group structure constants f^{pqr}, etc.):

$$[Y^p, Y^a] = if^{pqr}Y^r + if^{pqa}X^a$$

If we act with v^j on the equation it gives:

$$0 = 0 + if^{pqa}X^a v \rightarrow f^{pqa} = 0$$

Thus,

$$[Y^p, Y^a] = if^{pqr}Y^r$$

Thereby forming a subalgebra (an invariant subgroup which is not spontaneously broken).

Similarly, the X^a transform as

$$[Y^p, X^a] = if^{pab}X^b + if^{paq}Y^q = if^{pab}X^b$$

Thus, the X'^s form a representation of the invariant subalgebra.

Breaking subgroups

Let's now consider patterns of breaking for specific cases of interest.

Case 1: SO(3)

Suppose we have $SO(3)$ with φ'^s in defining rep: $\varphi = \begin{pmatrix} \varphi^1 \\ \varphi^2 \\ \varphi^3 \end{pmatrix}$. Since $SO(3)$ is real we are already working with real φ to get:

$$(t^a)_{ij} = -i\varepsilon_{aij}$$

Where

$$t^1 = -i\begin{pmatrix} 0 & 0 & 0 \\ 0 & 0 & 1 \\ 0 & -1 & 0 \end{pmatrix}, \quad t^2 = -i\begin{pmatrix} 0 & 0 & -1 \\ 0 & 0 & 0 \\ 1 & 0 & 0 \end{pmatrix}, \quad t^3 = -i\begin{pmatrix} 0 & 1 & 0 \\ -1 & 0 & 0 \\ 0 & 0 & 0 \end{pmatrix}$$

Suppose we have

$$\langle 0|\varphi^i|0\rangle = \begin{pmatrix} 0 \\ 0 \\ v_3 \end{pmatrix}$$

then $t^1 \cdot v \neq 0$, $t^2 \cdot v \neq 0$, and $t^3 \cdot v = 0$. The generator t^3 generates a $SO(2)$ algebra (x-y rotations leave v invariant. So we have the breaking $SO(3) \to SO(2)$. If we have an arbitrary result

$$\langle 0|\varphi^i|0\rangle = \begin{pmatrix} v_1 \\ v_2 \\ v_3 \end{pmatrix}$$

We can reconfigure the fields such that only $v_3 \neq 0$, with the same result as before. Alternatively, we can simply solve the expression in terms of the unit vector defined by:

$$\hat{n} = \frac{1}{\sqrt{v_1^2 + v_2^3 + v_3^3}}(v_i, v_2, v_3),$$

And then have

$$(v_1 t^1 + v_2 t^2 + v_2 t^3) \cdot v = 0$$

So, $SO(3)$ breaks to $SO(2)$, and only $SO(2)$.

***Case 2: SU*(2)**

For $SU(2)$ let's begin with the standard complex $\varphi = \begin{pmatrix} \varphi^1 \\ \varphi^2 \end{pmatrix}$ and complex generator representation with the matrices:

$$\sigma^1 = \begin{pmatrix} 0 & 1 \\ 1 & 0 \end{pmatrix}, \quad \sigma^2 = \begin{pmatrix} 0 & -i \\ i & 0 \end{pmatrix}, \quad \sigma^3 = \begin{pmatrix} 1 & 0 \\ 0 & -1 \end{pmatrix},$$

and consider $v = \begin{pmatrix} 0 \\ v \end{pmatrix}$, then all σ's give $\sigma \cdot v \neq 0$, so no invariant subgroup, everything is broken. For a more general argument, consider that any 2×2 Hermitian matrix can be diagonalized and must be traceless, so transformation can always give:

$$\varphi \to \varphi = a\sigma^3$$

and since σ^1 and σ^2 don't commute with σ^3, we must have breaking.

Now consider $SU(2)$ with real representation: $\varphi = \begin{pmatrix} \varphi^1 \\ \varphi^2 \\ \varphi^3 \end{pmatrix}$. This leads to the same derivation as for seen in case (1), but here we have the breaking $SU(2) \to U(1)$. If we do this approach, however, care must be taken in considering the adjoint representation given by:

$$\tilde{\varphi} = \sum_{i=1}^{3} \varphi^i \sigma^i,$$

where the adjoint has returned us back to a 2×2 Hermitian matrix formulation, with arguments like above to show breaking.

***Case 3: SU*(3)**

Let's start with a real representation with φ^i's, where $i = 1 \ldots 8$. The adjoint is given by:

$$\tilde{\varphi} = \sum_{i=1}^{8} \varphi^i \lambda_i,$$

where λ_i are the Gell-Mann matrices with $[\lambda_i, \lambda_j] = 2if^{ijk}\lambda_k$. Previously we found:

$$\delta\varphi^i = i(\theta \cdot t)^i_j \varphi^j(x)$$

and t's in adjoint representation are:

$$(t^i)^j{}_k = -if^{ijk}.$$

If we now compute

$$\delta\tilde{\varphi} = \sum_{i=1}^{8} \delta\varphi^i \lambda_i = \left[\frac{i\theta \cdot \lambda}{2}, \tilde{\varphi}\right]$$

In the above we can choose $\tilde{\varphi}$ to be diagonal since it is Hermitian, to get the form

$$\tilde{\varphi} = v_3\lambda_3 + v_8\lambda_8 + aI,$$

It's also traceless, so we drop the last term as well (or $a = 0$). If both v_3 and v_8 are arbitrary, this creates an invariant subalgebra after breaking:

$$SU(3) \to U(1) \times U(1).$$

If only one of the $\{v_3, v_8\}$ is arbitrary (the other zero, say), then we get for invariant subalgebra after breaking:

$$SU(3) \to SU(2) \times U(1).$$

Note that, in the Gauge Field analysis we don't just get "$SU(2)$" or "$SU(3)$", rather, we get them in pairs, usually denoted 'L' (left) and 'R' (right) to indicate the duality (and imply its nature is a chirality, which will be the case). We get, for an $SU(3)$ gauge field:

$$SU(3)_L \times SU(3)_R.$$

Suppose the Left field 'breaks' we then have a field gauge:

$$U(1) \times SU(2) \times SU(3)$$

There is no mixing with the "$SU(3)_R$" gauge part, so its just written as $SU(3)$, and the subscripting for 'L' is dropped on the left side as well … in reference to $SU(3)$. However, note that symmetry breaking brings us to $SU(2)$ (which does not break further), and that, as before, we can split to $SU(2)_L$ and $SU(2)_R$. This is very close to the gauge description of the Standard Model.

Case 4: SU(N)

There are many discussions exploring higher order gauge theories, including those indicated by various forms of "Grand Unification" such as $SU(5)$. In these efforts the odd gauge product of the Standard Model, explainable by way of symmetry breaking like indicated previously, is thought to have been the result of a 'unified' field theory, such as $SU(5)$, that occurred at early times (high temperature) and then 'broke'. The causality inferred is a deeper unified theory exists that breaks (when the Universe expands and cools sufficiently, say). In Book 7 [7] of this series, Emanator Theory describes how the odd product gauge is directly projected as is. The projection is to the gauge of the standard model $U(1) \times SU(2) \times SU(3)$ for as many generations as possible within the allowed number of constants in the theory (invariants under propagation), this turns out to be the three generations of light matter observed. There then remains the prospect of projecting a different breaking (attributable to the dark matter), such as sterile neutrino's, to complete the count on invariants of the theory (22). For purposes of this Series, therefore, further discussion of $SU(N)$ will be left to other sources.

Let's now return to the generating functional formalism, for generating Green's functions, and apply it to the multiple species gauge-field situation, starting with the definition for the effective action potential:

$$Z[J^i] = \int \mathcal{D}\varphi^i e^{i\int[\mathcal{L}(\varphi^i)+\Sigma_i J^i\varphi^i]d^4x} = e^{W[J^i]}$$

where $W[J^i]$ is the generator of connected Green's Functions as before. Note that in this construction, unlike before, the measure $\mathcal{D}\varphi^i$ is not infinite because the group space is compact (true for all Lie Groups). We are interested in the vacuum expectation of the fields, $\Phi^i(x)$, in the presence of external sources:

$$\frac{\delta}{i\delta J'(x)}W = \frac{1}{Z[J]}\frac{\delta Z}{i\delta J'(x)} \equiv \Phi^i(x) \quad \rightarrow \quad \Phi^i(x)\Big|_{J=0} \equiv \langle 0|\varphi^i(x)|0\rangle = v^i$$

$$\frac{\delta^2 W}{i\delta J^i(x) i\delta J^j(x)} = \frac{1}{Z[J]}\frac{\delta^2 Z}{i\delta J^i(x) i\delta J^j(x)} - \frac{1}{Z[J]^2}\frac{\delta Z}{i\delta J^i}\frac{\delta Z}{i\delta J^j}\Bigg|_{j=0}$$
$$= \langle 0|T\varphi^i(x)\varphi^i(y)|0\rangle - v^i v^i = \langle 0|T\big(\varphi^i(x) - v^i\big)\big(\varphi^j(x) - v^j\big)|0\rangle$$

Note that W automatically generates the shifted Greens function useful for the LSZ perturbation expansion.

Now consider

$$\frac{\delta W}{i\delta J(x)} = \Phi^i(x) \rightarrow \quad \Phi^i = \Phi^i[J]$$

Suppose we want to invert to a form $J^i = J^i[\Phi]$, this can be done by use of a functional Legendre Transformation and in doing so we find the effective action (Γ). Let's begin by simply considering the Legendre transformation of W to something 'Γ' to get:

$$i\Gamma[\Phi] = W[J] - i\int d^4x J^i(x)\Phi^i(x)$$

$$\frac{i\delta\Gamma}{\delta\Phi^i(x)} = \int dy \frac{\delta W}{\delta J^k(y)}\frac{\delta J^k(y)}{\delta\Phi^i(x)} - iJ^i(x) - i\int d^4y \frac{\delta J^k(y)}{\delta\Phi^i(x)}\Phi^k(y)$$
$$= -iJ^i(x)$$

So,

$$\frac{\delta\Gamma}{\delta\Phi^i(x)} = -J^i(x)$$

When we set $J'^s = 0$ we get

$$\frac{\delta\Gamma}{\delta\Phi^i(x)} = 0 \rightarrow \Phi^i(x) = \widehat{\Phi}^i(x)$$

Which then gives the form:

$$\frac{\delta W}{\delta J^i(x)}\Bigg|_{J^i=0} = \widehat{\Phi}^i(x) = \langle 0|\varphi^i(x)|0\rangle = v^i.$$

As we will now see, Γ is a generating functional for the 1.P.I. representation. Let's adopt a standard notational compaction before proceeding

Notational compaction

$$\Phi^i \equiv \Phi^i(x_i),\ \ \frac{\delta W}{\delta J^i} \equiv i\Phi^i,\ \ \frac{\delta \Gamma}{\delta \Phi^i} \equiv -J^i,$$

Actual shorthand: integration over x implied wherever appropriate, as is the placement of $\delta(x_i - x_j)$. Let's start by taking successive derivatives $\frac{\delta}{\delta\Phi^j}$:

$$i\delta^i_{\ j} = \frac{\delta^2 W}{\delta J^i \delta J^k}\frac{\delta J^k}{\delta \Phi^j} = \langle 0|T\varphi^i(x_i)\varphi^k(x_k)|0\rangle \frac{\delta^2\Gamma}{\delta\Phi^k(x_k)\delta\Phi^i(x_i)}$$

Since,

$$\langle 0|T\varphi^i(x_i)\varphi^k(x_k)|0\rangle = G^{(2)}_{ik}$$

we have:

$$\frac{\delta^2\Gamma}{\delta\Phi^i\delta\Phi^j} = i\left[G^{(2)}_{ij}\right]^{-1}.$$

Now for another derivative:

$$0 = \frac{\delta^3 W}{\delta J^i\delta J^k\delta J^p}\frac{\delta J^p}{\delta\Phi^\ell}\frac{\delta J^k}{\delta\Phi^j} + \frac{\delta^2 W}{\delta J^i\delta J^k}\frac{\delta^2 J^k}{\delta\Phi^j\delta\Phi^l}$$

So,

$$-iG^{(3)}_{ikp}\left[G^{(2)}_{pl}\right]^{-1}\left[G^{(2)}_{kj}\right]^{-1} = G^{(2)}_{ik}\frac{\delta^3\Gamma}{\delta\Phi^j\delta\Phi^\ell\delta\Phi^k}$$

This generalizes:

$$\Gamma = -i\sum_n \frac{1}{n!}\int \Gamma^{(n)}_{i1\ldots in}(x_1,\ldots,x_n)\Phi^{i1}(x_1)\ldots\Phi^{in}(x_n)dx_1,\ldots,dx_n$$

and

$$\Gamma^{(n)}_{i1\ldots in}(x_1,\ldots,x_n) = \langle 0|T\varphi_{i1}(x_1)\ldots\varphi_{in}(x_n)|0\rangle_{\ 1.P.I.}$$

Suppose we have the multi-species gauge symmetry described, where:

$$\varphi^i \to \varphi^i + \delta\varphi^i,\quad \mathcal{L}(\varphi^i + \delta\varphi^i) = \mathcal{L}(\varphi^i),\quad \delta\varphi^i = i(\theta\cdot t)^i_j\varphi^j(x),$$

we then have

$$Z[J^i] = \int \mathcal{D}\varphi\, e^{i\int[\mathcal{L}(\varphi^i)+\sum_i J^i\varphi^i]d^4x}$$

and for the gauge transformed form, if we take $\varphi^i \to \varphi^i + \delta\varphi^i = \varphi^i + i(\theta\cdot t)^i_j\varphi^j$, which can also be written $\varphi^i[\delta^i_j + i(\theta\cdot t)^i_j]$, we can then substitution in the exponential. The problem is the impact of a transformation $[\delta^i_j + i(\theta\cdot t)^i_j]$ on the measure. It turns out, by direct summation on the infinitesimals, that the Lie symmetry provides a unitary transformation, thus the Jacobian is 1 and the measure stays the same. We can therefore write:

$$Z[J^i]\big|_{\varphi^i \to \varphi^i + \delta\varphi^i} = \int \mathcal{D}\varphi\, e^{i\int\left[\mathcal{L}(\varphi^i) + \Sigma_i J^i[\delta_{ij}^{\square} + i(\theta\cdot t)_{ij}^{\square}]\varphi^j\right]d^4x}$$

$$= Z\left[J^i + i(\theta\cdot t)^i{}_j J^j\right].$$

Thus

$$Z[J^i] = Z\left[J^i + i(\theta\cdot t)^i{}_j J^j\right],$$

from which it follows that:

$$W[J^i] = W\left[J^i + i(\theta\cdot t)^i{}_j J^j\right],$$

$$\Gamma[\Phi^i] = \Gamma\left[\Phi^i + i(\theta\cdot t)^i{}_j \Phi^j\right]$$

$$V_{eff}[\hat{\Phi}^i] = V_{eff}\left[\hat{\Phi}^i + i(\theta\cdot t)^i{}_j \hat{\Phi}^j\right]$$

Thus,

$$i(\theta\cdot t)^i{}_j \hat{\Phi}^j \frac{\partial V_{eff}}{\partial \hat{\Phi}^i} = 0$$

If $\frac{\partial V_{eff}}{\partial \hat{\Phi}^i} = 0$ for some $\hat{\Phi}^i = v^i \neq 0$, then these are the previous v^i. Otherwise, we have the broken condition $(t^a)^i_j v^j \neq 0$ for some $t^{a's}$, and those $t^{a's}$ are broken.

Another way to express the gauge invariance is to work with the Γ relation, where:

$$\frac{\delta\Gamma}{\delta\theta^a} = 0 \to \int d^4x \Phi^i(x)(t^a)^i_j \frac{\delta\Gamma}{\delta\Phi^i(x)} = 0$$

In the latter form we've captured the Ward Identities in one of the standard forms. For an alternate form, consider taking the derivative with respect to $\Phi^k(y)$

$$\int d^4x\left[\delta^i_k \delta(x-y)(t^a)^i_j \frac{\delta\Gamma}{\delta\Phi^j(x)} + \Phi^i(x)(t^a)^i_j \frac{\delta\Gamma^2}{\delta\Phi^j(x)\delta\Phi^k(y)}\right]$$

And then let $\Phi^i = v^i$:

$$\int d^4x v^i (t^a)^i_j \Delta'^{-1}_{Fjk}(x-y) = 0$$

where Δ'^{-1}_{Fjk} is the full inverse propagator. In momentum space:

$$v^i(t^a)^i_j \hat{\Delta}'^{-1}_{ij}(0) = 0.$$

In this latter form we have a statement of Goldstone's theorem valid to all orders.

Ward identities without symmetry breaking gives relations between Green's functions with the same number of arguments. If there is symmetry breaking, the Ward identities relate Green's functions with different numbers of arguments.

Goldstone model

Let's consider the Goldstone model in detail:

$$\mathcal{L} = \partial_\mu \varphi^* \partial^\mu \varphi - \mu^2 |\varphi|^2 - \frac{\lambda}{4!} |\varphi|^4$$

Where we choose the potential term to be:

$$V = \mu^2 |\varphi|^2 + \frac{\lambda}{4!} |\varphi|^4$$

And get:

$$\frac{\partial V}{\partial \varphi^*} = 0 \quad \rightarrow \quad \mu^2 + \frac{2\lambda}{4!} |\varphi|^2 = 0.$$

There are two solutions, $\varphi = 0$ and $|\varphi|^2 = \frac{-12\mu^2}{\lambda}$, and the latter solution only exists for $\mu^2 < 0$. Suppose we have $\mu^2 < 0$, then we can write the minimal potential (vacuum) solution as:

$$\varphi_{vac} = \sqrt{\frac{-12\mu^2}{\lambda}}.$$

Note that this vacuum is degenerate due to the following symmetry:

$$\varphi \rightarrow e^{i\theta} \varphi, \;\; \varphi^* \rightarrow e^{-i\theta} \varphi^*,$$

thus the general vacuum solution is:

$$\varphi_{vac} = \sqrt{\frac{-12\mu^2}{\lambda}} e^{i\theta}.$$

Let's now consider real and imaginary terms separately, and examine a specific choice of vacuum. Suppose

$$\varphi = \frac{1}{\sqrt{2}} (\varphi_1 + i\varphi_2)$$

and we choose a vacuum

$$\langle 0|\varphi_1|0\rangle = \sqrt{\frac{-4!\,\mu^2}{\lambda}} \equiv v, \quad \langle 0|\varphi_2|0\rangle = 0.$$

Thus breaking the symmetry. Now define

$$\left.\begin{aligned} \varphi_1' &= \varphi_1 - v \\ \varphi_2' &= \varphi_2 \end{aligned}\right\} \rightarrow \langle 0|\varphi_i'|0\rangle = 0$$

And rewrite the Lagrangian:

$$\mathcal{L} = \frac{1}{2}\left[\left(\partial_\mu \varphi_1'\right)^2 + \left(\partial_\mu \varphi_2'\right)^2\right] - \frac{\lambda}{4!}v^2\varphi_1'^2 - \frac{\lambda}{4!}v\left(\varphi'^2_1 + \varphi_2'^2\right)\varphi_1'$$
$$- \frac{\lambda}{4!\,4}\left(\varphi_1'^4 + \varphi_2'^4\right).$$

With the Lagrangian in this form, there is no mass term for $\varphi_2'^2$, it becomes the Goldstone Boson, and there is a new (cubic) interaction term.

We have

$$[Q, \varphi] = -\varphi, \quad [Q, \varphi^*] = \varphi^*$$

thus

$$[Q, \varphi_1] = -i\varphi_2, \quad [Q, \varphi_2] = i\varphi_1$$

thus

$$\langle 0|[Q, \varphi_2]|0\rangle = i\langle 0|\varphi_1|0\rangle \neq 0 \quad \rightarrow \quad Q|0\rangle \neq 0,$$

And we meet the condition for spontaneous symmetry breaking. Because we know that $\langle 0|Q|n\rangle\langle n|\varphi_2|0\rangle$ is nonzero, then $\langle n|\varphi_2|0\rangle$ is nonzero, thus φ_2 is associated with a Goldstone boson.

Let's consider the gauge symmetry indicated by $e^{i\theta}$, e.g., the rotations. For this we have
$(t^a)^i_j\langle 0|\varphi_i|0\rangle$ vectors that ae zero eigenstates of the mass matrix. The 't' is the generator of rotations and is represented by:

$$t = -i\begin{pmatrix} 0 & 1 \\ -1 & 0 \end{pmatrix}, \qquad \langle 0|\varphi^j|0\rangle = \begin{pmatrix} v \\ 0 \end{pmatrix}.$$

So,

$$(t)\langle 0|\varphi|0\rangle = \begin{pmatrix} 0 \\ iv \end{pmatrix}$$

which indicates a zero eigenvector in the direction of φ_2.

So far we have used tree approximation for $\langle 0|\varphi|0\rangle = \binom{v}{0}$, etc. Corrections beyond tree approximation need the effective potential. This begs the question -- Do we "break φ" and implement perturbation theory before or after spontaneous symmetry breaking? The easiest way to maintain gauge invariance, and not have to redo the effective potential calculation at every order of perturbation, is to have perturbation expansion on Feynman diagram loop order. This works because loop order does correspond with a perturbation expansion in terms of a small constant factor (dimensionful) multiplying the Lagrangian in the Action.

Consider the standard Lagrangian in terms of kinetic and potential energy terms:

$$\mathcal{L} = T - V,$$

When evaluated as part of the Action, suppose the Lagrangian is dimensionful and we have terms involving:

$$\frac{1}{a}\mathcal{L}.$$

This means we will have

$$V \to \frac{1}{a}V, \;\; \Delta_f \to a\Delta_f,$$

in the associated Feynman diagrams of the quantum field theory. For a given Feynman diagram, the power of 'a' in the expression will be given by the number of internal lines 'I' and the number of vertices 'V':

$$power(a) = I - V.$$

Recall the equation for the number of independent momenta in a diagram gives a relation between loop count 'L' and the aforementioned parameters:

$$L = I - V + 1.$$

Thus, loop count, a gauge invariant feature, can be used for cutoff since it is equivalent to a perturbation cutoff in the power of 'a':

$$power(a) = L - 1.$$

We have the occurrence of parameter a as ħ in the path integral:

$$Z = \int \mathcal{D}\varphi \, e^{iS/\hbar} = e^{W/\hbar}.$$

From this we see that if W is expanded to 0-loops, it is zeroth order in ħ, and at 1-loop order it is first order in ħ.

Example: (real scalar field with mass and interaction)
We have:

$$\mathcal{L} = \frac{1}{2}\left(\partial_\mu \varphi\right)^2 - \frac{1}{2}\mu^2\varphi^2 - \frac{\lambda}{4!}\varphi^4.$$

In the perturbation expansion we have the relation that the effective potential is 'i' times the sum of all 1.P.I diagrams with zero external momenta. We have a choice in the perturbation theory to decide what if the 'free' part and what is the 'interaction' part, we choose the interaction to be:

$$U = \frac{1}{2}\mu^2\varphi^2 + \frac{\lambda}{4!}\varphi^4.$$

Thus,

$$U''(\varphi) = \mu^2 + \frac{(4 \cdot 3)\lambda}{4!}\varphi^2.$$

Now make Feynman Rules:

external lines : $\longrightarrow\limits_{p}$ $i\Delta_F(p) = \dfrac{i}{p^2 - m^2 + i\epsilon}$ $\left.\right\}$ for choice of u

internal lines : $\longrightarrow\limits_{k}$ $\int \dfrac{d^4k}{(2\pi)^4} \dfrac{i}{k^2 - m^2 + i\epsilon}$ above $\Rightarrow m \Leftrightarrow$

vertex : $\overset{k_1}{\longrightarrow}\bullet\overset{k_2}{\longleftarrow}$ $-i\mu^2(2\pi)^4 \delta(k_1 + k_2)$

$\asymp$ $\diagdown\!\!\!\!\diagup$ $-i\lambda(2\pi)^4\delta(k_1 + k_2 + k_3 + k_4)$

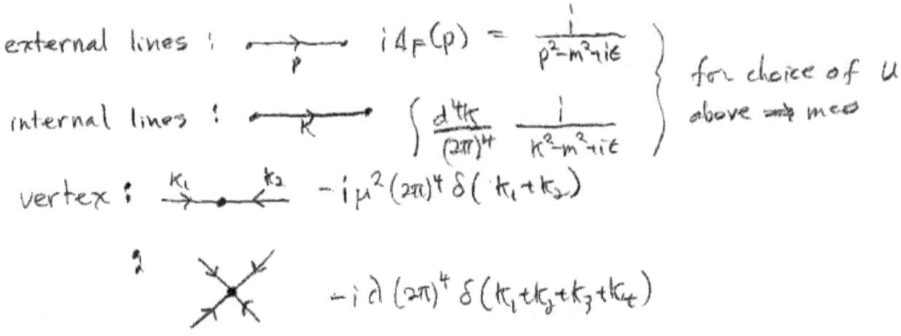

We want V_{eff} that generalizes beyond the tree level potential U to account for quantium corrections. For tree approximation analysis pertaining to ground state, i.e. Vacuum, relied on $\partial u/\partial\varphi = 0$. Extending to quantum analysis recall that we have the progression to a functional formulation as follows:

$$\mathcal{L}(\varphi) \to Z[J] \to W[J] \to \Gamma[\Phi]$$

where

$$\frac{\delta W}{\delta J} = i\Phi \quad \to \quad \frac{\delta\Gamma}{\delta\Phi^i} = -J^i.$$

$$\Gamma[\Phi] = -i\sum_n \frac{1}{n!} \int d^4x_1 \dots d^4x_n \Gamma^{(n)}_{i1\dots in}(x_1 \dots x_n)\Phi^{i1}(x_1) \dots \Phi^{in}(x_n)$$

where $\Gamma^{(n)}_{i1\dots in}(x_1 \dots x_n) = \langle 0|T\varphi^{i1}(x_1) \dots \varphi^{in}(x_n)|0\rangle$. Treating φ^i as constants:

$$\Gamma[\varphi^i] = -i\sum_n \frac{1}{n!} \int d^4x_1 \dots d^4x_n \Gamma^{(n)}_{i1\dots in}(x_1 \dots x_n)\varphi^{i1}(x_1) \dots \varphi^{in}(x_n)$$

$$= -(2\pi)^4\delta_4(o)V_{eff}(\varphi)$$

where

$$V_{eff}(\varphi) = i\sum_n \frac{1}{n!} \Gamma^{(n)}_{i1\dots in}(0 \dots 0)\,\varphi^{i1}(x_1) \dots \varphi^{in}(x_n)$$

and:

$$\frac{\partial V_{eff}}{\partial\varphi^i} = 0$$

yields the vacuum. Diagrammatically, a general 1-loop has:

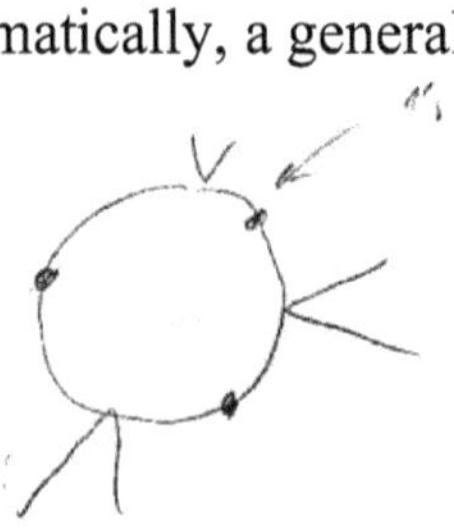

where the arrow point to a "mass insertion". We could simplify the diagram by using:

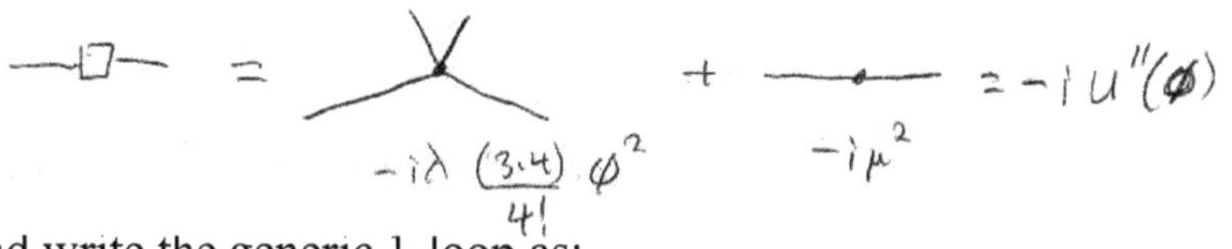

and write the generic 1-loop as:

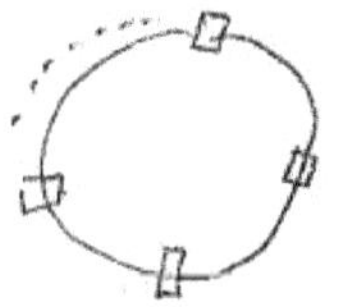

from this we get for the effective potential V_{eff}:

$$V^{(1)}{}_{eff} = i\sum_n \int \frac{d^4k}{(2\pi)^4}[-iU''(\varphi)]^n \left(\frac{i}{k^2+i\epsilon}\right)^n \left(\frac{1}{2n}\right)$$

where the $\left(\frac{i}{k^2+i\epsilon}\right)$ terms are the massless propagators, and there is a $\left(\frac{1}{2n}\right)$ symmetry factor (the symmetries on the circle, e.g., n cyclic permutations and 2 reflections). We can do the sum (a reverse Taylor series expansion) to get:

$$V^{(1)}{}_{eff} = \frac{i}{2}\int \frac{d^4k}{(2\pi)^4} ln\left[1 - \frac{U''[\varphi]}{k^2+i\epsilon}\right]$$

We have (hyper)spherical symmetry, so let's shift the measure accordingly and also Euclideanize:

$$d^4k \to id^4k_E = 12\pi^2 k_E^3 dk_E.$$

The integral divergence is now dealt with by momentum cut-off:

$$V^{(1)}{}_{eff} = \frac{\pi^2}{(2\pi)^4}\int_0^\Lambda k_E^3 dk_E ln\left[1 + \frac{U''(\varphi)}{k_E^2}\right]$$

So

$$V^{(1)}{}_{eff} = \frac{1}{64\pi^2}\{2\Lambda^2 U''(\varphi) - [U''(\varphi)]^2 ln\Lambda^2 + [U''(\varphi)]^2\left(lnU''(\varphi) - \frac{1}{2}\right)\}$$

We can then write the expansion of the potential to 1-loop, with addition of renormalization counter terms as usual:

$$V_{eff} = V_{eff}^{(0)}(\varphi) + V_{eff}^{(1)}(\varphi) + \cdots c.t.'s$$

Or

$$V_{eff} = U(\varphi) + V_{eff}^{(1)}(\varphi) + \cdots$$

As we have seen, there are three renormalization counter terms (from wavefunction, mass and coupling renormalization), thus we expect to have:

$$\mathcal{L} = \left(\frac{1}{2}\partial_\mu\varphi\right)^2 - \frac{1}{2}\mu^2\varphi^2 - \frac{\lambda}{4!}\varphi^4 + \frac{1}{2}A\left(\partial_\mu\varphi\right)^2 - \frac{1}{2}B\varphi^2 - \frac{1}{4!}C\varphi^4.$$

Recall that $\frac{1}{2}A\left(\partial_\mu\varphi\right)^2$ is the part of the correction for the propagator (free theory), so does not effect V_{eff}, Thus, for V_{eff} to 1-loop order:

$$V_{eff} = U(\varphi) + V_{eff}^{(1)}(\varphi) + \frac{1}{2}B\varphi^2 - \frac{1}{4!}C\varphi^4.$$

and B and C are chosen to cancel. We want to connect to the classical limit, so want:

$$\frac{\partial^2 V_{eff}}{\partial\varphi^2}\Big|_{\varphi=0} = \mu^2, \quad \frac{\partial^4 V_{eff}}{\partial\varphi^4}\Big|_{\varphi=0} = \lambda.$$

So,

$$B = -\frac{1}{64\pi^2}\frac{\lambda}{2}\left(\Lambda^2 - \mu^2 \ell n\left(\frac{\Lambda^2}{\mu^2}\right)\right), \quad C = \frac{1}{64\pi^2}\frac{3\lambda^2}{2}\left[\ell n\frac{\Lambda^2}{\mu^2} - 1\right]$$

And to 1-loop:

$$V_{eff} = \frac{1}{2}\mu^2\varphi^2 + \frac{\lambda}{4!}\varphi^4$$
$$+\frac{1}{64\pi^2}\left\{\left(\frac{\lambda}{2}\varphi^2 + \mu^2\right)^2 ln\left(1 + \frac{\lambda\varphi^2}{2\mu^2}\right) - \frac{3\lambda^2}{8}\varphi^4 - \frac{\lambda}{2}\mu^2\varphi^2\right\}$$

The above loop expansion can be trusted when φ is small, and then the loop correction is a "correction" to classical result. We can now identify the mass: $\frac{\partial^2 V_{eff}}{\partial\varphi^2}\Big|_{\varphi_{min}}$=mass. These ideas are expanded further in [50].

Note that if we start with a theory that is massless $\mu^2 = 0$, we cannot renormalize as indicated above, at $\varphi = 0$, and use instead:

$$\frac{\partial^2 V_{eff}}{\partial\varphi^2}\Big|_{\varphi=0} = 0, \quad \frac{\partial^4 V_{eff}}{\partial\varphi^4}\Big|_{\varphi=M} = \lambda(M).$$

where M is arbitrary and $\lambda(M)$ provides a renormalization group equation. Classically we have

$$V_{eff}\Big|_{\varphi_{min}=\varphi_c} = 0,$$

At 1-loop V_{eff} we can go negative. To see this consider e small so that we can ignore λ^2 terms (recall that $\lambda = \mathcal{O}(e^4)$), thus:

$$V_{eff} \cong \frac{\lambda}{4!}\varphi_c^4 + \frac{3e^4}{64\pi^2}\varphi_c^4\left(ln\frac{\varphi_c^2}{M^2} - \frac{22}{6}\right)$$

Let's now test for location of the minimum:

$$0 = \frac{dV_{eff}}{d\varphi_c} = \frac{\lambda}{3!}\varphi_c^3\left[1 + \frac{9e^2}{8\pi^2\lambda}\left(ln\frac{\varphi_c^2}{M^2} - \frac{22}{6}\right)\right]$$

So, we have the classical solution $\varphi_c = 0$, and the other solution, often denoted $\varphi_c = \langle\varphi\rangle \neq 0$. With the non-classical solution we see spontaneous symmetry breaking due to radiative corrections that did not exist at classical level. What is the mass of the field? Recall:

$$\Gamma^{(2)}(p) = (p^2 - m^2)(1 + \Sigma(p))$$

And

$$V_{eff} = \cdots + \frac{1}{2}V_{eff}''(\langle\varphi\rangle)(\varphi - \langle\varphi\rangle)^2 \rightarrow \; V_{eff}''(\langle\varphi\rangle) = m^2$$

Thus

$$m^2 = \frac{3e^4}{8\pi^2}\langle\varphi\rangle^2$$

Similar analysis for the photon gives $m_A = e\langle\varphi\rangle$. So massless scalar electrodynamics is neither massless nor electrodynamics. Furthermore, we started with no mass, thus no mass scale, and now, at the regularization step of renormalization, after symmetry breaking, we have $\langle\varphi\rangle = \langle 0|\varphi|0\rangle$ with $dim[\langle\varphi\rangle] = \dim\,[M]$. (This is because the regularization, momentum cutoff, etc., breaks scale invariance.) Suppose we proceed with choosing the renormalization parameter $M = \langle\varphi\rangle$, then we get

$$\lambda = \frac{33e^4}{8\pi^2}$$

And we've traded a dimensionless parameter (λ) for a dimensionful parameter $\langle\varphi\rangle$ (known as dimensional transmutation):

$$(e, \lambda) \rightarrow \; (e, \langle\varphi\rangle).$$

5.3 Higgs Mechanism

To consider the Higgs Mechanism we begin with a complex scalar field with potential as before:

$$\mathcal{L} = \partial_\mu\varphi^*\partial^\mu\varphi - \mu^2|\varphi|^2 - \frac{\lambda}{4!}|\varphi|^4$$

and now have:

$$\mu^2 < 0, \quad |\varphi|_{min} = \sqrt{\frac{-12\mu^2}{\lambda}} \equiv \frac{v}{\sqrt{2}}.$$

Let's split the complex scalar field and set vacuum values as shown in previous examples:

$$\varphi = \frac{1}{\sqrt{2}}(\varphi_1 + i\varphi_2),\quad \langle 0|\varphi_1|0\rangle = v,\quad \langle 0|\varphi_2|0\rangle = 0.$$

Let's shift to new field (measure transitive invariance allows):

$$\varphi'_1 = \varphi_1 - v,\qquad \varphi'_2 = \varphi_2$$

Then:

$$\mathcal{L} = \frac{1}{2}\left[\left(\partial_\mu\varphi'_1\right)^2 + \left(\partial_\mu\varphi'_2\right)^2\right] - \frac{\lambda v^2}{4!}\varphi_1'^2 - \frac{\lambda v}{4!}(\varphi_1'^2 + \varphi_2'^2)\varphi'_1$$
$$- \frac{\lambda}{4\cdot 4!}(\varphi_1'^4 + \varphi_2'^4)$$

and for this form of the Lagrangian we see that we have φ'_2 massless and $m_{eff} > 0$ for φ'_1.

We now take a different approach, beginning with a polar decomposition:

$$\varphi = |\varphi|e^{i\theta} = \frac{1}{2}\rho e^{i\theta} \rightarrow \langle 0|\rho|0\rangle = v$$

And to get the same as before, we need: $\langle 0|\theta|0\rangle = 0$. Considering just the free theory we then arrive at:

$$\mathcal{L}_{free} = \frac{1}{2}\left[\left(\partial_\mu\rho\right)^2 + \rho^2\left(\partial_\mu\theta\right)^2\right].$$

Note this no longer makes sense unless there is spontaneous symmetry breaking since as is it doesn't have a propagator term for θ, but a quartic term instead. Recall from previous analysis that the spontaneous symmetry breaking could be represented via $\rho' = \rho - v$, so let's do this along with $\varphi' = \frac{\varphi}{v}$ to have the convenient reformulation:

$$\mathcal{L} = \frac{1}{2}\left[\left(\partial_\mu\rho'\right)^2 + \left(\partial_\mu\theta'\right)^2\right] - \frac{\lambda v^2}{4!}(\rho')^2 + \frac{\lambda}{2v^2}(\rho'^2 + 2v\rho')(\partial_\mu\theta')$$
$$- \frac{\lambda}{4\cdot 4!}(\rho' + 4v)\rho'^3$$

Recall that the Goldstone Boson is related to the field that is shifted by phase changes: θ'.

So far we've considered the Higgs Lagrangian where spontaneous symmetry breaking exists, and indicated some interesting transformations in the Lagrangian theory (from which to then quantize). Let's now add a gauge field (related to classical electromagnetism) by introduction of a four-vector potential A_μ, known as the Coleman-Weinberg model, where:

$$\partial_\mu\varphi \rightarrow \left(\partial_\mu + ieA_\mu\right)\varphi.$$

As before, let's start with the φ_1, φ_2 (Complex Cartesian) decomposition to get:

$$\mathcal{L} = \frac{1}{2}\left[\left(\partial_\mu\varphi_1 - eA_\mu\varphi_2\right)^2 + \left(\partial_\mu\varphi_2 - eA_\mu\varphi_1\right)^2\right] - \frac{1}{4}F_{\mu\nu}F^{\mu\nu}$$
$$-\frac{1}{2}\mu^2(\varphi_1^2 + \varphi_2^2) - \frac{\lambda}{4\cdot 4!}(\varphi_1^2 + \varphi_2^2)^2$$

As before, let's have spontaneous symmetry breaking that amounts to the shift:

$$\varphi_1 = \varphi_1' + v; \quad \varphi_2 = \varphi_2'$$

Except let's now only do substitutions and only write the Lagrangian to quadratic level (the particle spectrum is determined by the quadratic part of the Lagrangian):

$$\mathcal{L}_{quadratic} = \frac{1}{2}\left[\left(\partial_\mu\varphi_1'\right)^2 + \left(\partial_\mu\varphi_2'\right)^2\right] - \frac{\lambda v^2}{4!}\varphi_1'^2 - \frac{1}{4}F_{\mu\nu}F^{\mu\nu}$$
$$+\frac{1}{2}e^2v^2A_\mu A^\mu + evA_\mu\partial^\mu\varphi_2'$$

At this point the last two terms spoil what we hope to describe in terms of spontaneous symmetry breaking as it appears we have a photon mass term $\frac{1}{2}e^2v^2A_\mu A^\mu$ and a coupling term $evA_\mu\partial^\mu\varphi_2'$ that results in the operators for A_μ and φ_2' no longer being diagonal in the mass matrix. Obviously a maneuver is needed to bring this into a proper form, and that is done with the gauge transformation freedom we've introduced. Let's choose the gauge:

$$A_\mu = B_\mu - \frac{1}{ev}\partial_\mu\varphi_2',$$

to get:

$$\mathcal{L}_{quadratic} = \frac{1}{2}\left(\partial_\mu\varphi_1'\right)^2 + \frac{1}{2}\left[\left(\partial_\mu B_\nu\right)^2 + \left(\partial_\nu B_\mu\right)^2\right]$$
$$-\frac{\lambda v^2}{4!}\varphi_1'^2 + \frac{1}{2}e^2v^2B_\mu B^\mu$$

We've now arrived at a particle spectrum with a massive scalar φ_1', where

$$\varphi_1' \to m = \sqrt{\frac{\lambda v^2}{12}}$$

We also have a massive vector boson (the photon part of the theory) with

$$B_\mu \to M = ev$$

Note that at this point we appear to have eliminated the Goldstone Boson φ_2' by introduction of the gauge field. Let's make this explicit by doing a polar transform on the gauged quadratic-level Lagrangian above, to get:

$$\mathcal{L} = \frac{1}{2}\left[\left(\partial_\mu\rho\right)^2 + \rho^2\left(\partial_\mu\theta + eA_\mu\right)^2\right] - \frac{1}{2}\mu^2\rho^2 - \frac{\lambda}{4\cdot 4!}\rho^4 - \frac{1}{4}F_{\mu\nu}F^{\mu\nu}$$

Alternatively, we could have started with the original Coleman-Weinberg model and switched to polar coordinates first to get:

$$\mathcal{L} = -\frac{1}{4}F_{\mu\nu}F^{\mu\nu} + \frac{1}{2}\left(\partial_\mu\rho'\right)^2 + \frac{1}{2}(\rho' + v)^2\left(\frac{1}{v}\left(\partial_\mu\theta'\right) + eA_\mu\right)^2$$
$$-\frac{1}{2}\mu^2\rho'^2 - \frac{\lambda}{4\cdot 4!}\rho'^4$$

And if we then do the gauge transformation similar to before:

$$A_\mu = B_\mu - \frac{1}{ev}\left(\partial_\mu\theta'\right)$$

We again eliminate the Goldstone boson arising from the symmetry breaking by a particular form of Gauge, known as the Unitary gauge.

Renormalizability

So far the theory is starting to look like one that might provide a massless photon (no massive boson), but in the unitary gauge the theory appears to not be renormalizable. Consider the Green's function for the aforementioned vector boson with $M = ev$, where we have Green's function:

$$[(\partial^2 - M^2)\eta^{\mu\nu} - \partial^\mu\partial^\nu]\Delta_{f\,\nu\rho}(x) = \delta^\mu{}_\rho\delta_4(x)$$

Whose Fourier Transform gives the Feynman propagator:

$$\widetilde{\Delta}_{f\,\nu\rho}(x) = -\left(\frac{\eta^{\mu\nu} - \frac{P^\mu P^\nu}{M^2}}{P^2 - M^2 + i\varepsilon}\right) \rightarrow \frac{1}{M^2} = \mathcal{O}(1)\ \ as\ P^2 \rightarrow \infty$$

Since the vector propagator goes as $\mathcal{O}(1)$ it does not provide any convergence factor, where general power-counting arguments then indicate non-renormalizability. This was the end of the matter until it was realized that almost miraculous cancellations could occur. Part of the problem is the unitary gauge appears manifestly nonrenormalizable. Consider instead the Landau gauge for the photon:

$$\widetilde{\Delta}_{f\,\nu\rho}(x) = -\left(\frac{\eta^{\mu\nu} - \frac{p^\mu p^\nu}{p^2}}{p^2 + i\varepsilon}\right) \rightarrow \frac{1}{p^2}\ \ as\ p^2 \rightarrow \infty,$$

which now behaves as a (renormalizable) scalar propagator. This suggests the R-gauges examined by t'Hooft (R for Renormalizable).

R-gauges

Let's start with the Lagrangian studied thus far $\mathcal{L}(\varphi_1', \varphi_2', A_\mu)$, and introduce the gauge-fixing term $f(\varphi_2', A_\mu)$ where:

$$\mathcal{L}' = \mathcal{L}(\varphi_1', \varphi_2', A_\mu) - f(\varphi_2', A_\mu), \quad f(\varphi_2', A_\mu) = -\frac{1}{2\xi}\left(\partial^\mu A_\mu + \xi e v \varphi_2'\right)^2.$$

The cross-term from the gauge-fixing directly cancels the terms in the Lagrangian that took us off mass eigenstate previously. If we now regard the kinematic part of the Lagrangian (grouping interaction terms separately):

$$\mathcal{L}' = \mathcal{L} - f = \frac{1}{2}\left[\left(\partial_\mu \varphi_1'\right)^2 + \left(\partial_\mu \varphi_2'\right)^2\right] - \frac{\lambda v^2}{4!}\varphi_1'^2 - \frac{1}{4}F_{\mu\nu}F^{\mu\nu} + \frac{\xi}{2}e^2 v^2 (\varphi_2')^2 + \frac{1}{2}e^2 v^2 A_\mu A^\mu + \frac{1}{2\xi}\left(\partial^\mu A_\mu\right)^2 + Interactions$$

Where the interaction terms are simply grouped as such. In this form we have for propagators:

$$A_\mu: -\frac{1}{P^2 - M^2 + i\varepsilon}\left(\eta^{\mu\nu} - (1-\xi)\frac{P^\mu P^\nu}{P^2 - \xi M^2}\right), \qquad M = ev$$

$$\varphi_2': \quad -\frac{1}{P^2 - \xi M^2 + i\varepsilon}$$

$$\varphi_1': \quad \frac{1}{p^2 - m^2 + i\varepsilon}$$

Consider A_μ with $\xi \to \infty$, this returns to the Unitary gauge seen previously. If we take $\xi \to 0$, we see that A_μ behaves asymptotically like the Landau gauge, which is renormalizable. Thus, the T'Hooft R-gauges interpolate between obvious renormalizable theories and physical theories. If gauge symmetry can be demonstrated, then we can thereby show overall renormalizability. An obstacle in this process is showing that the S-matrix is independent of the R-gauge. In particular, note that the φ_2' propagator has poles that depend on gauge parameter ξ, so terms with φ_2' should not appear in the S-matrix.

Let's now consider what happens to a renormalizable theory when you break it spontaneously. Consider a general theory with different types of vertices, and scalar and spin-1/2 (fermionic) fields. For the ith type of vertex let's denote the number of scalar lines by b_i, the number of fermion lines by f_i, and the number of derivatives by α_i. For a general graph we then have the collection of numbers:

$$n_i$$

The number of i type vertices.

And

$$IB$$

The number of internal scalar lines.

And

$$IF$$

The number of internal fermion lines.

And

$$B$$

The number of external scalar lines.

And

$$F$$

The number of external fermion lines.

From which we calculate the superficial degree of divergence from the topological identities:

$$B + 2(IB) = \sum_i n_i b_i$$

$$F + 2(IF) = \sum_i n_i f_i$$

And the Kirckoff's Law (loop count) type relation:

$$L = IB + IF - \sum_i n_i + 1$$

And since each loop involves a momenta integral $\int d^4p$ we then have the level of divergence of a graph given by:

$$D = 4L - 2(IB) - (IF) + \sum_i n_i \alpha_i$$

$$D = 4 - B - \frac{3}{2}F + \sum_i n_i \left(b_i + \frac{3}{2}f_i + \alpha_i - 4\right)$$

Where

$$\delta_i = \left(b_i + \frac{3}{2}f_i + \alpha_i - 4\right)$$

Is known as the index of divergence. Generally speaking, $\delta_i = 0$ will be characteristic of a renormalizable theory (such as for φ^4 theory), while $\delta_i < 0$ will be indicative of a super-renormalizable theory (φ^3 theory), and $\delta_i > 0$ will be indicative of a non-renormalizable theory (φ^6 theory).

In unit units of mass

$$[\varphi] = 1, \qquad [\psi] = \frac{3}{2}, \qquad [\partial] = 1$$

And the dimension of the i-th type vertex is $b_i + \frac{3}{2}f_i + \alpha_i = \delta_i + 4$.

Thus, the dimension of the coupling coefficient g_i in the i-th vertex will go as: $[g_i] = -\delta_i$. Thus, a dimensionless coupling gives $\delta_i = 0$, indicating renormalizability, while positive coupling gives superrenormalizability. Let's consider what this means for the renormalizability of a theory after spontaneous symmetry breaking. What results is $\varphi^4 \to (\varphi + v)^4$, which involves the terms with φ^4, φ^3, … each of the terms from the spontaneous breaking being more renormalizable (known as soft breaking).

Non-abelian gauge theories (Yang-Mills)

Let's now consider more than two fields (the complex scalar field). Recall that for complex field, with the minimal substitution bringing in a gauge field for electromagnetism, we had

$$\partial_\mu \to \partial_\mu + ieA_\mu$$

which corresponds to a transformation on the field which is an irreducible representation of the Lie group U(1). We now want to generalize to a set of fields transforming under gauge transformation as an irreducible representation of a Lie group in the more general sense (with examples like SU(2) and SU(3) to be considered in detail). The discussion that follows draws upon the Yang & Mills (1954) paper [51] and by the follow-up paper by Utiyama (1956) [52].

Suppose we have a set of fields φ^i, which we also denote as a column vector $\tilde{\varphi}$, which transforms under some irreducible representation of a Lie group G:

$$\tilde{\varphi}' = U(g)\tilde{\varphi}, \quad g \in G.$$

Suppose we have a Lagrangian, in the standard form examined thus far, such that under transformation:

$$\mathcal{L}(\tilde{\varphi}, \partial_\mu \tilde{\varphi}) = \mathcal{L}(\tilde{\varphi}', \partial_\mu \tilde{\varphi}).$$

If g remained a global choice of gauge $(\partial_\mu \tilde{\varphi})' = U(g)(\partial_\mu \tilde{\varphi})$, then we have the desired invariance of the Lagrangian. We want $g = g(x)$ to be a function of spacetime, however, for our gauge field generalization, which means

$$(\partial_\mu \tilde{\varphi})' \neq U(g)(\partial_\mu \tilde{\varphi}).$$

If we generalize the previous form $\partial_\mu \to \partial_\mu + ieA_\mu$ to include a general Lie (gauge) group we have:

$$D_\mu^{ij} \tilde{\varphi}_j = \left[\delta^{ij} \partial_\mu + ig(t^a)^{ij} A_\mu^a(x)\right] \tilde{\varphi}_j.$$

To have $\mathcal{L}' = \mathcal{L}$ we need $(D_\mu \tilde{\varphi})' = U(g(x))(D_\mu \tilde{\varphi})$. In shortened notation we have:

$$D_\mu \tilde{\varphi} = \left[\partial_\mu + igt^a A_\mu^a(x)\right] \tilde{\varphi}.$$

For $(D_\mu \tilde{\varphi})'$ we have:

$$(D_\mu \tilde{\varphi})' = [\partial_\mu + igt^a A'^a_\mu(x)]U\tilde{\varphi} = U[\partial_\mu + igt^a A^a_\mu(x)]\tilde{\varphi}$$

And since

$$\partial_\mu(U\tilde{\varphi}) = (\partial_\mu U)\tilde{\varphi} + U(\partial_\mu\tilde{\varphi}) = U[U^{-1}\partial_\mu U + \partial_\mu]$$

We have:

$$U^{-1}\partial_\mu U + igU^{-1}(t^a A'^a_\mu)U = igt^a A^a_\mu$$

Or

$$t^a A'^a_\mu = U(t^a A^a_\mu)U^{-1} + \frac{i}{g}(\partial_\mu U)U^{-1}$$

Using the summed notation given by: $A'^{\square}_\mu = \sum_a t^a A'^a_\mu$, we then have:

$$A'^{\square}_\mu = UA^{\square}_\mu U^{-1} + \frac{i}{g}(\partial_\mu U)U^{-1}$$

If U is constant, then $\partial_\mu U = 0$ and

$$U(g)t^a U(g)^{-1} = [D(g)^{-1}]^{ab}t^b,$$

where $[D(g)]^{ab}$ is real and unitary, thus orthogonal: $[D(g)^{-1}]^{ab} = [D(g)]^{ba}$. So, we can write:

$$t^a A'^a_\mu = [D(g)]^{bc}t^b A^c_\mu \rightarrow A'^a_\mu = [D(g)]^{ac}A^c_\mu.$$

Thus, the A's transform as representatives of the group.

To proceed we want to make a kinetic energy term for the $A's$ so as to make them dynamic. But before doing this, let's first consider the infinitesimal analysis. Suppose:

$$U(g) = I + i\theta^a t^a + O(\theta^2)$$

Then at first order in θ:

$$A'^a_\mu = A^a_\mu + f^{abc}A^b_\mu \theta^c - \frac{1}{g}\partial_\mu\theta^a, \quad [t^a, t^b] = if^{abc}t^c$$

The generators of the group in the adjoint representation are:

$$(T^a)^{bc} = -if^{abc}.$$

This permits us to rewrite as:

$$A'^a_\mu = A^a_\mu - \frac{1}{g}[\partial_\mu\theta^a + ig(T^a)^{bc}A^b_\mu\theta^c] = A^a_\mu - \frac{1}{g}[D_\mu\theta]^a$$

For electromagnetism we had θ equal to a constant and only one gauge field that transformed as $A'^{\square}_\mu = A^{\square}_\mu$ (a massless field). Now there is the column vector of vector fields A^a_μ, for which we can define more than the one $F_{\mu\nu}$ that existed for electromagnetism. Here we have the distinct generators given by t^a (or T^a in the adjoint representation), so we define the non-trivial tensor-field group scalar:

$$\tilde{F}_{\mu\nu} = t^a F^a_{\mu\nu},$$

where the tensor field group elements transform under gauge change according to:

$$F'^{a}_{\ \mu\nu} = D^{ab} F^{b}_{\ \mu\nu}.$$

From this it is clear that we can write a gauge invariant scalar

$$F^{a}_{\ \mu\nu} F^{a\,\mu\nu} = C\, tr \tilde{F}_{\mu\nu} \tilde{F}^{\mu\nu},$$

where C is a constant that depends on the normalization of the generators (if $tr(t^a t^b) = 2\delta^{ab}$ then $C = 1/2$). We've worked bottom-up with the generalization to A'^{a}_{μ} and top-down with the generalization to get $\tilde{F}_{\mu\nu}$. Again, paralleling the electromagnetism construct, we are inclined to guess the remaining relation to complete the generalization as (the column vector for A^a_μ is: $\tilde{A}^{\square}_{\mu}$):

$$\tilde{F}_{\mu\nu} = \partial_\mu \tilde{A}^{\square}_{\nu} - \partial_\nu \tilde{A}^{\square}_{\mu} \quad ???$$

To test this guess we need to show that

$$\tilde{F}'_{\mu\nu} = U(g) \tilde{F}_{\mu\nu} U(g)^{-1}.$$

We compute the following:

$$\begin{aligned} \tilde{F}'_{\mu\nu} = U(g)\big(\partial_\mu \tilde{A}^{\square}_{\nu} - \partial_\nu \tilde{A}^{\square}_{\mu}\big)U(g)^{-1} &+ \big[(\partial_\mu U)U^{-1}, U\tilde{A}^{\square}_{\nu}U^{-1}\big] \\ &- \big[(\partial_\nu U)U^{-1}, U\tilde{A}^{\square}_{\mu}U^{-1}\big] \\ &+ \frac{i}{g}\big[(\partial_\mu U)U^{-1}, (\partial_\nu U)U^{-1}\big] \end{aligned}$$

Thus

$$\tilde{F}'_{\mu\nu} = U(g)\tilde{F}_{\mu\nu}U(g)^{-1} - ig\big[\tilde{A}^{\square}_{\mu}, \tilde{A}^{\square}_{\nu}\big],$$

which is not the desired $\tilde{F}'_{\mu\nu} = U(g)\tilde{F}_{\mu\nu}U(g)^{-1}$ relation, but it is now clear that the following generalization will work:

$$\tilde{F}_{\mu\nu} = \partial_\mu \tilde{A}^{\square}_{\nu} - \partial_\nu \tilde{A}^{\square}_{\mu} + ig\big[\tilde{A}^{\square}_{\mu}, \tilde{A}^{\square}_{\nu}\big]$$

Now that we know a reasonable kinetic term upon introduction of the Yang-Mills non-abelian gauge fields, let's examine the generalized, pure Yang-Mills, Lagrangian that results:

$$\mathcal{L}(\tilde{\varphi}, \partial_\mu \tilde{\varphi}) \to \mathcal{L}(\tilde{\varphi}, D_\mu \tilde{\varphi}) - \frac{1}{4} F^{a}_{\ \mu\nu} F^{a\,\mu\nu}$$

Pure Yang-Mills is nontrivial (unlike electromagnetism), with cubic and quartic interaction terms. The first hint of such complications comes from the Euler-Lagrange derivation of the equations of motion for the Lagrangian, which give:

$$D_\mu F^{a\,\mu\nu} = j^{a\,\nu}_{matter} \to D_\nu j^{\nu}_{matter} = 0$$

This follows from the Euler-Lagrange result:

$$\partial_\mu F^{a\,\mu\nu} + g f^{abc} F^{b\,\lambda\nu} A^c_\lambda + ig\Pi^{\nu i}(t^a)^i_{\ j}\varphi^j = D_\mu F^{a\,\mu\nu} - j^{a\,\nu}_{matter} = 0,$$

where

$$\Pi^{\nu i} = \frac{\partial \mathcal{L}}{\partial(\partial_\nu \varphi^i)}.$$

Thus, we have the relation $D_\nu j^\nu_{matter} = 0$, not $\partial_\nu j^\nu_{matter} = 0$, which means that the total conserved current is no longer just due to matter motion – the Yang-Mills fields (photons in electromagnetism U(1) case) here carry charge.

5.4 Yang-Mills Quantization

Let's now proceed to doing the quantization for the Yang-Mills theories:

First: The Lagrangian is determined, as in the preceding.

Second: Shift from Lagrangian to Hamiltonian form, then go from Poisson brackets to commutators to quantize. In doing this the primary and secondary constraints are identified.

Third: The secondary constraints are generators of gauge transformations of the fields. Note that this means that the non-abelian generalization of Gauss's Law ($D_\mu F^{a\,\mu\nu} = 0$) provides a generator for infinitesimal gauge transformations.

Fourth: Now shift the quantization to the path integral formulation, knowing that the Yang-Mills Lagrangian describes a theory with gauge invariance, and adopting the Fadeev-Popov prescription to cure the infinity that results if not handled properly.

Fifth: Part of the Fadeev-Popov prescription introduces a matrix determinant that needs to be represented in exponentiated form. This can be done if the variables used are Grassmannian. What results, however, is the introduction of an infinite set of closed loops involving a scalar particle with fermion statistics (known as a ghost). The Grassmanian variables for the ghosts never appear as external lines, have zero mass, and couple to A via a derivative term. This might all sound like a hopeless mess, but consider that every ghost loop introduces a factor of minus one for fermionic-type cancellations, as well as having power-counting renormalizability if the non-ghost theory is renormalizable.

Sixth: Using the BRS method, the resulting quantized Yang-Mills theory will be shown to be renormalizable. In doing this, use is made of dimensional regularization. An incompatibility arises with the Ward Identities in this process, however, due to coupling terms dependent on Dirac's γ^5 matrix, which is defined using 4-dimensions. The resulting anomalies due to γ^5 are known as the chiral anomalies.

Seventh: The chiral anomalies for various gauge theories of relevance are examined. Charge relations among particles are predicted in some Yang-Mills theories, in agreement with observation, such that anomaly terms are eliminated.

Eighth: Instanton elements and a sophisticated vacuum structure is indicated for QCD, so this is examined as well.

The Yang-Mills Lagrangian
Starting with:

$$\mathcal{L}(\tilde{\varphi}, \partial_\mu \tilde{\varphi}) \rightarrow \mathcal{L}(\tilde{\varphi}, D_\mu \tilde{\varphi}) - \frac{1}{4} F^a_{\mu\nu} F^{a\,\mu\nu}$$

and

$$D^{ab}_\mu = \delta^{ab} \partial_\mu + ig(t^c)^{ab} A^c_\mu(x) = I\partial_\mu + ig t^a A^a_\mu = I\partial_\mu + ig\tilde{A}_\mu,$$

where the notation $t^a A^a_\mu = \tilde{A}_\mu$ is adopted. Recall the set of scalar fields φ^a, can be written as a column vector $\tilde{\varphi}$, so here we have the vector field generalization. We've determined that the field tensor is defined by:

$$\tilde{F}_{\mu\nu} = \partial_\mu \tilde{A}^{\square}_\nu - \partial_\nu \tilde{A}^{\square}_\mu + ig\left[\tilde{A}^{\square}_\mu, \tilde{A}^{\square}_\nu\right]$$

And

$$F^{a\,\mu\nu} = \left(\partial_\mu A^a_\nu - \partial_\nu A^a_\mu\right) - g f^{abc} A^b_\mu A^c_\nu$$

The shift from Lagrangian to Hamiltonian form
In the Hamiltonian formulation we have the conjugate momenta (time index t and 0 are the same):

$$\Pi^{a\mu} = \frac{\partial \mathcal{L}}{\partial(\partial_t \varphi^a)} = -F^{a\,0\mu},$$

where a is the Group Index, and μ is the Lorentz index. We then have $\Pi^{a0} = 0$ identically since F is antisymmetric, e.g., $\Pi^{a0} = 0$ is a primary constraint. Working with the Hamiltonian form (indices $i, j \in 1,2,3$):

$$\mathcal{H} = \int d^3x \left[\Pi^{a\mu} \partial_t A^a_\mu - \mathcal{L}\right]$$

$$= \int d^3x \left[\Pi^{aj}\Pi^a_j + \frac{1}{4} F^a_{ij} F^{a\,ij} - A^a_0 D_j F^{a\,j0}\right],$$

We want to require $\Pi^{a0} = 0$ to remain true at all time, thus we have the Poisson Bracket relation:

$$\{\Pi^{a0}, \mathcal{H}\} = 0 \rightarrow D_j F^{a\,j0} = 0,$$

where

$$D_j F^{a\,j0} = 0$$

describes a secondary constraint (akin to Gauss's Law $\nabla \cdot E = 0$).

Gauss's Law is the generator of infinitesimal gauge transformations

Since

$$\{D_j F^{a\,j0}, \mathcal{H}\} = 0,$$

We don't generate further constraints. Also, the secondary constraint is first class in that $\{\Pi^{a0}, D_j F^{a\,j0}\} = 0$ as well. In such a circumstance the secondary constraints are generators of gauge transformations of the fields. For infinitesimal transformation with $D_j F^{a\,j0}$ we have:

$$T = \left\{ \int d^3y\, \theta^a(y) D_j^{ab}(y) F^{b\,j0}(y), A^c{}_k(x) \right\}$$

$$= \int d^3y\, \theta^a(y) D_j^{ab}(y) \{F^{b\,j0}(y), A^c{}_k(x)\}$$

$$T = -\int d^3y\, \theta^a(y) D_k^{ac}(y) \delta_3(x-y) = \theta^a(x) D_k^{ca}(x) = -g\delta A^c{}_k(x)$$

Thus, Gauss's Law and its generalization, is the generator of infinitesimal gauge transformations.

Path Integral Quantization for a Gauge Theory via the Fadeev-Popov Method

To shift our Quantization of Yang-Mills to the Path Integral setting, for ease of performing certain calculations, such as with the Feynman prescription, we need to have a well-defined path integral. This is provided by the Fadeev-Popov method, which removes a group measure infinity factor that arises from the gauge invariance. Recall the Path Integral form

$$Z = \int \mathcal{D}A_\mu e^{iS[A]}$$

in terms of the Action $S[A]$. Let's now choose a gauge by imposing a set of conditions:

$$f^a[A^a{}_\mu, x] = 0.$$

If f^a is the Lorentz gauge condition, then we have: $\partial_\mu A^{a\,\mu}(x) = 0$. If f^a is the Coulomb gauge condition, then we have: $\partial_j A^{a\,j}(x) = 0$. And, if $A^a{}_3(x) = 0$, we have axial gauge. For whatever choice of gauge, we then group into gauge equivalent classes, with resulting structure a bundle whose base consists of gauge inequivalent A^μ's, with fiber the gauge transformation, and a choice of gauge condition corresponding to a cross-section. To proceed, let's assume we have found a set of $f's$ (note that in general non-abelian theories such a set may not be obtainable, but not an issue for the SU(2) and SU(3) theories, etc. that are of direct relevance to the Standard Model).

Fadeev-Popov Prescription

For each f define

$$\Delta_f[A_\mu]^{-1} = \int \mathcal{D}u(x)\delta(f(A_\mu^{u(x)})),$$

where $u(x)$ is a local gauge transform acting as a power $A_\mu^{u(x)}$ with structure as indicated in the previous infinitesimal analysis. Consider now the action on a gauge transformed A_μ by this mechanism according to $u_0(x)$, we then have:

$$\Delta_f[A_\mu^{u_0}]^{-1} = \int \mathcal{D}u(x)\delta(f(A_\mu^{u_0 u})).$$

If we then define $u' = u_0 u$ and make use of transitive invariance of the integration measure:

$$\Delta_f[A_\mu^{u_0}]^{-1} = \int \mathcal{D}u'(x)\delta(f(A_\mu^{u'})) = \Delta_f[A_\mu]^{-1}.$$

Thus, we find that the entity Δ_f so defined is gauge invariant. Rewriting we have a gauge-invariant expression for a unity operation:

$$1 = \Delta_f[A_\mu]\int \mathcal{D}u(x)\delta(f(A_\mu^{u(x)})),$$

Let's use this expression in out naïve path-integral formulation to properly recognize aspects of the measure present:

$$Z = \int \mathcal{D}u(x) \int \mathcal{D}A_\mu\, \Delta_f[A_\mu]\, \delta\left(f\left(A_\mu^{u(x)}\right)\right) e^{iS[A]}$$

From this form it is then argued that we can choose a gauge transformation such that A'_μ makes the inner integral above no longer explicitly dependent on $u(x)$, thus the outer integral is simply the infinite factor of $\int \mathcal{D}u(x)$, which is simply dropped going forward in the Fadeev-Popov prescription (a constant factor to be absorbed in any renormalization later as needed, so consistent with later actions). Thus we have:

$$Z = \int \mathcal{D}A_\mu\, \Delta_f[A_\mu]\, \delta\left(f\left(A_\mu^{u(x)}\right)\right) e^{iS[A]}.$$

So, we've effectively introduce a new element over what was indicated in the naïve Lagrangian with the improperly managed gauge, that element being $\Delta_f[A_\mu]$. Let's now calculate this new element of the formalism (with minor change in notation):

$$\Delta_f^{-1}[A_\mu] = \int \mathcal{D}u(x)\delta(f\left(A_\mu^{u(x)}\right))$$

Note that in Z we must have the $f(A_\mu) = 0$ and $f\left(A_\mu^{u(x)}\right) = 0$ conditions be met for general $A_\mu^{u(x)}$, which means that $\int \mathcal{D}A_\mu$ must integrate over A_μ's that include the identity, thus we can write:

$$\left(A^a{}_\mu\right)^u = A^a{}_\mu - \frac{1}{g} D_\mu^{ab}\theta^b.$$

If written this way, the integral over $\mathcal{D}u(x)$ becomes an integral over θ^b's in terms of the infinitesimal parameters of $u(x)$. Thus we can write:

$$\Delta_f{}^{-1}[A_\mu] = \int \mathcal{D}\theta(x)\delta(f\left(A^a{}_\mu - \frac{1}{g}D_\mu^{ab}\theta^b\right)).$$

Recall now the relationship:

$$\int dx\delta(f(x)) = \frac{1}{f'(x)}$$

Thus,

$$\Delta_f{}^{-1}[A_\mu] = det\left[\frac{\partial f^a\left(D_\mu\theta(x)\right)}{\partial\theta^b(y)}\right]_{\theta=0}^{-1} \quad \rightarrow \quad \Delta_f = \det M,$$

where

$$M^{ab}(x,y) = \frac{\partial f^a\left(D_\mu\theta(x)\right)}{\partial\theta^b(y)}$$

and $\det M$ is known as the Fadeev-Popov determinant.

In Lorentz gauge,

$$f^a = \partial_\mu A^{a\,\mu}(x) = 0,$$

and

$$M^{ab}(x,y) = \frac{\partial\left(\partial^\mu D_\mu^{ac}\theta^c(x)\right)}{\partial\theta^b(y)} = \partial^\mu D_\mu^{ab}(x)\delta(x-y).$$

And since $D_\mu^{ab} = \delta^{ab}\partial_\mu + ig(t^c)^{ab}A_\mu^c(x)$, we see that Δ_f will have a dependence on $A_\mu^c(x)$ in the integral.

In axial gauge,

$$f^a = A^a{}_3(x) = 0,$$

Thus

$$M^{ab}(x,y) = \frac{\partial(D_3^{ac}\theta^c(x))}{\partial\theta^b(y)} = D_3^{ab}(x)\delta(x-y) = \partial_3\delta^{ab}\delta(x-y)$$

So, for axial gauge there is clearly no $A_\mu^c(x)$ dependence, thus no Fadeev-Popov determinant complication (and no Fadeev-Popov ghosts in what follows).

In Coulomb gauge $\partial_j A^{a\,j}(x) = 0$, which is sufficient to determine the gauge uniquely in abelian gauge field, but not with the non-abelian examined here (known as Gribov ambiguity).

Thus, we can now write:

$$Z = \int \mathcal{D}A_\mu \; \det M \;\; \delta\left(f^a(A_\mu)\right) e^{iS[A]}.$$

Let's focus on the Yang Mills Action with no interactions,

$$S[A] = -\frac{1}{4}\int d^3x \; F^a_{\;\mu\nu} F^{a\,\mu\nu}$$

Which is already a familiar form, leaving only the need to move the $\det M \;\; \delta\left(f^a(A_\mu)\right)$ factors into the exponent (into the Action), to get a standard path integral form to work with. This means the $\det M$ and $\delta\left(f^a(A_\mu)\right)$ need to be rewritten in exponentiated form.

Since Z is independent of gauge choice, we can replace $\delta\left(f^a(A_\mu)\right)$ by a more general gauge condition $\delta\left(f^a(A_\mu) - \sigma^a\right)$:

$$Z = \int \mathcal{D}A_\mu \; \det M \;\; \delta\left(f^a(A_\mu) - \sigma^a\right) e^{iS[A]}.$$

From this form we have:

$$\int \mathcal{D}\sigma \; Z \, e^{-\frac{i}{2\alpha}\int d^4x \, \sigma^a \sigma^a} = \int \mathcal{D}A_\mu \; \det M \;\; e^{i\int d^4x \left[\mathcal{L}_{YM} - \frac{1}{2\alpha} f^a f^a\right]},$$

where

$$\mathcal{L}_{YM} = -\frac{1}{4} F^a_{\;\mu\nu} F^{a\,\mu\nu}, \quad \mathcal{L}_{GF} = -\frac{1}{2\alpha} f^a f^a.$$

Let's proceed with Lorentz gauge, where $f^a = \partial_\mu A^{a\,\mu}(x) = 0$. Then

$$M^{ab}(x,y) = \partial^\mu D^{ab}_\mu(x)\delta(x-y)$$
$$= \partial^\mu\left[\delta^{ab}\partial_\mu + ig(t^c)^{ab} A^c_\mu(x)\right]\delta(x-y)$$
$$M^{ab}(x,y) = \left[-\Box\,\delta^{ab} + g f^{abc}\partial^\mu A^c_\mu\right]\delta(x-y)$$

Recall the relation $\det M = e^{-Tr \ln M}$ and start from the factored form

$$M^{ab}(x,y) = (-\Box)\left[\delta^{ab} + g f^{abc}\,\Box^{-1}\,\partial^\mu A^c_\mu\right]\delta(x-y),$$

Thus focus on the term

$$Tr \ln M \cong Tr\; ln\left[\delta^{ab} + g f^{abc}\,\Box^{-1}\,\partial^\mu A^c_\mu\right].$$

Splitting the propagation (effectively using the Chapman-Kolmogorov property of the path integral formulation) we can rewrite:

$$Tr \ln M \cong -\sum_n \frac{(-ig)^n}{n} tr\left\{\int dx_1 \dots dx_n k(x_1 - x_2)\partial_\mu^{x_2} \tilde{A}^\mu_{\square}(x_2) \dots k(x_n - x_1)\partial_\mu^{x_1} \tilde{A}^\mu_{\square}(x_1)\right\}$$

where $\boxdot^{-1} = k$. This describes loop terms that have a (-1) factor for each loop as with fermion loops. In essence, the Fadeev-Popov determinant generates an infinite set of closed loops with a scalar particle with fermion statistics – referred to as a Fadeev-Popov ghost. So, to manage the $\det M$ term we are introducing a set of fictitious particles that have fermionic statistics as indicated. Interestingly, the Gaussian integral, for boson statistics has for $\det M$ expression:

$$(\det M)^{-1} = \int \mathcal{D}\varphi_1 \mathcal{D}\varphi_2 \, e^{i\int \varphi_1^a(x) M^{ab} \varphi_2^b(y) dxdy},$$

while the similar form but with Grassman variables gives:

$$(\det M) = \int \mathcal{D}c\mathcal{D}\bar{c} \, e^{i\int \bar{c}^a(x) M^{ab}(x) c^b(y) dxdy},$$

Which is what is needed to get the $\det M$ factor. In Lorentz gauge, the explicit Grassmannian representation indicates:

$$\int \bar{c}^a(x)\, M^{ab}(x) c^b(y) dxdy = \int \bar{c}^a(x)\, \partial^\mu D_\mu^{ab} c^b(y) d^4x$$

$$= -\int \partial^\mu \bar{c}^a(x) D_\mu^{ab} c^b(y) d^4x = -\int \left[\partial_\mu \bar{c} \partial^\mu c + ig\partial_\mu \bar{c} \tilde{A}^\mu_{\square} c\right] d^4x$$

From this we see the $c's$ never appear as external lines, they have zero mass, and they couple to $\tilde{A}^\mu_{\square}$ as a derivative term. We can now write the Fadeev-Popov ghost term that results from the determinant as:

$$\mathcal{L}_{FP} = \bar{c}^a(x) M^{ab}(x) c^b(y)$$

The Path integral is then written:

$$Z = \int \mathcal{D}A_\mu \mathcal{D}c\mathcal{D}\bar{c} \, e^{i\int d^4x[\mathcal{L}_{YM} + \mathcal{L}_{GF} + \mathcal{L}_{FP}]}$$

We can now read off the Feynman rules for the gauge field propagator:

$$iD_{F\,\mu\nu}^{\;ab}(p) = \left[-i\left(\eta_{\mu\nu} - \frac{p_\mu p_\nu}{p^2}\right)\frac{1}{(p^2 + i\varepsilon)} - i\alpha \frac{p_\mu p_\nu}{p^2(p^2 + i\varepsilon)}\right]\delta^{ab}$$

And for the ghost propagator:

$$iD_F^{ab}(p) = \frac{-i}{(p^2 + i\varepsilon)}\delta^{ab}$$

Now that we have the propagators (for internal and external lines) let's consider he interaction terms that arise:

$$\mathcal{L} = \mathcal{L}_0 + \mathcal{L}_I,$$

where $\mathcal{L}_0$ is the free part of the Lagrangian giving the propagators just derived, and $\mathcal{L}_I$ is the interaction Lagrangian:

$$\mathcal{L}_I = gf^{abc}(\partial^\mu A^{a\nu}_{\square})A^b_\mu A^c_\nu - \frac{1}{4}g^2 f^{abc} f^{ade} A^{b\mu}_{\square} A^{c\nu}_{\square} A^d_\mu A^e_\nu$$
$$- gf^{abc}\partial_\mu \bar{c}^b A^{a\mu} c^c$$

There are, thus, three types of interactions:

$$gf^{abc}\left[(p_1 - p_2)_\lambda \eta_{\mu\nu} + \begin{array}{c}(p_2 - p_3)_\mu \eta_{\nu\lambda} + (p_3 - p_1)_\nu \eta_{\mu\lambda}\\ + \text{cyclic perm.}\end{array}\right]$$

$$-ig^2\Big[f^{abe}f^{cde}(\eta_{\lambda\mu}\eta_{\rho\nu} - \eta_{\lambda\nu}\eta_{\rho\mu})$$
$$+ f^{cbe}f^{ade}(\eta_{\lambda\mu}\eta_{\rho\nu} - \eta_{\lambda\nu}\eta_{\rho\lambda})$$
$$+ f^{ace}f^{bde}(\eta_{\lambda\rho}\eta_{\mu\nu} - \eta_{\lambda\nu}\eta_{\mu\rho})\Big]$$

$$gf^{abc} p_{1\mu}$$

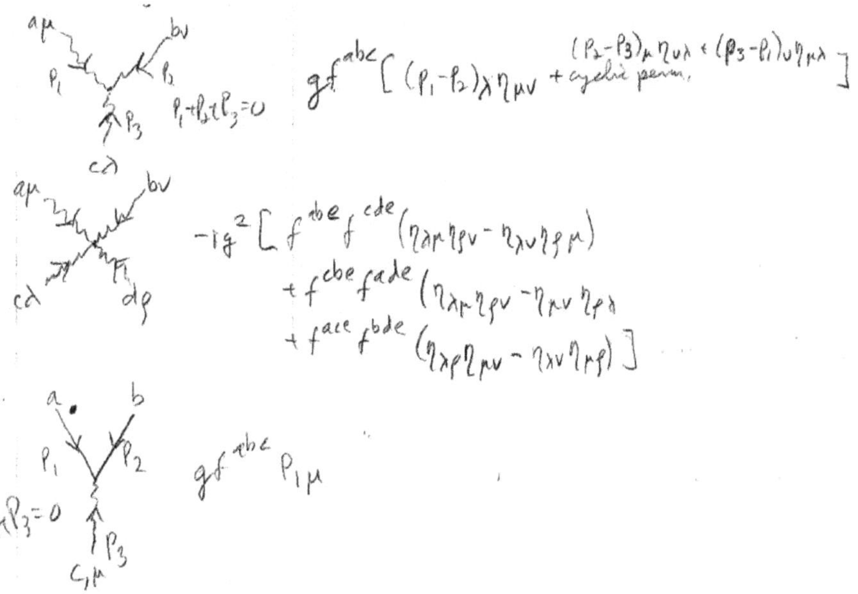

Note that every ghost loop has factor minus one for partial cancellation, as well as power counting super-renormalization.

The next step in completing the abelian gauge theory is to establish the Ward Identities (and from there attempt renormalization). However the effort to establish the Ward identities in the context of non-Abelian gauge theory, where they are known as the Taylor-Slavnov identities, was originally a very complex derivation. This was greatly simplified in later derivations by Becchi, Rouet, and Stora (BRS) in their papers [53] and [54].

The Action in the Yang-Mills quantization thus has form:

$$S_{YM} + S_{GF} + S_{FP} = \int d^4x[\mathcal{L}_{YM} + \mathcal{L}_{GF} + \mathcal{L}_{FP}]$$

where, for notational convenience, dropping the simple measure factors:

$$S_{YM} = -\frac{1}{4}\int F^{a}{}_{\mu\nu}F^{a\,\mu\nu}, \quad S_{GF} = -\frac{1}{2\alpha}\int f^a f^a. \quad S_{FP} = -\int \bar{c}^a(x)\, M^{ab}(x)c^b(y)$$

The problem is that under gauge transformation $A_\mu \to A_\mu + D_\mu\theta$, the S_{GF} and S_{FP} terms give contributions. The innovation of BRS was to look at gauge transformations where $\theta = c\lambda$ (where λ is a Grassmannian constant). Consider the following:

$$\delta S_{GF} = -\frac{1}{\alpha}\int f^a \frac{\partial f^a}{\partial A_\mu^b}\delta A_\mu^b = -\frac{1}{\alpha}\int f^a \frac{\partial f^a}{\partial A_\mu^b} D_\mu^{bd} c^d \lambda$$

Recall that:

$$M^{ab}(x,y) = \frac{\partial f^a\left(D_\mu\theta(x)\right)}{\partial\theta^b(y)} = \frac{\partial f^a(\delta A_\mu^b)}{\partial\theta^b(y)} = \frac{\partial f^a}{\partial A_\mu^b}\frac{\partial A_\mu^b}{\partial\theta^b(y)} = \frac{\partial f^a(x)}{\partial A_\mu^b(y)} D_\mu^b(y)$$

Thus,

$$\delta S_{GF} = -\frac{1}{\alpha}\int f^a M^{ab} c^b \lambda$$

If we now choose an appropriate change of $\bar{c}^a$ to eliminate this term from δS_{FP}:

$$\delta S_{FP} = -\int \delta\bar{c}^a(x)\, M^{ab}(x)c^b(y) - \int \bar{c}^a(x)\,\delta[M^{ab}(x)c^b(y)].$$

So, if we choose $\delta\bar{c}^a(x) = -\frac{1}{\alpha}f^a\lambda$, then the first integral cancels as desired, leaving the integral with $\delta[M^{ab}(x)c^b(y)]$. If we want this last term to be zero we are asking for:

$$0 = \delta[M^{ab}c_b] = \delta\left[\frac{\partial f^a(x)}{\partial A_\mu^b(y)} D_\mu^b c_b\right] = \frac{\partial f^a(x)}{\partial A_\mu^b(y)\partial A_\nu^c(z)}(D_\nu^b c_b\lambda)D_\mu^b c_b + \frac{\partial f^a(x)}{\partial A_\mu^b(y)}\delta\left[D_\mu^b c_b\right]$$

Note that the first term on the right contracts a symmetric term with an antisymmetric term, thus it is identically zero. Thus, to have $\delta[M^{ab}(x)c^b(y)] = 0$ we need $\delta\left[D_\mu^b c_b\right] = 0$, the latter we get by choosing our last transformation relation, that for $c(x)$:

$$\delta\left[D_\mu^b c_b\right] = 0 \to \delta c_a = -\frac{g}{2}f^{abc}c_b c_c\lambda.$$

Details in [55].

Using BRS, show Yang-Mills Theories are Renormalizable

Starting with the Yang-Mills Lagrangian in the form used by BRS, with Lorentz gauge chosen ($f^a = \partial_\mu A^{a\mu}(x) = 0$):

$$Z = \int \mathcal{D}A_\mu \mathcal{D}c\mathcal{D}\bar{c} \exp\left(i \int d^4x\, \mathcal{L}\right)$$

where

$$\mathcal{L} = -\frac{1}{4} F^a{}_{\mu\nu} F^{a\,\mu\nu} - \frac{1}{2\alpha}\left(\partial_\mu A^{a\mu}\right)\left(\partial_{\mu\nu} A^{a\nu}\right) + \bar{c}^a \partial^\mu D^{ab}_\mu c^b$$

And where we have established invariance of the Lagrangian under the following (BRS) transformation:

$$\delta A^a_\mu(x) = D^{ab}_\mu c^b(x)\lambda$$

$$\delta \bar{c}^a(x) = -\frac{1}{\alpha}\partial_\mu A^{a\mu}(x)\lambda$$

$$\delta c^a = -\frac{g}{2} f^{abc} c_b(x) c_c(x)\lambda$$

And since the measure is invariant, we then have Z is invariant. This means that under BRS transformation we have:

$$\langle 0|T(A+\delta A)\ldots(c+\delta c)\ldots(\bar{c}+\delta\bar{c})\ldots|0\rangle = \langle 0|TA\ldots c\ldots\bar{c}\ldots|0\rangle.$$

Thus

$$\langle 0|T\delta A\ldots c\ldots\bar{c}\ldots|0\rangle + \langle 0|TA\ldots\delta c\ldots\bar{c}\ldots|0\rangle + \langle 0|TA\ldots c\ldots\delta\bar{c}\ldots|0\rangle = 0.$$

The identities (like Ward identities from abelian) are known as the Slavnov-Taylor identities.

Consider the example of $\delta_{BRS}\langle 0|TA^a_\mu(x)\bar{c}^b|0\rangle = 0$:

$$0 = \langle 0|T\delta A^a_\mu(x)\bar{c}^b|0\rangle + \langle 0|TA^a_\mu(x)\delta\bar{c}^b|0\rangle$$

So,

$$0 =$$

$$0 = \langle 0|TD^{ad}_\mu c^d(x)\lambda\bar{c}^b|0\rangle + \left\langle 0\middle|TA^a_\mu(x)\left(-\frac{1}{\alpha}\partial_\mu A^{b\mu}(x)\lambda\right)\middle|0\right\rangle$$

Thus, upon factoring out the λ (a Grassmannian variable so anticommuting) we have the relation:

$$\langle 0|TD^{ad}_\mu c^d(x)\bar{c}^b|0\rangle + \frac{1}{\alpha}\langle 0|TA^a_\mu(x)\partial_\mu A^{b\mu}(x)|0\rangle.$$

Let's focus on the longitudinal part of A^a_μ and rewrite the derivatives:

$$\langle 0|T\partial^\mu D^{ad}_\mu c^d(x)\bar{c}^b(y)|0\rangle + \frac{1}{\alpha}\frac{\partial}{\partial x^\mu}\frac{\partial}{\partial y^\nu}\langle 0|TA^{a\mu}(x)A^{b\nu}(y)|0\rangle.$$

Solving for the term on the left-hand-side in a general way, note that it satisfies the form:

$$\left\langle 0\middle|T\left[\partial^\mu D^{\square}_\mu c(x)\right]^a \mathcal{F}\middle|0\right\rangle = \int \mathcal{D}A_\mu\mathcal{D}c\mathcal{D}\bar{c}\,\left[\partial^\mu D^{\square}_\mu c(x)\right]^a \mathcal{F}\, e^{iS},$$

where $\mathcal{F}$ is any grouping of fields. From this form we then have:

$$\left\langle 0\middle|T\left[\partial^{\mu}D_{\mu}^{\square}c(x)\right]^{a}\mathcal{F}\middle|0\right\rangle = \int \mathcal{D}A_{\mu}\mathcal{D}c\mathcal{D}\bar{c}\ \mathcal{F}\frac{\delta}{i\delta\bar{c}^{a}(x)}e^{iS}$$

$$= \int \mathcal{D}A_{\mu}\mathcal{D}c\mathcal{D}\bar{c}\ \frac{i\delta}{\delta\bar{c}^{a}(x)}\mathcal{F}\ e^{iS} = \left\langle 0\middle|T\frac{i\delta}{\delta\bar{c}^{a}(x)}\mathcal{F}\middle|0\right\rangle$$

Thus, we see that:

$$\left\langle 0\middle|T\partial^{\mu}D_{\mu}^{ad}c^{d}(x)\bar{c}^{b}(y)\middle|0\right\rangle = i\delta^{ab}\delta_{4}(x-y).$$

So,

$$\frac{1}{\alpha}\frac{\partial}{\partial x^{\mu}}\frac{\partial}{\partial y^{\nu}}\langle 0|TA^{a\mu}(x)A^{b\nu}(y)|0\rangle = -i\delta^{ab}\delta_{4}(x-y).$$

This is the same result as in QED. In Fourier space:

$$\langle 0|TA^{a\mu}(x)A^{b\nu}(y)|0\rangle = i\int\frac{d^{4}p}{(2\pi)^{4}}\,e^{-ip\cdot(x-y)}\widetilde{\Delta}_{F}(p)_{\mu\nu}^{ab}\,.$$

Thus,

$$p^{\mu}p^{\nu}\widetilde{\Delta}_{F}(p)_{\mu\nu}^{ab} = -\alpha\delta^{ab}$$

and

$$\widetilde{\Delta}_{F}(p)_{\mu\nu}^{ab} = \left[\widetilde{\Delta}_{F}(p)_{\mu\nu}^{ab}\right]_{\mathcal{T}} - \alpha\frac{p_{\mu}p_{\nu}}{p^{2}(p^{2}+i\varepsilon)}$$

Where $\mathcal{T}$ denotes the transverse part of the propagator. Thus, the longitudinal part of the gluon propagator does not get renormalized (as in QED).

Let's now turn to a proof of renormalizability for Yang-Mills theories given the tools developed, including the relation established above where

$$\left\langle 0\middle|T\left[\partial^{\mu}D_{\mu}^{\square}c(x)\right]^{a}\mathcal{F}\middle|0\right\rangle + \left\langle 0\middle|T\frac{i\delta}{\delta\bar{c}^{a}(x)}\mathcal{F}\middle|0\right\rangle = 0$$

Let:
E_{v} equal the number of external gluon lines.
E_{g} equal the number of external ghost lines.
I_{v} equal the number of internal gluon lines.
I_{g} equal the number of internal ghost lines.
L equals the number of loops
V_{3} equals the number of 3-point gluon vertices.
V_{4} equals the number of 4-point gluon vertices.
V_{g} equals the number of $\bar{c}Ac$ vertices.

The topological identities are:

$$3V_{3} + 4V_{4} + V_{g} = E_{v} + 2I_{v}$$
$$2V_{g} = E_{g} + 2I_{g}$$

$$L = I_v + I_g - (V_3 + V_4 + V_g) + 1$$

And the degree of divergence is given by (some vertices have ∂ terms):

$$D = 4L - 2(I_v + I_g) + (V_3 + V_g).$$

Thus,

$$D = 4 - (E_v + E_g)$$

as in QED. And, as with QED, we can now enumerate the possibly divergent graphs. For the ghost propagator we have an apparent $D = 2$ divergence, but because an extra momentum is carried by the external line we have an effective $D = 1$. Similarly, the ghost 4-point vetex is not $D = 0$, but effectively $D = -1$. The same argument hold for the cAc vertex, which appears as $D = 1$ but has effective $D = 0$ (for logarithmic divergence). The gluon propagator is $D = 2$, the gluon 3-point vertex is $D = 1$, and 4-point vertex is $D = 0$. As with QED, the divergent terms are precisely the interaction terms in the Lagrangian that are renormalized. So we see how to arrive at a theory that is finite as before, but now there is the problem that gauge invariance must be maintained. For the Yang-Mills generalization of the QED-type renormalization the difficult part is showing that gauge invariance is respected. To start this process, we must first show that we have a regularization process that respects the gauge invariance. To this end dimensional regularization is used.

Consider the bare Lagrangian in the form arrived at, with explicit invariance under BRS transformations. If we proceed with the regular renormalization of wavefunction, etc., we get:

$$A^a_{\mu\,B} = {Z_3}^{1/2} A^a_\mu$$

$$c^a{}_B = \tilde{Z}^{1/2} c^a$$

$$\bar{c}^a{}_B = \tilde{Z}^{1/2} \bar{c}^a$$

Now, however, we have two types of 3-point interaction vertices. The 3-gluon vertex will have a g_B along with three renormalization factors for the 3 A^a_μ's, thus we would usually have the relation for g_B:

$$g_B = \frac{Z_{YM}}{{Z_3}^{3/2}} g_{YM},$$

where g_{YM} is the coupling constant from the $F_{\mu\nu}$ term. However we must also consider the ghost 3-point function, for which we have an alternate form for g_B:

$$g_B = \frac{X}{{Z_3}^{1/2}\tilde{Z}} g_{ghost},$$

where g_{ghost} is the coupling constant from the D_μ^{ab} operator. From these two relations we have a restriction that follows from the Ward identities such that we get the constraint:

$$\frac{Z_{YM}}{Z_3} = \frac{X}{\tilde{Z}}.$$

Convention is to choose:

$$g_B = \frac{X}{{Z_3}^{1/2}\tilde{Z}} g.$$

In the process of working with the bare Lagrangian we can now see a problem – we have the aforementioned five divergent graphs, but we have only 3 independent renormalization parameters. In what we have just written, with the added Yang-Mills terms, we've introduced two extra renormalization constants: $\tilde{Z}$ and X. To clarify, let's rewrite the Lagrangian in terms of the renormalization quantities:

$$\mathcal{L} = -\frac{1}{4} Z_3 F^a{}_{\mu\nu} F^{a\,\mu\nu} - \frac{1}{2\alpha}(\partial_\mu A^{a\mu})(\partial_\nu A^{a\nu}) + \tilde{Z}\bar{c}^a \partial^\mu D_\mu^{ab} c^b$$

$$D_\mu^{ab} = \delta^{ab}\partial_\mu + \frac{X}{\tilde{Z}} g f^{abc} A_\mu^c(x)$$

$$F^{a\,\mu\nu} = (\partial_\mu A_\nu^a - \partial_\nu A_\mu^a) - \frac{X}{\tilde{Z}} g f^{abc} A_\mu^b A_\nu^c$$

From this form we see a renormalized BRS transformation is indicated to maintain gauge invariance in this more general formulation:

$$\delta A_\mu^a(x) = \tilde{Z} D_\mu^{ab} c^b(x)\lambda$$

$$\delta \bar{c}^a(x) = -\frac{1}{2\alpha}\partial_\mu A^{a\mu}(x)\lambda$$

$$\delta c^a = -X\frac{g}{2} f^{abc} c_b(x) c_c(x)\lambda$$

Rewriting the renormalized Lagrangian with renormalized terms separated:

$$\mathcal{L} = -\frac{1}{4} F^a{}_{\mu\nu} F^{a\,\mu\nu} - \frac{1}{2\alpha}(\partial_\mu A^{a\mu})(\partial_{\mu\nu} A^{a\nu}) + \bar{c}^a \partial^\mu D_\mu^{ab} c^b + \mathcal{L}_{c.t.}$$

$$\mathcal{L}_{c.t.} = -\frac{1}{4}(Z_3 - 1)(\partial_\mu A_\nu^a - \partial_\nu A_\mu^a)^2 + \frac{1}{2}(Z_{YM} - 1)(\partial_\mu A_\nu^a - \partial_\nu A_\mu^a) f^{abc} A_{\square}^{b\mu} A_{\square}^{c\nu}$$

$$+(\tilde{Z} - 1)\bar{c}^a \partial^\mu \partial_\mu c^a - \frac{1}{4}(\frac{Z_{YM}{}^2}{Z_3} - 1) g^2 f^{abc} f^{ade} A_{\square}^{b\mu} A_{\square}^{c\nu} A_\mu^d A_\nu^e - (X - 1) g \bar{c}^a \partial_\mu (f^{abc} \partial_\mu A^{b\mu} c^c)$$

We've arrived at a gauge invariant regularization scheme in the BRS formalism. Three terms will now be made finite:
(1) Use Z_3 to make the transverse part of the gluon propagator finite.
(2) Use X to make the gluon-ghost (cAc) term finite.
(3) Use $\tilde{Z}$ to make the ghost propagator finite.

Having made the above three terms finite, we must then show five other terms are finite in order to prove renormalizability:
(1) The longitudinal part of the gluon propagator.
(2) The gluon 3-vertex
(3) The gluon 4-vertex
(4) Composite operator from $\delta A_\mu^a(x)$ term
(5) Composite operator from δc^a term

Let's start by showing convergent terms involving $\delta A_\mu^a(x)$. Consider

$$\delta\langle A_\mu^a \mathcal{F}\rangle = 0 \quad \rightarrow \quad \langle\, (\delta A_\mu^a)\mathcal{F}\rangle + \langle A_\mu^a \delta\mathcal{F}\rangle = 0,$$

where $\mathcal{F}$ is any arbitrary collection of fields. We need to show the general $\langle\, (\delta A_\mu^a)\mathcal{F}\rangle$ term is finite. By operating on the equation with ∂_μ and using the equations of motion for ghosts:

$$\langle 0|T(\partial^\mu \delta A_\mu^a)\mathcal{F}|0\rangle = \left\langle 0\left|T\frac{i\delta}{\delta \bar{c}^a(x)}\mathcal{F}\right|0\right\rangle$$

On the left-hand-side only the fields $\bar{c}^a$ and $\bar{c}^a A_\mu^a$ are potentially divergent, but on the right-hand-side we see that $\mathcal{F} = \bar{c}^a$ → RHS=i1, while $\mathcal{F} = \bar{c}^a A_\mu^a$ → RHS=$i\langle 0|A_\mu^a|0\rangle = 0$. Thus, the Composite terms with $\delta A_\mu^a(x)$ are finite.

Next, consider the composite terms involving δc^a, which contain 2 c^a's, so must pair with a field entry $\mathcal{F}$ with at least 2 $\bar{c}^a$'s, and in doing so arrives at a $D = 0$ divergence that must be considered. So we need only consider whether $\langle 0|T\delta c^a \bar{c}^b \bar{c}^c|0\rangle$ is finite. Using BRS invariance:

$$0 = \delta\langle 0|Tc^a \bar{c}^b \bar{c}^c|0\rangle = \left\langle 0\left|T\left(-\frac{g}{2}L^a\lambda\right)\bar{c}^b\bar{c}^c\right|0\right\rangle$$

$$+\left\langle 0\left|Tc^a\left(-\frac{1}{2\alpha}\partial_\mu A^{b\mu}\lambda\right)\bar{c}^c\right|0\right\rangle + \left\langle 0\left|Tc^a\bar{c}^b\left(-\frac{1}{2\alpha}\partial_\mu A^{c\mu}\lambda\right)\right|0\right\rangle$$

where,

$$L^a = Xf^{abc}c_b c_c.$$

Removing the λ (taking care with signs since they are Grassmannian):

$$-g\langle 0|TL^a\bar{c}^b\bar{c}^c|0\rangle + \frac{1}{\alpha}\langle 0|Tc^a\partial_\mu A^{b\mu}\bar{c}^c|0\rangle - \frac{1}{\alpha}\langle 0|Tc^a\bar{c}^b\partial_\mu A^{c\mu}|0\rangle = 0$$

Under dimensional regularization the latter two terms are made finite, thus $\langle 0|TL^a\bar{c}^b\bar{c}^c|0\rangle$ must be finite, the composite operator L^a must be finite.

Next, consider BRS invariance for the following:

$$0 = \delta\langle 0|TA^{a\mu}\bar{c}^b|0\rangle = \langle 0|T(\delta A^{a\mu})\bar{c}^b|0\rangle + \langle 0|TA^{a\mu}\delta\bar{c}^b|0\rangle$$

$$0 = \langle 0|T(\delta A^{a\mu})\bar{c}^b|0\rangle + \left\langle 0\middle|TA^{a\mu}\left(-\frac{1}{2\alpha}\partial_\mu A^{b\mu}\lambda\right)\middle|0\right\rangle$$

We can rewrite this as:

$$\partial_\mu\langle 0|TA^{a\mu}A^{b\mu}|0\rangle = -\langle 0|T(\delta A^{a\mu})\bar{c}^b|0\rangle$$

Where the right-hand-side is finite as shown previously, and so the left-hand-side must also be finite, but the left-hand-side is the longitudinal part of A. Thus we have proven finiteness for the longitudinal part as well. To show finiteness for the 3-vertex consider:

$$0 = \delta\langle 0|T\bar{c}^bA^{b\mu}A^{c\nu}|0\rangle,$$

All resulting terms are of known type and finite, except for the term

$$\langle 0|T\delta\bar{c}^bA^{b\mu}A^{c\nu}|0\rangle,$$

which must be finite, but since it can be rewritten (using the BRS transform):

$$\left\langle 0\middle|T\left(-\frac{1}{2\alpha}\partial_\mu A^{a\mu}(x)\lambda\right)A^{b\mu}A^{c\nu}\middle|0\right\rangle = -\frac{1}{2\alpha}\lambda\partial_\mu\langle 0|TA^{a\mu}A^{b\mu}A^{c\nu}|0\rangle$$

Thus, we see that the 3-vertex is finite. A similar derivation then shows the same for the 4-vertex. Thus, Yang-Mills is renormalizable.

Anomalous Ward Identities

The anomalous Ward Identities are those for which no regularization exists. We are using dimensional regularization, so the question becomes where does this fail? The answer falls mainly on the element from the Dirac theory that relies on 4 dimensions, the $\gamma_5 = \gamma_0\gamma_1\gamma_2\gamma_3$ term. Anomalies having to do with couplings that involve γ_5 are known as chiral anomalies.

Chiral Symmetries

Let's start with SU(3) since it is easier than Weinberg-Salam. For Quantum Chromodynamics, that results from the Yang-Mills nan-Abelian gauge field theory with gauge group SU(3), we will derive the following Lagrangian (in ta later section):

$$\mathcal{L} = -\frac{1}{4}F^a{}_{\mu\nu}F^{a\,\mu\nu} + \sum_i \bar{\psi}_i^a(i\gamma\cdot D^{ab} - m_i)\psi_i^a,$$

$$D_\mu^{ab} = \delta^{ab}\partial_\mu + ig(t^c)^{ab}A_\mu^c = \delta^{ab}\partial_\mu + ig\sum_i \left(\frac{\lambda^c}{2}A_\mu^c\right)^{ab},$$

where ψ_i^a are the quark fields, and A_μ^a are the color gluons, $a = 1,2,3$ is the color index, and i is the flavor index (u,d,s,c,b,t).

Let's now consider the $m_i \cong 0$ approximation for the common low mass quarks (useful for the u, d quarks that have very low mass in perturbation theory):

$$\mathcal{L}_{m=0} = -\frac{1}{4}F^a{}_{\mu\nu}F^{a\,\mu\nu} + \sum_{i=u,d} \bar{\psi}_i^a(i\gamma\cdot D_{\square}^{ab})\psi_i^a,$$

Let's now examine the symmetry groups of $\mathcal{L}_{m=0}$:

$$\psi_i^{\square} \to \left(\psi_i^{\square}\right)' = \left(e^{-i\theta^\alpha\frac{\tau^\alpha}{2}}\right)_{ij}\psi_j^{\square},$$

which shows invariance under SU(2), and

$$\psi_i^{\square} \to \left(\psi_i^{\square}\right)' = e^{-i\theta}\ \psi_j^{\square},$$

which shows invariance of the Lagrangian under U(1). Thus, there is the familiar SU(2)×U(1). Notice however that the following invariance also exists (requires the mass term to be zero or approximately so since this term breaks the invariance by not having a γ factor, which is necessary to have the sign change/conjugation, for invariance:

$$\psi_i^{\square} \to \left(\psi_i^{\square}\right)' = \left(e^{-i\theta^\alpha\frac{\tau^\alpha}{2}\gamma_5}\right)_{ij}\psi_j^{\square},$$

which shows invariance under $SU(2)_A$, and

$$\psi_i^{\square} \to \left(\psi_i^{\square}\right)' = e^{-i\theta\gamma_5}\ \psi_j^{\square},$$

which shows invariance of the Lagrangian under $U(1)_A$. Thus, the approximate Lagrangian given by $\mathcal{L}_{m=0}$ is invariant under SU(2)×U(1) × $SU(2)_A$ × $U(1)_A$. The axial symmetry so indicated Is not seen in nature, even in low-mass approximation, thus is thought to be spontaneously broken. The flavor SU(2)×U(1) symmetry is then partly broken by $m_u \neq m_d$.

From the breaking of $SU(2)_A$ × $U(1)_A$ we expect 4 massless pseudoscalar bosons, but we get only 3 (π^+, π^-, π^0), that match with the SU(2) portion, leaving the U(1) part missing (known as the U(1) problem, this will be solved in the more complete Yang-Mills analysis by analysis of instantons). Even when we consider all the pseudoscalar bosons in the more general three-flavor model (with u, d, and s quarks), giving $SU(3)_A$ × $U(1)_A$, there is still the U(1) problem to resolve.

Chiral Anomaly

We will find in this section that the axial symmetry of classical electromagnetism is broken in quantum electrodynamics due to the Ward identities. The chiral anomaly was first described by Alder-Bell-Jackiw [56,57] in 1969, when is was used to explain the anomalous decay of neutral pions. In the process of obtaining the decay width the theory requires the number of quarks to be specified. There is agreement with the observed three generations of quarks (no more, no less). Thus, an important constraint on the standard model is forced by resolution of the chiral anomaly. In this section the constraint requiring total generation charge sum to zero is shown, while the constraint on generation number is left for more detailed review elsewhere [20,21].

Let's now consider quarks (hadrons) in more detail, starting with restating the Dirac equation from the Lagrangian formulation:

$$\mathcal{L} = \bar{\psi}(i\gamma \cdot \partial - m)\psi$$

gives equations of motion:

$$(i\gamma \cdot \partial - m)\psi, \quad \bar{\psi}(i\gamma \cdot \overleftarrow{\partial} - m)$$

which have chiral symmetry:

$$\psi \rightarrow e^{i\theta\gamma_5}\psi, \quad \bar{\psi} \rightarrow e^{-i\theta\gamma_5}\bar{\psi}$$

which gives rise to a Noether current:

$$J_{5\,\mu} = \bar{\psi}\gamma_\mu\gamma_5\psi, \quad \partial^\mu J_{5\,\mu} = 2im\bar{\psi}\gamma_5\psi.$$

If $m = 0$ (Weyl spinors), we have the conserved current $J_{5\,\mu}$, with $\partial^\mu J_{5\,\mu} = 0$.

Generalize to non-Abelian gauge space

Let's now consider a general gauge symmetry:

$$\psi \rightarrow e^{i\theta^a t^a}\psi, \quad \bar{\psi} \rightarrow e^{-i\theta^a t^a}\bar{\psi}$$

and get the Noether currents:

$$J_\mu^a = \bar{\psi}\gamma_\mu t^a\psi \equiv \bar{\psi}_i\gamma_\mu(t^a)^{ij}\psi_j$$

$$J_{5\,\mu}^a = \bar{\psi}\gamma_\mu\gamma_5 t^a\psi$$

for which there are associated conserved charges:

$$Q^a = \int d^3x J_0^a, \quad Q_5^a = \int d^3x J_{5\,0}^a.$$

The Q^a's generate an SU(N) group:

$$[Q^a, Q^b] = if^{abc}Q^c \rightarrow SU(N)$$

while the Q_5^a do not:

$$\left[Q_5^a, Q_5^b\right] = if^{abc}Q^c$$

(the product of pseudoscalars is a scalar). As with the generators of pure boost or rotations in the Lorentz group lacking closure separately, here the we will combine the charges so as to get two $SU(N)$'s. Let's define (L)eft and (R)ight chirality charges:

$$Q_R^a \equiv Q^a + Q_5^a, \quad Q_L^a \equiv Q^a - Q_5^a,$$

for which we have:

$$[Q_R^a, Q_R^b] = i\varepsilon^{abc} Q_R^c$$
$$[Q_L^a, Q_L^b] = i\varepsilon^{abc} Q_L^c$$
$$[Q_R^a, Q_L^b] = 0$$

which is $SU(N)_L \times SU(N)_R$, and explains why this (pairing) always appears in gauge field theories.

We can define two projection operations in terms of the chirality gamma, γ_5, where:

$$P_L = \frac{1}{2}(1 - \gamma_5), \quad P_R = \frac{1}{2}(1 + \gamma_5).$$

Thus

$$\psi = (P_L + P_R)\psi = \psi_L + \psi_R$$

This is convenient because now:

$$Q_R^a = \int d^3x \bar{\psi}\gamma_\mu(1 + \gamma_5)\psi = 2\int d^3x \bar{\psi}_R \gamma_\mu \psi_R$$

and

$$\bar{\psi}\gamma_\mu\psi = \bar{\psi}_L\gamma_\mu\psi_L + \bar{\psi}_R\gamma_\mu\psi_R$$

So, we can now write:

$$\mathcal{L} = \sum_i [\bar{\psi}_{iL} i\gamma \cdot \mathcal{D}\psi_{iL} + \bar{\psi}_{iR} i\gamma \cdot \mathcal{D}\psi_{iR}]$$

If massless the action of the left and right SU(N)'s act separately, but not massless, and the weak interaction symmetry only acts on the Left-handed quarks.

Ward Identities involving chirality

We have that

$$[Q_5^a, \psi_i(x)] = -i(t_a)_{ij}\psi_j = (\delta\psi)_i$$

and

$$[J_5^a{}_0(x), \psi_i(x)] = (\delta\psi)_i(x)\delta_3(\vec{x} - \vec{y}).$$

and

$$\partial^\mu J_5^a{}_\mu = \sum_{i,j} (m_i + m_j)\,\bar{\psi}_i\gamma_5(t^a)^{ij}\psi_j$$

We can use this to derive a set of Ward identities:

$$\frac{\partial}{\partial x_\mu}\langle 0|TJ_{5\ \mu}^{a}(x)\psi_i(y)\dots\psi_j(z)|0\rangle = \langle 0|T\partial^\mu J_{5\ \mu}^{a}(x)\psi_i(y)\dots\psi_j(z)|0\rangle$$
$$+\langle 0|T[J_{0\ 5}^{a}(x),\psi_i(y)]_{x^0=y^0}\dots\psi_j(z)|0\rangle$$
$$+\dots+\langle 0|T\psi_i(y)\dots[J_{0\ 5}^{a}(x),\psi_j(z)]_{x^0=z^0}|0\rangle$$

Now consider the case of only one axial (abelian) current, in which case all of the commutators in the above are zero and we have the derivative operation distributing without added terms. Working with

$$J_\mu(x)=\bar{\psi}\gamma_\mu\psi,\ \ J_{5\ \mu}(x)=\bar{\psi}\gamma_\mu\gamma_5\psi,$$

and from Dirac equation:

$$\partial^\mu J_\mu=0,\ \ \partial^\mu J_{5\ \mu}(x)=2im\bar{\psi}\gamma_5\psi$$

So,

$$\frac{\partial}{\partial x_\mu}\langle 0|TJ_\mu(x)J_\nu(y)J_{5\ \lambda}(z)|0\rangle=0$$

and

$$\frac{\partial}{\partial z_\mu}\langle 0|TJ_\mu(x)J_\nu(y)J_{5\ \lambda}(z)|0\rangle=2m\langle 0|TJ_\mu(x)J_\nu(y)i\bar{\psi}(z)\gamma_5\psi(z)|0\rangle$$

Let's now examine $\langle 0|TJ_\mu(x)J_\nu(y)J_{5\ \lambda}(z)|0\rangle$ directly by making use of Fourier Transforms:

$$\int dxdydze^{k_1x+k_2y+pz}\langle 0|TJ_\mu(x)J_\nu(y)J_{5\ \lambda}(z)|0\rangle$$
$$\equiv(2\pi)^4\delta(k_1+k_2+p)T_{\mu\nu\lambda}(k_1,k_2,p)$$

and from prior results we then have by the Ward Identity:

$$k_1^\mu T_{\mu\nu\lambda}=0,\qquad k_2^\nu T_{\mu\nu\lambda}=0,\quad p^\lambda T_{\mu\nu\lambda}=2imT_{\mu\nu}$$

Considered as a free theory, the composite operator $\langle 0|TJ_\mu(x)J_\nu(y)J_{5\ \lambda}(z)|0\rangle$ corresponds to:

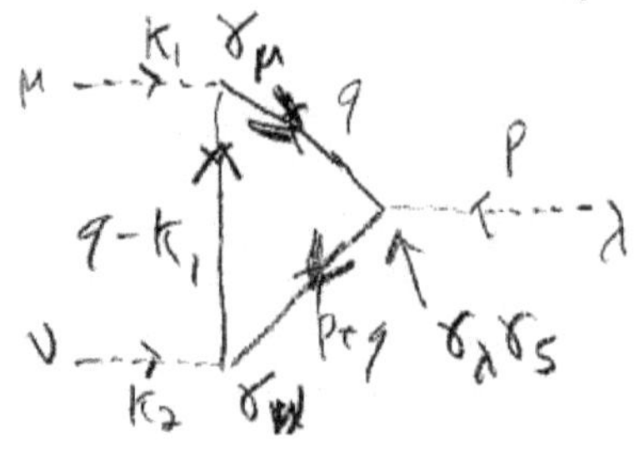

Thus,

$$T_{\mu\nu\lambda} = (-1)\int \frac{d^4q}{(2\pi)^4} Tr\left[\gamma_\mu \frac{i}{\boldsymbol{q}-\boldsymbol{k_1}-m}\gamma_\nu \frac{i}{\boldsymbol{q}-\boldsymbol{p}-m}\gamma_\lambda\gamma_5 \frac{i}{\boldsymbol{q}-m} + S\right]$$

where S is a similar term for the loop going the other way, and slash notation it replaced with bold font. From this we see that:

$$p^\lambda T_{\mu\nu\lambda} = 2imT_{\mu\nu} + \Delta^1_{\mu\nu} + \Delta^2_{\mu\nu},$$

where

$$\Delta^1_{\mu\nu} = i\int \frac{d^4q}{(2\pi)^4} Tr\left[\gamma_\mu \frac{1}{\boldsymbol{q}-\boldsymbol{k_1}-m}\gamma_\nu\gamma_5 \frac{1}{\boldsymbol{q}-m} + \gamma_\nu \frac{1}{\boldsymbol{q}-\boldsymbol{k_2}-m}\gamma_\mu\gamma_5 \frac{1}{\boldsymbol{q}-\boldsymbol{p}-m}\right]$$

$$\Delta^2_{\mu\nu} = i\int \frac{d^4q}{(2\pi)^4} Tr\left[\gamma_\mu \frac{1}{\boldsymbol{q}-\boldsymbol{k_1}-m}\gamma_\nu\gamma_5 \frac{1}{\boldsymbol{q}-\boldsymbol{p}-m} + \gamma_\nu \frac{1}{\boldsymbol{q}-\boldsymbol{k_2}-m}\gamma_\mu\gamma_5 \frac{1}{\boldsymbol{q}-m}\right]$$

We can write

$$\Delta^1_{\mu\nu} = i\int \frac{d^4q}{(2\pi)^4}\left[f_{\mu\nu}(q+k_2) - f_{\mu\nu}(q)\right],$$

$$f_{\mu\nu} = Tr\left[\frac{1}{\boldsymbol{q}-\boldsymbol{k_2}-m}\gamma_\mu \frac{1}{\boldsymbol{q}-\boldsymbol{p}-m}\gamma_\nu\gamma_5\right]$$

In the present form, with the split between $\Delta^1_{\mu\nu}$ and $\Delta^2_{\mu\nu}$ as indicated, we have two integrals that are linearly divergent. Notice, however, that the Δ's superficially appear as the difference in two terms that are related by a shift in integration variable. The suggests how a fix might be made. consider the example:

$$I(a) = \int_{-\infty}^{\infty} dx[f(x+a) - f(x)]$$

$$= \int_{-\infty}^{\infty} dx\left[\sum_{n=1} \frac{1}{n!} f^{(n)}(x)a^n\right] = \int_{-\infty}^{\infty} dx\left[f'(x)a + \frac{1}{2}f''a^2 + \cdots\right]$$

Thus,

$$I(a) = a[f(\infty) - f(-\infty)] + \frac{1}{2}a^2[f'(\infty) - f'(-\infty)] + \cdots$$

If $\int dxf(x)$ is at most logarithmically divergent, then as $f(x) \to 0\ as\ |x| \to \infty$, and $I(a) = 0$.
If $\int dxf(x)$ is linearly divergent, then $f(x) \to constant\ as\ |x| \to \infty$, and

$$I(a) = a[f(\infty) - f(-\infty)]$$

Let's now generalize this result to n-dimensional Euclidean space:

$$I^n(a) = \int_{-\infty}^{\infty} d^n x \left[\sum_j a_j \partial^j f(x) + \cdots \right] = \int_{S=\infty} d\vec{S} \cdot \vec{a} f(x)$$

$$= \lim_{R\to\infty} \frac{\vec{a} \cdot \vec{R}}{|R|} f(R) S_{N-1}(R)$$

For Minkowski:

$$\int_{-\infty}^{\infty} d^4x [f(x+a) - f(x)] = 2i\pi^2 \lim_{R\to\infty} \vec{a}_\mu \cdot \vec{R}^\mu R^2 f(R)$$

Thus,

$$\Delta^1_{\mu\nu} = i \lim_{|q|\to\infty} (2i\pi^2) a \cdot q q^2 f_{\mu\nu}(q) = i \lim_{|q|\to\infty} (2i\pi^2) \frac{a \cdot q}{q^2} Tr[\boldsymbol{q}\gamma_\mu \boldsymbol{q}\gamma_\nu\gamma_5]$$

where

$$Tr[\boldsymbol{q}\gamma_\mu \boldsymbol{q}\gamma_\nu\gamma_5] = q^\alpha q^\beta Tr[\gamma_\alpha\gamma_\mu\gamma_\beta\gamma_\nu\gamma_5] = q^\alpha q^\beta(-4i\varepsilon_{\alpha\mu\beta\nu}) = 0$$

with the latter zero due to $q^\alpha q^\beta$ being symmetric and $\varepsilon_{\alpha\mu\beta\nu}$ being antisymmetric, their contraction then giving zero. Thus we have zero at lowest order in the expansion for $\Delta^1_{\mu\nu}$, so we need to go to one order higher. doing so we then get at lowest (non-zero) order:

$$\Delta^1_{\mu\nu} = -\frac{i}{8\pi^2} \varepsilon_{\mu\nu\alpha\beta} k_2^\alpha p^\beta, \quad \Delta^2_{\mu\nu} = \frac{i}{8\pi^2} \varepsilon_{\mu\nu\alpha\beta} k_1^\alpha p^\beta$$

We thus have:

$$p^\lambda T_{\mu\nu\lambda} = 2imT_{\mu\nu} + \frac{i}{8\pi^2} \varepsilon_{\mu\nu\alpha\beta}(k_1^\alpha - k_2^\alpha) p^\beta$$

We should have the term $2imT_{\mu\nu}$ to have agreement with Ward Identity indicated at the start, so there is still a problem with the anomalous second term on the right. It turns out the anomalous term, partly cured by the integration shift above, is still an illusion, but clarifying what representation will give consistency with the Ward Identity (in this circumstance) is how to resolve the ambiguity. Note that instead of performing the integration according to a free theory Feynman diagram analysis with the specific diagram indicated, we could have chose an alternate diagram:

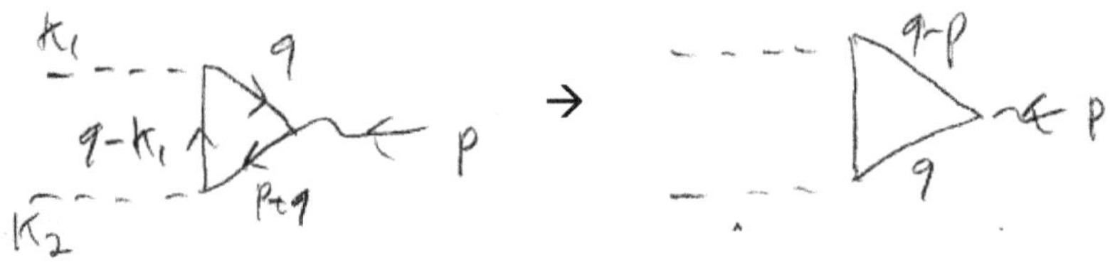

Thus, we could have an arbitrary change of variables such that $q \to q + xk_1 + yk_2$. The question then becomes, can we choose the x and y such that $p^\lambda T_{\mu\nu\lambda} = 2imT_{\mu\nu}$? The answer is yes, but in so doing we will lose the other Ward identity $k_1^\mu T_{\mu\nu\lambda} = 0$. In other words, we can't solve all of the Ward identities, the anomaly can't be eliminated in general.

Focusing on

$$f_{\mu\nu} = Tr\left[\gamma_\mu \frac{1}{\boldsymbol{q} + \boldsymbol{p} - m}\gamma_\nu\gamma_5\frac{1}{\boldsymbol{q} - \boldsymbol{k_2} - m}\right]$$

This can be rewritten as:

$$f_{\mu\nu}(q) = \frac{1}{[(q+p)^2 - m^2][(q-k_2)^2 - m^2]} \times Tr\{\gamma_\mu(\boldsymbol{q} + \boldsymbol{p} + m)\gamma_\nu\gamma_5(\boldsymbol{q} - \boldsymbol{k_2} + m)\}$$

and

$$f_{\mu\nu}(q) = \frac{1}{[(q+p)^2 - m^2][(q-k_2)^2 - m^2]} 4i\varepsilon_{\mu\alpha\nu\beta}(q+p)^\alpha(q-k_2)^\beta$$

Since the [...][...] part is fourth-order in q, and the only term with q in $\varepsilon_{\mu\alpha\nu\beta}(q+p)^\alpha(q-k_2)^\beta$ is first order since antisymmetry will eliminate other terms, this leaves linear divergence for the Δ's. Using the prior results for linearly divergent integrals we can then write

$$p^\lambda T_{\mu\nu\lambda} = 2miT_{\mu\nu} + \frac{i}{4\pi^2}\varepsilon_{\mu\nu\alpha\beta}k_1^\alpha k_2^\beta$$

and evaluating $k_1^\mu T_{\mu\nu\lambda}$ we get:

$$k_1^\mu T_{\mu\nu\lambda} = \frac{i}{8\pi^2}\varepsilon_{\mu\nu\alpha\beta}k_1^\alpha k_2^\beta$$

So, depending on how we evaluate the loop momenta we get different results. But the final result stands, if we get one Ward identity to work it is at the cost of another Ward identity working.

The anomaly can't be eliminated, as was first noticed by Adler-Bell-Jackiw [56,57]. Is the anomaly worse for higher order graphs (with higher order Ward identities)? No, it turns out additional vertices make the graph more convergent and the anomaly goes away – the only divergent graphs are the "triangle" graphs.

The anomaly in the triangle graphs is due to γ_5 terms, thus does not arise in electromagnetism but will arise in non-Abelian gauge theories (GUT theories) more generally, including the Weinberg-Salam (electroweak) theory. In GUT theories (including Weinberg-Salam) we have vertices like:

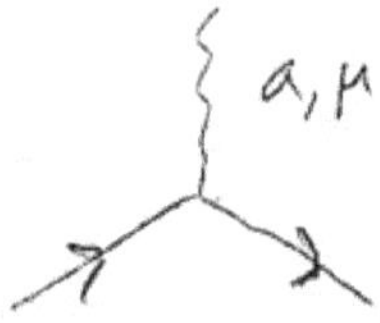

where

$$\bar{\psi}_a i\gamma \cdot \mathcal{D}\psi_b = \bar{\psi}_a[i\gamma \cdot \partial - ig(t^c)_{ab}\gamma \cdot A]\psi_b$$

where $(t^c)_{ab}$ is the matrix for generators in quark representation, and the diagram shows the coupling for $-igt^a\gamma_\mu$. For the coupling to axial current we have $-igt^a\gamma_\mu\gamma_5$. Returning to the problematic triangle diagram, consider

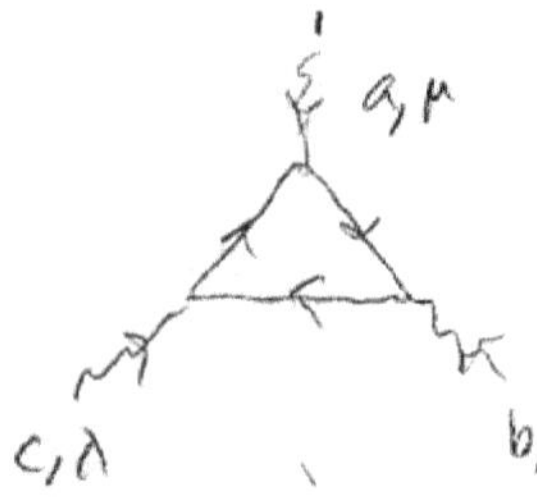

Evaluation of this diagram will result in a factor $Tr[t^a\{t^b, t^c\}]$, which is a group theoretic factor, and this is the factor that must be zero to eliminate the anomaly for a particular GUT theory and thereby have a consistent field theory quantization. Here are some cases where this term is zero:

(1) We see that the vector-like quarks have left and right chiralities that couple the same way to gluons, thus they will cancel each other. Thus, vector-like quarks are anomaly free.

(2) Quarks belonging to real representatives of groups will be anomaly free, since $Tr[t^a\{t^b, t^c\}] = 0$ is trivially satisfied.

(3) Weinberg-Salam: $SU(2)_L \times U(1)$. $U(1)$ is not 'safe' as in real, and $SU(2)_L$ not vector-like, so does not bode well. However, cancellation of anomalies does take place generator-by-generator, if the quark and lepton charge at a particular generation sum to zero. So, at the first generation, we have -1 electron charge, 0 electron-neutrino charge, up quarks with charge 2/3 and 3 colors, and down quark with -1/3 charge and 3 colors, for total charge 0. It was in the anomaly cancellation calculation indicated that the charm quark was first predicted.

5.5 Quantum Chromodynamics (QCD) Overview

Quantum Chromodynamics follows directly from quantum field theory methods developed for Quantum Electrodynamics. It turns out no fundamentally new principles are needed to understand the strong interactions. Instead the old principles (causality, Lorentz Invariance, unitarity, Gauge Invariance, Renormalizability) are again brought to bear, but with entirely new results, such as asymptotic freedom and confinement.

Quantum Chromodynamics was halfway solved at the outset, since there was already the quark model from fitting hadron resonances to the Eightfold Way by Gell-Mann [58]. In this description there were no isolated quarks. There followed the idea of $SU(3)$ color, denoted $SU(3)_C$, which was consistent with the quark model if color singlets were required $\{\bar{q}q$ and $qqq\}$. The example of $qqq = uuu$ in the case of the Δ^{++} resonance, was inconsistent with Fermi statistics, but consistent if the three u's different colors to form an $SU(3)$ singlet. A $SU(N)$ color model with $N = 3$ colors was also indicated by scattering experiments where:

$$\Gamma(\pi^0 \to \gamma\gamma) \propto N^2$$

and

$$\sigma(e^+e^- \to \{u, d, \ldots\}) \propto N$$

where $\{u, d, \ldots\}$ denotes the hadrons.

Quantum Chromodynamics was probably more than halfway done at the outset if you factor in that it was known what its main challenge would be – the need for asymptotic freedom. This need was recognized early from work on the Parton model where it was determined that the electroproduction scaling on ep→hadrons scattering, with eeq vertex and proton momentum 'p' went as:

$$\frac{1}{|q^2|} f\left(\frac{q^2}{2p \cdot q}\right).$$

This behavior is consistent with no heavy resonances, specifically, it is consistent with masses becoming irrelevant at large $|q^2|$. This is equivalent to saying that the strong interaction is weak at short distances, or that it is asymptotically free. A related experimental observation was quark confinement (Regge trajectories and quarkonia). So once Quantum Chromodynamics could explain asymptotic freedom, and then the observed hadron spectrum, cross-sections, structure functions, scaling violations, etc., it would be accepted. The objective in what follows is to explain asymptotic freedom, for the other areas of exploration there are numerous references [55].

So, we need to construct a theory of colored quarks with local $SU(3)$ symmetry:

$$q = \begin{pmatrix} q^1 \\ q^2 \\ q^3 \end{pmatrix}, \qquad q(x) \to \Omega(x)\, q(x), \qquad \Omega(x) \in SU(3)$$

To construct a gauge – invariant kinetic term for quark, note that

$$\partial_\mu(\Omega q) = (\partial_\mu \Omega)\, q + \Omega \partial_\mu q$$

So define

$$D_\mu q = \partial_\mu q = ig\Delta_\mu q$$

Where $\Delta_\mu(x)$ is Hermitian traceless 3x3 matrix:

$$\Delta_\mu(x) \rightarrow \Omega\Delta_\mu\Omega^{-I} + \frac{1}{ig}(\partial_\mu\Omega)\Omega^{-I}$$

Then $D_\mu q \rightarrow \Omega D_\mu q$ (e.g., transforms covariantly) and $\overline{q}iDq$ is gauge invariant. If we choose a basis for our eight gluon fields:

$$A_\mu = A_\mu^a T^a \quad tr\, T^a T^b = \frac{1}{2}\delta^{ab}$$

Since gluons are also dynamical we need a kinetic term, as with previous discussion, lets start differently here:

$$[D_\mu, D_\nu]q = \{-ig(\partial_\mu A_\nu - \partial_\nu A_\mu) - g^2[A_\mu, A_\nu]\}q$$

So define:

$$\mathrm{F}_{\mu\nu} = \frac{1}{-ig}[D_\mu, D_\nu] = (\partial_\mu A_\nu - \partial_\nu A_\mu) - ig[A_\mu, A_\nu]$$

Then

$$\mathrm{F}_{\mu\nu} \rightarrow \Omega\mathrm{F}_{\mu\nu}\Omega^{-1}$$

and $tr\mathrm{F}^{\mu u}\mathrm{F}_{\mu\nu}$ is gauge invariant. Let's write this in terms of gluon fields A_μ^a, but first a reminder on structure function notation and conventions:

$$[T^a T^b] = iC^{abc}T^c \rightarrow C^{abc} = -2i\, tr[T^a T^b]T^c$$

where C^{abc} is totally unsymmetric. Thus

$$F_{\mu\nu}{}^a = 2\, tr(\mathrm{F}_{\mu\nu}T^a) = \partial_\mu A_\nu^a - \partial_\nu A_\mu^a + gC^{abc}A_\mu^b A_\nu^c$$

and

$$-\frac{1}{2}tr\mathrm{F}_{\mu\nu}\mathrm{F}^{\mu\nu} = -\frac{1}{4}\mathrm{F}_{\mu\nu}^a\mathrm{F}^{\mu\nu a}$$

Suppose there are N quark flavors q_u with $u = 1,2 \ldots N$. The most general gauge invariant, invariant renormalizable Lagrangian is (slash notation on D is dropped):

$$\mathcal{L} = -\frac{1}{2}\mathrm{Z}tr\mathrm{F}_{\mu\nu}\mathrm{F}^{\mu\nu} + \overline{q}_L AiDq_L + \overline{q}_R BiDq_R - \overline{q}_L Mq_R - \overline{q}_R M^\dagger q_L$$
$$+\frac{1}{2}\mathrm{Z}'\epsilon_{\mu\nu\sigma\sigma}tr\mathrm{F}^{\mu\nu}\mathrm{F}^{\lambda\sigma}$$

where A, B, M are N x N matrices acting on flavor indices. A, B are Hermitian and

$$q_{L,R} = \frac{1}{2}(1 \pm \gamma_5)q.$$

Redefinitions are then typically done to get to the standard form described previously. The first step is to show that:

$$\epsilon_{\mu\nu\sigma\sigma}tr\mathrm{F}^{\mu\nu}\mathrm{F}^{\lambda\sigma} = \partial^\mu K_\mu.$$

As a total derivative the term is usually dropped in the classical theory, and that is how we will proceed now, but in the quantum theory this matter must be revisited in relation to the matter of instantons and global topological constants (winding number) that can exist.

Next

$$A_\mu \rightarrow \mathrm{Z}^{1/2}A_\mu, \quad g \rightarrow \mathrm{Z}^{-1/2}g$$

is used to put the renormalization constant Ƶ in the usual position. Since A and B are Hermitian we can diagonalize according to:

$$A \to {U_L}^{\dagger} A U_L, \quad B \to {U_R}^{\dagger} B U_R$$

with U_L and U_R unitary. Then rescale q_L and q_R such that

$$\overline{q}_L A i D q_L + \overline{q}_R B i D q_R \to \overline{q}_L i D q_L + \overline{q}_R i D q_R = \overline{q} i D q.$$

Finally, for the mass terms, the mass matrices can be diagonalized by:

$$M' \to {U_L}^{\dagger} M'^{U_R},$$

with U_L and U_R unitary. Thus we arrive at the standard form:

$$\mathcal{L} = -\frac{1}{2} tr F_{\mu\nu} F^{\mu\nu} + \sum_n \overline{q}_n i D q_n - m_n \overline{q}_n q_n$$

Regarding this simplified form for the Lagrangian we can see the following symmetries:

1. Invariance under CPT.
2. Flavor symmetry.
3. Chiral Symmetry.
4. U(1) problem.

These symmetries are not completely destroyed by weak interaction radiative corrections even though these violate C, P, CP. These corrections generate operators of Dimension ≤ 4 ⇒ Absorbed into most general Z, or Dimension > 4 ⇒ suppressed by powers of $\frac{1}{M}$. Things would have been different if we had chosen the strong interaction gauge group to be $SU(3)_L \times SU(3)_R$.Then parity is conserved but not by the most general Lagrangian. Thus weak radiative corrections would generate parity violation in the strong interaction at O(α). In general, operators of dimension > 4 in the Lagrangian are expected to be suppressed by inverse powers of a large mass scale.

Interaction vertices

The interaction terms in the QCD Lagrangian for the eight gluons:

$$\overline{q} i g \gamma^{\mu} A_{\mu}^{a} T^{a} q$$

Since gluons carry color change, they have self-couplings, so we have 3-vertex and a 4-vertex from:

$$tr\left(\left(\partial_{\mu} A_{\nu} - \partial_{\nu} A_{\mu}\right) - ig\left[A_{\mu}, A_{\nu}\right]\right)^2$$

One gluon exchange

In the classical theory for the Lagrangian indicated thus far, what is the force between static quarks? This is generated by the tree graph and differs from the coulomb potential result from QED only by a group theoretic factor:

$$E(r) = \frac{g^2}{4\pi r} (T_1^a)_{ki} (T_2^{a*})_{lj}$$

We can write

$$T_1^a T_2^{a*} = \frac{1}{2}[(T_1^a + T_2^{a*})(T_1^a + T_2^{a*}) - T_1^a T_1^a - T_2^a T_2^{a*}]$$

Consider the interaction between a quark and an antiquark, 3x3=8+1, there is a singlet channel and an octet channel:

$$(T^a T^a)_{singlet} = \left(\frac{8}{3}\right)\left(\frac{1}{2}\right) = \frac{4}{3}$$

$$(T^a T^a)_{octet} = 3$$

Thus

$$E(r) = \frac{g^2}{4\pi r}\begin{cases} -\frac{4}{3} & singlet \\ +\frac{1}{6} & octet \end{cases}$$

and $q\overline{q}$ attract in the singlet channel and repel in the octet channel

Chapter 6 Standard Model

The Standard Model and the Extended Standard Model
According to the Standard model of Particle Physics, the Universe consists of point-like spin-1/2 fermions that reside in two families: the Leptons and the Quarks. These particles exist in a geometry with a gauge field, where they interact through that gauge field according to local gauge invariance and indirectly through the geometry according to general relativity. Even more indirectly, the gravitational interaction via mass has mass itself determined through exchange of spin-0 Higgs particles.

The local gauge invariance of the Standard model is:

$$U(1)_\gamma \otimes SU(2)_L \otimes SU(3)_C.$$

Let's consider each part separately:

$U(1)_\gamma$→ gives rise to the electromagnetic force between charged particles.

$SU(2)_L$→ gives rise to the force between left-handed particles.

$SU(3)_C$→ gives rise to the strong force that operates on particles (hadrons) that carry color charge (three types).

In Book 5 [5] we see that each gauge group listed above derives from the general local gauge invariant group form:

$$SU(N)_L \otimes SU(N)_R.$$

For $N = 1$, we have the trivial case (no left or right): $U(1)$.
For $N = 2$, we have the asymmetric case of only left: $SU(2)_L$.
For $N = 3$, we have the symmetric case of left-right, or simply a fixed ratio: $SU(3)$.

In Book 5 we see that there is mixing between the $U(1)$ and $SU(2)$ parts to give the actual electroweak theory observed:

$$U(1) \otimes SU(2) \rightarrow \; U(1)_\gamma \otimes SU(2)_L.$$

The mixing introduces universal (globally gauge invariant) 'mixing angles' that number among the fundamental parameters of the theory (along with particle masses and a few other constants).

In Book 7 [7] we see that Emanator Theory predicts the form of the local gauge invariance as a general form, to be exactly that observed: $U(1)_Y \otimes SU(2)_L \otimes SU(3)_C$. Emanator theory also predicts a theory governed by 22 parameters, not 19 as currently listed for the Standard Model.

As strange as the form $U(1)_Y \otimes SU(2)_L \otimes SU(3)_C$ might seem, it is so far only applied to the interaction element of te theory. Let's now consider an interaction with what? The answer resides in the representations of the indicated groups (the left-handed particles interacting according to $SU(2)_L$ will be seen as interacting visa the weak force, etc.). So, saying there is a local gauge invariance is to say that there will be particles according to the irreducible, independent, representations of these groups. Thus, for any chosen representation there will be a collection of particles predicted. This is precisely what is observed experimentally. Thus, we not only know the groups $SU(2)_L$ and $SU(3)_C$, say, we also know their specific representations to obtain particle numbers and groupings observed experimentally.

So, the Standard Model is actually indicating the Model pair of Local Gauge Field and Representation $\mathcal{R}$:

$$\{U(1)_Y \otimes SU(2)_L \otimes SU(3)_C;\ \mathcal{R}\},$$

and the Standard Model then says "times three", where the representation has three generations or copies:

$$\{U(1)_Y \otimes SU(2)_L \otimes SU(3)_C;\ 3 \times \mathcal{R}\}.$$

For the fundamental Leptons and Quarks this is shown in tabular form with the symbols for the particles as:

	electroweak left-handed Leptons	electro-only right-handed Leptons	Electro-only Quarks
1st Generation	$[e^-{}_L, \nu_{e,L}(m=0)]$	$[e^-{}_R, ---]$	d, u
2nd Generation	$[\mu^-{}_L, \nu_{\mu,L}(m=0)]$	$[\mu^-{}_R, ---]$	s, c
3rd Generation	$[\tau^-{}_L, \nu_{\tau,L}(m=0)]$	$[\tau^-{}_R, ---]$	b, t

Note that there are no right-handed neutrinos in the Standard Model and the left-handed neutrinos are massless. We already have ample evidence that the left-handed neutrinos are massive, so we know the Standard model will extend in this regard, which leaves the supposedly non-existent right-handed neutrinos. There probably are right-handed neutrino but if "electro-only" and having no charge, what results is a particle completely decoupled from the other matter, except gravitationally, precisely what describes dark matter. More on this will follow once we go from the 19 parameter model to the hypothesized 22 parameter model, as this suggests limits on extensions to the theory, as does cosmological data, which suggests a possible 4th neutrino, and no more, scenario consistent with the cosmological evolution, unless the other neutrinos are very massive (all probably the case).

With the Standard Model quantum field theory structure, with scatter results according to the indicated local gauge field and particle representation, we find additional conservation rules. One of these rules describes the conservation of Lepton number at each generation, in other words:

$$L_e = N(e^-) + N(\nu_e) - N(e^+) + N(\bar{\nu}_e)$$

is a constant, and the same for constants for the muon and tau generations. There is also conservation of baryon number and the familiar conservation of charge.

The Standard Model is a renormalizable quantum field theory, and while this poses no difficulty with the $SU(2)$ elements alone or $SU(3)$ alone, here we have a product gauge and, what's more, it has handedness with use of $SU(2)_L$ not $SU(2)$. Not surprisingly, anomalous terms arise in he renormalization effort and for the renormalization to work, additional constraints are imposed on the structure of the theory. Most notable is that at a generational level across both leptons and baryons there should be a charge sum of zero. This is observed and was used to predict missing quarks (charm) in the early discovery process.

The structure identified by the standard model and the high decimal number agreement on key results from quantum field theory together provide a theory that can't be displaced directly by anything better, and only minimally augmented (with addition of right-handed neutrino), so trying to find a unified theory appears impossible from 'within' the theory, and Godel's incompleteness theorem also suggests the hopelessness of such a task. This, then, leaves no choice but to look for an encompassing theory that projects the quantum theory with the Standard Model structure in its entirety. And this is what is attempted with emanator theory (Book 7 [7]).

The Standard Model can't explain the three generations of matter, the origin of the local gauge group with its odd product form, the

dimensionless constant alpha, and certainly can't explain the extension to massive(light) neutrinos, or the possible extension to massive (dark) right-handed neutrinos. In Emanator Theory (Book 7 [7]) we have:
(1) The local gauge $U(1) \otimes SU(2) \otimes SU(3)$ is projected by the theory. Why it should have three generations and why it should by asymmetric with use of $SU(2)_L$ is due to those choices providing the maximal packing of the Light matter sector (for maximal complexity information flow, an aspect of the MIE hypothesis), within the constraint of a 22-parameter emanation process.
(2) The 22 parameter, constants of the motion (emanation), result helps to constrain the particle representations such as to stay at 22-parameters, yet have maximally complex interaction.
(3) The MIE Hypothesis indicates the given generational structure given the constraint of working with the indicated product algebra. It also suggests any 'fine-tuning' on gravitational constant G might be dominated by the dark matter sector and its contribution to the evolution of the universe. Such fine-tuning would be most powerful if it indicated a fractal scale invariance property (see fractal G discussion in [59]).

Before considering the 22-parameter Emanator theory prediction of an extended Standard Model, let's first recount the 19-parameter Standard Model, which I'll separate into four groups:

(I) 9 Yukawa coupling constants (masses) for the charged fermions
(II) 5 constants for Weinberg Angle and the CKM matrix (with three mixing angles and CP-violating phase)
(III) 3 Constants for electromagnetic coupling (alpha), for strong interaction (g3), and strong CP-violating phase ($\theta_3 \approx 0$).
(IV) 2 Higgs parameters: Mass and Vacuum Expectation

If we allow for the left-handed neutrinos to have mass, then we get 3 more masses and another 4 constants for the PMNS matrix (three mixing angles and a CP-violating phase):

(V) Extended model: 7 more constants → We, thus, have 26 parameters.

Let's update out table with this extended version of the theory:

	electroweak left-handed Leptons	electro-only right-handed Leptons	Electro-only Quarks
1st Generation	$[e^-{}_L, \nu_{e,L}(m \neq 0)]$	$[e^-{}_R, ---]$	d, u
2nd Generation	$[\mu^-{}_L, \nu_{\mu,L}(m \neq 0)]$	$[\mu^-{}_R, ---]$	s, c
3rd Generation	$[\tau^-{}_L, \nu_{\tau,L}(m \neq 0)]$	$[\tau^-{}_R, ---]$	b, t

There is now a problem. In order to maintain renormalizability, if there are left-handed neutrinos with mass there must be right handed neutrinos with mass [60] So, just how sure are we about neutrinos having mass? There is strong evidence not only for neutrino mass, but for neutrino family dynamics (here seen as oscillation between neutrino mass states). We'll see more evidence of generational family kinematics/dynamics in a later section. Neutrino family dynamics was first seen in measurements indicating that neutrinos spontaneously changed flavor (the Solar neutrino experiments [61]), which not only indicates they have mass, but allows us to determine the mass differences.

Clearly the Extended Standard Model needs to be extended further, to allow for the massive right-handed neutrino that is hypothesized to have no weak interaction like its right-handed electron cousin (and no electric interaction since no electric charge and no strong interaction since no color charge, thus 'dark' matter). Such a right-handed neutrinos (with no charge) can act as their own antiparticle (which is consistent with formation of "Bright" supermassive Black Holes in the early universe as seen with Webb). Furthermore, in addition to the Dirac mass relation to the left-handed neutrino it also has a Majorana mass term not tied to the Higgs mechanism [62] What results is that instead of mass $e^-{}_L$ equal to mass $e^-{}_R$ here we have

$$m_{\nu_{e,L}} \propto \frac{1}{m_{\nu_{e,R}}},$$

which is known as the see-saw mechanism [63] Thus the very low-mass left-handed neutrinos indicate very large mass right-handed neutrinos. Large mass neutrinos are an excellent candidate for cold dark matter, precisely what is needed to complete the Cosmological Standard Model.

Given the renormalization constraints and the observation of neutrino mass we arrive at the following updated Model:

	electroweak left-handed Leptons	electro-only right-handed Leptons	Electro-Quarks
1st Generation	$[e^-_L, \nu_{e,L}(m \ll m_e)]$	$[e^-_R, \nu_{e,R}(m \gg m_e)]$	d, u
2nd Generation	$[\mu^-_L, \nu_{\mu,L}(m \ll m_e)]$	$[\mu^-_R, \nu_{\mu,R}(m \gg m_e)]$	s, c
3rd Generation	$[\tau^-_L, \nu_{\tau,L}(m \ll m_e)]$	$[\tau^-_R, \nu_{\tau,R}(m \gg m_e)]$	b, t

which is described by the 26 parameter theory indicated for massive left-handed neutrinos (if we assume that the right handed neutrino masses can be determined from the left-handed neutrino masses).

The standard counting to arrive at the 19 parameters (or, now, 26) includes $\theta_{QCD} \cong 0$ as a parameter of theory and its nearness to exactly zero is often referred to as the Strong CP problem. Going forward this is removed as a concern. The 'fine-tuning' that selects $\theta_{QCD} = 0$ is considered to be the same as that which selects the generational number, to arrive at 22 parameters while respecting the local gauge field constraint. Thus $\theta_{QCD} = 0$ is part of the MIE optimal selection.

Also included in the 19 parameter count is the U(1) gauge coupling g_1 and the SU(2) gauge coupling g_1, and these coupling are related to alpha by:

$$\alpha = \frac{1}{4\pi} \frac{{g_1}^2 {g_2}^2}{{g_1}^2 + {g_2}^2}$$

Now, the emanation's projection of the quantizable Lagrangian theory involves various constants (22) in the resulting quantization. In that process, however, alpha already exists, in fact, from the derivation of alpha, we see that it establishes maximal perturbation starting at the level of the emanation process itself. Thus, alpha doesn't number among the 22, and the relation between coupling constants involving alpha means only one of those coupling constants is independently specifiable, so only one should be counted. At this juncture we've reduced the count on 26 parameters to 24.

Notice in the prior counting, to arrive at the count of 26, we extended the count to be the same with dark neutrinos included since the dark neutrino mass was assumed to have some fixed (inverse) relation to the light neutrino mass. Let's not make this assumption, but retain the same mixing angle matrix, to now have 3 more masses in the theory (adding to the 24 count we now have 27).

In order to reduce the 27 parameter count to 22 we need to remove an 'overcount' of 5. Such an overcount would occur if there was a family-relationship generationally, for all masses. Consider such a mass relationship for the electron-muon-tau family, it is already known to exist and is known as the Koide relation. The Koide relation [64] was first observed for the three massive leptons currently known:

$$\frac{m_e + m_\mu + m_\tau}{\left(\sqrt{m_e} + \sqrt{m_\mu} + \sqrt{m_\tau}\right)^2} = \frac{2}{3}$$

To a lesser extent this relation is satisfied for the quarks as well, particularly for the three most massive, where the value is 0.6695. The problem with a simple application to the quark masses is that they are dependent on energy scale. A theoretical explanation for the Koide relation describes how this relation might exist for the masses of a given generation (or family group) [65]. As mentioned with neutrino oscillations, their oscillations involve the existence of such a family relationship in the quantum field theory implementation, so the hypothesis that it exists for each column in the table is not such a stretch. There are five columns of (independent) masses (the right handed electrons not independent of the left handed electrons removes them from the count). Thus, the five groups of three represent five groups with only two independent mass parameters, and the overall parameter count is thereby reduced by 5 from 27 to 22, as needed to be in agreement with Emanator theory. Thus Emanator theory, with the MIE hypothesis, would select for family group dynamics as observed with the neutrinos since it is what allows the maximal particle set within the 22-parameter constraint.

Chapter 7. Thermal Quantum Field Theory

In this chapter we consider quantum field theory where even in flat space-time there are alternate vacua. First we consider the thermal Green's function that results if we shift to complex time. We then consider the motion of a constantly accelerating observer in the flat space-time's vacuum. The observer's constant acceleration is along a congruence of a Killing Vector field of the (flat) spacetime. The observer sees a thermal spectrum of particles. If instead we consider the vacuum chosen by the accelerating (Rindler space-time) observer, versus that of the inertial (Minkowski space-time) vacuum, we find that again a thermal spectrum of particles is observed (from the accelerated observer's perspective).This will provide an initial exposure to how thermality and complex time are inextricably linked (this is explored further in Book 6 [6]).

7.1 Thermal Quantum Field Theory

7.1.1 The Propagator and the Partition Function

Indications of a connection between Quantum Mechanics and Statistical Mechanics (likewise, quantum field theory and statistical field theory) have occurred in prior examples, let's now explore this in detail. Time consisting of a real parameter that labels causal events is familiar and is part of the dynamical description in a special role from the earliest classical mechanical descriptions. From special relativity we learn time has different forms (parametrizations) and in the context of spacetime, is simply another coordinate variable (albeit with differently signed signature). Once demoted to being a coordinate variable, even part of the time (no pun intended), we can ask about complex time in that context. The metric signature will change from Lorentzian to Euclidean if we switch to pure imaginary time, for example, where all of the Euclidean-based sums will be well-defined. In essence, we will find that the connection between Quantum Mechanics and Statistical Mechanics is that they share the same analytic time, one referencing the real-part of time in a standard dynamical context (Quantum Mechanics) while the other references the imaginary part of time in a standard equilibrium thermodynamics description of the system. The analyticity of time overall will be maximally extended in the manner that gives rise to the Feynman propagator. (which will embed the proper causality into the theory).

As usual, when a convenient mathematical connection is used a lot (Planck's original introduction of his constant and quantization [67]; or use of Vector potential in electromagnetism, proven to be real by the Aharonov-Bohm Effect [68) we should consider what it means if analytic time really 'exists'. And it probably does, a supporting reason for this is that bound state descriptions, such as the solution for the hydrogen atom, so easy in standard Schrodinger or Heisenberg analysis, if attempted with the path integral formalism that works everywhere else (when other methods don't), is found to only work describing bound states if in curved spacetime with analytic time. So analytic time is needed to fully utilize the path integral formalism. What does it mean to have complexified time? From the Unruh analysis of the accelerated observer [10], one possibility is that this indicates a universal thermality according to acceleration. In the analysis, imaginary time is periodic and that periodicity defines the inverse temperature of the system. what results for the accelerated observer (in a standard quantum vacuum) is the appearance of a thermal flux from direction accelerating away from, with thermal spectrum at temperature:

$$T = \frac{\hbar a}{2\pi c k_B},$$

where $\hbar$ is Planck's constant, a is the acceleration, c is the speed of light, and k_B is Boltzmann's constant. For an acceleration of $1m/s^2$ we have a temperature $T = 4.06 \times 10^{-21}K$. The strongest acceleration we usually feel is due to Earth, 9.8 m/s^2, so we are bathed in a thermal bath at temperature $T = 3.98 \times 10^{-20}K$ from this effect (so drowned out by the CMBR). As explored in [6], in cosmological and Black Hole analysis complexified time provides a bridge to the thermodynamics of the system, which allows a thermal quantum gravity solution to be described (but not a quantum gravity solution, especially since such a solution might not exist). Once we've adjusted to the notion of analytic time, its natural to ask for the maximal analytic extension and if there is more than one choice (there is) the choice consistent with other aspects of the physics (such as causality) can be adopted at this juncture, and that is what is done with the aforementioned Feynman propagator. Analytic time provides a more complete and interconnected physical description, but it also constrains time and other aspects even more than before. So we now need to know why we have a theory with analytic time in addition to the odd local gauge group that is a product group of U(1), SU(2), and SU(3), and the set of 19 (or 22?) constants, etc. To have a deeper understanding of time, and the rest, requires an enveloping formalism to the structured formalism currently seen, and that is what is explored in Book 7 [7], with a synopsis provided in Appendix D.

7.1.2 The Propagator with Complexified Time

From the first appearance of the Schrodinger equation it was noted that changing time to imaginary time, $t \to -i\tau$, would give the diffusion equation. Let's now consider complexification of time when working with the propagator.

7.1.2.1 Direct Substitution

Recall that the propagator is based on the unitary evolution operator that is defined by the Hamiltonian for the system, or by the Action on paths:

$$K(t,q,q') = \langle q|e^{itH/\hbar}|q'\rangle = \int_{\chi(0)=q'}^{\chi(t)=q} \mathcal{D}\chi e^{iS[\chi]/\hbar},$$

with paths parametrized by $\chi(t)$, that start at q' and end at q. Let's now shift from $t \to i\tau$ in this context:

$$K_E(\tau,q,q') = \langle q|e^{-\tau H/\hbar}|q'\rangle = \int_{\chi(0)=q'}^{\chi(\tau)=q} \mathcal{D}\chi e^{-S_E[\chi]/\hbar},$$

where now the integrand is real and well-defined. The partition function is then simply given by:

$$\int dq K_E(\tau,q,q') = tr[e^{-\tau H/\hbar}]$$

where the temperature is $T = \hbar/\tau k_B = 1/\beta$. In standard notation ($k_B = 1$) using for partition function $Z(\beta)$ and free energy $F(\beta)$ we then have:

$$Z(\beta) = tr[e^{-\beta H}] = e^{-\beta F} = \oint_{\chi(0)=\chi(\tau)} \mathcal{D}\chi e^{-S_E[\chi]/\hbar}$$

Note that to have the correspondence with the definition of partition function we identified the ends of the paths, or equivalently, we've shifted to integration on periodic paths with period $\tau = \hbar\beta$. Let's now apply this to the harmonic oscillator fundamental case to see if it makes sense. For the Euclideanized harmonic oscillator, we have:

$$K_E(\hbar\beta,q,q) = \sqrt{\frac{m\omega}{2\pi\hbar\sinh(\hbar\omega\beta)}}\exp\left[-\frac{2m\omega q^2}{\hbar}\frac{\sinh^2(\hbar\omega\beta/2)}{\sinh(\hbar\omega\beta)}\right]$$

$$Z(\beta) = \int dq K_E(\tau, q, q') = \frac{1}{2\sinh\left(\frac{\hbar\omega\beta}{2}\right)} = \sum_n \exp\{-\beta E_n\},$$

where

$$E_n = \hbar\omega\left(n + \frac{1}{2}\right).$$

For low temperature, $\beta \to \infty$, and $F(\beta) \to E_0 = \frac{1}{2}\hbar\omega$, as expected. This simple substitution works out, so let's now consider complex time in more detail. In particular, is time analytic?

7.1.2.2 Full analyticity – Wick rotation

We've seen that the direct substitution $t \to -i\tau$ provides interesting connections. If we arrive at this change more formally in terms of analytic time, we will have better understanding. First, as regards a Lorentzian spacetime with integrations (the Action) described as integrals in real time, it was observed that complexified time being analytic allowed for a change in the contour of integration, effectively a rotation from the real axis by 90 degrees about the origin to turn the integration into an integral along the imaginary axis, i.e., we've achieved $t \to -i\tau$ by way of a "Wick rotation". There are different ways to do this in the context of the quantum propagator, however, but only one encodes the causal structure consistently, the Feynman propagator.

Before continuing with analysis of the Feynman propagator, consider the significance of analyticity and the Wick rotation is not only that a 1-dimensional time and (N-1)-dimensional spatial dynamics problem can be turned into a N-dimensional statics problem, allowing for Euclideanized path integrals that are convergent and well-defined to obtain system integral solutions (that can be analytically carried back to the Lorentzian system representation, thereby making the path integrals well-defined via analyticity). We can also take this in the other direction, an intractable N-dimensional statics problem might be more easily solvable as a (N-1)-dimensional dynamics problem.

7.1.3 Green's Function – Feynman Propagator – Choice of analytic extension

Let's follow the discussion of analyticity given by [68] and consider the Feynman propagator given in their notation by:

$$G_F(x^\alpha, t; x'^\alpha, t') = \begin{cases} i\sum_n \frac{\psi_n(x^\alpha)\psi_n(x'^\alpha)}{2\omega_n} \exp[-i\omega_n(t-t')], & t > t' \\ i\sum_n \frac{\psi_n(x^\alpha)\psi_n(x'^\alpha)}{2\omega_n} \exp[i\omega_n(t-t')], & t < t' \end{cases}$$

To effect the substitution by $t \to -i\tau$ by analytically extending off the real line, there must be a rotation as indicated in Figure 1, which requires analyticity in quadrants Two and Four as shown in shade.

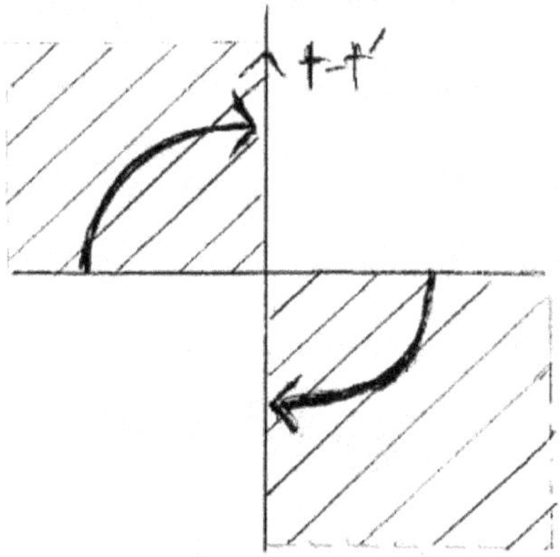

Here's the new Euclideanized Green's function G_E (or Euclideanized Feynman propagator):

$$G_E(x^\alpha, t; x'^\alpha, t') = \begin{cases} i\sum_n \frac{\psi_n(x^\alpha)\psi_n(x'^\alpha)}{2\omega_n} \exp[-\omega_n(\tau-\tau')], & \tau > \tau' \\ i\sum_n \frac{\psi_n(x^\alpha)\psi_n(x'^\alpha)}{2\omega_n} \exp[\omega_n(\tau-\tau')], & \tau < \tau' \end{cases}$$

which is well-defined and unique given the fall-off condition to zero for large$|\tau - \tau'|$.

The choice of shaded regions allows the "Wick rotation" shown, and this convention for analytic extension is usually captured in the mathematics (e.g., not diagrammatically as shown) by introduction of a small imaginary part to ω_n, or in momentum representation (taking a Fourier Transform), to $p^2 - m^2$. Let's shift to the standard Feynman propagator form with this in mind. Staring with the standard Green's function to the Klein Gordon equation we have the standard solution

$$G(x, y) = \frac{1}{(2\pi)^4} \int d^4p \frac{e^{-ip(x-y)}}{p^2 - m^2 \pm i\varepsilon}$$

where the $\pm i\varepsilon$ denotes various choice of deformation to the integration contour to have a well-defined propagator, and these solutions are different. We want the Feynman propagator convention allowing the "Wick rotation" indicated, thus, we want:

This is equivalent to the definition in terms of the limit as the $i\varepsilon$ contour deformation goes to zero:

$$G_F(x,y) = \lim_{\varepsilon\to 0} \frac{1}{(2\pi)^4} \int d^4p \frac{e^{-ip(x-y)}}{p^2 - m^2 + i\varepsilon}.$$

7.1.4 Thermal Green's Function

Let's now consider a system at a temperature $T = 1/\beta$ and get the associated Thermal Green's Function. If we have a temperature and using the grand canonical ensemble formalism (see [6]) we can write the expectation value of any operator A as:

$$\langle A \rangle_\beta = \frac{Tr[\exp(-\beta H)\, A]}{Tr[\exp(-\beta H)]}$$

The thermal Green's function would then be:

$$G_T(x,y) = i\langle T\varphi(x)\varphi(y)\rangle_\beta.$$

The solution has the same terms as before, but now has a Bose-Einstein statistics contribution:

$$G_T(x^\alpha, t; x'^\alpha, t') = \begin{cases} i\sum_n \frac{\psi_n(x^\alpha)\psi_n(x'^\alpha)}{2\omega_n} \begin{Bmatrix} (1+n_B)\exp[-i\omega_n(t-t')] \\ +n_B \exp[i\omega_n(t-t')] \end{Bmatrix}, & t > t' \\ i\sum_n \frac{\psi_n(x^\alpha)\psi_n(x'^\alpha)}{2\omega_n} \begin{Bmatrix} (1+n_B)\exp[i\omega_n(t-t')] \\ +n_B \exp[-i\omega_n(t-t')] \end{Bmatrix}, & t < t' \end{cases}$$

where

$$n_B = \frac{1}{(\exp(\omega_n\beta) - 1)}$$

As before, we can analytically continue to obtain the more manageable Euclideanized form, and this will correspond to the analytic continuation under conditions requiring periodicity in imaginary time with period β:

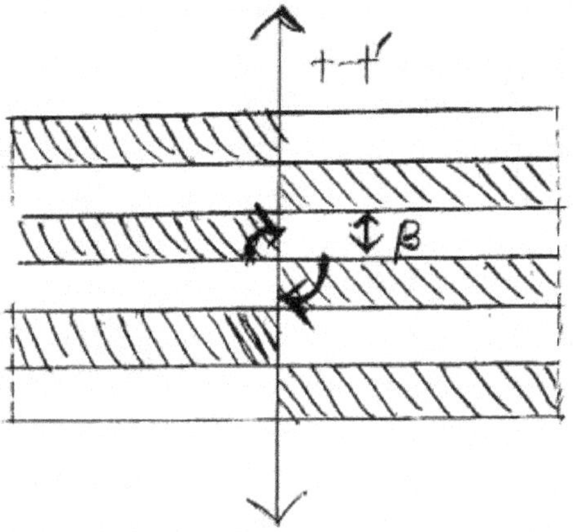

This will be useful in describing the Black holes thermodynamics analysis that is used in [6].

7.2 Complexification of time

If we have a quantum system that is in equilibrium with its environment, and where nontrivial exchange is possible in energy and particles with that environment, the 'Euclideanization' of the spacetime described above brings us to a Grand-Canonical Ensemble formulation for the thermal state given by:

$$\Phi_{\beta,\mu} = \frac{e^{-\beta(H-\mu N)}}{Z}, \quad Z = Tr\left[e^{-\beta(H-\mu N)}\right]$$

where H is the Hamiltonian operator of the system, N is the number operator, μ is the chemical potential, and Z is the partition function (from which the entire system thermodynamics can be derived).

Non-interacting particles: Bose-Einstein and Fermi-Dirac statistics

Suppose we have non-interacting particles with energy and number density given by (harmonic oscillator form):

$$H = \sum_n \varepsilon_n a_n{}^{\dagger} a_n, \quad N = \sum_n a_n{}^{\dagger} a_n$$

where $\{a_n{}^{\dagger}, a_n\}$ are the standard creation and annihilation operators, which for Bosons will obey the standard commutation relations:

$$[a_j, a_k{}^{\dagger}] = \delta_{jk}, \qquad [a_j, a_k] = 0, \qquad [a_j{}^{\dagger}, a_k{}^{\dagger}] = 0$$

The mean number of Bosons molecules in mode n is then n_B:

$$n_B = Tr\left[a_n{}^{\dagger} a_n \Phi_{\beta,\mu}\right] = \frac{1}{\exp(\beta[\varepsilon_n - \mu]) - 1},$$

where $\varepsilon_n > \mu$ is required to have a positive mode number.

For Fermions the commutator relations above get replaced with anticommutators, and we now have:

$$n_F = Tr\left[a_n{}^\dagger a_n \Phi_{\beta,\mu}\right] = \frac{1}{\exp(\beta[\varepsilon_n - \mu]) + 1},$$

there is no longer a problem $\varepsilon_n < \mu$ in this situation to stay consistent with positive occupation numbers (we have antimatter).

7.3 Laws of Thermodynamics are recovered

7.3.1 Definition of Heat Bath

Let's consider a total system Hamiltonian H consisting of the system Hamiltonian H_S and of a heat bath Hamiltonian H_B, and an interaction Hamiltonian between System and Heat Bath, H_{SB}. Thus:

$$H = H_S + H_B + H_{SB}.$$

The Hamiltonian gives the unitary time evolution operator as

$$U(t) = T\{\exp -i \int_0^t dt' H(t')\}$$
$$\equiv \lim_{\delta t \to 0} e^{-i\delta t H(t)} e^{-i\delta t H(t-\delta t)} \ldots e^{-i\delta t H(\delta t)} e^{-i\delta t H(0)}$$

(the starting point form many path integral formulations). Instead of a wavefunction, with mixing, and with Euclideanization, we have a system density matrix that is acted on by the unitary evolution operator:

$$\rho(t) = U(t)\rho(0)U^\dagger(t), \quad \frac{\partial \rho(t)}{\partial t} = -i[H(t), \rho(t)].$$

Let's consider the mean energy change in the heat reservoir:

$$\frac{\partial \langle H_B(t) \rangle}{\partial t} = \frac{\partial}{\partial t}\langle H_B(t) - \mu N \rangle + \mu \frac{\partial}{\partial t}\langle N \rangle = F + P,$$

where $F = \frac{\partial}{\partial t}\langle H_B(t) - \mu N \rangle$ is the heat flow that enters the reservoir and $P = \mu \frac{\partial}{\partial t}\langle N \rangle$ is the power that enters the reservoir. The separation into F and P parts is indicated by the definition of entropy in the next section.

7.3.2 Entropy

Recall the Boltzmann entropy formula in terms of system probabilities (Shannon Entropy) is:

$$S = -k_B p \ln p,$$

here we take $k_B = 1$ and switch to the density matrix formulation, to get the standard von Neumann entropy form relevant to our situation:

$$S = -Tr\{\rho \ln \rho\}$$

and we then have:

$$\partial_t S = -\partial_t Tr\{\rho \ln \rho\} = 0$$

under unitary evolution. Now suppose we split the density matrix into a non-thermal system part (non-system parts traced out) and a thermal system part $\Phi_{\beta,\mu}$ as indicated previously (thereby ignoring reservoir information). Denote the effective description by $\rho_S \otimes \Phi_{\beta,\mu}$. The entropy difference (divergence) that describes the difference between the full density matrix and the non-thermal state density matrix, the information loss by such an approximation, is then $(\rho || \rho_S) \otimes \Phi_{\beta,\mu}$ and the entropy of this can be written

$$S[(\rho || \rho_S) \otimes \Phi_{\beta,\mu}] = S[\rho_S] - S[\rho] + \beta \langle H_B - \mu N \rangle + \ln Z$$

Thus (showing k_B):

$$\partial_t S[(\rho || \rho_S) \otimes \Phi_{\beta,\mu}] = \partial_t S[\rho_S] + \frac{F}{k_B T}.$$

This allows us to write:

$$\partial_t S[\Phi_{\beta,\mu}] = \beta \partial_t \langle H_B - \mu N \rangle = \frac{F}{k_B T}$$

7.3.3 First Law of Thermodynamics

Let's start with the implications of number conservation. Global number conservation gives:

$$[N_S + N_B, H] = 0,$$

while local number conservation gives:

$$[N_B, H_B] = 0, \quad [N_S, H_S] = 0.$$

Taken together we then know that the thermal coupling terms satisfies:

$$[N_S + N_B, H_{SB}] = 0,$$

which means that particles are either in the system or in the bath (the coupling does not change their total number). If there were a coupling number, this would be the same as the change in that number being zero. There is a coupling energy term, however, so let's make change in the coupling energy over time equal to zero. This means that:

$$\partial_t \langle H_{SB} \rangle = i \langle [H_S + H_B, H_{SB}] \rangle = 0.$$

In terms of the F, P variables introduced earlier:

$$F = i \langle [H_S - \mu N_S, H_{SB}] \rangle, \qquad P = i\mu \langle [N_S, H_{SB}] \rangle,$$

and

$$\partial_t \langle H_S \rangle = \langle \partial_t H_S \rangle - (F + P),$$

where $\langle \partial_t H_S \rangle$ is power entering the system, and F, P describe energy flows leaving the system. Thus, the first law of thermodynamics is shown for the system.

7.3.4 Second Law of Thermodynamics

Let's return to the product notation for the effective density of states given our (typical) missing information of a system: $\rho_S \otimes \Phi_{\beta,\mu}$. Let's now consider the real system to have precisely this tensor product form as initial state:

$$\rho(0) = \rho_S(0) \otimes \Phi_{\beta,\mu}.$$

We can now write change in entropy as:

$$\Delta S(t) = S\big[(\rho(t) || \rho_S(t)) \otimes \Phi_{\beta,\mu}\big] = S[\rho_S(t)] - S[\rho(t)] + \frac{Q}{k_B T}$$

where heat is

$$Q = Tr\{(H_B - \mu N)\rho(t)\} - Tr\{(H_B - \mu N)\rho(0)\}.$$

Note that the expression above for change in entropy is in terms of a relative entropy, which is always positive of zero. Thus,

$$\Delta S(t) \geq 0,$$

the Second Law of Thermodynamics.

7.4 Wightman axioms and the Osterwalder-Schrader reconstruction theorem

For quantum theory there are a variety of prescriptions for making the path integral formulation well-defined. The one most compact mathematically is to work with Euclideanization with use of analytic time. In this process, however, we are speaking of a quantum description, where an integral description is given, for which we Wick rotate to arrive at statistical description where the integral is well-defined and can be solved, and then Wick rotating back with our solution. How general is this procedure? How general is the Wick rotation? What does this suggest?

The Wightman axioms [69] answer how general such a construction can be is very general as long as the key axiom is satisfied: "If the supports of two fields are spacelike separated, then the fields either commute or anticommute". Note that this burdens the commutator relations with imposing the causality relation unlike in local quantum field theory. The Wightman axioms require the standard spin statistics relations. A true spinor representation is incompatible with the Euclideanization maneuvers used in the Wightman axioms, so not directly generalizable in that sense (but suggested of a variety of ways to make some consistent for fermions. The Wightman framework also can't, technically, work with gauge theories. It makes sense there would be a complications for Non-Abelian, but there's even a complication for Abelian gauge theories because gauge theories allow an indefinite norm (incompatible with the positive definite norm requirement for a Hilbert space). This is trivially overcome with appropriate choice of gauge (Coulomb), thus falling under the purview of the Wightman axioms in that form.

The Field Theories that can be 'Wick' rotated appear to be a large enough family to cover many cases of interest (if not all approximately) but this requires that we look at the analytic continuation Wick rotation maneuver more closely. In doing so we arrive at the Osterwalder-Schrader reconstruction theorem [70], where we find a set of rules for when a (well-defined) Euclidean quantum field theory can be Wick-rotated not a Wightman quantum field theory. Equivalence of the technical conditions for the Osterwalder-Schrader theories and Wightman quantum field theory axioms was shown in 1994.

7.5 Accelerated Observer

7.5.1 Particle detector in 4D Minkowski

Let's consider a particle detector in 4D Minkowski. Let the detector consist of an idealized point particle with internal energy levels labeled by the energy E, coupled via a monopole interaction with scalar field ϕ. Suppose the detector moves along a world line described by $x^{\mu}(\tau)$ where τ is the detectors proper time. The detector-field interaction is then described by the interaction term in the Lagrangian given by:

$$\mathcal{L} = cm(\tau)\phi[x(\tau)]$$

Where c is a small coupling constant, $m(\tau)$ is a monopole moment operator, and ϕ is the scalar field. For a general trajectory the detector will not remain in its ground state E_0 but will undergo a transition to $E > E_0$, while the field will make a transition to an excited state $|\psi\rangle$. To first order in perturbation theory the transition amplitude is:

$$ic\langle E,\psi| \int_{-\infty}^{\infty} m(\tau)\phi[x(\tau)]d\tau \,|0_M, E_0\rangle$$

$$= ic\langle E|m(0)|E_0\rangle \int_{-\infty}^{\infty} e^{i(E-E_0)\tau} \,\langle\psi|\phi(x)|0_m\rangle d\tau$$

and first order transitions only occur to the state $|\psi\rangle = |1_k\rangle$, so we need:

$$\langle 1_k|\phi(x)|0_m\rangle = \int d^3\,k' \,(16\pi^3\omega')^{-1/2}\langle 1_k|a_{k'}^{t}|0_M\rangle e^{-i\vec{k}'\cdot\vec{x}+i\omega t}$$

$$= (16\pi^3\omega')^{-1/2}e^{-i\vec{k}\cdot\vec{x}+i\omega t}$$

So,

$$\langle 1_k|\phi(x)|0_m\rangle = (16\pi^3\omega)^{-1/2}\exp\{-i\vec{k}\cdot\vec{x} + i\omega t\},$$

but $\vec{x}$ above is not an independent variable, instead being determined by the detectors trajectory. Suppose that trajectory is an inertial world line:

$$\vec{x} = \vec{x}_0 + \vec{v}t = \vec{x}_0 + \vec{v}\tau(1 - v^2)^{-1/2}$$

Thus, we get

$$(16\pi^3\omega)^{-1/2}\, e^{-\vec{k}\cdot\vec{x}_0} \int_{-\infty}^{\infty} e^{i(E-E_0)\tau}\, e^{i\tau(\omega-\vec{k}\cdot\vec{v})(1-v^2)^{-1/2}}\, d\tau$$

$$= C\,\delta\left(E - E_0 + (\omega - \vec{k}\cdot\vec{v})(1-v^2)^{-1/2}\right)$$

where C is a constant. Since the argument of δ is always > 0, the transition amplitude vanishes. Also, the transition is forbidden on energy conservation grounds - a direct consequence of Poincare invariance.

If we consider a more complicated trajectory, it would not yield a δ function, thus transitions are possible.

Transition to all possible E and ψ:

$$c^2 \sum_E |\langle E|m(0)|E_0\rangle|^2 \int_{-\infty}^{\infty} d\tau \int_{-\infty}^{\infty} d\tau'\, e^{-i(E-E_0)(\tau-\tau')}\, G^+(x(\tau), x(\tau'))$$

$$= c^2 \sum_E |\langle E|m(0)|E_0\rangle|^2\, \mathcal{F}(E - E_0)$$

where $\mathcal{F}(E - E_0)$ is the detector response function, and is independent of the details of the detector. G^+denotes the Wightman Green function:

$$G^+(x, \tau') = \langle 0|\phi(x)\phi(x')|0\rangle$$

$\mathcal{F}$ represents the bath of "particles" that the detector effectively experiences as a result of its motion. The remaining factor represents the selectivity of the detector to detecting this bath. For detector trajectories in Minkowski space that satisfy

$$G^+\big(x,(\tau), x,(\tau')\big) = g(\Delta\tau)$$

for some function g and $\Delta\tau = \tau - \tau'$, the system is invariant under time translation in the reference frame of the detector. This means that the detector is in equilibrium with the ϕ field, so that the number of quanta absorbed by the detector per unit τ is constant. The aforementioned inertial trajectory satisfies above form and yields no particles production (detection rate is constant at zero). For the uniformly accelerated detector, which also satisfies the above form, we obtain a(constant) thermal background. Detector "equilibrium"/detection between the accelerated detector and the ϕ field in the Minkowski vacuum state $|0_M\rangle$ is the same as that which would have been achieved had the detector remained unaccelerated, but immersed in a bath of thermal radiation at the temperature $T = \frac{acceleration}{2\pi k_B}$. Note that the same conclusion can also be drawn from an examination of the thermal Green's function for an inertial detector: The vacuum Green's function for a uniformly acceleration detector is the same as thermal Green's function for an inertial detector.

Observer Vacuum induces Lenz's Law backreaction against non-Inertial motion

As the detector accelerates, its coupling to the ϕ field causes the emission of quanta, which produces a resistance against the accelerating force. The work done by the external force to overcome this resistance supplies the missing energy that feeds into the field via the quanta emitted from the detector, and also into the detector which simultaneously makes upward transition. But as far as the detector is concerned, the net effect is the absorption of thermally distributed quanta.

Consider an inertial detector moving in a quantum field in a many-particle state. Then G^+ is replaced by:

$$\left\langle {}^1n_{\vec{k}_1}, {}^2n_{\vec{k}_2}, \cdots \middle| \phi(x)\phi(x') \middle| {}^1n_{\vec{k}_1}, \cdots \right\rangle$$
$$= G^+(x,x') + \sum_i {}^i n u_{\vec{k}_i}(x) u^*_{\vec{k}_i}(x') + \sum_i {}^i n u^*_{\vec{k}_i}(x) u_{\vec{k}_i}(x')$$

For $v = 0$, $\mathcal{F}(E - E_0) \propto \theta(E - E_0)$ shows that the absorption of a single quantum of mass m by the detector will not occur unless the energy level spacing $E - E_0$ in the detector is at least equal to the particle rest energy m.

For $\vec{v} \neq 0$:

$$\mathcal{F}(E - E_0) = \mathcal{F}(E') \propto \int_{E'^-}^{E'^+} n_k \, dk \qquad E'^{\pm} = E' \sqrt{\frac{1 \pm v}{1 \mp v}}$$

As the detector moves through the isotropic bath of radiation, a particular transition with energy $E - E_0$ will not select quanta from only one mode, but a whole range, varying from blue-shfled lower energy modes in the forward direction to redshfled higher energy modes in the backward direction. In general there is no simple relation between $\langle N_i \rangle$ and the particle number as measured by a detector, even if it is freely falling.

Use of a detector to describe particles usually involves the Wightman function G^+, so let's examine it further. Consider a spacetime asymptotically static to past and future. Let's analyze G^+ constructed using the in vacuum evaluated in the out region. This generally is a complicated function of $\vec{x}, \vec{x}'$. Consider the simple case of a homogeneous universe, such as the asymptotically static, spatially flat, FRW model, G^+ is then invariant under spatial translations and rotations. Starting with

$$G_{in}^+(x, x') = \langle in, 0|\phi(x)\phi(x')|0, in\rangle = \int u_k^{in}(x) u_k^{in*}(x') d^{n-1}k$$

Due to spatial homogeneity u_k^{out} and u_k^{in} remain as plane waves, thus

$$\left(u_k^{in}, u_k^{out}\right) \propto \delta_{kk'}, \left(u_k^{in}, u_k^{out*}\right) \propto \delta_{-kk'}$$

So, in terms of a Bogoliubov transformation:

$$u_{\vec{k}}^{in}(x) = \alpha_k u_{\vec{k}}^{out}(x) + \beta_k u_{-\vec{k}}^{out*}(x)$$

Thus,

$$\mathcal{F} \propto |\beta_k|^2 \theta(E - m)$$

implies the interpretation of $|\beta_k|^2$ as the number of quanta created in mode k when compared with $\vec{v} = 0$ of prior analysis.

In the 2D (conformably flat) spacetime we have $\bar{\phi}(x) = \Omega^{(2-n)/2}(x)\phi(x) = \phi(x)$ for $n - 2$. Thus, particle production takes place only when the conformal symmetry is broken by the presence of a mass, which provides a length scale for the theory. The production process can be regarded as caused by the coupling of the spacetime expansion to the quantum field via the mass.

The total uncertainty in the particle number over a time interval Δt is:

$$\Delta N \gtrsim (m\Delta t)^{-1} + |A|\Delta t,$$

where the first term on the right-hand side is the particle number uncertainty due to Heisenberg's energy-time uncertainty relation (a dynamical relation), and the second term is for particle production. A is the average particle production rate in $\tau = \Delta t$.

When attributing the production of field quanta to the changing of gravitation field, it seems natural to describe the particles themselves as being produced *during* the period of expansion. But, when spacetime is curved, no natural definition of particles is available. In spite of this, because of the special symmetry of FRW spacetime, one can identify a privileged class of observers, the comoving observers, who see the universe as expanding precisely isotropically. One might then identify particles in the expansion region with the excitation of comoving particle detectors.

7.5.2 Adiabatic Vacuum = Minkowski Vacuum using WKB solutions

Consider the mode equation:

$$\frac{d^2}{d\eta^2}F_k(\eta) + \omega_k^2(\eta)F_k(\eta) = 0, \;\; \omega_k^2(\eta) = k^2 + C(\eta)m^2.$$

The formal WKB solutions are then:

$$F_k = (2\omega_k)^{-1/2} \exp\left[-i\int^{\eta} \omega_k(\eta')d\eta'\right]$$

$$W_k^2(\eta) = \omega_k^2(\eta) - \frac{1}{2}\left(\frac{\ddot{\omega}_k}{\omega_k} - \frac{3}{2}\frac{\dot{\omega}_k^2}{\omega_k^2}\right)$$

Zeroth order approximation:

$$W_k^{(0)}(\eta) \equiv \omega_k(\eta)$$

Working with our K(100) example:

$$\omega_k^2 = k_x^2 + \left(k_y^2 + k_z^2 + m^2\right)e^{2gt}$$

Introduce slowness parameter: $\eta \to \eta/T$:

$$\frac{d^2x}{d\eta^2} + T^2\omega^2 x = 0 \to T^2\omega_k^2(t) = T^2k_x^2 + T^2\left(k_y^2 + k_z^2 + m^2\right)e^{2gtT},$$

For Large T:

$$\omega_k \to \omega_k \cong \left(k_y^2 + k_z^2 + m^2\right)e^{2gt}$$

So, as $t \to \infty$:

$$F_k \sim \exp\left(-i\left(k_y^2 + k_z^2 + m^2\right) e^{2gt}/g\right)$$

which matches the asymptotic behaviour of $H_{ik}^{(2)}(\mathrm{Z})$, thus F_k is positive frequency with respect to the adiabatic vacuum. The adiabatic definition of vacuum agrees with the Minkowski positive frequency definition. An inertial particle detector registers no particles, to any finite adiabatic order, in this vacuum, it also registers strictly no particles with respect to the Minkowski vacuum.

From the adiabatic analysis we know that as $t \to \infty$ the proper Bessel function that matches the asymptotic behaviour is $H_{ik}^{(2)}(\tau)$ and yet, remaining arbitrariness in the adiabatic vacuum allows for a term which vanishes to infinite order in the adiabatic parameter (such as if involving factor of e^{-k}). Is it possible to construct a different Bessel function that differs by factors of e^{-k}? Here is an example:

$$2J_{-ik}(\tau) = e^{-k\pi}H_{ik}^{(1)}(\tau) + e^{k\pi}H_{iK}^{(2)}(\tau).$$

A vacuum based on J_{-ik} at early times is a valid adiabatic vacuum, furthermore, if we claim it evolves into $H_{ik}^{(2)}$ at late times we have particle production that is finite:

$$\sum_k |\beta|^2 = \sum_k \left(\frac{1}{e^{(2\pi/g)|k_x|} - 1}\right) \neq \infty,$$

Which suggests that a unitary transformation might exist between the early and late time vacua (a Bogoliubov transformation).

7.5.3 Kasner space-time with Kasner Vacuum

The Kasner solutions of the vacuum Einstein equations correspond to universes which expand along two spatial directions and contract along a third. They have metric

$$ds^2 = -d\tau^2 + \tau^{2p_1}dx^2 + \tau^{2p_2}d\psi^2 + \tau^{2p_3}d\zeta^2$$

With $\tau, x, \psi,$and $\zeta\epsilon(-\infty, \infty)$ are characterized by the three numbers p_1, p_2andp_3. In order to satisfy the vacuum Einstein equations the exponent satisfy

$$p_1 + p_2 + p_3 \; and \; p_1^2 + p_2^2 + p_3^2 = 1$$

The degenerate kasner spacetime is the solution with$p_1 = 1$ and $p_2 = p_3 = 0,$

Henceforth labeled K_{100}. It is more easily treated in a different coordinate system, namely

$$t = \tfrac{1}{g}In(g\tau), x = \tfrac{1}{g}x, \; y = \psi \; and \; z = \zeta$$

The metric is then

$$ds^2 = e^{2gt}(-dt^2 + dx^2) + dy^2 + dz^2$$

It is easily seen that K_{100} is a region of Minkowski spacetime (henceforth M^4).

The K_{100} coordinates are related to the Minkowski coordinates by

$X = g^{-1}e^{gt}\sinh(gx),\; T = g^{-1}e^{gt}\cosh(gx),\; Y = y, and \; Z = z$

yielding the familiar metric

$$ds^2 = -dT^2 + dX^2 + dY^2 + dZ^2$$

However, the transformation only maps into the region of M^4 with$T > |X|$. If K_{100} is shown as a subsection of M^4 using the Penrose diagram representation , the difference between their Cauchy surfaces is seen to only depend on contributions at sections of future null infinity.The form of the coordinate relation also shows that surfaces of constant t in K_{100} provide a foliation by hyperbolic sheets in the upper wedge of M^4.

For Minkowski space with Cartesian coordinates:

$$dS^2 = dT^2 - dX^2 - dY^2 - dZ^2$$

Using Kasner (100) coordinates (where $Y = y, Z = z$):

$$X = \pm g^{-1}e^{\pm gt}\sinh(gx), \quad T = \pm g^{-1}e^{\pm gt}\cosh(gx)$$
$$dX = \pm e^{\pm gt}\cosh(gx)\,dx + e^{\pm gt}\sinh(gx)\,dt$$
$$dT = \pm e^{\pm gt}\sinh(gx)\,dx + e^{\pm gt}\cosh(gx)\,dt$$

Thus

$$(dT)^2 = e^{\pm gt}\sinh^2(gx)\,dx^2 + e^{\pm 2gt}\cosh^2(gx)\,dt^2 \pm 2e^{\pm 2gt}\sinh(gx)\cosh(gx)\,dxdt$$
$$(dX)^2 = e^{\pm gt}\cosh^2(gx)\,dx^2 + e^{\pm 2gt}\sinh^2(gx)\,dt^2 \pm 2e^{\pm 2gt}\sinh(gx)\cosh(gx)\,dxdt$$

which leads to:

$$S^2 = e^{\pm 2gt}(dt^2 - dx^2) - dy^2 - dz^2$$

where the positive case gives future (x-t) light wedge (a 2D light-cone product with y-z plane) and negative gives past light wedge.

Spacelike surfaces of constant t provide a foliation of the upper quadrant: $\mathcal{U}$. The data on some t=constant surface will allow prediction of future dynamics in $\mathcal{U}$. Using the K(100) metric

$$dS^2 = e^{2gt}(dt^2 - dx^2) - dy^2 - dz^2$$

let's now examine the scalar field equation (Klein-Gordon equation) in this circumstance:

$$(\Box + m^2)\varphi = 0$$

$$g_{\mu\nu} = \begin{pmatrix} e^{2gt} & & & \\ & -e^{2gt} & & \\ & & -1 & \\ & & & -1 \end{pmatrix}$$

where

$$\Box = \frac{1}{\sqrt{-g}}\partial_\mu\left(\sqrt{-g}\,g^{\mu\nu}\partial_\nu\right), \quad g = -e^{4gt}, \quad \sqrt{-g} = e^{2gt}$$

Thus

$$\Box = \frac{1}{e^{2gt}}\frac{\partial}{\partial t}\left(e^{2gt}(e^{-2gt})\frac{\partial}{\partial t}\right) + \partial_i\left(g^{ik}\partial_k\right)$$
$$= \frac{1}{e^{2gt}}\frac{\partial^2}{\partial t^2} + (-e^{-2gt})\frac{\partial^2}{\partial x^2} - \frac{\partial^2}{\partial y^2} - \frac{\partial^2}{\partial z^2}$$

Let $\varphi = f_{\vec{k}}(t)\exp(i\vec{k}\cdot\vec{x}) = f_{\vec{k}}(t)\exp(ik_x x + ik_y y + ik_z z)$, then

$$\frac{d^2 f_{\vec{k}}}{dt^2} + \left[k_x^2 + (k_y^2 + k_z^2 + m^2)e^{2gt}\right]f_{\vec{k}} = 0$$

Let $\tau = g^{-1}\left(k_y^2 + k_z^2 + m^2\right)^{1/2}e^{gt}$ and $k = |k_x|/g$, then

$\frac{d\tau}{dt} = \left(k_y^2 + k_z^2 + m^2\right)^{1/2}$, $\frac{d^2 f}{dt^2} = \frac{d}{d\tau}\left(\frac{df}{d\tau}\frac{d\tau}{dt}\right)\frac{d\tau}{dt}$, $\frac{d^2 f_{\vec{k}}}{dt^2} = \frac{d^2 f}{d\tau^2}\left(\frac{d\tau}{dt}\right)^2 + \frac{df}{d\tau}\left(\frac{d^2\tau}{dt^2}\right)$

So,

$$\frac{d^2 f}{d\tau^2}(g\tau)^2 + \frac{df}{d\tau}(g^2\tau) + [(gk^2) + (g\tau)^2]f = 0$$

$$\tau^2\frac{d^2 f}{d\tau^2} + \tau\frac{df}{d\tau} + (\tau^2 + k^2)f = 0$$

And in the last form we have the standard Bessel equation, where asymptotic behaviour indicates the first and second Hankel functions as solutions:

$$f_k = C_1 H_{ik}^{(1)}(\tau) + C_2 H_{ik}^{(2)}(\tau),\ \ k = \frac{|k_x|}{g},\ \ \tau = g^{-1}\left(t_z^2 + k_y^2 + m^2\right)^{1/2} e^{gt}.$$

From the asymptotics:

$t \to -\infty, \frac{d^2 f_k}{dt^2} + k_x^2 f_k = 0 \to \quad f_k \sim \exp(\pm i|k_x|t)$

$t \to +\infty, \frac{d^2 f_k}{d\tau^2} + f_k = 0 \to \quad f_k \sim \exp(\pm i\tau)$

we see that for adiabatic regularization we will want an approximate solution of the wave equation that reduces to the positive frequency (generalized) WKB form.

For x large, the various Bessel functions behave as follows:

$J_n(x) \sim \sqrt{\frac{2}{\pi x}}\cos\left(x - \frac{n\pi}{2} - \frac{\pi}{4}\right)$

$Y_n(x) \sim \sqrt{\frac{2}{\pi x}}\sin\left(x - \frac{n\pi}{2} - \frac{\pi}{4}\right)$

$I_n(x) \sim \frac{e^x}{\sqrt{2\pi x}}$

$K_n(x) \sim \frac{e^{-x}}{\sqrt{2\pi x}}$

$H_n^{(2)}(x) = J_n(x) - iY_n(x) \sim \sqrt{\frac{2}{\pi x}}(\cos\alpha - i\sin\alpha) = \sqrt{\frac{2}{\pi x}}e^{-i\alpha}$

$H_n^{(2)}(x) = J_n(x) + iY_n(x) \sim \sqrt{\frac{2}{\pi x}}e^{i\alpha}$

$$\begin{cases} H_n^{(2)}(x) \sim \sqrt{\frac{2}{\pi x}}\exp\left(-i\left[x - \frac{n\pi}{2} - \frac{\pi}{4}\right]\right) \\ H_n^{(1)}(x) \sim \sqrt{\frac{2}{\pi x}}\exp\left(i\left[x - \frac{n\pi}{2} - \frac{\pi}{4}\right]\right) \end{cases}$$

So,

$$H_{ik}^{(2)}(\tau) \sim \sqrt{\frac{2}{\pi\tau}} \exp\left(-k\frac{\pi}{2}\right) \exp\left(i\frac{\pi}{2}\right) \exp(-i\tau)$$

is consistent with the adiabatic requirement at large τ (note exponential suppression for high k modes). Let's now compare this with the natural definition of positive frequency in some local frame of reference. Starting with

$$H_v^{(2)}(z) = -\frac{e^{v\pi i/2}}{\pi i} \int_{-\infty}^{\infty} e^{-iz\cos t - vt dt} \quad , \qquad -1 < Rev < 1$$

Consider: $v = ik$, $z = \beta[T^2 - X^2]^{1/2} = \frac{\beta}{g} e^{gt}$, $\beta = \left[k_z^2 + k_y^2 + m^2\right]^{1/2}$, then:

$$x = \frac{1}{g} tanh^{-1}(X/_T) = \frac{1}{g} sinh^{-1}(X \,/\, (z/\beta)) = \frac{1}{g} cosh^{-1}(T/(z/\beta))$$

If we make the substitution:

$$\mathfrak{z} = \mathfrak{z}' + x, \;\; d\mathfrak{z} = d\mathfrak{z}'$$

$H_{ik}^{(2)}(z) = i\pi^{-1} e^{\pi k/2} \int_{-\infty}^{\infty} d\mathfrak{z}' \exp\left\{-i\beta[T^2 - X^2]^{1/2}(\cosh \mathfrak{z}' \cosh x + \sinh \mathfrak{z}' \sinh x)\right\} e^{-ik\mathfrak{z}}$

$$= i\pi^{-1} e^{-\pi k/2} \int_{-\infty}^{\infty} d\mathfrak{z}' \exp[-i\beta \cosh \mathfrak{z}' \, T - i\beta \sinh \mathfrak{z}' \, X] \, e^{-ik\mathfrak{z}}.$$

Let $\omega_{\vec{p}} = \beta \cosh \mathfrak{z}'$, $P_1 = -\beta \sinh \mathfrak{z}'$, $P_y = k_y$, $P_z = k_z$, then:

$$\omega_{\vec{p}} = (P_1^2 + P_2^2 + P_3^2 + m^2)^{1/2}$$

$$H_{ik}^{(2)}(z) = i\pi^{-1} e^{-\pi k/2} \int_{-\infty}^{\infty} d\mathfrak{z}' \, e^{-i\omega_p T} e^{i\vec{p}\cdot\vec{x}} e^{-ik\mathfrak{z}'}$$

We therefore find that $H_{ik}^{(2)}(\tau)$ consists of a superposition of modes which are positive frequency with respect to Minkowski time T, thus providing the same vacuum as the Minkowski Vacuum. In this situation, a comoving observer (a Minkowski Observer) will see no particles in this vacuum if there are none to begin with (no particle creation, the Bogoliubov transformation only has alpha coefficients nonzero, e.g., only mixing of positive modes amongst themselves, no mixing between negative and positive modes is occurring). Furthermore as $t \to \infty, z \to \infty$, $H_{iH}^{(2)}(Z)$ → positive frequency with respect to the adiabatic vacuum $|0>$. So, the adiabatic vacuum and the local (Minkowski) positive frequency vacuum are in agreement. This hints at a solution mesh consisting of local patches with local vacuum definitions and particle production flux at the 'boundaries' (where the approximation of locally flat breaks down sufficient for particle production). More formal axiomatic/algebraic formulations show that such an intuitive solution does exist (in Globally Hyperbolic space-time, e.g., where Cauchy data can be specified in an initial-value specification).

7.5.4 Killing Vectors

Starting with the Minkowski metric, $dS^2 = +dT^2 - dX^2 - dY^2 - dZ^2$, consider the coordinate transformation:

$$X = g^{-1}e^{gt}\sinh(gx),\ \ T = g^{-1}e^{gt}\cosh(gx),\ \ T^2 - X^2 = g^{-2}e^{2gt},\ \ X/T = \tanh(gx)$$

Which results in the Kasner (100) space-time metric:

$$dS^2 = e^{2gt}(dt^2 - dx^2) - dy^2 - dz^2$$

K (100) is imbedded in Minkowski as shown here:

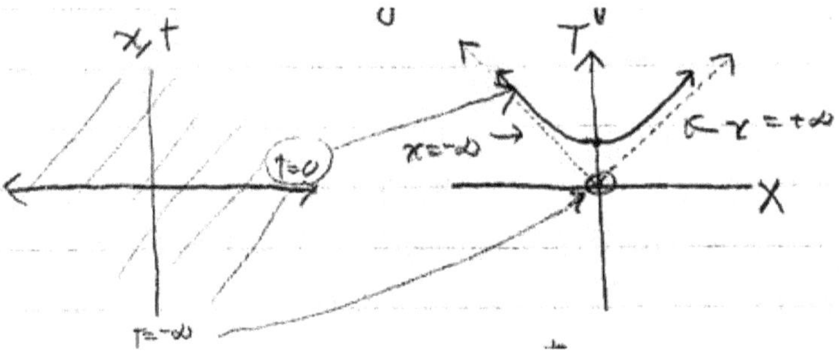

Definition of Killing Vector

A vector $\mathfrak{z}^\mu$ whose norm is invariant under symmetry transformations (such as those associated with translations, boosts, and rotations in Minkowski space-time) must satisfy (covariant form of equation for flat spacetime):

$$0 = \mathfrak{z}^\mu{}_{,\rho}g_{\mu\sigma} + \mathfrak{z}^\nu{}_{,\sigma}g_{\rho\nu} + \mathfrak{z}^\mu\frac{\partial g_{\rho\sigma}}{\partial x^\mu}$$

Killing vectors for K(100) and Minkowski

Since we know the metric is diagonal and only time-dependent (for K(100) case), we get:

$$0 = \mathfrak{z}^\mu{}_{,\rho}g_{\sigma\sigma} + \mathfrak{z}^\nu{}_{,\sigma}g_{\rho\rho} + \mathfrak{z}^0\frac{\partial g_{\rho\sigma}}{\partial x^0}$$

For the Minkowski metric with no time-dependence this becomes:

$$0 = \mathfrak{z}^\mu{}_{,\rho}g_{\sigma\sigma} + \mathfrak{z}^\nu{}_{,\sigma}g_{\rho\rho}$$

Thus:

$$\rho = \sigma:\ \mathfrak{z}^\rho{}_{,\rho} = 0 \rightarrow \mathfrak{z}^\rho = C, \qquad \rho = 0,1,2,3\ ,$$

where $\mathfrak{z}^\rho$ is constant for each of $\rho = 0,1,2,3$, these are the translations. And,

$$\rho = 0, \sigma = i:\ \ \mathfrak{z}^0{}_{,i} - \mathfrak{z}^i{}_{,0} = 0 \rightarrow \mathfrak{z}^0 = \chi_i, \mathfrak{z}^i = \chi_0, \qquad i = 1,2,3\ ,$$

these are the boost transformations. And,

$$\rho = i, \sigma = j: \ \mathfrak{z}^i{}_{,j} - \mathfrak{z}^j{}_{,i} = 0 \rightarrow \mathfrak{z}^i = \chi_j, \mathfrak{z}^j = -\chi_i, \qquad i,j = 1,2,3 \ , \quad i \neq j,$$

these are the rotation transformations.

The vectors corresponding to these transformation:
There are 4 translation vectors: ∂_μ.
There are 3 boost vectors: $\chi_i \partial_0 + \chi_0 \partial_i$.
There are 3 rotation vectors: $\chi_i \partial_0 + \chi_0 \partial_i$.

Some terminology:
Diffeomorphism: a bijective, bicontinuous, C^r map.

Isometry: a diffeomorphism that carries the metric into itself:

$$g(x, y)|_p = g(\varphi_* X \varphi_* Y)|_{\varphi(p)}$$

Killing vector field: a local one-parameter group of isometries.

The mapping between K(100) and Minkowski is not bijective, therefore it is not a diffeomorphism.
Note FRW and deSitter have a particle horizon whereas K(100) doesn't. In deSitter the vacuum is unspecified by the isometries to within a one parameter group of vacua. Boundary conditions on the possible Minkowski vacua yield only one vacuum.

7.5.5 K(100) observer has 10 KV fields (see Appendix C for details)
The K(100) observer describes 10 killing vector fields (which correspond directly to those of Minkowski), the extension of these vector beyond the coordinate singularity at spatial infinity $x \rightarrow \infty$ in K(100) coordinates is direct. We know that the affine parameter is finite for any geodesic in K(100) space that intercepts the light cone. The extension of these geodesics beyond K(100) space immediately returns us to the Minkowski space time. The 10 killing vector fields given in K(100) space are continuous along a given geodesic and correspond directly with the Minkowski KV fields in the K(100) region. The extension of the KV fields to all of Minkowski spacetime then follows trivially from the theory of differential equations since we must thereby have a unique solution and we know of the KV – Minkowski fields. Thus the extension is simply to the KV field as it known in Minkowski.

It is possible to describe a deformation of a K(100) hypersurface (a hyperbola is chosen) to a Minkowski hypersurface (T=0 is chosen). For massive fields we need only assume that $\varphi(x) \to 0$ at spatial infinity to get a well defined rotation. So, we need to identify the class of states on the Minkowski hypersurface that have zero Cauchy data on the Kasner(100) hypersurface $\Sigma_{k(100)}$:

$$\left(P_{\mathfrak{z}}\right)_{k100} = \int_{t=0} d\Sigma n_\mu T_\nu^\tau \mathfrak{z}^\nu(x) = 0$$

In operator formation (upon quantifying the theory) we have

$$\left(\hat{P}_{\mathfrak{z}}\right)_{K100} |0>_{K100} = 0$$

where $\left(\hat{P}_{\mathfrak{z}}\right)_{K100}$ annihilates the K(100) vacuum. For massive fields we simply have:

$$\left(P_{\mathfrak{z}}\right)_{Mink} = \left(P_{\mathfrak{z}}\right)_{k100} = 0,$$

and since

$$\left(\hat{P}_{\mathfrak{z}}\right)_{Mink} |0>_{Mink} = 0,$$

for the Poincare vacuum for Minkowski, we must have:

$$|0>_{K100} = |0>_{Mink}$$

for massive fields. For massless fields we start as before, $\left(P_{\mathfrak{z}}\right)_{K100} = 0 \Rightarrow \left(\hat{P}_{\mathfrak{z}}\right)_{K100} |0>_{K100} = 0$, but now we have:

$$\left(P_{\mathfrak{z}}\right)_{Mink} = \int_{\mathcal{J}^+} d\Sigma_{null}\, n_\mu T_\nu^\mu \mathfrak{z}^\nu(x) + \left(P_{\mathfrak{z}}\right)_{K100} \neq 0$$

Thus, $\left(P_{\mathfrak{z}}\right)_{Mink} |\Psi> \neq 0$ in general. So, $|0>_{K(100)} \neq |0>_{Mink}$ for massless fields, the difference being any number of null particles on Σ_{Mink} going to positive null infinity, $\mathcal{J}^+$, yielding an entire class of possible $|\Psi>$ that give $|0>_{K(100)}$ in the K(100) region.

<u>Field Theory</u>

First consider the classical modal decomposition of the wavefunction in terms of outgoing and ingoing modes defined in terms of the Minkowski Cauchy surface $\mathcal{C}_A$:

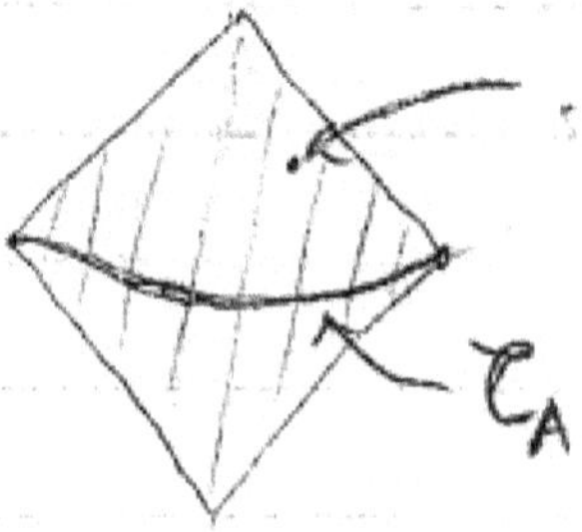

The upper arrow indicates a point in Minkowski where the wavefunction field is specified (as it is for all of Minkowski). The lower arrow indicates the $\mathcal{C}_A$ hypersurface. Similarly, we have for the K(100) Cauchy surface $\mathcal{C}_B$:

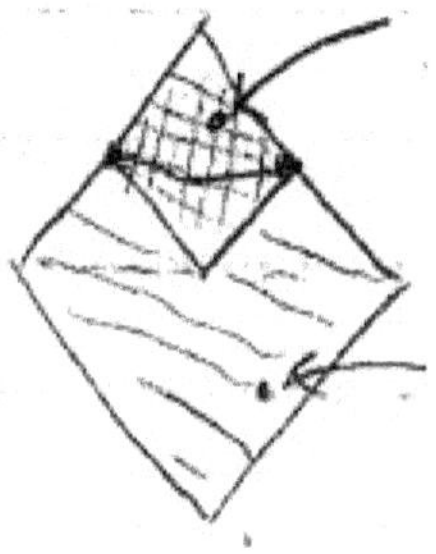

The upper arrow indicates a point in K100 (also Minkowski) where the wavefunction field is specified (as it is for all of Kasner100). The lower arrow indicates a point in Minkowski, that is not in K100, where the Minkowski wavefunction field is specified. The $\mathcal{C}_B$ hypersurface of K100 is also shown.

First, let's study the scalar product and field defined in Minkowski space:

$$(\varphi_1, \varphi_2) = -i \int_\Sigma \varphi_1(x) \overleftrightarrow{\partial_\mu} \varphi_2{}^*(x) d\Sigma^\mu$$

where for convenience a modal decomposition of φ in terms of ingoing and outgoing modes is made (for Minkowski space):

$$\varphi_{Mink} = \sum_k (a_k f_k + b_k g_k + h.c.)$$

Where $h.c.$ stands for Hermitian conjugate. We have for the complete orthonormal $f's$ and $g's$:

$$(f_i, f_j) = \delta_{ij}, \qquad (f_i^*, f_j^*) = -\delta_{ij}, \qquad (f_i, f_j^*) = 0$$

with the same for g_k.

It can be shown that the scalar product is invariant under deformations of the Cauchy surface Σ to another Cauchy surface Σ' (by the completeness properties of Cauchy data on a given surface). Lets' deform $\mathcal{C}_A$ to have a portion the same as $\mathcal{C}_B$:

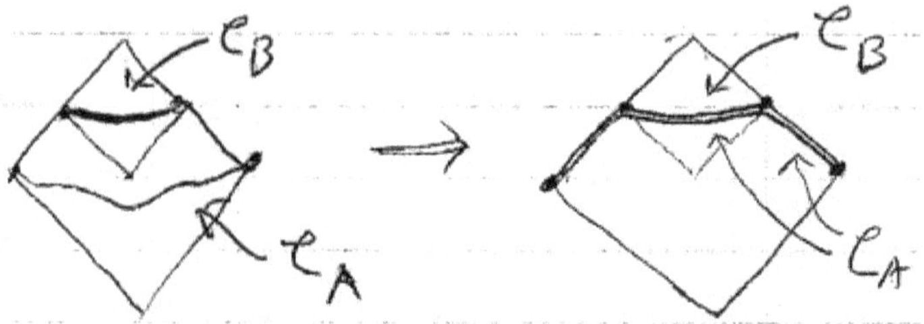

In this way we divide the scalar product integral defined over Cauchy surface C_A into an integral over surface $\mathcal{C}_B$ and $\mathcal{J}^+$.

7.5.6 Milne

Consider a field and surface product defined for Milne space (still using Minkowski coordinates) with modal decomposition (f_k, g_k)

$$\varphi_{Milne} = \sum_k (c_k f_k + d_k g_k + h.c.)$$

with

$$(\varphi_1, \varphi_2)_{Milne} = \int_{\mathcal{C}_B} \varphi_1(x) \overleftrightarrow{\partial_\mu} \varphi_2{}^*(x) d\Sigma^\mu.$$

For massive fields:

$$\varphi(\mathcal{J}^+)_{m \neq o} = 0$$

in which case,

$$(f_i, \varphi)_{\mathcal{J}^+} = 0 \quad \rightarrow \quad (f_i, f_k)_{\mathcal{J}^+} = 0\,, \ \ m \neq 0$$
$$(g_i, \varphi)_{\mathcal{J}^+} = 0 \quad \rightarrow \quad (g_i, g_k)_{\mathcal{J}^+} = 0\,, \ \ m \neq 0$$

Thus, in conclusion:

<u>For massive fields:</u> $a_k = c_k, \quad b_k = d_k$.

<u>For massless fields:</u> $a_k = c_k$, and since

$$(g_i, \varphi_{Mink})_{\mathcal{C}_B, m=0} = 0 \ \rightarrow \ d_k = 0.$$

Thus at $\mathcal{C}_B$:

$$\varphi_{Milne} = \sum_k (c_k f_k + h.c.)$$

And the Milne vacuum is specified by

$$a_k|0>_{Milne} = 0$$

for all a_k. To relate this to the Mink vacuum we write any Mink state as a tensor product

$$|S>_{Mink} = |S_f>_{Mink} \otimes |S_g>_{Mink}$$

where $|S>$ is a general state, $|S_f>_{Mink}$ is an ingoing quantum state, and $|S_g>_{Mink}$ is an outgoing quantum state. So, the Fock space for Mink out vacuum (free scalar wave) can be expressed as the tensor product of Fock spaces for incoming and outgoing quanta. It is now possible to see that the Milne vacuum will agree with a variety of Minkowski vacua, each of what has $a_k|0>_{Mink}= 0$, with any $|S_g>_{Mink}$ allowed.

Defining properties of Milne:

(1) Isomorphic to the future light cone of Minkowski.
(2) There is a simple coordinate transformation to Minkowski like the Rindler transform.
(3) The resulting Milne metric is time dependent in the Milne coordinate system.
(4) Physically, Milne observers are related to each other by boosts relative to a common origin. Proper time for a given Milne observer then ranges from $-\infty < t < \infty$ for which the constant t hypersurface, in Minkowski space, is a hyperbola in the upper quadrant. Of course, translations on the constant t hypersurface in Milne coordinates corresponds to a boost in Minkowski coordinates, the orbit of which is the aforementioned hyperbola.

So, it would seem that field theory in CST has a nonlocal feature. This would seem reasonable and yet direct application of the mathematical machinery quickly grinds to a halt because of this.

Recap of the Kasner (100) and Milne spacetimes

Kasner (100) metric:

$$ds^2 = e^{+2gt}(dt^2 - dx^2) - dy^2 - dz^2$$

By the coordinate transformation:

$$X = g^{-1}e^{gt}\sinh(gx), Y = y, \quad Z = z, \qquad T = g^{-1}e^{gt}\cosh(gx)$$

we arrive at a sector of Minkowski:

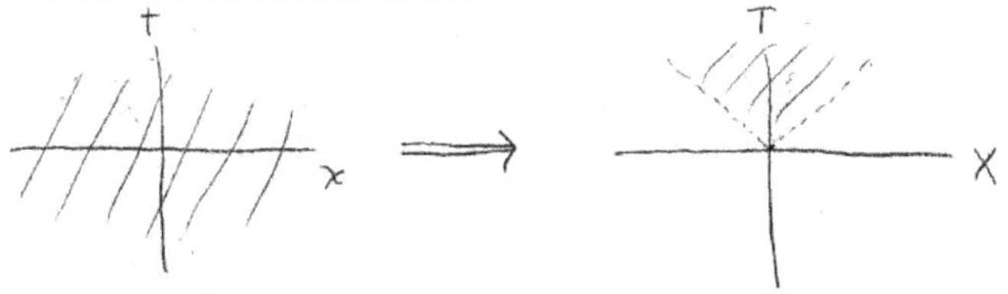

where spacetime surfaces of constant t in Kasner(100) provide a foliation of hyperbolae in the upper (null) quadrant of Minkowski.

Milne metric (equivalent to FRW with $a(t) = t, K = -1, t > 0$):

$$ds^2 = -dt^2 + t^2[d\chi^2 + \sinh^2\chi\,(d\theta^2 + \sin^2\theta d\varphi^2)]$$

with the coordinate transformation $r = t\sinh\chi, \theta = \theta, \ \varphi = \varphi, t' = t\cosh\chi$, becomes a sector of Minkowski:

$$ds^2 = -dt'^2 + r^2(d\theta^2 + \sin^2\theta d\theta^2)$$

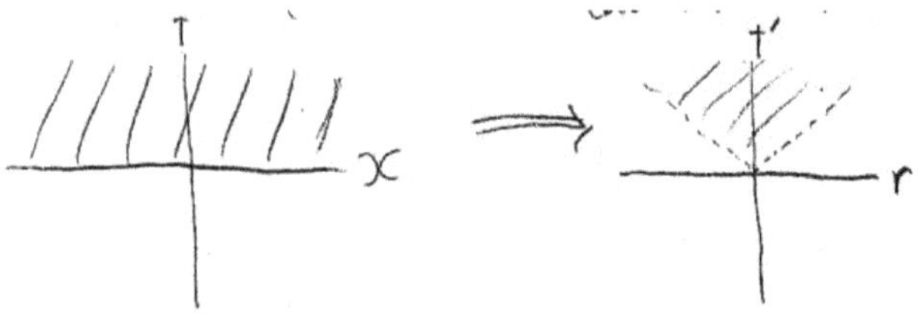

7.5.7 Field theory in the K(100) and Milne spacetimes.

I. Properties of the spacetimes

A. Kasner (100)

i. The k(100) spacetime is embedded in the future light cone of one of the Minkowski Cartesian coordinates.

ii. The K(100) metric is $dS^2 = e^{2gt}(dt^2 - dx^2) - dy^2 - dZ^2$

iii. The coord. Trans. Relating K(100) to its Mink. Imbeddings is $X = g^{-1}e^{2gt}\sinh(gx), T = g^{-1}e^{2gt}\cosh(gx), Y = y, Z = Z$

B. Milne

i. Milne spacetime in imbedded in the future light core of Minkowski

ii. The Milne metric: $dS^2 = -dt^2 + t^2[dx^2 + \sinh^2 x\,(d\theta^2 + \sin^2\theta d\varphi^2)]$

iii. The coord. Trans.: $r = t\sin x, t' = t\cosh x\ , \theta = \theta, \varphi = \varphi$

II. Quantum Field Theory in K(100)

A. The Klein-Gordon equation is $(\Box + m^2)\varphi = 0$ $\varphi(x) = f_{\vec{x}}(t)\exp(i\vec{k}\cdot\vec{x})$, we get $\ddot{f}_{\vec{k}} + [k_x^2 + (k_z^2 + m^2)e^{2gt}]f_{\vec{k}}$

i. When $a = (k_x^2 + k_z^2 + m^2) \neq 0$: $f_{\vec{x}} = c, H_{ik}^{(2)}(\tau) + C_2 H_{ik}^{(2)}, \tau = \frac{1}{g}(k_x^2 + k_z^2 + m^2)$

ii. When $a = (k_x^2 + k_z^2 + m^2) \neq 0$: $f_{\vec{x}} = a, e^{-ik_x x} + a_2 e^{i|k_x|t}e^{-ik_x x}$

B. Adiabatic analysis reveals that either the J_{-ik} or $H_{ik}^{(2)}$ decomposition is an acceptable adiabatic vacuum.

C. K(100) has 10K.V.'s one, and only one, of which is timelike. Thus a unique prescription exists for defining positive frequency.
$\mathcal{L}_{(\partial/\partial t)}\Psi^t = -i\omega\Psi^t \rightarrow \Psi^t =$
$f\left(g^{-1}e^{gt}\sinh(gx)\exp\left\{-\omega\left[\frac{1}{g}e^{gt}\cosh(gx)\right]\right\}\omega\right)$

i. $H_{ik}^{(2)}$ is only composed of terms satisfying the pos. freq. lie. Equation analogically for $H_{ik}^{(1)}$ $(a \neq 0)$; $H_{ik}^{(2)}$ can be expressed in terms of Mink. Mode position

a. $k_x = 0$, Ψ^t modes are composed of only $e^{-i\omega_p T}\cos(P_x x)$ modes, X,T Mink coord.'s.

b. $k_x \neq 0$, Ψ^t has a full modal departure $\{\cos(P_x X)\ and \sin(P_x X)\}$

ii. When a=0, the Ψ^t (plane wave) modes are composed from a subspace of the associated Minkowski modal Hilbert space.

iii. Relations between the Mink and M(100) modal decomp. For either a=0, are derived.

D. By deforming the surfaces of integration (from $\nabla_\mu[T^{\mu v}\mathfrak{z}_v] = 0, \mathfrak{z}_v a\ K.V)$ we are able to show the equivalent of conserved quanitions – quantum operations – vacuum, between K(100) and Mink for the $m \neq 0$ case. For $m \neq 0$ we find a similar relations between K(100) and a subsector of Mink. Quantum operations yielding an equiv. class of Mink vacua to the K(100) vacuum.

E. Equivalence of vacua that are defined by any of the possible timelike K.V.'s is established for both the Mink. Case and the K(100) case.

III. QFT in Milne

A. Solutions to the Klein-Gordon equation : $U_k(x,) = e^{-\eta}T_{KJ}^{(-)}(X)Y_J^{\mu}(\theta,\varphi)\Psi_k(\eta)$

i. $\pi_{KJ}^{(-)}(x) = \left\{\frac{1}{2}\pi k^2(k^2+1)\dots(k^2+J^2)\right\}^{-1/2}\sinh^J x\left(\frac{d}{d\cosh x}\right)^{1+J}\cos kx$

a. The relation $\pi_{KJ}^{(-)}(x) = e^{-i\pi/4}\sinh^J x\, P_{(ik-1/2)}^{\left(J+\frac{1}{2}\right)}(\cosh x) is\ shown$

ii. $\frac{d^2 x_k}{d\eta^2} + \{k^2 + e^{2\eta} m^2\} x_k = 0 - x_k(\tau) = C_1 H_{ik}^{(1)}(\tau) + C_1 H_{ik}^{(2)}(\tau)$

B. As before the adiabatic analysis yields either $H_{ik}^{(2)} or\, J_{-k}$ modal decompositions.

C. Milne has 10K.V.'s, only one timelike

i. $\mathcal{L}_{\left(\frac{\partial}{\partial t}\right)} \Psi^t = -i\omega \Psi^t - \Psi^t = f(t \sinh x) \exp\{-i\omega t \cosh x\}\, \omega(\theta, \varphi)$

a. $m = 0\,, x_k = d_1 e^{-i|k|}$ ingoing modes, (completeness and wave-product type modes used)

b. $m \neq 0, u_k{}'s$ composed of all modes.

ii. Construction of $\Psi^t\ from\ u_k{}'s$ requires $\pi(x) e^{-ikx}$ to be grouped such that $\pi(x) e^{-ikx} = \sinh^{-1} x$

D. Integral definition of conserved quantities over spacelike hypersurfaces used as in II.D reaffirm the Fock space relations between Milne and Minkowski for both the $m = 0\ and\ m \neq 0$ cases.

7.6 The Bogoliubov Transform

Given the two field decomposition:

$$\varphi(x) = \sum_i \left[a_i u_i(x) + a_i^\dagger u_i^*(x)\right]$$

And

$$\varphi(x) = \sum_i \left[\bar{a}_i \bar{u}_i(x) + \bar{a}_j^\dagger \bar{u}_j^*(x)\right]$$

We can write

$$\bar{u}_j = \sum_i \left(\alpha_{ji} u_i + \beta_{ji} u_i^*\right),$$

or, conversely

$$u_i = \sum_i \left(\alpha_{ji} u_i + \beta_{ji} u_j^*\right)$$

These relations are the Bogoliubov transformation and α_{ji}, β_{ji} are the Bogoliubov coefficients:

$$\alpha_{ji} = (\bar{u}_j, u_j), \quad \beta_{ji} = -(\bar{u}_j, u_j^*)$$

So, $a_i = \sum_j \left(\alpha_{ji} \bar{a}_j + \beta_{ji}^* \bar{a}_j^\dagger\right)$ and $a_j = \sum_i \left(\alpha_{ji}^* a_j - \beta_{ji}^* a_i^\dagger\right)$ [71,72].

Now, in regards to Rindler spacetime, we with use the Unruh method to choose the extension of Right Rindler modes into the left region such that the resulting mode given on a Cauchy surface for all of Minkowski is analytic and bounded in the lower half complex $\bar{u}$ plane ($\bar{u}$ for regular Mink mull coordinates). The extended Rindler modes:

$$\omega_k^{(1)} = \left({}^{R}u_k + e^{-\pi\omega/a}\, {}^{L}u_{-k}^{*}\right)$$

and

$$\omega_k^{(2)} = \left({}^{R}u_{-k}^{*} + e^{-\pi\omega/a}\, {}^{L}u_k\right)$$

Where ${}^{R}u_k = (4\pi\omega)^{-1/2}\, e^{ik\varepsilon - i\omega\eta}$ in Right

$= 0$ in Left

${}^{L}u_k = (4\pi\omega)^{-1/2} e^{ik\varepsilon + i\omega\eta}$ in Left

$= 0$ in Right

$\varphi = \sum_i \left[a_k \bar{u}_k + a_k^{\dagger} \bar{u}_k^{*}\right]$ Mink.

$$\varphi = \sum_k \left(b_K^{(1)}\, {}^{L}u_k + b_K^{(1)\dagger}\, {}^{L}u_k + b_K^{(2)}\, {}^{R}u_k + b_K^{(2)\dagger}\, {}^{R}u_k^{*}\right)$$

Given the above analyticity properties we have:

$$\varphi = \sum_k \left[2\sinh\left(\frac{\pi\omega}{a}\right)\right]^{-1/2} \left(d_k^{(1)}\left[e^{\pi\omega/2a}\, {}^{R}u_k e^{-\pi\omega/2a}\, {}^{L}u_{-k}^{*}\right] + d_k^{(2)}\left[e^{-\pi\omega/2a}\, {}^{R}u_k^{*} e^{\pi\omega/2a}\, {}^{L}u_k\right]\right) + h.c.$$

Now, in calculating the Bogoliubov transform from a Minkowski positive frequency to Rindler modes we use the Unruh relation to our advantage:

$$b_K^{(1)} = [2\sinh(\pi\omega/a)]^{-1/2}\left[e^{\pi\omega/2a} d_k^{(2)} + e^{-\pi\omega/2a} d_{-k}^{(1)t}\right]$$

$$b_K^{(2)} = [2\sinh(\pi\omega/a)]^{-1/2}\left[e^{\pi\omega/2a} d_k^{(1)} + e^{-\pi\omega/2a} d_{-k}^{(2)t}\right]$$

So,

$$\left\langle 0_M \middle| b_k^{(1,2)\dagger} b_k^{(1,2)} \middle| 0_M \right\rangle = \frac{e^{-\pi\omega/a}}{2\sinh(\pi\omega/a)} = \left(e^{2\pi\omega/a} - 1\right)^{-1}$$

${}^{R}u_k$ is not analytic at the origin, and neither is the Unruh combination, however Fourier transforms are capable of transforming (taking as argument) non-analytic functions (consider the step-function), and in a distribution sense, at least, everything is well defined.

Furthermore, even though $^R_\square u^\square_k$ is not in L^2, we may still speak of its Fourier transform, just as a Fourier transform of one of its own basis modes yields a delta function (a distributionally well-defined object). Thus, at this junction we could brute force Fourier transform $^R_\square u^\square_k$ and proceed as is, but let's consider the problem from a different prospective for a moment.

We may surmount the difficulties in the modal analysis in a different way – if a specific extension is not pleasing. The alternative is not to speak of Rindler plane waves at all but to use wavepackets. Since these wavepackets can be chosen to have bounded support with sufficiently fast die off (in some Gaussian wavepacket parameter, say), then we can arrange to have no support at Rindler $x = -\infty$ where the nonanalyticity problems arise, but this still isn't quite right since an arbitrary Minkowski state may have support at the origin. Of course, we could continue along this tack by having some boundary condition at Rindler $x = -\infty$, while that at $x = \infty$ is the bounded support $\rightarrow u(x = \infty) = 0$. However, this presents many difficulties being an additional complication on the quantization on the half-line problem that is similar to he moving mirror problem, and not discuss further here.

Considering $^R_\square u^\square_k$, lets find

$$^R_\square u^\square_j = \sum_i (\alpha_{ji} u_i + \beta_{ji} u_i^*)$$

where $u_i's$ are Minkowski modes. Thus, we need to find the Bogoliubov coefficients:

$$\alpha_{ji} = (^R_\square u^\square_j, u_i), \quad \beta_{ji} = (^R_\square u^\square_j, u_i^*)$$

where the scalar product has been defined on a spacelike hyperplane and follows from the existence of a second order partial differential equation (the Klein-Gordon equation):

$$(\varphi_1, \varphi_2) = -i \int_t \varphi_1(x) \overleftrightarrow{\partial_t} \varphi_2^*(x) d^3 x$$

The $u_i's$ and $^R_\square u^\square_j's$ have been normalized with respect to the scalar product respective to Cauchy surfaces in their spacetimes:

$$u_{\vec{k}} = [2\omega(2\pi)^3]^{-1/2} e^{i\vec{k}.\vec{x} - i\omega t} \rightarrow (u_{\vec{k}}, u_{\vec{k}'}) = d^3(\vec{k} - \vec{k}')$$

In 1 + 1 Mink. space we have $ds^2 = dt^2 - dx^2 = d\bar{u}d\bar{v}$, thus

$$u_k = [4\pi\omega]^{-1/2}\, e^{ikx - i\omega t}\,, \quad (u_k, u_{k'}) = \delta(k - k')$$

Here we are also considering a massless field, so $\omega = |k|$:

For $k > 0$:

$$u_k = (4\pi\omega)^{-1/2}\, e^{-i\omega(t-x)} = (4\pi\omega)^{-1/2} e^{-i\omega\bar{u}}\,, \quad \begin{matrix} t - x = \bar{u} \\ t + x = \bar{v} \end{matrix}$$

For $k < 0$

$$u_k = (4\pi\omega)^{-1/2}\, e^{-i\omega\bar{v}}$$

As for Rindler modes let's first sort out the coordinates:

$$t = a^{-1}\, e^{a\xi} \sinh a\eta \quad u = \eta - \xi$$

$$x = a^{-1}\, e^{a\xi} \cosh a\eta \quad v = \eta - \xi$$

So,

$\bar{u} = t - x = -a^{-1}\, e^{a\xi}[\cosh a\eta - \sinh a\eta] = -a^{-1}\, e^{a\xi}\, e^{-a\eta} = -a^{-1}e^{-au}$

$\bar{v} = a^{-1}e^{-av}$

Modal solution are

$$f_k = (4\pi\omega)^{-1/2}\, e^{ik\xi \mp i\omega\eta}$$

With the upper sign for Right Rindler and the lower sign for Left Rindler.

Define ${}^{R}_{\square}u_k^{\square} = (4\pi\omega)^{-1/2}\, e^{ik\xi - i\omega\eta} \quad in\ R$

$= 0 \quad in\ L$

Etc.

Converting the ${}^{R}_{\square}u_k^{\square}(\xi, \eta) = {}^{R}_{\square}u_k^{\square}(x, t)$ we get:

For $k < 0$:

$${}^{R}_{\square}u_k^{\square} = (4\pi\omega)^{-1/2}\, e^{i\omega(\eta - \xi)} = (4\pi\omega)^{-1/2}\, e^{-i\omega u}$$

$$= (4\pi\omega)^{-1/2}(-a^{-1}\, e^{-au})^{\frac{i\omega}{a}}\, (-a)^{i\omega/a}$$

$${}^{R}_{\square}u_k^{\square} = (4\pi\omega)^{-1/2}\, (-a)^{i\omega/a}\, (\bar{u})^{i\omega/a}$$

For $k < 0$:

$${}^{R}_{\square}u_k^{\square} = (4\pi\omega)^{-1/2}\, e^{-i\omega v} = (4\pi\omega)^{-1/2}\, (a^{-1}\, e^{av})^{-\frac{i\omega}{a}}\, (a)^{-i\omega/a}$$

$${}^{R}_{\square}u_k^{\square} = (4\pi\omega)^{-1/2}\, (a)^{-i\omega/a}\, (\bar{v})^{-i\omega/a}$$

Let's concern ourselves with the $k > 0$ ${}^{R}_{\square}u_k^{\square}$ modes for now:

$$\alpha_{ji} = -i \int_0^\infty \left\{(4\pi\omega)^{-1/2} (-a)^{i\omega/a}\, \bar{u}^{i\omega/a}\right\} \overleftrightarrow{\partial}_t \left\{(4\pi\omega')^{-1/2} \begin{bmatrix} e^{-i\omega'\bar{u}} \\ e^{-i\omega'\bar{v}} \end{bmatrix}\right\} dx$$

Where the column matrix is associated with the conditions $\begin{bmatrix} k' > 0 \\ k' < 0 \end{bmatrix}$. Let's consider the $k' > 0$ integral first and split according to $\alpha_{ji} = \{\alpha_{ji}\}_{k'>0} + \{\alpha_{ji}\}_{k'<0}$:

$$\{\alpha_{ji}\}_{k'>0} = \frac{-i(-a)^{i\omega/a}}{4\pi\sqrt{\omega\omega'}} \int_0^\infty \Big\{(-\omega')(-x)^{i\omega/a} e^{-i\omega'(-x)} - \left(\frac{i\omega}{a}\right)(-x)^{i\omega/a-1} e^{-i\omega'(-x)}\Big\} dx$$

This requires solving

$$I_p = \int_0^\infty (-x)^{i\omega/a-p}\, e^{-i\omega'(-x)} dx$$

Let $t = i\omega' x$:

$$I_p = \int_0^{-i\infty} \left(\frac{t}{i\omega'}\right)^{i\omega/a-p} e^{-t} \frac{dt}{(-i\omega')} = -(i\omega')^{i\omega/a-(p+1)} \int_0^{-\infty} t^{\left(\frac{i\omega}{a}-p\right)}\, e^{-t}\, dt$$

Since $\Gamma(z) = \int_0^\infty t^{z-1}\, e^{-t}\, dt$ we get:

$$\{\alpha_{ji}\}_{k'<0} = \frac{i(-a)^{i\omega/a}}{4\pi\sqrt{\omega\omega'}} (-1)^{i\omega/a} \left[(i\omega')^{i\omega/a} \Gamma\left(\frac{i\omega}{a} + 1\right) + \left(\frac{i\omega}{a}\right)(i\omega')^{i\omega/a-2}\, \Gamma\left(\frac{i\omega}{a}\right)\right]$$

So,

$$\alpha_{ji} = \frac{\left(1 + e^{\pi\omega/2a}\right) a^{i\omega/a}}{4\pi\sqrt{\omega\omega'}} (i\omega')^{i\omega/a} \left[\left(\frac{\omega}{a}\right)\left(1 - \frac{1}{\omega'^2}\right) - i\right] \Gamma\left(\frac{i\omega}{a}\right).$$

Next, we need the β Bogoliubov coefficient:

$$\beta_{ji} = -\left({}^{R}_{\square}u_j^{\square}, u_i^*\right)$$

where ${}^{R}_{\square}u_k^{\square} = \begin{cases} (4\pi\omega)^{-1/2}\, e^{ik\xi - i\omega\eta} & \text{in Right} \\ 0 & \text{in Left} \end{cases}$ and $\begin{matrix} \bar{u} = t - x \\ \bar{v} = t + x \end{matrix}$, thus

$${}^{R}_{\square}u_k^{\square} = \begin{Bmatrix} (4\pi\omega)^{-1/2} (-a)^{i\omega/a} \bar{u}^{i\omega/a} \\ (4\pi\omega)^{-1/2} (a)^{-i\omega/a}\, \bar{v}^{i\omega/a} \end{Bmatrix}$$

defined in the In Right and with $\begin{pmatrix} k > 0 \\ k < 0 \end{pmatrix}$. Thus

$$u_k^* = \begin{Bmatrix} (4\pi\omega)^{-1/2}\, e^{+i\omega\bar{u}} \\ (4\pi\omega)^{-1/2}\;\, e^{+i\omega\bar{v}} \end{Bmatrix}$$

And since

$$\beta_{jk} = i\int_{\Sigma} {}^{R}u_k^{*}\, \overleftrightarrow{\partial}_t u_k\, dx, \qquad \omega_j = j, \omega_k = k \text{ when } j > 0, k > 0$$

For $j > 0, k > 0$:

$$\beta_{jk} = i\int_0^\infty dx\, \left(4\pi\omega_j\right)^{-1/2} (-a)^{i\omega_j/a} \left(4\pi\omega_j\right)^{-1/2}$$

$$\times \left[\bar{u}^{i\omega/a}\, e^{-i\omega\bar{u}}\, (-i\omega_k) - \bar{u}^{i\omega/a}\, e^{-i\omega\bar{u}} \left(\frac{i\omega j}{a}\right)\right]_{t=0}$$

Thus

$$\beta_{jk} = \frac{i(-a)^{ij/a}}{4\pi\sqrt{jk}} \int_0^\infty dx\, \{(-x)^{ij/a}\, e^{ikx}\, (-ik) - (-x)^{ij/a-1}\, e^{ikx}\, (ij/a)\}$$

$$= \frac{-(-a)^{ij/a}}{4\pi} \int_0^\infty dx\; e^{ikx} \left[\frac{1}{x}\left(\frac{j}{k}\right)^{1/2} \frac{1}{a} - \left(\frac{k}{j}\right)^{1/2}\right] (-x)^{ij/a}$$

This Bogoliubov transformation will convert a ${}^{R}u$ into ${}^{M}u$ modes, on $-\infty < x < \infty$, while a ${}^{M}u$ mode will be down converted to a ${}^{R}u$ mode on its domain $0 < x < \infty$.

Thus, mode-by-mode, we can convert ${}^{R}u$ to ${}^{M}u$. So a $|0_R\rangle$ state can be expressed as a Minkowski state, and $\langle 0_M|a_R^\dagger a_R|0_M\rangle$ has a thermal spectrum.

The regions of Minkowski spacetime complementing $T > |X|$and $T > |X|$("time-reversed" K_{100}) are the Rindler wedges. Since K_{100} and Rindler spacetime (henceforth $\mathcal{R}$) have trivial coordinate relations with Minkowski spacetime in the $Y - Z$plane, those two dimensions are dropped in the remaining analysis. The Rindler wedges, in the $X - T$plane, are covered by coordinates $\eta \in (-\infty, +\infty)$ $\varepsilon \in (-\infty, +\infty)$,and a discrete coordinate Q $= \pm 1$ which distinguishes the right and left wedges. These are related to the Minkowski coordinates via

$$T = g^{-1} e^{g\xi} \sinh(g\eta)$$
$$X = Qg^{-1} e^{g\xi} \cosh(g\eta)$$

with g a constant (acceleration). The Rindler metric is thus

$$ds^2 = e^{2g\xi}(-d\eta^2 + d\xi^2)$$

in either wedge.

The Minkowski spacetime and its sections are maximally symmetric, i.e. they have ten Killing vector fields [14], Restricted to the $X - T$plane, the three independent Killing vector fields are the time translation

$$\frac{\partial}{\partial T} = \begin{cases} e^{-gt}\left(\cosh(gx)\frac{\partial}{\partial t} - \sinh(gx)\frac{\partial}{\partial x}\right) & in\ K_{100,} \\ e^{-g\varepsilon}\left(\cosh(g\eta)\frac{\partial}{\partial \eta} - \sinh(g\eta)\frac{\partial}{\partial \xi}\right) & in\ \mathcal{R}, \end{cases}$$

The spatial translation

$$\frac{\partial}{\partial X} = \begin{cases} e^{-gt}\left(\cosh(gx)\frac{\partial}{\partial x} - \sinh(gx)\frac{\partial}{\partial t}\right) & in\ K_{100,} \\ Qe^{-g\xi}\left(\cosh(g\eta)\frac{\partial}{\partial \xi} - \sinh(g\eta)\frac{\partial}{\partial \eta}\right) & in\ \mathcal{R}, \end{cases}$$

and

$$X\frac{\partial}{\partial T} + T\frac{\partial}{\partial X} = \begin{cases} \frac{\partial}{\partial x} & in\ K_{100,} \\ \frac{\partial}{\partial \eta} & in\ \mathcal{R}, \end{cases}$$

A boost in the X-direction

7.7 Physical Distinction among Alternative Vacuum States in Flat Spacetime

Even in flat spacetime, the states of a quantized field can be described via a variety of inequivalent Fock-space representations, associated with different congruences of inertial or non-inertial observers. But it appears possible to distinguish among the possibilities on physical grounds: field positive- and negative-frequency eigenfunctions might be required to be well-defined and regular throughout the spacetime, so that the states can be attained by evolution from regular data in the remote past. This criterion distinguishes the familiar Minkowski-coordinate construction from that corresponding to the diverging congruence of observers whose world lines trace out a degenerate-Kasner subspace of Minkowski spacetime, for example. It also draws a physical distinction between the Minkowski-coordinate Fock-space states and those associated with a congruence of uniformly accelerated observers (Rindler observers), since the latter states cannot be represented as any combinations of the former. This analysis of alternative descriptions of a quantized field may extend to more general classes of observers, and to more general spacetime geometries as well.

7.7.1 Introduction

The Fock space of states central to ordinary flat-spacetime quantum field theory is determined by the global timelike Killing vector field present in Minkowski space. In more general spacetimes there are inequivalent vacua leading to different Fock space representations for a quantum field. Indeed, such Fock representations appear even in at Minkowski spacetime: they are associated with congruences of observers other than those which delineate the familiar Minkowski coordinates. The problems of defining vacuum states, Fock spaces, and particles have occupied researchers in curved-spacetime quantum field theory since the inception of the subject [73]. The significance of these problems even in at spacetime was highlighted in the seminal work of Fulling [74], Hawking [8], Davies [75], and Unruh [10], since elaborated upon by many authors [76].

In flat spacetime, however, it appears possible to select on physical grounds a particular vacuum state and Fock space from among those which can be defined by various congruences of observers. This might be done by requiring field positive and negative-frequency eigenfunctions to be well-defined and regular (i.e. at least C2) throughout the spacetime, so that the states can be said to evolve from regular data in the distant past. Note that this criterion is applied to the positive frequency eigenfunctions obtained directly from the defining Lie-derivative eigenvalue equation. Superpositions of these fundamental positive-frequency modes are still regarded as positive frequency functions but need not satisfy the regularity criterion (just as for Fourier series). As we show in detail in this work, such a regularity criterion suffices, for example, to distinguish the field-theory description constructed by observers tracing out a degenerate-Kasner [77,78] subspace of Minkowski spacetime from that obtained in Minkowski coordinates. It also clarifies the physical distinction between the vacuum/particle definitions associated with uniformly accelerated (Rindler) observers and those of inertial observers.

There are three common approaches to the construction of Fock-space representations for quantized-field states in the general case: the C*-algebra, complex structure, and \positive/negative-frequency splitting" methods [79]. The C*-algebra 126 and complex-structure approaches afford great generality, but require more elaborate formalism than is called for here. Instead we shall rely on the more familiar positive/negative-frequency splitting method. This gives rise to a Fock-space representation by using a timelike vector field and a foliation of spacetime. On each folium positive- and negative-frequency field modes are defined via a Lie-derivative eigenvalue equation, the Lie-derivative being taken along the given vector field and evaluated at the folium. (This is a well-known and straightforward generalization of the Minkowski-coordinate time derivative.) The field decomposed into these modes on any given folium defines creation and annihilation operators with which a vacuum and Fock-space states are constructed as usual. The Fock spaces obtained by implementing this procedure on different hypersurfaces of a given foliation in general need not be unitarily equivalent. However, if the congruence of observers moves along the integral curves of a Killing vector field, then there is a single vacuum and Fock space associated with the hypersurfaces of the foliation. As the well-known example of Fulling shows [74], in a region of spacetime with two linearly independent Killing vector fields, the vacua and Fock spaces associated with these two Killing vector fields may be unitarily inequivalent. The set of Fock spaces thus obtained is intimately connected with the vector field and foliation, i.e., the congruence of observers involved. The inequivalence of these representations for different congruences of observers constitutes an ambiguity in vacuum and particle definitions.

The observer congruences associated with the degenerate-Kasner [78,79] and the Rindler [80] subspaces of Minkowski spacetime are convenient illustrations of the use of our physical criterion to clarify the ambiguity. The former consists of all observers moving with uniform speeds in a single spatial direction, their trajectories filling the forward light cones of all points of the plane perpendicular to their motion. (These Kasner observers do not move along integral curves of a Killing vector field.) The latter consists of the well-known congruence of observers moving with constant acceleration in a single spatial direction; their trajectories fill the 127 region (\Rindler wedges") outside the light cones of all points of a plane perpendicular to their direction of motion. The Killing vector field integral curves followed by the Rindler observer trajectories make them a favorite choice for modelling non-inertial observers.

Our analysis shows that while the Kasner observers can generate a Fock space representation of a quantized field with vacuum different from the familiar Minkowski vacuum, our physical criterion will single out the Minkowski vacuum in the degenerate-Kasner subspace, just as it does for a field theory on the full Minkowski spacetime. Likewise we find clear distinctions between the Fulling vacuum state obtained via positive/negative-frequency splitting using the Rindler congruence, and the ordinary Minkowski vacuum state. For example, while the Minkowski vacuum can be described as an excited state in the Fulling-Rindler Fock space (exhibiting the famous \acceleration radiation"), the Fulling vacuum cannot be represented as any state in the inertial observers' (Minkowski) Fock space. But the Fulling vacuum and associated states are based upon positive frequency field functions which are singular on the Rindler horizon bifurcation event, and have discontinuous first derivatives at the horizontal regular events in the spacetime. Hence our criterion selects the inertial Fock space as the appropriate physical framework for the field theory, even when accelerated or more general observers are involved. (Of course the Fulling vacuum and associated states remain of physical significance, e.g., in spacetimes with boundaries imposed by accelerating mirrors [81].) Furthermore, our criterion could prove useful in more general spacetimes, where no timelike Killing vector field exists (a related criterion was considered in [82]).

The requirement that states be obtainable from regular data in the remote past differentiates the Minkowski vacuum and Fock-space structure from the ostensibly similar Rindler-spacetime constructions. The Rindler positive-frequency functions have oscillatory singular behavior on the past and future Rindler horizons, and singular behavior at the horizon bifurcation event. The Rindler-observer congruence cannot be extended beyond the horizons and remain both timelike and Killing. At the horizons physical quantities such as stress-energy expectation values, in Fulling-Rindler Fock-space states, are divergent [81]. But these are ordinary points of the M4 spacetime. Hence regular evolution from regular data in the distant past could not be expected to put a field into these states. The Minkowski vacuum and Fock-space states do not suffer from these drawbacks, and may thus be distinguished as the physical Hilbert space of the quantized field. Hence, for example, in treating the interaction between a field and an observer on an arbitrary world line, it would be appropriate to represent the state of the field using the Fock space constructed with a congruence of inertial (Minkowski) observers.

These results show that the familiar Minkowski-coordinate Fock-space representation of the states of a quantized field stands out on simple physical grounds from the diverse array of alternative representations which can be constructed even in at spacetime. The Minkowski-coordinate timelike Killing vector field (time translation) used to define positive- and negative-frequency normal modes extends over the entire

M4 manifold without singularity; the resulting normal-mode functions are regular on the entire spacetime; renormalized stress-energy expectation values in the associated Fock-space states likewise exhibit no singularities anywhere on M4. Hence these states of the quantized field can be obtained by evolution from regular data in the remote past. Alternative constructions of vacuum states and Fock spaces may fail to satisfy one or more of these criteria.

The degenerate-Kasner geometry traced out by the conformal-Killing observers, moving with all possible velocities perpendicular to a fixed spacelike plane, admits a variety of choices of vacuum state and positive-frequency modes for a quantized field. Inequivalent choices can be made at different \times," giving rise to particle production [83]." But a unique choice can be made corresponding to the Minkowski-coordinate construction carried out on the full spacetime. Then no particles are produced, and the theory is at least for massive fields exactly equivalent to the familiar Minkowski-space theory.

The Fulling-Rindler vacuum and Fock space associated with a congruence of uniformly accelerated observers is a well-known example of an alternative representation of a quantized field. But these field states exhibit a variety of pathologies: they are unitarily inequivalent to the Minkowski-coordinate Fock space; they cannot be represented as countably finite combinations of Minkowski states; and they give rise to singularities in many operator expectation values. In particular, since these states give rise to stress-energy divergences on the light cones which are the horizons of the Rindler coordinates, they cannot arise via regular evolution from the remote past. Hence it is the Minkowski Fock-space states, rather than these, which should represent the physical states of the field in interaction with observers on arbitrary world lines.

In general spacetimes, lacking the extensive isometry structure of M4, choosing an appropriate vacuum state and Fock space from the infinity of possibilities is problematic. But the criteria considered here of extensibility over the entire spacetime, of nonsingularity on the entire spacetime (except, of course, at actual physical singularities of the geometry), of evolution from the remote past (or initial singularity), may prove useful in these more general circumstances as well.

7.7.2 Functional Schrodinger Approach

We consider a real scalar field of mass *m* with Lagrangian density

$$\mathcal{L}(x) = -\frac{\sqrt{-g}}{2}\left(g^{\mu\nu}\partial_\mu\varphi\partial_{\nu\varphi} + [m^2 + \xi\mathcal{R}]\varphi^2\right)$$

where $\mathcal{R}$ is the Ricci scalar and ξ the curvature-coupling constant. We employ both the Heisenberg and Schrôdinger pictures. To establish notation we begin with a description of the Fock space representation in the Heisenberg picture and then describe the functional-Schrödinger approach in the Schrödinger picture. For flat spacetime $\mathcal{R} = 0$, and the field equation corresponding to the above Lagrangian is the familiar Klein-Gordon equation. The conserved inner product is given by

$$(\varphi_{1,}\varphi_{2,}) = -i\int_\Sigma \varphi_1^{\overleftrightarrow{\partial}}\mu^{\varphi_2}{}^*\sqrt{-g_\Sigma}d\Sigma^\mu,$$

where $d\Sigma^\mu$ is the surface element on the spacelike hypersurface Σ. For the Heisenberg picture a complete set of orthonormal mode solutions, denoted $ui(x)$, are used to describe φ:

$$\varphi^{(x)} = \sum_q [a_q u_q + a_q^+ u_q^*].$$

Canonical quantization of the theory is implemented by imposing the usual commutation relations, $\left[a_{q\,,a_{q'}^+}\right] = \delta_{qq'}$ etc. The vacuum is defined by $a_q|0\rangle = 0$ for all q .A different complete set of solutions, distinguished by overbars, likewise yields

$$\varphi(x) = \sum_k [\bar{a}_k\bar{u}_k + \bar{a}_k^+\bar{u}_k^*],$$

for which the new vacuum is defined by $\bar{a}_k|\bar{0}\rangle = 0$. Since the sets of orthonormal modes are complete they can be expressed in terms of one another:

$$\bar{u}_k = \sum_k (\alpha_{kq}u_q + \beta_{kq}u_q^*),$$

$$u_q = \sum_k (\alpha_{kq}^*\bar{u}_k - \beta_{kq}\bar{u}_k^*).$$

The Bogoliubov coefficients relating the mode functions are

$$\alpha_{q^k} = (\bar{u}_q, u_k) \; and \; \beta_{q^k} = -(\bar{u}_q, u_k^*)\,,$$

and the Bogoliubov transformations for the creation and annihilation operators are

$$a_q = \sum_k (\alpha_{kq}\bar{a}_k + \beta_{kq}^*\bar{a}_k^+)$$

$$\bar{a}_k = \sum_q (\alpha_{kq}^* a_q + \beta_{kq}^*\bar{a}_k^+),$$

and their Hermitian conjugates.

In the functional-Schrôdinger approach [84] wave functionals are used to represent quantum states of the field. The functionals give probability amplitudes for the values of a complete set of commuting observables, one such set being the values of the field on a spacelike hypersurface (known as the "field-coordinate" representation). The dynamics in the functional Schrodinger approach is implemented on a specified foliation of the spacetime. Given a foliation, labeled by s, the evolution of the wave functional between hypersurfaces is described by

$$\left(H(s) - i\tfrac{\partial}{\partial s}\right)\Psi[\varphi(\Sigma), s] = 0$$

the functional Schrödinger equation. The Hamiltonian $H(s)$ is obtained by integration of the field energy density over the hypersurface $\Sigma(s)$. The *canonical* stress-energy tensor is used in the integration, since this leads to the canonical Hamiltonian [84]. (If the gravitational stress-energy tensor which appears in the Einstein field equations were used, the canonical commutation relations would be satisfied only in the instance of minimal coupling, i.e., with curvature coupling $\xi = 0$.) In the gauge = $g_{oi} = g^{oi} = 0$ the parameter s may be identified with a time coordinate x^0. The appropriate form of $H(x^0)$ as a functional differential operator then follows upon the replacement

$$\nabla_{0\varphi} \equiv \pi = i g_{00}\sqrt{-g}\frac{\delta}{\vec{x}};$$

this implements the canonical commutation relation between the conjugate momentum π and the field φ in the field-coordinate representation. Finally this yields

$$H(x^\circ) = \int_{\Sigma x^\circ} T_{00} g^{00}\sqrt{-g}\, d^3$$

$$= \frac{1}{2}\int_{\Sigma(x^\circ)} g^{00}\sqrt{-g}\left\{-\left(\frac{g_{00}}{\sqrt{-g}}\right)^2 \frac{\delta^2}{\delta\varphi\vec{x}^2} - g_{00}\left[g^{ij}\nabla_{i\varphi}\nabla_{i\varphi} - (m^2 + \xi R)\varphi^2\right]\right\} d^3x$$

for the Hamiltonian operator.

For a massless scalar field in (1 + 1)-dimensional Minkowski space, with a foliation by constant-T slices, this Hamiltonian is very simple:

$$H(T) = \frac{1}{2}\int_{-\infty}^{\infty} dX\left(-\frac{\delta^2}{\delta_\varphi(X)^2} + (\nabla\varphi)^2\right)$$

Since the hypersurface-orthogonal vector field given by $\partial/\partial T$ is Killing, the Hamiltonian is independent of *T*. For this foliation the arguments of the wave functionals are the field values$\varphi(X, T = constant)$. But it proves simpler here to express the theory in terms of an alternative set of observables, i.e., m terms of a "spatial mode" expansion of the field values. The "spatial modes" are eigenfunctions of the spatial part of the Klein-Gordon operator. The field configurations are thus expanded, e.g.,

$$\varphi(X, T = 0) = \pi^{-1/2}\int_0^\infty dk\left[Y_k^{(+)}\cos(kX)\right]$$

The inverse of this transformation is

$$Y_k^{(+)} = \pi^{-1/2}\int_{-\infty}^{+\infty} \varphi(X)\cos(kX)dX$$

$$Y_k^{(-)} = \pi^{-1/2}\int_{-\infty}^{+\infty} \varphi(X)\sin(kX)dX$$

For convenience we use real mode functions, so that all quantities are real. The index $\sigma = \pm$ on the $Y_k^{(\sigma)}$identifies modes symmetric and antisymmetric about $X = 0$. In terms of the amplitudes $Y_k^{(\sigma)}$ then, Hamiltonian (2.22) takes the form

$$H_M \frac{1}{2}\int_0^\infty dk \sum_{\sigma=\pm}\left(-\frac{\delta^2}{\delta Y_k^{(\sigma)2}} + k^2 Y_k^{(\sigma)^2}\right),$$

a sum of simple-harmonic-oscillator Hamiltonians. The annihilation and creation operators for these oscillators are given by

$$a_k^{(\sigma)} = \left(\frac{k}{2}\right)^{1/2} K_k^{(\sigma)} + \left(\frac{1}{2k}\right)^{1/2}\frac{\delta}{\delta Y_k^{(\sigma)}}$$

and its Hermitian conjugate. The wave functional for the Minkowski vacuum state, in terms of the$Y_k^{\pm}$, is just the product of the oscillators' ground-state wavefunctions:

$$\Psi_0^{(M)}[\{Y_k^\sigma\}, T] = \aleph(T)\exp\left[-\frac{1}{2}\int_0^\infty dk\, k\left(Y_k^{(+)^2} + Y_k^{(-)^2}\right)\right]$$

where $\aleph(T) = \prod_k \sigma^{(k/\pi)^{1/4}} e^{-ikT/2}$ consists of a normalization constant and an infinite time-dependent phase, corresponding to the zero-point energy in each mode. We can write

$$\varphi = \sum_{k=-\infty}^{\infty}\left(a_k u_{L,k} + a_k^\dagger u_{L,k}^* + b_{kR,k}^u + b_{kuR^*,k}^\dagger\right) \qquad (5.35)$$

(The discrete ``box`` description will be converted to a continuum description as needed.) Analyticity properties (in the Rindler coordinates) imply that the combinations.

$$\bar{\bar{u}}_k^{(1)} = \left[2sinh\left(\frac{\pi w}{g}\right)\right]^{-1/2}\left(e^{-\pi w}/2g u_R^*, -K + e^{\pi w/2g} uL, k\right)$$

(5.36a)

$$\bar{\bar{u}}_k^{(2)} = \left[2sinh\left(\frac{\pi w}{g}\right)\right]^{-1/2}\left(e^{\pi w}/2g u_R^*, k + e^{-\pi w/2g} u_L^*, -k\right)$$

(5.36B)

Are positive-frequency functions in the inertial (Minkowski) description, i.e these are superpositions of Minkowski-spacetime positive-frequency solutions. The field can also be expanded in terms of $\bar{u}_k^{(1)}$ *and* $\bar{u}_k^{(2)}$:

$$\varphi \sum_{k=-\infty}^{\infty}\left(\bar{a}_k \bar{u}_k^{(1)} + \bar{b}_{k\bar{u}_k^{(2)} + \bar{a}_k^\dagger\left(\bar{u}_k^{(1)}\right)}^{\;*} + \bar{b}_k^\dagger\left(\bar{u}_k^{(2)}\right)\right)$$

(5.37)

The inner products $(\varphi, u_{R,k})$ and $((\varphi, u_{L,k}))$ give the relations

$$a_k = \left[2sinh\left(\frac{\pi w}{g}\right)\right]^{-1/2}\left(e^{-\pi w}/2g \bar{a}_k, -K + e^{\pi w/2g} \bar{b}^\dagger_{-k}\right)$$

(5.38a)

$$b_k = \left[2sinh\left(\frac{\pi w}{g}\right)\right]^{-1/2}\left(e^{-\pi w}/2g \bar{b}_k, -K + e^{\pi w/2g} \bar{a}^\dagger_{-k}\right)$$

(5.38b)

Here the conditions a_k I0›=0, for all *k,* define the fulling-Rindler vacuum state 0› and the conditions $\bar{a}_k$I$\bar{0}$› = $\bar{b}_k$I$\bar{0}$› = 0 define the Minkowski vacuum I$\bar{0}$›.

The Boguliubov transformation (4.4) can be written

$$\bar{a}_{k=a_k} cosh(\theta_k) - b^\dagger_{-k} sinh(\theta_k)$$

(5.39a)

$$\bar{b}_{k=b_k} cosh(\theta_k) - a^\dagger_{-k} sinh(\theta_k)$$

(5.39b)

With θ_k define via

$$cosh\theta_k = \left(1 - e^{-2\pi w/g}\right)^{-1/2}$$

(5.39c)

This can be represented formally as the operator relations

$$\bar{a}_k = G^{-1} a_k G(0)$$

(5.40a)

$$\bar{b}_k = G^{-1} b_k G(0)$$

(5.40b)

Where

$$G(0) = exp\left[\sum_k \theta_k\left(a_k{}^b - k - b^{\dagger}{}_{-k}a_k^{\dagger}\right)\right]$$

(5.40c)

Is formally the unitary transformation between Fulling-Rindler and Minkowski Fock-space bases. But because this definition infinitely many terms G may not actually exist as a well-defined operator on the Fock space built from operations of $a^{\dagger,}s$ on I0›. While the groupings $G^{-1}bG$ exist, the same cannot be assumed for G. again formally, G should effect the transformation.

$$\bar{0}\rangle = G^{-1}(0)I0\rangle = \exp\left\{\sum_k\left[-In\,(cosh\theta_k) + tanh\theta_k\left(a_k^{\dagger}b_k^{\dagger}\right)\right]I0\rangle.\right.$$

(5.41)

If the $\sum_k$ In cosh (θ_k) terms in (4.7) diverges to +∞, this implies $\langle 0I\bar{0}\rangle = 0$,i.e., the two vacua are orthogonal. In such an instance G(0) would not exist as an operator on the Minkowski Fock space. Since the field theory considered here is actually constructed on an unbounded domain, a continuum limit on the sum over k introduces an infinite spatial-volume factor that multiplies the integral of in cosh (θ_k), and the sum is indeed infinite. What this indicates is that the formal transformation G is not actually a unitary operator- a situation frequently encountered in systems with an infinite number in systems with an infinite number of degrees of freedom [85].

The inequivalence of the Fock spaces associated with the Rindler and Minkowski observes is further illustrated by considering explicit wave functionals for the states. A functional-Schrodinger treatment in Rindler spacetime is formally similar to that in Minkowski spacetime. Here the dynamical variables are the values of φ on the constant n. Equivalent variables are the ``spatial-mode`` amplitudes in the expression.

$$\varphi(\xi, Q, \eta = 0) = \pi^{-1/2}\int_0^{\infty}\left\{\left[u_q^{(+)}cos(q\xi) + v_q^{(+)}sin(q\xi)\right]\chi^{(+)}(Q)\right.$$
$$+\left\{\left[u_q^{(-)}cos(q\xi) + v_q^{(-)}sin(q\xi)\right]\chi^{(-)}(Q)\right.$$

Viz.,

$$u_q^{(+)} = \pi^{-1/2}\sum_{Q=\pm}\int_{-\infty}^{+\infty} d\xi\, cos(q\xi)\chi^{(\pm)}\varphi(\xi, Q)$$

(5.43a)

$$u_q^{(+)} = \pi^{-1/2}\sum_{Q=\pm}\int_{-\infty}^{+\infty} d\xi\, sin(q\xi)\chi^{(\pm)}\varphi(\xi, Q)$$

(5.43b)

With

$$\chi^{(\pm)}(Q) \equiv \begin{cases} +\frac{1}{\sqrt{2}}, if\ Q = +1; \\ +\frac{1}{\sqrt{2}}, if\ Q = -1. \end{cases}$$

(5.43c)

Here, as in sec5.2, the $\pm$ index identifies left-right symmetric and antisymmetric modes. The Hamiltonian takes the form

$$H_R = \frac{1}{2}\int_0^\infty dq \sum_{\sigma=\pm}\left(-\frac{\delta^2}{\delta u_q{}^{(\sigma)2}} + q^2 u_q{}^{(\sigma)2} - \frac{\delta^2}{\delta u_q{}^{(0)2}}\right).$$

(5.44)

This is a different operator than the H_M of equation (2.25), generating evolution in η rather than *t;* it is a sum of Hamiltonian for a different set of harmonic oscillators than H_M. Rindler annihilation and creation operators can be written

$$b_q^{(0)} = \left(\frac{q}{2}\right)^{1/2} u_q^{(\sigma)} + \left(\frac{1}{2q}\right)^{1/2} \frac{\delta}{\delta u_q{}^{(\sigma)}} \quad (5.45)$$

$$c_q^{(0)} = \left(\frac{q}{2}\right)^{1/2} u_q^{(\sigma)} + \left(\frac{1}{2q}\right)^{1/2} \frac{\delta}{\delta u_q{}^{(\sigma)}} \quad (5.45b)$$

And their Hermitian conjugates. The wave functional for the associated Fulling-Rindler vacuum state is

$$\psi_0{}^{(R)}\left[\left\{u_q^{(\sigma)}, v_q^{(\sigma)}, n\right]\right. = \widetilde{\mathcal{N}}(T) of\ Eq. (2.27)$$

Transformation between the Minkowski and Ridler and amplitudes $\left\{Y_k^{(\sigma)}\right\}$and $\left\{u_q^{(\sigma)}, v_q^{(\sigma)}\right\}$ takes place on the T=n=0 hypersurface , i.e,line; only there do two sets of amplitudes correspond t the same set of field vlues. Transformation in either direction is affected by substituting ether of the expansions. (4.8a) Or (2.23) into the inverse of the other, yielding

$$Y_k^{(+)} = \frac{\sqrt{2}}{\pi k}\int_0^\infty dq \sin\left(\frac{\pi q}{2g}\right)\left\{u_g^{(+)} \xi\left[\Gamma\left(1+\frac{iq}{g}\right)\left(\frac{k}{g}\right)^{-iq/g}\right]\right.$$
$$\left. -v_q^{(+)} \Re\left[\Gamma\left(1+\frac{iq}{g}\right)\left(\frac{k}{g}\right)^{-iq/g}\right]\text{-}\right\}$$

(5.47a)

$$Y_k^{(-)} = \frac{\sqrt{2}}{\pi k}\int_0^\infty dq\ cosh\left(\frac{\pi q}{2g}\right)\left\{u_q^{(-)} \Re\left[\Gamma\left(1+\frac{iq}{g}\right)\left(\frac{k}{g}\right)^{iq/g}\right] + v_q^{(-)} \Im\left[\Gamma\left(1\right.\right.\right.$$
$$\left.\left.\left. +\frac{iq}{g}\right)\left(\frac{k}{g}\right)^{-iq/g}\right]\right\}$$

(5.47b)

And

$$u_q^{(+)} = \frac{\sqrt{2}}{\pi g} cosh\left(\frac{\pi g}{2g}\right) \int_0^\infty dk Y_k^{(+)} \Re\left[\left(\frac{g}{k}\right)^{iq/g} \Gamma(iq/g)\right] \tag{5.48a}$$

$$u_q^{(+)} = \frac{\sqrt{2}}{\pi g} cosh\left(\frac{\pi g}{2g}\right) \int_0^\infty dk Y_k^{(+)} \Re\left[\left(\frac{g}{k}\right)^{iq/g} \Gamma(iq/g)\right] \tag{5.48b}$$

$$u_q^{(-)} = \frac{-\sqrt{2}}{\pi g} sinh\left(\frac{\pi g}{2g}\right) \int_0^\infty dk Y_k^{(-)} \Re\left[\left(\frac{g}{k}\right)^{iq/g} \Gamma(iq/g)\right] \tag{5.48c}$$

$$u_q^{(-)} = \frac{\sqrt{2}}{\pi g} cosh\left(\frac{\pi g}{2g}\right) \int_0^\infty dk Y_k^{(-)} \Re\left[\left(\frac{g}{k}\right)^{iq/g} \Gamma(iq/g)\right] \tag{5.48d}$$

Bogoliubov transformations akin to those of equations (4.4) and (4.5) follow from these relations, forms (2.26) and (4.10), and the chain rule, e.g.,

$$b_q^{(+)} = \int_0^\infty \left(\frac{q}{2k}\right)^{1/2} \frac{1}{2} \Re\left[\left(\frac{k}{g}\right)^{-iq/g} \Gamma\left(\frac{iq}{g}\right)\right] \left[exp\left(\frac{\pi q}{2g}\right) a_k^{(+)} + exp\left(-\frac{\pi q}{g}\right) a_k^{(+)\dagger}\right] dk, \tag{5.50}$$

And

$$a_k^{(+)} = \int_0^\infty \left(\frac{q}{2k}\right)^{1/2} \frac{1}{2} \{\Re\left[\left(\frac{k}{g}\right)^{-iq/g} \Gamma\left(\frac{iq}{g}\right)\right] \left[exp\left(\frac{\pi q}{2g}\right) b_q^{(+)} - exp\left(-\frac{\pi q}{g}\right) b_q^{(+)\dagger}\right]$$

$$+\Im\left[\left(\frac{k}{g}\right)^{-iq/g} \Gamma\left(\frac{iq}{g}\right)\right]\left[exp\left(\frac{\pi q}{2g}\right) c_q^{(+)} - exp\left(-\frac{\pi q}{g}\right) c_q^{(+)\dagger}\right]\} dq$$

The Minkowski vacuum $|\bar{0}\rangle$ can be described by a wave functional in the Rindler-spacetime formulaion, but not vice versa, i.e., the Rindler vacuum $|0\rangle$ cannot be described by a well defined wave functional in the Minkowski-spacetime formulation. The wave functional (2.27) and transformations (4.12) yield.

$$\psi_0^{(M)}\left[\left\{u_q^{(0)}, v_q^{(0)}\right\}, 0\right] = \mathcal{N}(0) exp\left\{-\frac{1}{2}\int_0^\infty dg \left[q\, tanh\left(\frac{\pi q}{2g}\right)\left(u_q^{(+)2} + v_q^{(+)2}\right) + q coth\left(\frac{\pi q}{2g}\right)\left(u_q^{(+)2} + v_q^{(+)2}\right)\right\} \tag{5.51}$$

The wave functional at nonzero ηcan be obtained , e.g., by applying to this the propagator corresponding to the Hamiltonian H_R. of course this is a different state than the Fulling-Rindler vacuum with wave functional (4.11): this state is ``squeezed`` relative to that, in the (-) modes, 11anti-squeezed`` in the (+) modes. But no corresponding description of the

Fulling-Rindler vacuum in the Minkowski formulation is even possible. Transformations (4.12b) applied to wave (4.11) yield

$$\psi_0^{(R)}\left[\left\{Y_k^{(\sigma)}, v_q^{(0)}\right\}, 0\right] =$$

$$\bar{\mathcal{N}}(0) exp\left(-\frac{1}{2\pi}\int_0^\infty \int_0^\infty dk\, dk`\left\{Y_k^{(+)} Y_{k`}^{(+)} \int_0^\infty coth\left(\frac{\pi z}{2}\right) cos\left[z\, ln\left(\frac{k}{k`}\right)\right] dz\right.\right.$$

$$\left.\left.+Y_k^{(-)} + Y_{k`}^{(-)} \int_0^\infty tanh\left(\frac{\pi z}{2}\right) cos\left[z\, ln\left(\frac{k}{k`}\right)\right] dz\right\}\right), \tag{5.52}$$

Where z is just q/g. Clearly this is no simple counterpart of Eq. (4.14). indeed, since the first integral over z is divergent, this transformed wave functional wave is not even well-defined. The Fulfilling-Rindler vacuum cannot be expressed in terms of Minkowski modes, hence not as any convergent combination of Minkowski Fock-space states.

The ``particle numbers`` in these vacua reflect the same asymmetry between the two formulations. The ``Rindler particle number`` of the Minkowski vacuum is the expectation value $\left\langle \bar{0} \middle| b_q^{(\sigma)} \dagger b_{q`}^{(\sigma)} \mathrm{I}\bar{0} \right\rangle$. This can be evaluated using the Bogoliubov transformations (4.13), and only terms with expectation values of the form $\left\langle \bar{0} \middle| a_k^{(\sigma)} \dagger a_{k`}^{(\sigma)\dagger} \mathrm{I}\bar{0} \right\rangle$. Contribute. Thus, for example, one obtains

$$\left\langle \bar{0} \middle| b_q^{(+)} \dagger b_{q`}^{(+)} \mathrm{I}\bar{0} \right\rangle = \int_0^\infty \frac{dk}{k} \frac{(qq`)^{1/2}}{4\pi^2 g^2} exp\left(-\frac{\pi(q+q`)}{2q}\right)$$

$$\mathrm{x}\left\{\Re\left[\left(\frac{k}{g}\right)^{-i(q+q`)/q} \Gamma\left(\frac{iq}{g}\right)\Gamma\left(\frac{iq}{g}\right)\Gamma\left(\frac{iq`}{g}\right)\right] + \Re\left[\left(\frac{k}{g}\right)^{i(q-g)/g} \Gamma\left(\frac{-iq}{g}\right)\Gamma\left(\frac{iq`}{g}\right)\right]\right\} \tag{5.53}$$

the integration over k can be performed, yielding

$$\left\langle \bar{0} \middle| b_q^{(+)} \dagger b_{q`}^{(+)} \mathrm{I}\bar{0} \right\rangle$$

$$= \frac{(qq`)^{1/2}}{2\pi g} exp\left(-\frac{\pi(q+q`)}{2\pi}\right) \mathrm{I}\Gamma\left(\frac{iq}{g}\right)\mathrm{I}^2[\delta(q+q`)+\delta(q-q`)]$$

$$= \frac{1}{e^{2kq/g}-1}\delta(q-q`),$$

Where the $\delta(q+q`)$term is discarded since q and q` range over nonnegative values. This is the famous ``thermal`` practice number, with ``temperature`` $k_B\Theta = G/(2\pi)$, the state $\mathrm{I}\bar{0}\rangle$ remaining however a pure quantum state. The ``Minkowski particle number`` of the Fulling-Rindler vacuum is a different matter entirely. Evaluating $\left\langle 0\middle| a_k^{(\sigma)\dagger} a_{k`}^{(\sigma)} |0\right\rangle$, using transformations (4.13b), one obtains in like manner

$$\left\langle 0\middle| a_k^{(\pm)\dagger} a_{k`}^{(\pm)} |\bar{0}\right\rangle = \frac{1}{\pi(kk`)^{1/2}} \int_0^\infty \frac{1}{e^{2\pi q/g-1}} cos \left[\frac{q}{g} log\left(\frac{k}{k`}\right)\right] \frac{dq}{g} \tag{5.55}$$

The integral over q is divergent for any values of k and k`. moreover, there results no well-defined ``diagonal`` form in k and k`, making an interpretation as a ``particle number`` problematic.

The inequivalence or asymmetry between the two vacuum states accords with the notion of the Minkowski vacuum as a ``minimum-energy`` state. Since it contains excitations (4.17), state $|\bar{0}\rangle$, making the latter a state of negative energy with respect to the Minkowski vacuum. No such state can be constructed within the Minkowski Fock space, at least without introducing singularities.

Chapter 8 QFT on CST

8.1 Overview

In this chapter we briefly consider quantum field theory on curved space-time. The first step n generalizing to the curved space-time setting is to redo the quantum mechanics and quantum field theory structures used in a covariant form. This is done in Section 8.3. In Section 8.3 another preparatory step is made – the introduction of useful space-time transforms and the compact diagrams (Penrose Diagram's) that result. A variety of topics are then briefly described with the covariant tools obtained, ranging from Covariant Quantization in Section 8.4 to Adiabatic Regularization in Section 8.12

8.2 Quantum Mechanics in Covariant Formulation

Classical Canonical Formulation

Given a Lagrangian $L(q, \dot{q})$ we have Action S and Hamiltonian $H(q, p)$:

$$S = \int_{t_1}^{t_2} dt L(q, \dot{q}), \quad H(q,p) = \sum_i p_i \dot{q}_i \; - L$$

where

$$p_i = \frac{\partial L}{\partial \dot{q}_i}.$$

Canonical Quantization

In canonical quantization the $q's$ and $p's$ become Hermitian operators acting in a Hilbert space and the $q's$ and $p's$ obey the canonical communication relations (with corresponding operator algebra revealed):

$$[q_i, q_j] = 0\,, [p_i, p_j] = 0\,, [q_i, p_j] = i\delta_{i,j}\,.$$

Note that a function of operators like $F(q, p)$, such as the Hamiltonian, now has an order dependency on $\{q, p\}$ and satisfies:

$$[q_i, F] = i\,\frac{\partial F}{\partial p_i}.$$

Now that the operator algebra is described, let's turn to the operator spectra. Consider the position operator q_i and recall the Dirac Bra-Ket notation for states:

$$q_i|q'\rangle = q_i|q'\rangle, \quad \langle q'|q''\rangle = \delta(q', q'')$$

and

$$\langle q'|p_i|q''\rangle = -i\frac{\partial\delta(q',q'')}{\partial q_i}, \quad \langle q'|p'\rangle = (2\pi)^{-n/2}\exp\left(i\sum p_i' q_i'\right).$$

Note that operations such as the derivative of a delta-function are well defined in distribution theory, and effectively reduce to resolution/definition in an integral context by way of integration by parts. If we have $F(q,p)$ with a definite ordering of factors:

$$\langle q'|F(q,p)|q''\rangle = F\left(q', -i\frac{\partial}{\partial q'}\right)\delta(q',q'')$$

Configuration space representation

In the configuration space representation the operators are represented by matrix elements based on the $|q'\rangle$, and the states are represented by functions. Consider a state $|\Psi\rangle$, it is represented by the Schrodinger wave function $\Psi(q') = \langle q'|\Psi\rangle$. An example being $\langle q'|p'\rangle$, which represents a particle of definite momentum. Similarly in the momentum space representation. So far the description has been kinematical, with time playing no role. The dynamical evolution of the system is governed by the Hamiltonian $H(q,p,t)$. The time-evolution may be described in several physically equivalent ways or "pictures":

Schrodinger picture

In the Schrodinger picture the fundamental observables q and p are not time dependent. Rather, the state vector (Ket) describing the system is time dependent. The fundamental dynamical equation is that of Schrodinger:

$$i\frac{d}{dt}|\Psi(t)\rangle = H(q,p,t)|\Psi(t)\rangle$$

(in general $H(t')$ and $H(t'')$ may not commute), which becomes in the configuration representation:

$$i\langle q'|d/dt|\Psi(t)\rangle = \int \langle q'|H(q,p,t)|q''\rangle\, dq''\langle q''|\Psi(t)\rangle$$

$$i\frac{\partial}{\partial t}\Psi(q',t) = H\left(q', -i\frac{\partial}{\partial q'}, t\right)\Psi(q',t)$$

i.e. Schrodinger's equation $i\frac{\partial}{\partial t}\Psi = \mathrm{H}\Psi$.

Time dependence of State Vector is a Unitary Transformation

If we consider the time dependence of the state vector as a unitary transformation (as it must be to conserve probability, etc.) we have time evolution according to:

$$|\Psi(t)\rangle = U(t,t_0)|\Psi(t_0)\rangle,$$

where to be norm preserving we must have $UU^\dagger = 1$ and also $U(t_0, t_0) = 1$. Thus

$$i\frac{d}{dt}(U(t,t_0)|\Psi(t_0)\rangle) = H(q,p,t)(U(t,t_0)|\Psi(t_0)\rangle)$$

or simply:

$$i\frac{d}{dt}U(t,t_0) = H(t)U(t,t_0)$$

Heisenberg picture

In the Heisenberg picture the state vector is time independent. The dynamical evolution of the system is expressed through the time dependence of the fundamental observable $q(t)$ and $p(t)$. Applying $U(t,t_0)^\dagger$ to the Schrodinger picture Ket we get a time independent Ket:

$$|\Psi_H\rangle = U(t,t_0)^\dagger|\Psi_s(t)\rangle = |\Psi_s(t_0)\rangle.$$

In order that measurable expectation values remain the same, the Heisenberg picture operator corresponding to the Schrodinger picture operator is:

$$F_H(t) = U(t,t_0)^\dagger F_S U(t,t_0).$$

Recall that $i\frac{\partial}{\partial t}\, U(t,t_0) = H(t)U(t,t_0)$, here $H(t) = H_s(t)$ as follows from the fact that we are in the time dependent state vector space of Schrodinger's fundamental dynamical equation. Using $i\frac{d}{dt}|\Psi(t)\rangle = H_s|\Psi(t)\rangle$ and taking the Hermitian conjugate:

$$-i\frac{d}{dt}|\Psi(t)\rangle = \langle\Psi(t)|H_s^\dagger$$

Thus

$$-i\frac{d}{dt}\langle\Psi(t_0)|U^\dagger(t,t_0) = \langle\Psi(t_0)|U^\dagger(t,t_0)H_s$$

So

$$i\frac{d}{dt}U^\dagger(t,t_0) = -U^\dagger(t,t_0)H_s$$

to give us the complementary relation. If $H_s(t)$ has no explicit time dependence, then

$$U(t,t_0) = \exp[-i(t,t_0)H]$$

and since here $[H,U] = 0$, then $H_H = H_s$.

Let's now consider an Heisenberg operator's time dependence:

$$i\frac{\partial}{\partial t}F_H = i\frac{d}{dt}\{U(t,t_0)^\dagger F_s U(t,t_0)\} + i\frac{\partial}{\partial t}F_H$$

$$= \left(i\frac{d}{dt}U(t,t_0)^\dagger\right)F_sU(t,t_0) + U(t,t_0)^\dagger\left(i\frac{d}{dt}F_s\right)U(t,t_0)$$

$$+U(t,t_0)^\dagger F_s\left(i\frac{d}{dt}U(t,t_0)\right) + i\frac{\partial}{\partial t}F_H$$

Using $i\frac{\partial}{\partial t}U(t,t_0) = H_sU(t,t_0)$ this simplifies to:

$$i\frac{d}{dt}F_H = -U^\dagger(t,t_0)H_sF_sU(t,t_0) + U(t,t_0)^\dagger F_sH_sU(t,t_0) + i\frac{\partial}{\partial t}F_H$$

or

$$i\frac{dF_H}{dt} = [F_H, H_H] + i\frac{\partial}{\partial t}F_H$$

which is Heisenberg's equation of motion.

Recall the Classical Poisson Bracket formulation for the equations of motion:

$$\frac{dF}{dt} = \{F, H\} + \frac{\partial}{\partial t}F$$

So, to transit to the quantum mechanics version we replace the bracket with $(-i)$ times the commutator. (Also note that this means that we need not define the Schrodinger picture before the Heisenberg picture as was done here as an example – the Heisenberg picture is just as fundamental in this regard.)

Interactions

In most of the quantum formulation, two systems interact through a term in the Hamiltonian that can be regarded as perturbative.

Interaction picture

In the Interaction Picture we break up the time dependence of the dynamics into that of the unperturbed Hamiltonian and that of the perturbation. The state vector evolves (as in the Schrodinger picture) but only under the influence of the interaction term in the Hamiltonian. The operators evolve (Heisenberg) under the influence of the unperturbed Hamiltonian. This picture is useful when interacting fields are considered.

Consider what canonical quantization of a system of independent real fields $\phi_a(x^\mu)$ entails: the μ index in x^μ refers to spacetime coordinate, while the index a refers to tensor or spinor indices and internal quantum numbers of the field multiplet. Consider bosons for now, and suppress

"a", and let's write a Lagrangian for $L[\phi, \partial_0\phi]$ where $\phi(x)$ analogous to $q_i(t)$, with $\vec{x}$ analogous to i (a continuous label instead of discrete). Thus

$$S[\phi] = \int_{t_1}^{t_2} dt\, L[\phi, \partial_0\phi]$$

where

$$L[\phi, \partial_0\phi] = \int dV_x\, \mathcal{L}\big(\phi(\vec{x},t), \partial\phi(\vec{x},t)\big), \quad \partial\phi = \partial_\mu\phi.$$

Note that $L[\phi, \partial_0\phi]$ has become a functional, so we now have functional variational analysis. Consider the momentum π conjugate to ϕ:

$$\pi(\vec{x},t) = \delta L[\phi, \partial_0\phi]_t / \delta\big(\partial_0\phi(\vec{x},t)\big)$$

where we now have the functional derivative. Recall that a functional derivative of $F(X)$, a functional of $X(x)$, is defined by the integral relation:

$$\delta F[X] = \int dV_x \frac{\delta F(X)}{\delta X(x)} \delta X(x)$$

where variation $\delta X(x)$ vanishes on boundary of integration. If $F(X) = \int dV_x\, f\big(X(x), \partial X(x)\big)$, then we can also write:

$$\delta F[X] = \left(\frac{\partial}{\partial X} f(X, \partial X) - \partial_\mu\left(\frac{\partial f(X, \partial X)}{\partial(\partial_\mu X)}\right)\right)\delta X(x)$$

Thus,

$$\frac{\delta F[X]}{\delta X(x)} = \frac{\partial}{\partial X} f(X, \delta X) - \partial_\mu\left(\frac{\partial f(X, \partial X)}{\partial(\partial_\mu X)}\right)$$

In particular:

$$\pi(x,t) = \frac{\delta L}{\delta\big(\partial_0\phi(\vec{x},t)\big)} = \frac{\partial L}{\partial\big(\partial_0\phi(\vec{x},t)\big)} - \partial_\mu\left(\frac{\partial \mathcal{L}}{\partial\big(\partial_\mu[\partial_0\phi(\vec{x},t)]\big)}\right)$$

where the last term is dropped because it only depends on first derivatives:

$$\pi(\vec{x},t) = \frac{\partial \mathcal{L}}{\partial\big(\partial_0\phi(\vec{x},t)\big)}$$

Also note that:

$$\frac{\delta X(\vec{x}',t)}{\delta X(\vec{x}',t)} = \frac{\partial X(\vec{x}',t)}{\partial X(\vec{x}',t)} = \delta(x', x)$$

The Lagrange (also known as the Euler-Lagrange) field equations then result upon imposing the usual boundary condition of zero variation on the boundary:

$$\partial_\mu\left(\frac{\partial\mathcal{L}}{\partial(\partial_\mu\phi)}\right)-\frac{\partial\mathcal{L}}{\partial\phi}=0$$

For the Hamiltonian we do the Lagendre Transform to phase space as before:

$$H[\phi,\pi]_t=\int dV_x\,\pi^a(\vec{x},t)\partial_0\phi_a(\vec{x},t)-L(\phi,\partial_0\phi)_t$$

(note that t-dependence is denoted by a subscript on some terms where this might not be apparent from the notational compression). We can now write:

$$\mathcal{H}=\pi^a\partial_0\phi_a-\mathcal{L}$$

Since we are working with bare fields, which by definition satisfy the field equation with unrenomalized masses (and coupling constants). We have the bare commutation relations:

$$[\phi_a(\vec{x},t),\phi_b(\vec{x}',t)]=0,[\pi^a(\vec{x},t),\pi^b(\vec{x}',t)]=0$$
$$[\phi_a(\vec{x},t),\pi^b(\vec{x}',t)]=i\delta_a^b\delta(\vec{x},\vec{x}'),$$

where there is a renormalization 'Z' factor with the delta functions in the renormalization. So, at the unrenormalized level, we have at a single time t :

$$[\phi_a(\vec{x}),F(\phi,\pi)]=i\frac{\delta}{\delta\pi^a(x)}F(\phi,\pi).$$

Continuing the parallel with the development of the quantum mechanics version, for the quantum field theory version we now consider a Schrodinger field representation using the eigenstates of $\phi(x)$ defined by:

$$\phi(\vec{x})|\phi'\rangle=\phi'(\vec{x})|\phi'\rangle$$

where $\phi'(\vec{x})$ is a c-number function (a real number function) while $\phi(\vec{x})$ denotes an operator, so the notation is starting to need a further iteration of development (to be done in a later section). For now, recall that if $|\psi\rangle$ is spanned by $|q'\rangle$ then

$$\langle q'|F(q,p)|\psi\rangle=F\left(q',-\mathrm{i}\frac{\partial}{\partial q'}\right)\langle q'|\psi\rangle.$$

Similarly, for $|\overline{\Psi}\rangle$ spanned by $|\phi'\rangle$ we have

$$\langle\phi'|F(\phi,\pi)|\overline{\Psi}\rangle=F\left(\phi',=\mathrm{i}\frac{\delta}{\delta\phi'}\right)\langle\phi'|\overline{\Psi}\rangle\equiv\overline{\Psi}[\phi']$$

In the Schrodinger picture (and dropping the overbar notation: $\overline{\Psi}\to\Psi$):

$$i\frac{d}{dt}|\Psi(t)\rangle=H(\phi,\pi)|\Psi(t)\rangle.$$

Note that $\{\phi,\pi\}$ depends on $\vec{x}$ but not t (since using Schrodinger picture), so:

$$i\frac{d}{dt}\Psi(\phi',t) = H\left(\dot{\phi}', -i\frac{\delta}{\partial\phi'}\right)\Psi(\phi',t)$$

In the Heisenberg picture:

$$i\frac{d}{dt}F(\phi,\pi;t) = [F(\phi,\pi;t), H(\phi,\pi;t)] + i\frac{\partial}{\partial t}F(\phi,\pi)$$

where the $\partial/\partial t$ derivation is only with respect to explicit t dependence in $F(\phi,\pi;t)$. Let's consider the conserved vector observable P^{μ}. To begin, multiply the Lagrange equations by $\partial_\mu\phi$

$$\partial_\mu\phi\left\{\partial_\nu\left(\frac{\partial\mathcal{L}}{\partial(\partial_\nu\phi)}\right) - \frac{\partial\mathcal{L}}{\partial\phi}\right\} = 0,$$

and using

$$\partial_\mu\mathcal{L} = \frac{\partial}{\partial(\partial_\nu\phi)}\partial_\mu\partial_\nu\phi + \frac{\partial\mathcal{L}}{\partial\phi}\partial_\mu\phi$$

obtain:

$$\partial_\mu\phi\partial_\nu\left(\frac{\partial\mathcal{L}}{\partial(\partial_\nu\phi)}\right) + \frac{\partial\mathcal{L}}{\partial(\partial_\nu\phi)}\partial_\mu\partial_\nu\phi - \partial_\mu\mathcal{L} = 0$$

or

$$\partial_\nu\left\{\partial_\mu\phi\left(\frac{\partial\mathcal{L}}{\partial(\partial_\nu\phi)}\right) - \delta_\mu^\nu\mathcal{L}\right\} = 0$$

We therefore have the standard result in terms of stress-energy tensor T_ν^μ:

$$\partial_\nu T_\mu^\nu = 0, \quad T_\nu^\mu = \partial_\nu\phi\left(\frac{\partial\mathcal{L}}{\partial(\partial_\mu\phi)}\right) - \partial_\nu^\mu\mathcal{L}$$

So,

$$\int dV_x\partial_\mu T_\nu^\mu = 0 \rightarrow \frac{d}{dt}\int dV_x T_\nu^0 = 0$$

by use of the Gauss divergence theorem, and we obtain:

$$P_\nu = \int dV_x T_\nu^0$$

where the sign is chosen such that $P_0 = H$:

$$P_0 = \int dV_x T_0^0 = \int dV_x\left\{\partial_\partial\phi\left(\frac{\partial L}{\partial(\partial_0\phi)}\right) - \mathcal{L}\right\} = \int dV_x\{\pi(x)\partial_0\phi - \mathcal{L}\}$$

As a special case of the Heisenberg field equation, suppose $F = f(\phi(x), \partial_i\phi(x), \pi(x))$ then

$$i\frac{d}{dt}F(\phi,\pi) = [f,H] = [f,P_0] = i\partial_\partial f$$

This is the 0-component of the more general relation:

$$i\partial_\mu f = [f, P_\mu]$$

The proof for the $\mu = i$ part: for functions that are power series in ϕ or π we focus on terms of the form:

$$[\pi(x)^n, P_i] = \int dV_x' \, \pi(\vec{x}')[\pi(\vec{x})^n, \partial_i' \phi(\vec{x}')] =$$

$$\int dV_x' \, \pi(\vec{x}') n\pi(\vec{x})^{n-1} \partial_i' \big(\delta(\vec{x}, \vec{x}')\big) = i\partial_i (\pi(x)^n)$$

In general

$$[\pi(x)^n, P_i] = \left[\pi(x)^n, \int dV_x' \, \pi(x')\partial_i' \phi(x')\right]$$

$$= \int dV_x' \, [\pi(x)^n, \pi(x')\partial_i' \phi(x')] = \int dV_x' \, \pi(x')[\pi(x)^n, \partial_i' \phi(x')]$$

Since

$$[\pi(x)^n, \partial_i' \phi(x')] = \partial_i' [\pi(x)^n, \phi(x')]$$

And recall the identity: $[A, BC] = ABC - BCA = B[A, C] + [A, B]C$, we can write:

$$[\pi(x)^n, \phi(x')] = -[\phi(x'), \pi(x)^n]$$

$$= -\pi[\phi(x), \pi^{n-1}] + [\phi(x), \pi(x)]\pi^{n-1}$$

$$= -\{\pi^n[\phi(x'), \pi^{n-1}] + \pi^2[\phi(x'), \pi^{n-2}] + \pi[\phi, \pi^{n-2}]\pi\}$$

So,

$$[\pi(x)^n, \phi(x')] = -\sum_{s=0}^{n-1} \pi^s \, [\phi(x), \pi]\pi^{n-s-1} = -i\delta(x, x') \sum_{s=0}^{n-1} \pi^{n-1}$$

$$= -n\delta(x, x')\pi^{n-1}$$

Thus

$$[\pi(x)^n, \partial_i \phi(x')] = -i\partial_i' (\vec{x}, \vec{x}') n\pi^{n-1}$$

and

$$[\pi(x)^n, P_i] = i \int dV_x' \, \pi(\vec{x}') n\pi(\vec{x})^{n-1} \partial_i' \big(\delta(\vec{x}, \vec{x}')\big) = +i\partial_i (\pi(x)^n)$$

From this general result, we can then show that:

$$i\partial_\mu f = [f, P_\mu]$$

as desired.

Infinitesimal translation

Let's obtain the $i\partial_\mu f = [f, P_\mu]$ result from invariance of the Lagrangian under infinitesimal translations with constant time hypersurfaces:

$$x^\mu \to x'^\mu = x^\mu + \delta x^\mu$$

$$\phi(x) \to \phi'(x) = \phi(x) + \delta_0 \phi(x)$$

where variations vanish on spatial boundary of the integration but not on the interior of the constant time hypersurfaces at t_1 and t_2. For infinitesimal translation:

$$\delta\phi = 0 \quad \to \quad \delta_0 \phi = -(\partial_\mu \phi)\delta x^\mu$$

Consider $f = f(\phi, \partial_\mu \phi)$ then

$$\delta_0 f = \frac{\partial f}{\partial \phi}\delta_0\phi + \frac{\partial f}{\partial(\partial_\mu \phi)}\partial^\mu \delta_0 \phi$$

$$= \frac{\partial f}{\partial \phi}\left(-(\partial_\mu \phi)\delta x^\mu\right) + \frac{\partial f}{\partial(\partial_\mu \phi)}\partial^\mu[-(\partial_v \phi)\delta x^v]$$

$$= -\left\{\frac{\partial f}{\partial \phi}\frac{\partial \phi}{\partial x^v} + \frac{\partial f}{\partial(\partial_\mu \phi)}\frac{\partial(\partial_\mu \phi)}{\partial x^v}\right\}\delta x^v = -(\partial_\mu f)\delta x^\mu$$

Thus

$$i\delta_0 f = [f, -P_\mu \delta x^\mu] \to -i(\partial_\mu f)\delta x^\mu = [f, -P_\mu]\delta x^\mu \to i(\partial_\mu f) = [f, P_\mu]$$

In agreement with the same result by way of the Gauss Divergence theorem. From this we also obtain the conserved current $\partial_\mu J^\mu = 0$ where:

$$J^\mu = \frac{\partial L}{\partial(\partial_\mu \phi_a)}\delta\phi_a - T_v^\mu \delta x^v$$

In Minkowski space the Lagrangian density of a free field is symmetric under space and time translations, which implies that energy and momentum are conserved. In curved spacetime, because of the presence in the Lagrangian density of the metric, we will find that for a free field (i.e., one influenced by gravitation alone), only the component of P_μ along the direction of an isometry of the spacetime is conserved.

Particles

The notation in this section and what follows is that of my PhD thesis [15] and papers references there.

Consider a single Hermitian free scalar field (such as the neutral pion) described by the Lagrangian density:

$$\mathcal{L} = \frac{1}{2}\left(\eta^{\mu v}\partial_\mu \phi \partial_v \phi - m^2\phi^2\right)$$

where the Euler-Lagrange equations give the Klein-Gordon Equation:

$$(\Box + m^2)\phi = 0.$$

We have for this Lagrangian:

$$\pi = \partial_0 \phi = \dot{\phi}$$

$$T_v^\mu = \frac{\partial L}{\partial(\partial_\mu \phi_a)}\partial_v \phi_a - \delta_v^\mu \mathcal{L}$$

$$P_v = \int dV_x T_v^0$$

$$P_0 = \int dV_x \left\{ \frac{\partial L}{\partial(\partial_0\phi)} \partial_0\phi - L \right\}$$
$$= \int dV_x \left\{ (\partial_0\phi)^2 - \frac{1}{2}(\partial_0\phi)^2 + \frac{1}{2}\left(\vec{\partial}\phi\right)^2 + \frac{1}{2}m^2\phi^2 \right\}$$
$$= \frac{1}{2}\int dV_x \left(\dot{\phi} + \left(\vec{\partial}\phi\right)^2 + m^2\phi^2 \right)$$
$$P_i = \int dV_x \left\{ \frac{\partial L}{\partial(\partial_0\phi)} \partial_i\phi - \delta_i^0\, \mathcal{L} \right\} = \frac{1}{2}\int dV_x \left(\dot{\phi}(\partial_i\phi) + (\partial_i\phi)\dot{\phi} \right)$$

(note, we've switched to a symmetrized form to be in agreement with General Relativity).

Let's define the scalar product as:
$$(f_1, f_2) = i\int dV_x \{ f_1^* \partial_0 f_2 - \partial_0 f_1^* f_2 \} \equiv i\int dV_x\, f_1^* \overleftrightarrow{\partial}_0 f_2$$
Let's now verify it is independent of time:
$$\partial_0 (f_1, f_2) = i\int dV_x \{ f_1^* \partial_0^2 f_2 - \partial_0^2 f_1^* f_2 \}$$
$$= i\int dV_x\, \{ f_1^* (\partial_i^2 - m^2) f_2 - (\partial_i^2 - m^2) f_1^* f_2 \}$$
$$= i\int dV_x \partial_i \{ f_1^* \partial_i f_2 - (\partial_i f_1^*) f_2 \} = i\{ f_1^* \partial_i f_2 - \partial_i f_1^* f_2 \}|_{surface} = 0.$$
Thus, (f_1, f_2) is independent of time.

Positive Energy solutions

Let's now obtain a complete set of positive energy solutions to the Klein-Gordon equation:
$$(\Box + m^2)\phi = 0$$
with
$$g_k(x) = (2\pi)^{-3/2}(2\omega_k)^{-1/2} \exp[i(kx - \omega_k t)], \qquad \omega_k = +(k^2 + m^2)^{1/2}$$
with orthonormality in this sense:
$$(g_{\vec{k}}, g_{\vec{k}'}) = \delta(\vec{k}, \vec{k}'), \qquad (g_{\vec{k}}, g_{\vec{k}'}^*) = 0.$$
Expanding $\phi(x)$ in g_k:
$$\phi(x) = \int d^3k \{ a(\vec{k}) g_{\vec{k}}(x) + a^\dagger(\vec{k}) g_{\vec{k}}^*(x) \}$$
We can now compute:
$$a(\vec{k}) = (g_k, \phi(x)) = i\int dV_x\, (g_k^* \pi - i\omega_k g_k^* \phi)$$
The commutation relation is then:

$[a(k), a^\dagger(k')] =$

$$\int dV_x dV_x' \{(g_k^*\pi - i\omega_k g_k^*\phi)(g_{k'}\pi + i\omega_{k'} g_{k'}\phi)$$
$$- (g_k\pi + i\omega_k g_k\phi)(g_{k'}^*\pi - i\omega_{k'} g_{k'}^*\phi)\}$$
$$= \int dV_x dV_x' \{(i\omega_{k'}\pi\phi - i\omega_k\phi\pi)g_{k'}g_k^* + (i\omega_k\pi\phi - i\omega_{k'}\phi\pi)g_{k'}g_k^*\}$$

Since

$$(g_{\vec{k}}, g_{\vec{k}'}) = i\int dV_x(-i\omega_{k'} g_k^* g_{k'} - (i\omega_k) g_k^* g_{k'})$$
$$= -i\int dV_x (g_k^* g_{k'})\,[i\omega_k + i\omega_{k'}] = \delta(\vec{k}, \vec{k}')$$

We can write:

$$[a(k), a^t(k')] = \int dV_x dV_x' \{[i\omega_k](\pi\phi - \phi\pi)$$
$$+ [i\omega_{k'}](\pi\phi - \phi\pi)\}(g_k^* g_{k'})$$
$$= -\int dV_x dV_x' \,[\pi(x)\phi(x') - \phi(x)\pi(x')]\,(i\omega_k + i\omega_{k'})g_k^*(x)g_{k'}(x')$$
$$= -i\int dV_x\,(i\omega_k + i\omega_{k'})g_k^*(x)g_{k'}(x)$$
$$= (g_{\vec{k}}, g_{\vec{k}'}) = \delta(\vec{k}, \vec{k}')$$

Note that $\delta(\vec{k}, \vec{k}')$ involves a Dirac delta not a Kronecker delta (as in the harmonic oscillator analogue from quantum mechanics). Similarly,

$$[a(k), a(k')] = (g_{\vec{k}}, g_{\vec{k}'}^*) = 0.$$

The Vacuum State

Let's consider the vacuum state $|0\rangle$ in the Schrodinger field representation:

$$P_0 = H = \frac{1}{2}\sum_k \omega_k \left(a_k^\dagger a_k + a_k a_k^\dagger\right) = \sum_k \omega_k \left(N_k + \frac{1}{2}\right)$$

Or, with normal ordering convention:

$$:P_0: = \sum_k \omega_k N_k$$

From modal solution in a cube (taken to infinite size) we have (as $L \longrightarrow \infty$):

$$\Delta k^i = 2\pi/L \rightarrow \sum_k B_k = \left({}^{L}/_{2\pi}\right)^{n-1} \sum_k \omega_k(\Delta^{n-1}k)B_k$$
$$\rightarrow V(2\pi)^{-(n-1)} \int d^{n-1}\,KB_{k},$$

So, $(2\pi)^{n-1}V^{-1}\sum_k \sim \int d^{n-1}k$. Thus, a comparison of the continuous and discrete mode expansion reveals:

$$(2\pi)^{-(n-1)/2}\, V^{1/2}\, a_k \sim a(k)$$

Thus, defining $N(k) \equiv a^t(k)a(k)$ we have

$$\sum_k N_k \sim \int d^{n-1}\, k\, N(k)$$

Thus, the evolution operator for a free particle system becomes:

$$U(t, t_0) = \exp[-i(t - t_0)H] = \exp\left[-i(t - t_0)\sum_k \omega_k N_k\right]$$

Thus,

$$U(t, t_0) = \exp\left[-i(t - t_0)\int d^{n-1}\omega_k a(k)^\dagger a(k)\right]$$

Whence, it follows from $|\psi_H\rangle = U(t, t_0)^\dagger |\psi_s(t)\rangle$ that

$$|0_H\rangle = U(t, t_0)^\dagger |0_s\rangle = U(t_0, t)|0_s\rangle = |0_s\rangle$$

Thus, the vacuum, $|0\rangle$, is independent of time, and $|0\rangle$ is the same for S and H.

Vacuum Energy

Let's now explore vacuum energy in the context of two parallel neutral conducting plates. Consider a massless neutral scalar field in the presence of those plates separated by distance a in the z-direction and one plate at $z = 0$. Assume "a" is sufficiently small that the vacuum energy outside the plates is small with respect to that between the plates, we then neglect the region outside the plates. E want to adopt the usual periodic boundary conditions on the plates, and this is done by requiring ϕ to be have the same value on each plate simultaneously in the rest frame of the plates. Let's impose periodic position in the x and y regions (each going from coordinate 0 to L, where $L \gg a$. Thus,

$$k_x = 2\pi\, L^{-1} i, \quad k_y = 2\pi\, L^{-1} j$$

where i, j are an integers, and

$$k_z = 2\pi\, a^{-1} \ell$$

where ℓ is an integer. The Volume is

$$V = aL^2.$$

Let $|0\rangle_a$ denote the vacuum, or no-particle state, of this system (and do not normal order to remove ground state energy):

$$\rho_0(a) = a^{-1}L^{-2}\ {}_a\langle 0|P_0|0\rangle_0 = a^{-1}L^{-2}\ {}_a\langle 0|\textstyle\sum_k \omega_k \left(N_k + \frac{1}{2}\right)|0\rangle_a$$

Recall

$$\omega_k = (k^2+m^2)^{1/2} = (k^2)^{1/2} = \left[\left(\frac{2\pi}{L}\right)^2 (i^2+j^2) + \left(\frac{2\pi}{a}\right)^2 \ell^2\right]^{1/2}$$

with i, j, ℓ integers. We can write:

$$\rho_0(a) = -\frac{1}{2}\, a^{-1} L^{-2} \lim_{\alpha\to 0} \frac{d}{d\alpha} \sum_{\vec{k}} e^{-\alpha\omega_k} = -\frac{1}{2}\, a^{-1} \lim_{\alpha\to 0} \frac{d}{d\alpha}\, S(\alpha, a)$$

where

$$S(\alpha, a) = L^{-2} \sum_{k} e^{-\alpha\omega_k}$$

Let, as $L\to\infty$, the relation $\sum_i \sum_j \longrightarrow L^2(2\pi)^{-2} \int dk^x\, dk^y$, with $k^2 = [(k^x)^2+(k^y)^2]$ be enforced:

$$S(\alpha, a) = (2\pi)^{-2} \sum_{\ell=\infty}^{\infty} \int_{-\infty}^{\infty} dk^x dk^y \exp\left\{-\alpha\left[k^2 + \left(\frac{2\pi}{a}\right)^2 \ell^2\right]^{1/2}\right\}$$

$$S(\alpha, a) = \pi^{-1} \sum_{\ell=1}^{\infty} \int_0^{\infty} dk\, k \exp\left\{-\alpha\left[k^2 + \left(\frac{2\pi}{a}\right)^2 \ell^2\right]^{1/2}\right\}$$

$$+ (2\pi)^{-1} \int_0^{\infty} dkk \exp[-\alpha k]$$

Define $F(\ell) \equiv \int_0^{\infty} dk\, k \exp\left\{-\alpha[k^2+(2\pi/a)^2\ell^2]^{1/2}\right\}$. Then we have:

$$\pi S(\alpha, a) = \sum_{\ell=1}^{\infty} F(\ell) + \tfrac{1}{2} F(0).$$

Recall the Euler-Maclaurin formula

$$\frac{1}{2}\, F(b) + \sum_{\ell=1}^{\infty} F(b+\ell) = \int_b^{\infty} F(\ell) d\ell - \sum_{m=1}^{\infty} \frac{B_{2m}}{(2m)!} F^{(2m-1)}(b)$$

Rewriting in terms of the Euler-Maclaurin formula:

$$\pi S(\alpha, a) = \int_0^{\infty} F(\ell) d\ell - \sum_{m=1}^{\infty} \frac{B_{2m}}{(2_m)!} F^{(2m-1)}(0)$$

Note, the coefficients B_{2m} are known as the Bernoulli numbers. Let's now solve the integral for $F(\ell)$ to get:

$$F(\ell) = \left(\frac{1}{\alpha}\frac{2\pi}{a}\,\ell + \frac{1}{\alpha^2}\right) \exp\left(-\alpha \frac{2\pi}{a}\,\ell\right).$$

Notice that $F(\ell)$ depends on ℓ only through ℓ/a, and that $F^{(1)}(0) = 0$, $F^{(3)}(0) = 2\alpha\left(\frac{2\pi}{a}\right)^3$, and $F^{(j)}(0) = \mathcal{O}(\alpha^2)$ for $j \geq 5$. Let

$$\int_0^\infty F(\ell)d\ell = aG(\alpha),$$

where $G(\alpha)$ doesn't depend on a. We can now write:

$$\pi S(\alpha, a) = aG(\alpha) + \frac{\pi^3}{45a^3}\alpha + \mathcal{O}(\alpha^2)$$

Thus,

$$\rho_0(a) = -\frac{1}{2}a^{-1}\lim_{\alpha\to 0}\frac{d}{d\alpha}S(\alpha, a) = \lim_{\alpha\to 0}\frac{d}{d\alpha}\left[-\frac{1}{2\pi}G(\alpha)\right] - \frac{\pi^2}{90a^4}$$

For Minkowski space we replace a by $L \to \infty$ and

$$\rho_0(\infty) = \lim_{\alpha\to 0}\frac{d}{d\alpha}\left[-\frac{1}{2\pi}G(\alpha)\right]$$

If we assume the physical vacuum energy density is zero in Minkowski space, then it is natural to define the vacuum energy density $\rho(a)$ between the plates by

$$\rho(a) = \lim_{\alpha\to 0}[\rho_0(\alpha, a) - \rho_0(\alpha, \infty)] = \frac{\pi^2}{90a^4}$$

Using vanishing boundary conditions instead of periodic we have $k^z = \pi\ell/a \to \pi^2$ replaces $(2\pi)^3$ in $F^{(3)}(0)$, also ℓ does not run over $\pm\ell$, only $+\ell$, this introduces a factor of $({}^1/_8)({}^1/_2)$.

One then obtains for a massless scalar field with vanishing boundary conditions

$$\rho(a) = \frac{-\pi^2}{1440a^4}$$

For the Electromagnetic field there are two independent polarization states, thus

$$\rho_{em}(a) = \frac{-\pi^2}{720a^4}.$$

Let's use the above derivation (with a convenient subtraction renormalization), which indicates a vacuum energy of the electromagnetic field in a cylindrical volume between two parallel plates, of area A, is $aA\rho_{em}(a)$. Since the energy increases as the plates are separated, an attractive force is exerted between the plates. Force per unit area in the z-direction is:

$$F/A = -\frac{d}{da}\left(a\,\rho_{em}(a)\right) = \frac{-\hbar c\pi^2}{240a^4}$$

(where a shift to explicit cgs units is made). Consider now two plates with separation and area:

$$a = 100\text{Å} = 10^{-8}\, m = 10^{-6}\, cm, A = 100cm^2.$$

We then have:

$$F = -\frac{(1.05 \times 10^{-27}\, erg\, sec)(3 \times 10^{10}\, cm/sec)\pi^2(100cm^2)}{240(10^{-6}\, cm)^4}$$

Recall that $erg = g \cdot cm^2/s^2$ and $10^7 \frac{erg}{cm} = 1N$:

$$|F| = 13\text{ N}.$$

Suppose

$$a = 10^{-5}\, cm, A = 100\, cm^2$$

then

$$|F| = 1.3 \times 10^{-3}\text{ N}.$$

So only noticeable when plates are very close, with $< 10^{-6}\, cm$ separation.

The Charged Scalar Field

Suppose you have two scalar fields, can they be rewritten as a complex scalar field?

$$\mathcal{L} = \frac{1}{2}\eta^{\mu\nu}\partial_\mu\phi_1\partial_\nu\phi_1 - \frac{1}{2}m^2\phi_1^2 + \frac{1}{2}\eta^{\mu\nu}\partial_\mu\phi_2\partial_\nu\phi_2 - \frac{1}{2}m^2\phi_2^2$$

$$= \frac{1}{2}\eta^{\mu\nu}\left(\partial_\mu(\phi_1 + i\phi_2)\partial_\nu(\phi_1 - i\phi_2)\right) - \tfrac{1}{2}m^2(\phi_1 + i\phi_2)(\phi_1 - i\phi_2)$$

Thus

$$\mathcal{L} = \eta^{\mu\nu}\partial_\mu\phi^\dagger\partial_\nu\phi - m^2\phi^\dagger\phi, \quad \phi = \frac{1}{\sqrt{2}}[\phi_1 - i\phi_2]$$

In this last form the Lagrangian is clearly invariant under the following transformation:

$\phi(x) \to \phi'(x) = exp(i\alpha)\phi(x) = \frac{1}{\sqrt{2}}\{(cos\,\alpha)\phi_1 + (sin\,\alpha)\phi_2 + i[(sin\,\alpha)\phi_1 - (cos\,\alpha)\phi_2]\}$

Thus

$$\begin{pmatrix}\phi_1'\\ \phi_2'\end{pmatrix} = \begin{pmatrix}\cos\alpha & \sin\alpha\\ -\sin\alpha & \cos\alpha\end{pmatrix}\begin{pmatrix}\phi_1\\ \phi_2\end{pmatrix}$$

Which is the classic rotation matrix $O(2) \simeq U(1)$, thus the gauge transformation is the simplest (nontrivial) case $U(1)$ and it can be represented as a rotation between the ϕ_1 and ϕ_2 state vectors in the 2-scalar field Lagrangian (recall the masses are the same). Since the gauge invariance is simpler for the complex representation let's use that:

$$\pi = \frac{\delta L}{\delta(\partial_0\phi)} = \partial_0\phi^\dagger;\ \pi^\dagger = \partial_0\phi$$

Thus $(\Box + m^2)\phi = 0$, $(\Box + m^2)\phi^\dagger = 0$, and we expand in terms of periodic plane waves (using a compete set: f_k, f_k^*):

$$(f_k, f_{k'}) = \delta_{k,k'}, (f_k, f_k^*) = 0$$

and

$$\phi(x) = \sum_k \{a_k f_k(x) + b_k^\dagger f_k^*(x)\}$$

$$\phi^\dagger(x) = \sum_k \{b_k f_k(x) + a_k^\dagger f_k^*(x)\}$$

Notice the coefficients oddly labeled with a's and b's – this convention is adopted for convenience later.

8.3 Curved space-time transforms and diagrams

Conformal transformations

Conformal transformations shrink or stretch the manifold, unlike coordinate transformations $x^\mu \to x^\mu$ which merely reliable the coordinates in some patch, leaving the geometry itself unchanged. A conformal transformation on the metric may be described by:

$$g_{\mu\nu}(x) \to \bar{g}_{\mu\nu}(x) = \Omega^2(x) g_{\mu\nu}(x)$$

Note the useful transformation:

$$\left[\Box + \frac{1}{4}\frac{(n-2)}{(n-1)}R\right]\phi \to \left[\bar{\Box} + \frac{1}{4}\frac{(n-2)}{(n-1)}\bar{R}\right]\bar{\phi}$$

$$= \Omega^{-(n+2)/2}\left[\Box + \frac{1}{4}\frac{(n-2)}{(n-1)}R\right]\phi$$

where $\bar{\phi} \equiv \Omega^{(2-n)/2}(x)\,\phi(x)$.

Penrose diagram for 2D Minkowski

Consider the metric

$$ds^2 = dt^2 - dx^2 = dudv$$

where $u = t - x, v = t + x$ and in the $\{u, v\}$ coordinate system the metric is:

$$g_{\mu\nu} = \frac{1}{2}\begin{bmatrix} 0 & 1 \\ 1 & 0 \end{bmatrix}$$

Let's now perform the 'compactifying' coordinate transformation:

$$u' = 2\tan^{-1} u, \quad v' = 2\tan^{-1} v$$

where $-\pi \leq u', v' \leq \pi$ (contrast with $-\infty < u, v < \infty$):

$$ds^2 = \frac{1}{4}\sec^2\left(\frac{1}{2}u'\right)\sec^2\left(\frac{1}{2}v'\right)du'dv'$$

Thus,

$$g_{\mu\nu}(u', v') = \frac{1}{8}\sec^2\left(\frac{1}{2}u'\right)\sec^2\left(\frac{1}{2}v'\right)\begin{bmatrix}0 & 1\\ 1 & 0\end{bmatrix}$$

Let's now perform a conformal transformation with

$$\Omega^2(x) = \left(\frac{1}{4}\sec^2\left(\frac{1}{2}u'\right)\sec^2\left(\frac{1}{2}v'\right)\right)^{-1}$$

So,

$$g_{\mu\nu}(u', v') \to \bar{g}_{\mu\nu}(u', v') = \frac{1}{2}\begin{bmatrix}0 & 1\\ 1 & 0\end{bmatrix} \to d\bar{s}^2 = du'dv'$$

Penrose conformal diagram

This is a device that enables the whole of an infinite spacetime to be represented as a finite diagram (compact manifold), by applying a conformal transformation to the metric structure.

The following Penrose Diagram results:

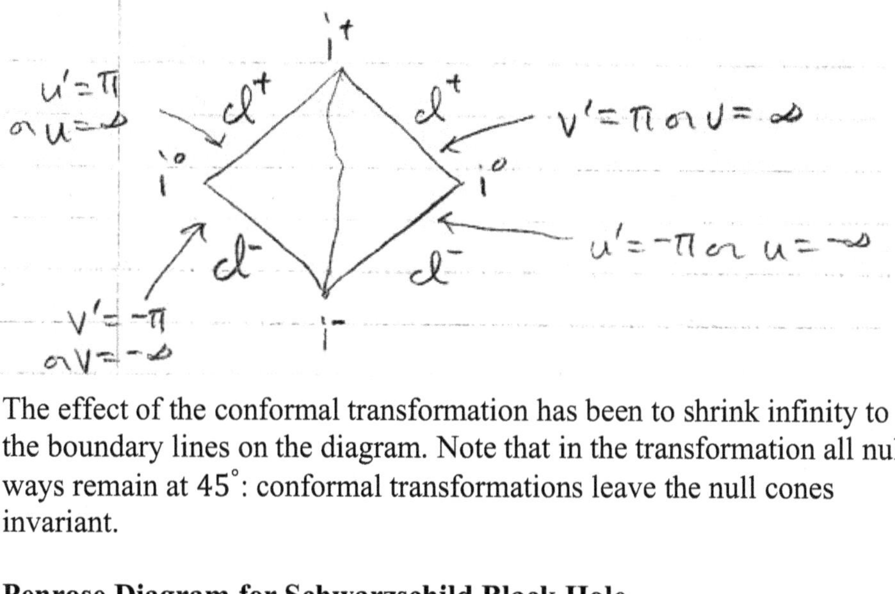

The effect of the conformal transformation has been to shrink infinity to the boundary lines on the diagram. Note that in the transformation all null ways remain at 45°: conformal transformations leave the null cones invariant.

Penrose Diagram for Schwarzschild Black Hole

Let's now repeat this conformal transformation procedure on the 4D Schwarzschild Black Hole metric to obtain the Penrose diagram for a Black Hole. Starting with the standard form of he Schwarzschild metric:

$$ds^2 = (1 - {}^{2M}/_{r})dt^2 - (1 - {}^{2M}/_{r})^{-1}dr^2 - r^2(d\theta^2 + \sin^2\theta d\phi^2)$$

We now shift to Kruskal coordinates, first using:

$$r^* = r + 2M\ln|(r/2M) - 1|$$

thus

$$dr^* = dr\left(1 + 2M\left\{\frac{\left(\frac{1}{2M}\right)}{\left(\frac{r}{2M}\right) - 1}\right\}\right) = dr\left(\frac{1}{1 - \frac{2M}{r}}\right)$$

and we have

$$ds^2 = \left(1 - \frac{2m}{r}\right)\left[dt^2 - dr^{*2}\right] - r^2(d\theta^2 + \sin^2\theta d\phi^2)$$

Next, introduce

$$u = t - r^*, \qquad v = t + r^*,$$

From which we get:

$$r^* = \frac{1}{2}(v - u) = r + 2M\ln\left|\left({}^r/_{2M}\right) - 1\right| \rightarrow \frac{1}{4M}(v - u) - \frac{r}{2M}$$
$$= \ln\left|\frac{r}{2M} - 1\right|$$

Thus,

$$-\left(1 - \frac{r}{2M}\right) = \exp(-r/2M)\exp\left(\frac{v - u}{4M}\right)$$

and using the trivial relation $\left(1 - \frac{2M}{r}\right) = -\frac{2M}{r}\left(1 - \frac{r}{2M}\right)$ we can then write:

$$ds^2 = \frac{2M}{r}\exp\left({}^{-r}/_{2M}\right)\exp\left(\frac{v - u}{4M}\right)dudv - r^2\left(d\theta^2 + \sin^2\theta d\phi^2\right)$$

Next, let's do the 'compactifying' transformation:

$$\bar{u} = -4M\ e^{-u/4M}, \qquad \bar{v} = 4M\ e^{v/4M}$$

thus

$$d\bar{u} = e^{-u/4M}du, \quad d\bar{v} = e^{v/4M}dv$$

to get:

$$ds^2 = (2M/r)\,e^{-r/2M}\,d\bar{u}d\bar{v} - r^2(d\theta^2 + \sin^2\theta d\phi^2)$$

Similar to the 2D-Minkowski derivation, the part of the metric consisting of

$$(2M/r)\,e^{-r/2M}\,d\bar{u}d\bar{v}$$

is conformal to 2D-Minkowski, however, unlike that derivation, $\bar{u}, \bar{v}$ are defined only in the quadrant given by $-\infty < \bar{u} \leq 0$ and $0 \leq \bar{v} < \infty$:

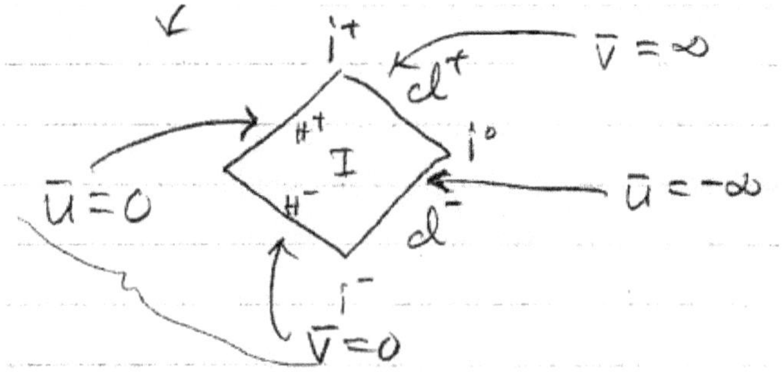

The indicated $\bar{u} = 0, \bar{v} = 0$ boundary occurs when $r = 2M, t = \pm\infty$, which is a singularity of the u,v coordinate system but not of $\bar{u}, \bar{v}$. Therefore, the spacetime may be analytically extended beyond the left-hand edge using the Kruskal coordinates $\bar{u}, \bar{v}$ defined over the whole plane:

$$-\infty < \bar{u}, \bar{v} < \infty$$

The resulting space-time is the maximally extended Kruskal manifold:

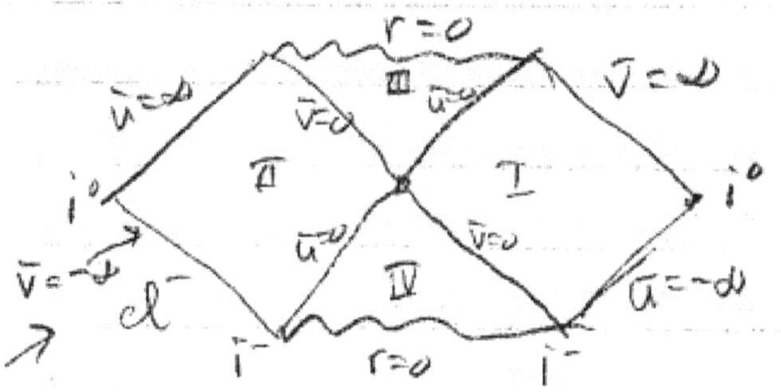

Note that the time direction formally reversed on the left-hand side $(t \rightarrow -t)$ (thus the description of falling into a black hole or emerging from a white hole).

The maximally extended Kruskal spacetime is everywhere (except at $r = 0$) a solution of the vacuum Einstein equation In the real would, where a black hole is more likely to form from the implosion of a star, the vacuum equations only apply to the region outside the star and so only a fragment of the figure will be relevant.

Spacetimes will geometric symmetries:

Symmetries can be described using Killing vectors z^{μ} which are solutions to

$$\mathcal{L}_z g_{\mu v}(x) = 0 \Leftrightarrow z_{\mu;v} + z_{v;\mu} = 0.$$

For symmetries associated with conformal flatness (when the spacetime is conformal to Minkowski space), the geometry admits a conformal killing vector field, which satisfies

$$\mathcal{L}_z g_{\mu v}(x) = \lambda(x) g_{\mu v}(x).$$

Consider the general Lagrangian density in curved spacetime:

$$\mathcal{L}(x) = \frac{1}{2}[-g(x)]^{1/2}\{g^{\mu\nu}(x)\phi(x)_{,\mu}\,\phi(x)_{,\nu} - [m^2 + \varepsilon R(x)]\phi^2(x)\}$$

with $S = \int \mathcal{L}(x)d^n x$ and $\delta S = 0$, we get the equations of motion:

$$[\Box_x + m^2 + \varepsilon R(x)]\phi(x) = 0$$

Note that attention has been focused on linear field theories, and even so there is no unique quantization scheme.

Generalization of scalar product

The following generalization of the scalar product is used, where the integration is over a constant τ hypersurface:

$$(f_1, f_2) \equiv i\int d^{n-1}x\,|g|^{1/2}\,g^{0\nu} f_1^*(\vec{x},t)\overleftrightarrow{\partial}_\nu f_2\,(\vec{x},t).$$

With this definition, if the $f's$ are solutions of the field equation $(\Box + m^2 + zR)\phi = 0$, then (f_1, f_2) is conserved.

In terms of a general spacelike hypersurface σ with future directed n^μ vector and hyper-surface element $d\sigma$:

$$(f_1, f_2) = i\int_\sigma d\sigma\,|g|^{1/2}\,n^\nu f_1^*\overleftrightarrow{\partial}_\nu f_2$$

Note that this formulation is conserved under deformation of $\sigma \to \sigma'$ and use of the Gauss' Divergence theorem:

$$(f_1, f_2)_{\sigma'} - (f_1, f_2)$$
$$= i\int_v d^n x\partial^\mu\left(|g|^{1/2} f_1^*\overleftrightarrow{\partial}_\mu f_2\right) = i\int_v d^n x\nabla^\mu\left(f_1^*\nabla_\mu f_2\right) = 0.$$

8.4 Covariant quantization

In covariant quantization Heisenberg quantum fields manifest themselves in the traditional way as operators defined on a single Hilbert space. The method has origins in Segal's work [21]. Segal's methods relied heavily on the existence and structure of classical solutions of the field equations. In the present context this implies that the space-time manifold must be globally hyperbolic in the sense of possessing a spatial hypersurface on which Cauchy data can be freely specified. Globally hyperbolic manifolds are necessarily of the form $R \times \Sigma$, where Σ is a three-space.

Anti-de Sitter (Ads)

Anti-de Sitter (Ads) is an example of a non-globally hyperbolic manifold, it possesses both closed timelike curved and a timelike boundary at spatial infinity. Ads is a homogeneous space of the group O(3, 2), but similarly, deSitter with a group oriented approach to quantization (SO(4, 1) for deSitter) misses thermal radiation associated with the event horizon of an inertial observer. The role of timelike infinity is not readily discussed in the group oriented approach.

The aim of the covariant approach is to construct a quantum field $\psi(x)$ satisfying the classical field equation:

$$\{\Box + u(x)\}\hat{\psi}(x) = 0$$

and the covariant commutation relation:

$$\left[\hat{\psi}(x), \hat{\psi}(x')\right] = -i\hbar\, \tilde{G}(x, x').$$

$\tilde{G}(x, x')$ is defined as the difference of advanced and retarded Green's functions, evolves classical Cauchy data specified on a Cauchy hypersurface Σ by:

$$\psi(x) = \int_{\Sigma} \tilde{G}(x, x')\, \overleftrightarrow{\partial}_{\mu}\psi(x')d\sigma^{\mu}(x')$$

(in this form it becomes clear that global hyperbolicity is an essential prerequisite).

Rigorous quantization schemes in a globally hyperbolic space-time attach considerable importance to Cauchy data of compact support. If our data has compact support a consequence of global hyperbolicity is that the Cauchy data on any Cauchy hypersurface will then posses compact support. For the case at hand, of a non-globally hyperbolic spacetime, initial data with compact support on one spacelike hypersurface will in general evolve in such a way that it becomes noncompact on many other spacelike hypersurface.

The universal covering space of AdS can be conformally mapped into half of the Einstein Static Universe. A globally hyperbolic spacetime and quantization therein follows the pattern set by Segal. It is possible to use this quantization, mapped black into AdS, to give an acceptable QFT in AdS (thus sidestepping the problem of how to fix the information passing into the space-time).

A spatial "periodicity" of the Einstein Static Universe induces an effective temporal periodicity. The specification of Cauchy data on $\{\tau = 0, 0 \leq \rho \leq \pi\}$ is equivalent to its specification on the pair of incomplete surfaces $\left\{\tau = 0,\ \rho < {}^{\pi}/_{2}\right\}$ and $\left\{\tau = \pi,\ \rho < {}^{\pi}/_{2}\right\}$ (consistent C^{∞} solution require proper boundary values). The set of solutions of this effective Cauchy surface in AdS generates a Hilbert space. To reconstruct the AdS solution from its Cauchy data we require the analog of $\hat{G}^T(x, x')$. G^T is obtained from the Einstein Static Universe G^E by restriction and mapping back (conformal relation). This completes quantization. The effect of the boundary conditions obtained by conformally mapping into ESU is to recirculate the energy, angular momentum, etc., lost to time-like infinity, resulting in a well defined, if rather unusual $\left(P_a = Q_a(\tau) + Q_a(\tau + \pi)\right)$, conservation law.

Particle production is possible whenever the Fourier decomposition of the external field contains frequencies $k_0 > \Sigma m_i \approx 10^{23}\ Hz$, where the sum has to be extended over all particles created in one elementary process. For present the expansion of the universe this phenomenon becomes pronounced only at distances of the order of 10^8 light years, which corresponds to frequencies of $10^{-25} Hz$ and is therefore a negligible source of particle creation. There is appreciable particle creation in first 10^{-20}sec of universe, however....

8.5 QFT in CST with minimal coupling

Let's consider minimal coupling for the free, neutral, scalar field from flat space-time:

$$\mathcal{L} = \frac{1}{2}\left(\eta^{\mu v}\, \partial_\mu \phi \partial_v \phi - m^2 \phi^2\right)$$

where scalar field has $\nabla_\mu \phi = \partial_\mu \phi$, and generalize to curved space-time:

$$\mathcal{L} = \frac{1}{2}\, |g|^{1/2} \left(g^{\mu v} \partial_\mu \phi \partial_v \phi - m^2 \phi^2\right)$$

Since conformal invariance is of interest we'll also study:

$$\mathcal{L} = \frac{1}{2}\, |g|^{1/2} \left(g^{\mu v} \partial_\mu \phi \partial_v \phi - m^2 \phi^2 - zR\phi^2\right),$$

where $R = g^{\mu v} R_{\mu v}$ is the scalar curvature and conformal invariance occurs when (derived later for 4D): $m = 0, z = 1/6$.

Conformal invariance occurs when:

$$S[\phi, g_{\mu v}] = S[\tilde{\phi}, \tilde{g}_{\mu v}]$$

where $\tilde{g}_{\mu v}(x) = \Omega^2(x)\, g_{\mu v}(x)$ and $\tilde{\phi}(x) = \Omega^{2p}(x)\phi(x)$.

An infinitesimal conformal transformation can be written:

$$\Omega^2(x) = 1 + \lambda(x)$$
$$\delta_0 g_{\mu v} = \tilde{g}_{\mu v} - g_{\mu v} = \lambda(x) g_{\mu v}(x)$$
$$\delta_0 \phi(x) = \tilde{\phi}(x) - \phi(x) = p\lambda(x)\phi(x)$$

Consider a conformal transformation with $p = -1/2, z = 1/6$. We have $\tilde{\phi}(x) = \left(1 - \frac{1}{2}\lambda\right)\phi(x)$ and:

$$\tilde{g}_{\mu\nu}(x) = (1+\lambda)g_{\mu\nu}(x) \rightarrow \ |\tilde{g}|^{1/2} = (1+2\lambda)|g|^{1/2}, \ \ g^{\mu\nu} \rightarrow \tilde{g}^{\mu\nu} = (1-\lambda)g^{\mu\nu}$$

Thus

$$\tilde{\Gamma}^{\mu}_{\nu\tau} = \Gamma^{\mu}_{\nu\tau} + \frac{1}{2}\left(\delta^{\mu}_{\tau}\partial_{\nu}\lambda + \delta^{\mu}_{\nu}\partial_{\tau}\lambda - g^{\mu\sigma}g_{\nu\tau}\partial_{\sigma}\lambda\right)$$

$$\tilde{R} = (1-\lambda)R + 3\Box\lambda$$

$$T^{\mu}_{\nu\tau} \rightarrow \Gamma^{\mu}_{\nu\tau} = \Gamma^{\mu}_{\nu\tau} + \frac{1}{2}\left(\partial^{\mu}_{\tau}\,\partial_{\nu}\lambda + \delta^{\mu}_{\nu}\partial_{\tau}\lambda - g^{\mu\sigma}g_{\mu\tau}\partial_{\sigma}\lambda\right)$$

Recall $R^{\mu}_{\nu\lambda\sigma} = \partial_{\sigma}\Gamma^{\mu}_{\nu\lambda} - \partial_{\lambda}\Gamma^{\mu}_{\nu\sigma} + \Gamma^{\tau}_{\nu\lambda}\Gamma^{\mu}_{\sigma\tau} - \Gamma^{\tau}_{\nu\sigma}\Gamma^{\mu}_{\lambda\tau}$ and $R_{\mu\nu} = R^{\lambda}_{\mu\lambda\nu}$, so, the general form with $m \neq 0$ becomes:

$$\tilde{\mathcal{L}} = \frac{1}{2}\,|\tilde{g}|^{1/2}\left(\tilde{g}^{\mu\nu}\partial_{\mu}\tilde{\phi}\partial_{\nu}\tilde{\phi} - m^2\tilde{\phi}^2 - \frac{1}{6}\,\tilde{R}\,\tilde{\phi}^2\right)$$

and in terms of infinitesimals to first order:

$$\tilde{\mathcal{L}} = \frac{1}{2}\,(1+2\lambda)|g|^{1/2}([1-\lambda]g^{\mu\nu}\partial_{\mu}$$

$$\left\{\left(1 - \frac{1}{2}\lambda(x)\right)\phi(x)\right\}\partial_{\nu}\left\{\left(1 - \frac{1}{2}\lambda(x)\right)\phi(x)\right\}$$

$$-\frac{1}{6}\{(1+\lambda)R + 3\Box\lambda\}\left(1 - \frac{1}{2}\lambda\right)^2\phi^2(x)$$

At first order $\left(1 - \frac{1}{2}\lambda\right)^2 \longrightarrow (1-\lambda)$, and we can write:

$$\tilde{\mathcal{L}} = \mathcal{L} - \frac{1}{2}\,|g|^{1/2}g^{\mu\nu}\{\phi\partial_{\mu}\lambda\partial_{\nu}\phi\} - \frac{1}{4}\,|g|^{1/2}\,(\Box\lambda)\phi^2$$

And using $\Box\lambda = \partial_{\mu}\left(|g|^{1/2}g^{\mu\nu}\partial_{\nu}\lambda\right)$, we have:

$$\tilde{\mathcal{L}} = \mathcal{L} - \partial_{\mu}\left(\frac{1}{4}|g|^{1/2}g^{\mu\nu}\phi^2\partial_{\nu}\lambda\right) = \mathcal{L}.$$

The total derivative gives rise to a surface term, which has no variation so is dropped, thus the conditions $p = -1/2, z = 1/6$ give rise to a conformal transformation as indicated. Let's now build a finite transformation from our infinitesimal trsnaormation:

$$\Omega^2(x) = \exp[\lambda(x)] \rightarrow \ \lambda = 2\ln\Omega$$

So,

$$\tilde{\mathcal{L}} = \mathcal{L} - \partial_{\mu}\left(\frac{1}{2}|g|^{1/2}g^{\mu\nu}\phi^2\partial_{\nu}\,ln\,\Omega\right)$$

From $T^{\mu\nu} = -|g|^{1/2} \frac{\delta S}{\delta g_{\mu\nu}(x)}$ and $S = \int d^n x \frac{1}{2} |g|^{1/2}\big(g^{\mu\nu}\partial_\mu\phi\, \partial_\nu\phi - m^2\phi^2 - zR\phi^2\big)$ we want to compute δS. Let's start with computing $\delta g^{\mu\nu}$, $\delta|g|^{1/2}$, and δR ($R = R^{\mu\nu} g_{\mu\nu}$):

$\delta g^{\mu\nu} = -g^{\mu p} g^{\nu\sigma} \delta g_{\rho\sigma}$

$\delta|g|^{1/2} = \frac{1}{2}|g|^{1/2} g^{\mu\nu} \delta g_{\mu\nu}$

$\delta R = \tilde{R} - R = -\lambda R + \frac{1}{2}\Box\lambda = R^{\mu\nu}\delta g_{\mu\nu} + \frac{1}{2}|g|^{-1/2}\partial_\mu\big(|g|^{-1/2} g^{\mu\nu}\partial_\nu\lambda\big)$

$\quad = -R^{\mu\nu}\delta g_{\mu\nu} + \frac{1}{2}|g|^{-1/2}\partial_\mu\big(|g|^{-1/2} g^{\mu\nu}\partial_\nu\big[2g^{\rho\sigma}\lambda g_{\rho\sigma} + 2\lambda g^{\rho\sigma} g_{\rho\sigma}\big]\big)$

$\quad = -R^{\mu\nu}\delta g_{\mu\nu} + g^{\rho\sigma} g^{\mu\nu}\big(\delta g_{\rho\sigma;\mu\nu} - \delta g_{\rho\mu;\sigma\nu}\big)$

So,

$$\delta S = \frac{1}{2}\int d^n x\ |g|^{1/2} \{\frac{1}{2} g^{\mu\nu}\delta g_{\mu\nu}\big(g^{\rho\sigma}\partial_\rho\phi\partial_\sigma\phi - m^2\phi^2 - zR\phi^2\big)$$
$$-\delta g_{\rho\sigma}\nabla^\rho\phi\nabla^\sigma\phi - z\big[-R^{\mu\nu}\delta g_{\mu\nu} + g^{\rho\sigma}g^{\mu\nu}\big(\delta g_{\rho\sigma;\mu\nu} - \delta g_{\rho\mu;\sigma\nu}\big)\big]\phi^2\}$$

Let $\delta g_{\mu\nu}$ and its first derivative be zero on boundary:

$\int d^n x\, |g|^{1/2} g^{\rho\sigma} g^{\mu\nu} \delta g_{\rho\sigma;\mu\nu}\phi^2 = \int d^n x\, |g|^{1/2}\, g^{\rho\sigma}\delta g_{\rho\sigma}\, \Box\phi^2$

$\int d^n x\, |g|^{1/2} g^{\rho\sigma} g^{\mu\nu} \delta g_{\rho\sigma;\mu\nu}\phi^2 = \int d^n x\, |g|^{1/2} g^{\rho\sigma} g^{\lambda\nu}\delta g_{\mu\nu}\, \nabla_\sigma\nabla_\lambda(\phi^2)$

Then,

$$\delta S = -\frac{1}{2}\int d^n x\, |g|^{1/2}\, T^{\mu\nu}\delta g_{\mu\nu}$$

with

$$T^{\mu\nu} = \nabla^\mu\phi\nabla^\nu\phi - \frac{1}{2}g^{\mu\nu}\nabla^\rho\phi\nabla_\rho\phi + \frac{1}{2}g^{\mu\nu}m^2\phi^2 - z\left(R^{\mu\nu} - \frac{1}{2}g^{\mu\nu}R\right)\phi^2$$
$$+z\big[g^{\mu\nu}\Box(\phi^2) - \nabla^\mu\nabla^\nu(\phi^2)\big]$$

From this result we can see that:

$$\nabla_\mu T^{\mu\nu} = 0,$$
$$T_\mu^\mu = 0 \text{ when } \xi = \frac{1}{6}, m = 0$$

Quantization follows by imposing canonical quantization rules.

8.6 Spatially flat isotropically changing metric

Let's now consider the spatially flat isotropically changing metric:

$$ds^2 = dt^2 - a^2(t)(dx^2 + dy^2 + dz^2)$$

In what follows we consider the situation where $a(t)$ has arbitrary time dependence except asymptotically, where it becomes constant for $t = \pm\infty$.

Let's consider what happens with the massless minimally coupled field ($z = 0$) in this situation:

$$\Box\phi = |g|^{1/2}\,\partial_\mu\left(|g|^{1/2}\,g^{\mu\nu}\,\partial_\nu\phi\right) = 0, \quad |g|^{1/2} = a^3$$

Thus,

$$\Box\phi = a^{-3}\,\partial_t\,(a^3\partial_t\phi) - a^{-2}\sum_i \partial_i^2\phi = 0.$$

Let's expand the field using:

$$\phi = \sum_{\vec{k}}\left\{A_{\vec{k}}\,f_{\vec{k}}(x) + A_{\vec{k}}^{\dagger}\,f_{\vec{k}}^{*}\,(x)\right\}$$

This is the formulation in a box with side L ($V = L^3$), with $L \longrightarrow \infty$ taken later:

$$f_{\vec{k}} = V^{-1/2}\,e^{i\vec{k}\cdot\vec{x}}\,\Psi_{\vec{k}}(\tau), \quad k^i = \frac{2\pi}{2}n^i, \quad k = |\vec{k}|, \quad \tau = \int^t a^{-3}(t')dt'$$

Thus,

$$\frac{\partial^2\Psi_k}{\partial\tau^2} + k^2a^4\Psi_k = 0.$$

Now, given the asymptotic behavior of $a(t) \to a$, a constant, we have as $t \to -\infty$ the Minkowski space field expansion given by:

$$(f_k, f_{k'}) = \delta_{k,k'}\ , \quad (f_k, f_{k'}^*) = 0$$

$$\phi(x) = \sum_k\left(a_k f_k(x) + a_k^{\dagger} f_k^*(x)\right)$$

$$\left[a_k, a_{k'}^{\dagger}\right] = \delta_{k,k}\quad , \qquad [a_k, a_{k'}] = 0$$

where the constant scale factor a_1 (as $t \to -\infty$) taken into account in the mode solutions:

$$f_{\vec{k}} \sim (Va_1^3)^{-1/2}\ (2\omega_{1k})^{-1/2}\exp\left[i\left(\vec{k}\cdot\vec{x} - \omega_{1k}t\right)\right], \quad \omega_{1k} = {k}/{a_1}$$

The initial Minkowski space has the metric above with $a = a_1$. The coordinates are rescaled by $x^{i'} = ax^i$ to arrive at the Minkowski metric and $x^{i'}$ is the physical or measured distance. Also

$$k^{i'} = k^i/a_1\ , \quad |\vec{k}'| = k/a_1 = \omega_{1k}\,.$$

Company $f_{\vec{k}} = V^{-1/2}\,e^{i\vec{k}\cdot\vec{x}}\,\Psi_{\vec{k}}(\tau)$ with the asymptotic form:

$$f_{\vec{k}} \sim (Va_1^3)^{-1/2}\ (2\omega_{1k})^{-1/2}\exp\left[i\left(\vec{k}\cdot\vec{x} - \omega_{1k}t\right)\right]$$

reveals the asymptotic form of $\Psi_{\vec{k}}(\tau)$, as $t \to -\infty$:

$$\Psi_k(\tau) \sim\ (2\omega_{ik}a_1^3)^{-1/2}\exp(-i\,\omega_{ik}a_1^3\tau),$$

where τ has any constant from integration absorbed into choice of time origin.

We can now patch through the Minkowski space relations that are conserved as we go to curved space. Minkowski space makes the initial calculation of those values or definitions well posed:

Scalar product is conserved:

$$(f_k, f_{k'}) = \delta_{k,k'}$$

Quantization commutators conserved:

$$\left[A_{\vec{k}}, A^{\dagger}_{\vec{k}'}\right] = \delta_{\vec{k},\vec{k}'}, \left[A_{\vec{k}}, A_{\vec{k}'}\right] = 0,$$

where $A_{\vec{k}}$ annihilates particles with momentum $\vec{k}/a$, energy ω/a, in initial space. WE also have from the Hamiltonian formulation that $\pi = a^2\partial_t\phi$, so we can now verify the field canonical commutation relations:

$$[\phi(x,t), \pi(x',t)] = a^3(t)\sum_{k,k'}\left[\left\{A_{\vec{k}}f_{\vec{k}} + A^{\dagger}_{\vec{k}}f^{*}_{\vec{k}}\right\}\partial_t\left\{\begin{matrix}A_{k'}f_{k'}(x') \\ +A^{\dagger}_{k'}f^{*}_{k'}(x)\end{matrix}\right\}\right]$$
$$-\partial_t\{\cdots\}\cdot\{\cdots\}$$

Thus

$$[\phi(x,t), \pi(x',t)] = a^3(t)\sum_{k}\{f_{\vec{k}}(x)\partial_t f^{*}_{\vec{k}}(x') - f^{*}_{\vec{k}}(x,t)\partial_t f_{\vec{k}}(x',t)\}.$$

To show that the right-hand-side expression in $f_{\vec{k}}$ is the delta function needed, consider h an arbitrary solution of $\Box h = 0$:

$$h(\vec{x},t) = \sum_{k}\{f_{\vec{k}}(\vec{x},t)(f_{\vec{k}}, h) - f^{*}_{\vec{k}}(\vec{x},t)(f^{*}_{\vec{k}}, h)\}$$

$$= -i\int d^3x'(t)\sum_{\vec{k}}\{f_{\vec{k}}(\vec{x},t)\partial_t f^{*}_{\vec{k}}(\vec{x}',t) - f^{*}_{\vec{k}}(\vec{x},t)\partial_t f_{\vec{k}}(\vec{x}',t)\}h(\vec{x}',t)$$

$$+i\int d^3x' a^3(t)\sum_{\vec{k}}\{f_{\vec{k}}(\vec{x},t)f^{*}_{\vec{k}}(\vec{x}',t) - f^{*}_{k}(\vec{x},t)f_{k}(\vec{x}',t)\}\partial_t h(\vec{x}',t)$$

From which we see that:

$$a^3(t)\sum_{k}\{f_{\vec{k}}(x)\partial_t f^{*}_{\vec{k}}(x') - f^{*}_{\vec{k}}(x,t)\partial_t f_{\vec{k}}(x',t)\} = \delta(\vec{x},\vec{x}')$$

and

$$a^3(t)\sum_{k}\{f_{\vec{k}}(x)f^{x}_{k}(x') - f^{*}_{k}(x)f_{k}(x')\} = 0$$

Thus,

$$[\phi(x,t), \pi(x',t)] = \delta(\vec{x},\vec{x}').$$

8.7 Generic Cosmological Particle Production

In the initial Minkowski Space $f_{\vec{k}}$ is a positive frequency solution of the field equations and $A_{\vec{k}}$ annihilates a particle at early times. Suppose we have no particles at early times: $A_{\vec{k}}|0\rangle = 0$, for all $\vec{k}$. The time development of $\Psi_{\vec{k}}(\tau)$ is governed by $\frac{d^2\Psi_k}{\partial\tau^2} + k^2a^4\Psi_k = 0$, which has two linearly independent solutions $\Psi_k^{(\pm)}(\tau)$ (and $a_2 = a(t)$ at this later time, and is approximately constant):

$$\Psi_{\vec{k}}^{(\pm)} \sim (2a_2^3\,\omega_{2k})^{-1/2}\exp\,(\mp\, i\omega_{2k}a_2^3\tau),$$

and using the standard modal decomposition:

$$\Psi_{\vec{k}}(\tau) = \alpha_k\Psi_k^{(+)}(\tau) + \beta_k\Psi_k{}^{(-)}(\tau)$$

we have

$$\Psi_{\vec{k}}(\tau) \sim (2a_2^3\,\omega_{2k})^{-1/2}\left[\alpha_k e^{-ia_2^3 i\omega_{2k}\tau} + \beta_k e^{ia_2^3 i\omega_{2k}\tau}\right].$$

The Wronskian of the second-order differential equation gives a conserved quantity:

$$\Psi_{\vec{k}}\partial_\tau\Psi_k^*\partial_\tau\Psi_k = i$$

where the constant is determined to be i using the early-time asymptotic form. At later times, in terms of the above modal decomposition, we have from this:

$$|\alpha_k|^2 - |\beta_k|^2 = 1.$$

If we write the future time asymptotic solution as:

$$\phi(x) = \sum_k \{a_{\vec{k}}g_{\vec{k}}(x) + a_{\vec{k}}^t\, g_{\vec{k}}^*(x)\}$$

with

$$g_{\vec{k}}(x) \sim (a_2^3)^{-1/2}(2\omega_{2k})^{-1/2}\exp\left[i\left(\vec{k}\cdot\vec{x} - \omega_{2k}t\right)\right]$$

and

$$a_{\vec{k}} = \alpha_k A_{\vec{K}} + \beta_k^*\, A_{-\vec{k}}^\dagger\,.$$

The last relation is known as a Bogoliubov transformation (originally arose in condensed matter physics). We can now verify consistency on the operator commutator relation:

$$\left[a_{\vec{K}}, a_{\vec{k}'}^\dagger\right] = \delta_{\vec{k},\vec{k}'}(|\alpha_k|^2 - |\beta_k|^2) = \delta_{\vec{k},\vec{k}'}$$

We can now see that if $a(t)$ is such that $|\beta_k|$ is non-zero, then particles are created by the changing scale factor of the universe [73]:

$$\langle N_{\vec{k}}\rangle_{t\to\infty} = \left\langle 0\middle|a_{\vec{k}}^\dagger a_{\vec{k}}\middle|0\right\rangle = |\beta_k|^2$$

When had no particles to start with:

$$\langle N_{\vec{k}}\rangle_{t\to\infty} = \langle 0|A_k^\dagger A_k|0\rangle = 0.$$

No Cosmological Particle Production when Conformally Invariant

$L = \frac{1}{2}\,|g|^{1/2}\left(g^{\mu\nu}\partial_\mu\phi\partial_\nu\phi - \frac{1}{6}\,R\,\phi^2\right)$ was shown to be invariant under a curved spactime conformal transformation:

$g^{\mu\nu} \to \tilde{g}_{\mu\nu}, \phi \to \tilde{\phi}$

$\tilde{g}_{\mu\nu}(x) = \Omega^2(x)g_{\mu\nu}(x)$

$\tilde{\phi}(x) = \Omega^{-1}(x)\phi(x)$

Conformal invariance of $S = \int d^4x\,L$ implies $\frac{\delta S}{\delta\phi} = \frac{\delta\tilde{S}}{\delta\phi} = \frac{\delta\tilde{S}}{\delta\tilde{\phi}}\Omega^{-1}$, so, using $|g|^{1/2} = \Omega^4|g|^{1/2}$:

$$\left(\Box + \frac{1}{6}R\right)\phi = \Omega^3\left(\tilde{\Box} + \frac{1}{6}\tilde{R}\right)\tilde{\phi}.$$

Oddly, even though the scalar field is massless and the conformal invariance is when there is no length (or mass) scale, we arrive at an equation of motion with an effective mass term. We can choose a conformal transformation such that $\tilde{R} = 0$, however, to arrive at a set of positive frequency solutions, that are also unchanging in the expansion of the conformally invariant universe, leading to no particle production (since there is no modal mixing, the Bogoliubov transformation is trivial): Since $\left(\tilde{\Box} + \frac{1}{6}\tilde{R}\right)\tilde{\phi} = 0$, and $ds^2 = dt^2 - a^2(t)(dx^2 + dy^2 + dz^2)$ (for spatially flat FRW), define:

$$\eta = \int^t a^{-1}(t')dt' \;\rightarrow\; ds^2 = a^2(t)(d\eta^2 - dx^2 - dy^2 - dz^2)$$

If we now choose the conformal transformation:

$$\Omega = a^{-1}(t) \;\Rightarrow\; \tilde{g}_{\mu\nu} = \eta_{\mu\nu} \;\Rightarrow\; \tilde{R} = 0 \;\Rightarrow\; \tilde{\Box}\tilde{\phi} = 0$$

and we arrive at the simple modal solutions:

$$\tilde{f}_{\vec{k}}(x) = V^{-1/2}(2k)^{1/2}\exp\left[i\left(\vec{k}\cdot\vec{x} - k\eta\right)\right]$$

The corresponding early time modes are:

$$f_{\vec{k}}(x) = a^{-1}\tilde{f}_k =$$

$$\left(Va^3(t)\right)^{-1/2}\left(2\omega_k(t)\right)^{-1/2}\exp\left[i\left(\vec{k}\cdot\vec{x} - \int^t \omega_k(t')\,dt'\right)\right]$$

which are precisely the positive frequency solution at early times. The $f_k(x)$ are also positive frequency at any time, since $a(t)$ is positive, so $\beta_k = 0$ at all times.

The result generalizes to FRW universes which are not spatially flat, and to massless fields of higher spin, as long as they obey conformally invariant free field equations → such as massless neutrinos, photons.

If you believe in gravitons, e.g., Grav as fundamentally field, and to be quantized…:
Gravitons are not excluded from production in a conformally invariant universe, however, since they are linearized perturbation constructs and do not satisfy a conformally invariant wave equation. Thus, gravitons are created by the change of $a(t)$ even in an isotropic conformally invariant universe.

8.8 Spin-statistics
A short derivation is now given for the dynamical spin-statistics relation, for more details see [87].

Note that we are working with a scalar field (spin-0) in the derivation thus far, where the $|\alpha_k|^2 - |\beta_k|^2 = 1$ relation follows purely as a consequence of the field equation. In general, β_k is not zero. Thus, a scalar spin-0 field at early times is dynamically constrained at later times to obey Bose-Einstein statistics (a dynamics constraint that enforces part of the spin-statistics relation). Let's write the late-time creation/annihilation operator commutator and anti-commutator relations:

$$\left[a_{\vec{k}}, a_{\vec{k}'}\right]_{\pm} = \left[\alpha_k A_{\vec{k}} + \beta_k^* A^{\dagger}_{-k}, \alpha_{k'} A_{\vec{k}'} + \beta_{k'}^* A^{\dagger}_{-k'}\right]_{\pm} =$$
$$\alpha_k \beta_{k'}^* \left[A_{\vec{k}}, A^{\dagger}_{-\vec{k}'}\right]_{\pm} + \alpha_{k'} \beta_k^* \left[A^{\dagger}_{-k}, A_{\vec{k}'}\right]_{\pm}$$
$$= (\alpha_k \beta_k^* \pm \alpha_k \beta_k^*) \delta_{\vec{k},-\vec{k}'}$$

$$\left[a_{\vec{k}}, a^{\dagger}_{\vec{k}'}\right]_{\pm} = \left[\alpha_k A_{\vec{k}} + \beta_k^* A^{\dagger}_{-\vec{k}}, \alpha_k^x A_k^{\dagger} \beta_k A_{-\vec{k}}\right]_{\pm} = |\alpha_k|^2 \left[A_{\vec{k}}, A^{\dagger}_{\vec{k}}\right]_{\pm} +$$
$$|\beta_k|^2 \left[A^{\dagger}_{-\vec{k}}, A_{\vec{k}}\right]_{\pm}$$
$$= (|\alpha_k|^2 \pm |\beta_k|^2) \delta_{\vec{k},\vec{k}'}$$

For Bose-Einstein (spin 0) we need $\left[a_{\vec{k}}, a_{\vec{k}'}\right] = 0$ and $\left[a_{\vec{k}}, a^{\dagger}_{\vec{k}'}\right] = (|\alpha_k|^2 - |\beta_k|^2) \delta_{\vec{k},\vec{k}'}$ and this is precisely what we get for the usual commutator forms in the above dynamical solution result.

8.9 Explicit Fock Space Analysis

Consider a Bogoliubov transformation describing a mode change with boson commutation rules dynamically consistent as indicated:

$$a_{\vec{k}} = \alpha_k A_{\vec{k}} + \beta_k^* A^\dagger_{-\vec{k}}, \qquad |\alpha_k|^2 - |\beta_k|^2 = 1$$

So, at early times $A_{\vec{k}}|0\rangle$ for all $\vec{k}$ and at late times $a_{\vec{k}}|0)$ for all $\vec{k}$, where the late-time vacuum is denoted with the curved brace: $|0)$ not $|0\rangle$. We want to find the late-time particle production that will result from $a^\dagger_{\vec{k}}$ acting on $|0)$.

Consider the amplitude $(0|n(\vec{k})n(-\vec{k})|0\rangle$, a state with n particles in $(\vec{k})$ and n in mode $(-\vec{k})$ is:

$$(0|n(\vec{k})n(-\vec{k})|0\rangle = (n!)^{-1}\left(0\left|\left(a_{-\vec{k}}\right)^n\left(a_{\vec{k}}\right)^n\right|0\right\rangle$$

Now, $a_{\vec{k}}|0\rangle = \beta_k^* A^\dagger_{-\vec{k}}\,|0\rangle = \beta_k^*(\alpha_k^*)^{-1} a^\dagger_{-\vec{k}}|0\rangle$, So,

$$(0|n(\vec{k})n(-\vec{k})|0\rangle = (n!)^{-1}(\beta_k^*/\alpha_k^*)^n\,(0|\left(a_{\vec{k}}\right)^n\left(a^\dagger_{-\vec{k}}\right)^n|0\rangle$$

$$= (n!)^{-1/2}(\beta_k^*/\alpha_k^*)^n\left(\left(n(-\vec{k})\right|\left(a^\dagger_{-\vec{k}}\right)^n\left|0\right\rangle$$

$$= (\beta_k^*/\alpha_k^*)^n(0|0\rangle$$

Thus,

$$\left(n(\vec{k}),n(-\vec{k})|0\right\rangle = (\beta_k^*/\alpha_k^*)^n(0|0\rangle,$$

And we see that particles are created in pairs of equal and opposite momenta. Let's refer to the number of pairs of momenta by n_j and write the paired particle production as:

$$(\{n_j(\vec{k}_j)\}|0\rangle = \prod_j (\beta_k^*/\alpha_k^*)^n(0|0\rangle$$

The early-time vacuum, when seen as late times, appears as:

$$|0\rangle = \sum_{\{n_j(\vec{k}_j)\}} |\{n_j(\vec{k}_j)\})\,(\{n_j(\vec{k}_j)\}|0\rangle$$

Let's now evaluate $|(0|0\rangle|$:

$$1 = \sum_{\{n_j(\vec{k}_j)\}} |(\{n_j(\vec{k}_j)\}|0\rangle|^2 = \left(\sum_{\{n_j(\vec{k}_j)\}}\prod_j \left|\beta_{k_j}/\alpha_{k_j}\right|^{2n_j}\right)|(0|0\rangle|^2$$

Recall the relationship:

$$\sum_{\{n_j(\vec{k}_j)\}}\prod_j x^{n_j} = \prod_j\left(\sum_{n_j=0}^{\infty} x^{n_j}\right)$$

Thus

$$1 = |\langle 0|0\rangle|^2 \prod_j \sum_{n_j=0}^{\infty} \left|\beta_{k_j}/\alpha_{k_j}\right|^{2n_j} = |\langle 0|0\rangle|^2 \prod_j \left(1 - \left|\beta_{k_j}/\alpha_{k_j}\right|^2\right)^{-1}$$
$$= |\langle 0|0\rangle|^2 \prod_j \left|\alpha_{k_j}\right|^2$$

So,

$$|\langle 0|0\rangle|^2 = \prod_j \left|\alpha_{k_j}\right|^{-2}$$

Thus

$$\left|\langle \{n_j\}|0\rangle\right|^2 = \prod_j \left(\left|\beta_{k_j}/\alpha_{k_j}\right|^{2n_j} \left|\alpha_{k_j}\right|^{-2}\right).$$

The probability of n particles in $\vec{k}$:

$$P_n(\vec{k}) = |\beta_k/\alpha_k|^{2n}\, |\alpha_k|^{-2}$$

This is consistent with

$$\sum_{n=0}^{\infty} P_n(\vec{k}) = 1$$

And also consistent with our prior result:

$$\langle N_{\vec{k}}\rangle_{t\to\infty} = \sum_{n=0}^{\infty} P_n(\vec{k}) = |\beta_k|^2$$

8.10 High Frequency black-body distribution

Let's derive the high-frequency black-body distribution. Let's rewrite the above relation for $P_n(\vec{k})$ with the dynamical relation to get:

$$P_n(\vec{k}) = |\beta_k/\alpha_k|^{2n}\, (1 - |\beta_k/\alpha_k|^2)$$

What is the asymptotic form of $|\beta_k/\alpha_k|^2$, $P_n(\vec{k})$, for large k? Let $a_<$ and $a_>$ denote the lesser and greater of a_1, a_2 from the prior discussion:

$$|\beta_k/\alpha_k|^2 \approx \exp[2\pi ks(a_>^2 - a_<^2)]/\exp[2\pi ks(a_>^2 + a_<^2)]$$
$$\approx \exp[-4\pi ksa_<{}^2 a_<^2] \approx \exp[-\mu k]$$

where $\mu \equiv 4\pi ksa_<^2$. In an expanding universe as discussed $a_2 = a_>, a_1 = a_<$, and suppose
$a_2 \gg a_1$ (such as with some inflationary models).The low frequency end of the spectrum is effectively red shifted away, since $|\beta_k/\alpha_k|^2 \approx \exp[-\mu k]$, and we have:

$$P_n(\vec{k}) = \exp[-n\mu k][1 - \exp[-\mu k]]$$

From this we have for the average number in mode $\vec{k}$:

$$\langle N_{\vec{k}} \rangle = |\beta_k|^2 = \frac{1}{\exp(\mu k) - 1}$$

The average particle density as $V \to \infty$:

$$\langle N \rangle = (2\pi^2 a_2^3)^{-1} \int_0^\infty dk k^2 \, [\exp(\mu k) - 1]^{-1}$$

Note: the form $P_n = \exp[-n\mu k][1 - \exp[-\mu k]]$ for the probability of observation of n particles in mode $\vec{k}$ is the same as for blackbody radiation with temperature:

$$T = (k_B \mu a_2)^{-1}.$$

Technically, the state $|0\rangle$ is actually a coherent superposition of states containing pairs of particles at late times, in contrast to black body radiation which is an incoherent mixture. Local observations would be unable to distinguish between this created radiation and strict black body radiation, however, because the correlation pairs of particles would have separations of a cosmological scale. In addition, interactions with other systems would also tend to destroy the correlation.

8.11 Adiabatic Vacuum

In the case of a FRW model universe with static in and out regions, the notion of particle is clear. If either the in or out vacuum states are chosen as the state of the quantum field, then a comoving detector will, over its entire world line, almost certainly fail to detect quanta in the high energy modes. So long as the mode frequency is much greater than the expansion rate, the probability of no detector response will remain very close to unity. However, for the lower modes, there will be quanta registered, signaling a breakdown of the approximation used to define the vacuum state.

If, on the other hand, there are no static in and out regions, then a method must be found of selecting those mode solutions of the field equation that come in some sense "closest" to the Minkowski space limit, Physically this might be envisaged as a construction that "least disturbs" the field by the expansion, i.e., results in a definition of particles for which there is minimal particle production by the changing geometry. With such a perturbative construct in mind, the mathematical description of adiabatic vacua with entail some sort of high-mass expansion of the field modes.

Adiabatic expansion of Green functions:
In the regularization of uv divergences only the high frequency field behavior is of interest. Because the high frequencies probe only the short distances, one is led to examine short distance approximations. Special interest attaches to the short distance behavior of Green functions such as $G_F(x,x')$ in limit $x \to x'$. Bunch and Parker (1979) [87] obtain an adiabatic expansion of G_F. Their approach does the following:

(1) Introduce Riemann normal coordinates. Expand $g_{\mu v}(x) = \eta_{\mu v} + \cdots$
(2) Define $\mathcal{G}_F(x,x') = \left(-g(x)\right)^{1/4} G_F(x,x'), \to \mathcal{G}_F(k)$
(3) Expand $\mathcal{G}_F(k)$ in normal coordinates, regain $\mathcal{G}_F(x,x')$, then $G_F(x,x')$

We thereby arrive at the DeWitt-Schwinger (DS) expansion:

$$G_F(x,x') = -i\Delta^{1/2}(x,x')(4\pi)^{-n/2} \int_0^\infty ids(is)^{-n/2} \exp[-im^2 s + (\sigma/2is)]F(x,x';is)$$

Since we have not imposed global boundary conditions on the Green function solution of the field equation the DS expansion does not determine the particular vacuum state in $iG_F(x,x') = \langle 0|T(\phi(x)\phi(x'))|0\rangle$. In particular, the "$i\varepsilon$" in the expansion of G_F only ensures that the DS equation represents the expectation value of a time ordered product of fields. (In flat space the $i\varepsilon$ gives additional information concerning the global nature of the states.) In a FRW spacetime, the Feynman Green function calculated as an expectation value in an adiabatic vacuum $|0^A\rangle$

$$iG_F^A(x,x') = \langle 0^A|T(\phi(x)\phi(x'))|0^A\rangle$$

should have an expansion of the form of G_F^{DS}. The vacuum $|0^A\rangle$ is defined in terms of a set of adiabatic posture frequency modes $u_{\vec{k}}$ given by

$$u_{\vec{k}} = \alpha_k^{(A)}(\eta)u_{\vec{k}}^{(A)} + \beta_k^{(A)}(\eta)u_{\vec{k}}^{(A)*}$$
$$\alpha_k^{(A)}(\eta_0) = 1 + O\left(T^{-(A+1)}\right)$$
$$\beta_k^{(A)}(\eta_0) = 0 + O\left(T^{-(A+1)}\right)$$

For some η_0, and so one can also write G_F^A as

$$iG_F^A(x,x') = \theta\left(x^0 - x^{0'}\right) \int d^{n-1}k\, u_{\vec{k}}(x) u_{\vec{k}}^*(x')$$
$$+ \theta\left(x^{0'} - x^0\right) \int d^{n-1}k u_{\vec{k}}^*(x)\, u_{\vec{k}}(x)$$

One can verify that two expansion of G_F^A agree to order A [87]. In particular, one can check that the "$i\varepsilon$" is necessary to guarantee the time ordering in both expansions.

The response of a comoving particle detector in FRW to Ath order vacuum remains unexcited with a probability that differs from unity only by terms of order $A + 1$. If the detector is not comoving there will generally be particle detection in the high frequency modes that falls to zero slower than T^{-A}.

8.12 Adiabatic regularization

Adiabatic regularization as applied to the free scalar field in a Robertsen-Walker universe:

$$ds^2 = dt^2 - a^2(t)d\vec{x}^2$$
$$L = \frac{1}{2}|g|^{1/2}(g^{\mu v}\, \partial_u\phi\partial_v\phi - m^2\phi^2 - \xi R\phi^2)$$
$$\Rightarrow (\Box + m^2 + \xi R)\phi = 0$$

Now, how do we construct the Hilbert space of state vectors for our system in the absence of an isometry or belling vector in the timelike direction?

In the absence of special symmetries which serve to specify a unique vacuum state, any space of state vectors which can reproduce any physically allowed set of expectation values may be used in setting up the quantum field theory.

One condition must be met in the space described above, the particle numbers in a given mode should not change if the rate of change of $a(t)$ is sufficiently slow. So, introduce slowness parameter: $a_T(t) \equiv a(t/T)$ and consider derivatives of order T^{-1}.

Adiabatic condition: The orthonormal solution, $f_{\vec{k}}(x)$ of the field equation, which will be used to expand the quantum field and to define the basis of the space of state vectors, should in lowest adiabatic order have the form:

$$f_{\vec{k}}(x) \sim (Va(t)^3)^{-1/2} \left(2\omega_k(H)\right)^{-1/2} \exp\left[i\left(\vec{k}\cdot\vec{x} - \int^t \omega_k(t')dt'\right)\right]$$

With $\omega_k(t) = [k^2/a(t)^2 + m^2]^{1/2}$

The adiabatic condition is satisfied at all times if $a(t)$ approaches a constant at early or late times and $f_{\vec{k}}(x)$ is taken to approach Minkowski positive frequency solution. Causality then motivates extending the adiabatic condition to hold when there are no asymptotically static regions. Now,

$$\phi(x) = \sum_{\vec{k}} \left\{A_{\vec{k}} f_{\vec{k}}(x) + A_{\vec{k}}^{\dagger} f_k^*(x)\right\}$$

Adiabatic vacuum state: $A_{\vec{k}}|0\rangle = 0$. The adiabatic expansion:

$$f_k = (2v)^{-1/2} a(t)^{-3/2} h_k(t) e^{i\vec{k}\cdot\vec{x}}$$

Chakraborty

$$\frac{d^2}{dt^2} h_k + \Omega_k^2 h_k = 0 \qquad \Omega_k^2 = \omega_k^2 + \sigma$$

$$\omega_k(t) = \left(k^2/a(t)^2 + m^2\right)^{1/2}$$

$$\sigma(t) = \cdots$$

$$\langle 0|\phi(x)\phi(x')|0\rangle = \sum_{\vec{k}} f_{\vec{k}}(x) f_{\vec{k}}^*(x')$$

$$= \frac{1}{2(2\pi)^3} [a(t)a(t')]^{-3/2} \int d^3 k e^{i\vec{k}\cdot(\vec{x}.\vec{x'})} h_k(t) h_k^*(t')$$

Now,

$Q^{(n)} \sim$ Term of adiabatic n order n in expansion of Q

$Q^{(n)} \sim$ Term in $Q^{(n)}$ which has smallest power of k^{-1}.

$Q_1{}^{(n+1)}$ has at least one more power of k^{-1} than $Q_1^{(n)}$

In lowest adiabatic order and for large K:

$\langle 0|\phi(x)\phi(x')|0\rangle \alpha \int d^3 k e^{i\vec{k}\cdot(\vec{x}-\vec{x}')} k^{-1} e^{-ik \int_{t'}^{t} a^{-1} dt''}$

When $\vec{x} = \vec{x}'$ and $t = t'$ diverges quadratically.

When $\vec{x} \neq \vec{x}'$ or $t = t'$ oscillating exp. make integral well defines as a distribution (introduce $e^{-\alpha k}$ will $\alpha \to 0^t$).

Divergences for large K are called UV divergences. In adiabatic regularization, the physically relevant finite expression is obtained from the formal one containing UV divergences by subtracting mode by mode each term in the adiabatic expansion of the integrand which contains at least one UV divergent part for arbitrary values of the parameters of the theory (in this case ξ and m).

Is a unique prescription subtracting a minimal number of terms such that a quantity quadratic in the fields will be continuous as a function of the independent parameters (here z, m; except in exceptional circumstances) ? → trace anomaly

Applying the adiabatic procedure to $\langle 0|\phi^2(x)|0\rangle$

$$h_k(t) \sim \omega_k^{-1/2}(t)\exp\left(-i\int^t \omega_k(t')dt'\right)$$
$$\langle 0|\phi(x)^2|0\rangle \sim (4\pi^2 a(t)^3)^{-1}\int_0^\infty dk\, k^2\, \omega_k(t)^{-1}$$

Notice $T^\mu_\mu = m^2\phi^2$ when $\xi = 1/6$ $(m = 0$ in limit)

But $\langle 0|T^\mu_\mu|0\rangle = m^2\langle 0|\phi^2|0\rangle$ does not imply (phys) equality:
$\langle 0|T^\mu_\mu|0\rangle_{phys} \neq m^2\langle 0|\phi^2|0\rangle_{phys}$!

Regularization is done on T^μ_ν prior to contraction, thus,

$$\langle 0|T^\mu_\mu|0\rangle_{phys} \neq m^2\langle 0|\phi^2|0\rangle_{phys} - m^2(4\pi^2 a(t)^3)^{-1}\int_0^\infty dk\, k^2\,(\omega_k(t)^{-1})^{(4)}$$

Take $m \to 0, \langle\phi^2\rangle$ has no divergences as $m \to 0$

Integral term yields constant value → trace anomaly or conformal anomaly for a free scalar field with $m \to 0$ and $\xi = 1/6$.

$$\langle 0|T^{\mu\nu}|0\rangle \sim \langle 0|T^{\mu\nu}|0\rangle^{(0)} + \langle 0|T^{\mu\nu}|0\rangle^{(2)}T^{-2} + \cdots$$
$$\nabla_\mu \langle 0|T^{\mu\nu}|0\rangle = 0$$

Thus, each adiabatic order variants on application of ∇_μ. Such $\langle 0|T^{\mu\nu}|0\rangle$ and terms subtracted at each adiabatic order all satisfy $\nabla_\mu\{\square\} = 0$ then!
$\nabla_\mu\langle 0|T^{\mu\nu}|0\rangle_{phys} = 0$

Consider the vector field:

$\xi^{\mu} = a(t)\delta_0^{\mu}$ and $E_z\, g_{\mu\nu}(x) = \lambda(x)\, g_{\mu\nu}(x)$
$E_z\, g_{\mu\nu} = \xi^{\alpha}\, \partial_{\alpha}\, g_{\mu\nu} + g_{\mu\alpha}\, \partial_{\nu}\xi^{\alpha} + g_{\alpha\nu}\, \partial_{\mu}\xi^{\alpha} = 2\nabla(\mu\xi\nu)$
And $\lambda(x) = 2\dot{a}(t)$
ξ satisfying the above is a conformal killing vector field:
$\nabla_{\mu}\left(\langle T^{\mu\nu}\rangle\xi_{\nu}\right) = \langle T^{\mu\nu}\rangle\nabla_{(\mu\xi\nu)} = \frac{1}{2}\lambda\langle T^{\mu\nu}\rangle g_{\mu\nu}$

$$\frac{1}{2}\int d^4x\sqrt{-g}\ \lambda\langle T_{\mu}^{\mu}\rangle = \int_{t_2} d^3x\ \sqrt{-g}\,\langle T^{o\nu}\rangle\xi_{\nu} - \int_{t_1} d^3\gamma\sqrt{-g}\,\langle T^{o\nu}\rangle\xi_{\nu}$$

$$\int d^4xa^3\acute{a}\langle T_{\mu}^{\mu}\rangle = \int_{+_2} d^3xa^4\langle T_{oo}\rangle - \int_{+_2} d^3xa^4\langle T_{oo}\rangle$$

$$\int_{t_1}^{t_2} d^4xa^3\dot{a}\langle T_{\mu}^{\mu}\rangle = a^4(t_2)\,\langle T_{oo}(t_2)\rangle - a^4(t_1)\,\langle T_{oo}(t_1)\rangle$$

$$\int_{t_1}^{t_2} d^4xa^3\dot{a}\langle T_{\mu}^{\mu}\rangle = \frac{1}{480\pi^2}\int_{t_1}^{t_2} dt\,\frac{d}{dt}\left(-\dot{a}^4 + a^2\dot{a}\dddot{a} + a\dot{a}^2\ddot{a} - \frac{1}{2}a^2\ddot{a}^2\right)$$
$$= g(t_2) - g(t_1)$$

$$g(t) = \frac{1}{480\pi^2}\left(-\dot{a}^4 + a^2\dot{a}\dddot{a} + a\dot{a}^2\ddot{a} - \frac{1}{2}a^2\ddot{a}^2\right)$$

Thus, $a^4(t)\,\langle T_{oo}(t)\rangle = g(t) + E$ $\quad$ E is a const

Thus, for $\xi = {}^1/_6, m \overrightarrow{\equiv} 0$

$$\langle T_{oo}(t)\rangle = \frac{1}{480\pi^2}\left(-\frac{\dot{a}^4}{a^4} + \frac{\dot{a}\dddot{a}}{a^2} + \frac{\dot{a}^2\ddot{a}}{a^3} - \frac{1}{2}\frac{\ddot{a}^2}{a^2}\right) + \frac{E}{a^4}$$

Symmetry $\langle T_1^1\rangle = \langle T_2^2\rangle = \langle T_3^3\rangle$
$\langle T_1^1\rangle = \frac{1}{3}\left(\langle T_{\mu}^{\mu}\rangle = \langle T_0^0\rangle\right)$
$$= \frac{1}{1440\pi^2}\left\{\frac{\dot{a}^4}{a^4} + \frac{2\dot{a}\dddot{a}}{a^2} - \frac{4\dot{a}^2\ddot{a}}{a^3} + \frac{3}{2}\frac{\ddot{a}^2}{a^2} + \frac{\ddot{a}^1}{a}\right\} - \frac{E}{3a^4}$$

$\langle T_{\mu\nu}\rangle = 0$ for $\mu \neq \nu$

No particle creation for $g = {}^1/_6, m = 0$

$\lambda\phi^4$ term affects ξ after renormalization, thus if initially we have a classically conformally invariant field later we don't, breaking conformal invariance and causing particle creation to occur.

Particle Concept

Two values of ε are of particular interest: (1) the minimally coupled case: $\varepsilon = 0$; (2) the conformally coupled case: $\varepsilon = \frac{1}{4}[(n-2)/(n-1)]$.

In the conformally coupled case, if $m = 0$ the action and hence the field equations are invariant under conformal transformations:

$$g_{\mu v}(x) \to \bar{g}_{\mu v}(x) = \Omega^2(x) g_{\mu v}(x)$$

If the field is assumed to transform as:

$$\bar{\phi}(x) = \Omega^{(2-n)/2}(x)\phi(x)$$

Curved spacetime generalizations of the Green function equations are easily obtainable:

$$[\square_x + m^2 + \varepsilon R(x)] G_f(x, x') = -[-g(x)]^{-1/2}\, \delta(x - x')$$

The difficulty arises in that the usual procedure does not uniquely specify the state $|0\rangle$ (to be explored in detail in Section) nor does it ensure that the solution has the properties of a time-ordered product. To fix the state $|0\rangle$ and impose the time ordering, boundary conditions must be imposed on the solution of the above equation. In Minkowski these boundary conditions take the form of a choice of contour in an integral. In curved spacetime the specification of boundary conditions is not so simple, and will depend on the global features of the particular case under consideration.

The equation arises as to which set of modes furnishes the "best" description of a physical vacuum, i.e., corresponds most closely to our actual experience of "no particles". This question cannot be answered as stated because it is necessary to specify also the details of the quantum measurement process that is used to detect the presence of quanta. In particular the state of motion of the measuring device can affect whether or not particles are observed to be present.

The special feature of Minkowski space is not that there is a unique vacuum (there is not), but that the conventional vacuum state as defined in terms of modes can be the agreed-upon vacuum for all inertial measuring devices, throughout the spacetime. This is because the vacuum defined by $a|0\rangle = 0$ is invariant under the Poincare group and so are the set of inertial observers in Minkowski space.

Thus, the particle concept does not have universal significance. Particles may register their presence on some detectors but not on others, there is an essential observer-dependent quality about them. Part of the reason for the nebulousness of the particle concept is it is actually defined to have a nebulous global nature → modes are defined on the whole of spacetime (or at least a large patch) so that a particular observer's specification of the field mode decomposition, hence number operator, will depend on his entire past history. To obtain a more objective probe of the state of a field one must construct locally–defined quantities, such as $\langle\psi|T_{\mu v}(x)|\psi\rangle$.

Appendix A. The Appearances of Alpha

The quantum domain is clearly part of the physics if Planck's constant is present in the governing equations. More subtle is that the quantum domain is part of the physics if the Fine Structure constant α (alpha) is present. Part of the subtlety is that alpha may be present but not be recognized as such. Beginning with classical physics, we can ask if alpha is revealed in some context (it is), then repeat for the physics disciplines that follow: electromagnetism, early atomic/quantum theory; modern quantum mechanics; quantum field theory and QED; and maximal perturbation parameter in chiral Emanation process. Clearly alpha will be revealed in many distinctive ways, but it is also informative when alpha is not part of the physics – this appears to be the case for "manifold physics" in the form of general relativity on a geometric manifold and statistical field theory on a neuromanifold (see [88] or Book 6 [6] for a description). The significance of this for measurement theory distinctions of quantum observable and classical apparatus will be discussed as well.

A.1 Universality constant C_∞ and its relation to alpha

The appearance of alpha in classical mechanics (including chaos) occurs for the classical trajectory "at the edge of chaos," where it is at an extremum of the non-chaotic motion of the system. It has been found [89] that the maximal perturbation from a stable state near the chaos boundary has universal maximal perturbation (involving the universal parameter C_∞, sometimes called the Myrberg constant or the second Feigenbaum constant).

Viewed as a maximal perturbation in a 2-D complexified 1-D variable (or dimension), we have a possible connection to alpha, since the latter is a maximal perturbation parameter in chiral trigintaduonion propagation (see Book 7[7]). This relation is found to exist, valid to 9 decimal places, even in the heuristic/approximate form derived in [7]. The relation is:

$$\alpha_{\square}^{-1} = \left(\sqrt{C_\infty}\right)^{29^*}, \quad 29^* \cong 29 + \left(\frac{4\pi}{72}\right)\left[1 + \left(\frac{\pi}{137 \cdot 29}\right)\left(\left(\frac{\pi}{72}\right) + \left(\frac{3}{72}\right)\right)\right],$$

where

$$C_\infty = 1.4011551890920506004 \ldots$$

Results in $\alpha_{\square}^{-1}$ having:

$$\alpha_{\square}^{-1} = 137.035999333370198263 \ldots$$

A.2 In electromagnetic coupling constant

The appearance of alpha in electromagnetism is hidden in plain sight – it is the electric coupling constant between two fundamental charge units:

$$\alpha = \frac{e^2}{4\pi}$$

(with appropriate choice of units).

A.3 A "kabbalistic" spectral-fit parameter

In early atomic experiments discrete spectra were observed for heated gases, such as hydrogen. In describing the spectral lines for Hydrogen, Balmer (1885) [90] was able to obtain an equation for describing the position of spectral lines and observed that, remarkably, only a single fit parameter was required. This result was extended to hydrogenic molecules under Rydberg, but still with a single fit parameters, now referred to as the Rydberg constant R_H. Once we obtain a modern quantum mechanics solution for the system we will find that the Rydberg constant consists of alpha, Planck's constant, the speed of light constant, and the mass of the electron:

$$\alpha \propto R_H.$$

A.4 Perturbation parameter in Modern Quantum Mechanics

In Modern Quantum Mechanics alpha arises in descriptions of charged matter as before, but now in the context of a perturbation analysis. The order of perturbation corresponds to an order of alpha, so a perturbation expansion in terms of orders of alpha can be done. Thus, here we see alpha in the role of a perturbation parameter (perhaps maximal in some context yet to be revealed, is will be shown in Emanator Theory [7]). Thus we have for some observable:

$$E = \sum\nolimits_n E^{(n)} \alpha^n, \ \alpha = \frac{e^2}{4\pi}.$$

A.5 Perturbation parameter in QED

In QED alpha enters a fundamental perturbation parameter. Upon renormalization alpha is seen as a running coupling constant in renormalization group analysis. Due to vacuum polarization effects in QED and QCD, the coupling constant actually depends on the length scale (energies) being probed. Thus, for QED coupling we have:

$$\alpha(r) = \frac{\alpha}{\left(1 - \frac{\alpha}{3\pi}\log(1 + \frac{\hbar}{m_e rc}\right)}$$

Where, as $r \to \infty$, $\alpha(r) \to \alpha$, the low-energy alpha limit (at largest value as running coupling constant) that is approximately what we measure in our low-energy experiments (low-energy when compared to GUT energy scale). Interestingly, the QCD running coupling 'constant' for the strong interaction appears to have very similar behavior:

$$\alpha_S(r) = \frac{\alpha}{\frac{b}{2\pi}\log(\frac{\hbar c}{\Lambda_c r}}.$$

It turns out the formalisms do have similar behavior as renormalizable theories operating in thc samc (four) dimensionality, and therefore have the same (universal) asymptotic running coupling constant behavior (a form of renormalization group universality that garnered the Nobel Prize).

A.6 Perturbation parameter in Emanator Theory

The maximum perturbation in Emanation Theory is first evaluated for a chiral trigintaduonion emanation where we take a norm=1 T_{base}=(A,B) and take the right product with T_{chiral} in the form T_{chiral}=(C,β), with product (A,B)(C,β) proven to be unit norm [91]. In Appendix D we will see that there are 137 independent octonion terms at the octonion sub-level of the new unit norm trigintaduonion that results, which leads to 137 independent terms at component level.

Recall that the exponential function (map) provides a well-defined 'lift' of a hypercomplex (Cayley) algebra from T to a complex Cayley product algebra C × T. Let's represent the system noise in the 137 independent 'noise channels' in the complex trigintaduonion space indicated. Let's assume that system noise (in an established emanation process) has settled, at component-level to be equipartitioned in both real and imaginary parts. Also assume that the total imaginary noise component is 'π' for maximal antiphase overall (to be justified later). This leads to a description of maximum transmittable chiral emanation noise. We have 137 terms with max unit norm each, for the real part, and for the imaginary part have a "phase angle" β such that 137β=π. The noise magnitude at octonionic-component level is then given by the right triangle with real part = 137 and angle β=π/137, thus maximum chiral emanation noise magnitude is:

$$H = \frac{137}{\cos\left(\frac{\pi}{137}\right)}.$$

The achiral emanation 'seen' is comprised of a sum over chiral emanations of the different types. For achiral emanation noise we have 29 "free" components, each with 137 independent terms. For maximum achiral emanation we thus have 137 x 29 independent terms that are built from the aforementioned chiral emanation terms (to make achiral). If we equipartition as before, with noise magnitude H, we have a "noise triangle" with magnitude (hypotenuse) H and with angle $\theta = \pi/137x29$. The imaginary part is then (H)sin(θ). As regards the H magnitude separated form (separating out the 'H' factor for now), we have for the imaginary part sin(θ)c. As before, we take maximal noise transmission when all the imaginary parts add to maximal antiphase. Given the equipartitioning assumption, we then simply have the factor 137x29:

$$\sin(\theta)\, c\, (137x29) = \pi \rightarrow c = \theta/\sin\theta.$$

The maximum real noise perturbation that the system can have is then α, where:

$$\alpha^{-1} = \frac{137}{\cos\beta}\cos\theta\frac{\theta}{\sin\theta}, \qquad where\ \beta = \frac{\pi}{137}\ and\ \theta = \frac{\pi}{137x29}$$

Thus,

$$\alpha^{-1} = 137.03599978669910 \ldots.$$

A.7 Nonappearance of alpha in manifold physics

If alpha and Planck's constant are taken as distinctive markers of quantum processes being relevant in the description, consider that such markers are absent in 'manifold-based physics'. This is consistent with quantum measurement theory if the manifold is deemed not quantum observation but classical apparatus. This means that gravitation and geometry (in 4 dimensions) is not quantum (e.g., no quantum gravity in 4D spacetime, but possible in 3D spacetimes where the AdS/CFT renormalizability properties exist, such as, possibly, at Black Hole horizons). No quantum gravity means that we resolve this gap in the theory, and the issues of objective reduction, by refencing a larger theory, referred to as emanator theory, which projects the {classical apparatus}⊗{quantum matter} that we see. In time, the maximal information emanation will shift (adiabatically) the description such that we see a new, slightly altered (optimal 'fit' on projection) result of{classical apparatus}⊗{quantum matter}, which may include the phenomenon we observe as wavecollapse. Here objective reduction is obtained by a ‘strain’ on the evolving {classical apparatus}⊗{quantum matter} projection eventually relieving that strain by adopting a new {classical apparatus}⊗{quantum matter} projection. So analogous to Penrose’s objective reduction criterion in terms of a one-graviton exchange relieving said geometry-field mismatch strain, effecting objective reduction [92]. Here, however, the strain is on the projection to a particular {classical apparatus}⊗{quantum matter} that shifts, and may even have a similar one-graviton cutoff for strain reduction, but here we speak of an effective one-graviton exchange (there being no true quantum gravity only a linear-order effective theory). Thus, the non-appearance of alpha in some areas of the physics is found to be just as interesting as the many odd ways it appears in other disciplines.

Appendix B. Math Review

Operator Mathematics

Products of operators

$(AB)\Psi = A[B\Psi]$

In general $AB \neq BA$ (just consider matrix representation).

The Commutator

$[A, B] = AB - BA$ is called the commutator.

Discrete orthogonal bases in $\mathcal{F}$: $\{u_i(r)\}$

Consider a countable set of functions of $\mathcal{F}$, labelled by a discrete index i:

The set $\{u_i(r)\}$ is orthonormal if $(u_i, u_j) = \int d^3r\, u_i^* u_j = \delta_{ij}$.
The set constitutes a basis if every $\Psi(\vec{r}) \in \mathcal{F}$ can be expanded in only one way in terms of the u_i :

$$\Psi(\vec{r}) = \sum_i c_i u_i(r).$$

Components of a wave function in the $\{u_i(r)\}$ basis

Project out:

$$(u_j, \Psi) = \left(u_j, \sum_i c_i u_i\right) = \sum_i c_i (u_j u_i) = \sum c_i \delta_{ij} = c_j$$

Thus,

$$c_j = (u_j, \Psi) = \int d^3r u_j^*(\vec{r})\, \Psi(\vec{r}).$$

The $\{c_i\}$ represent $\Psi(\vec{r})$ in the $\{u_j(\vec{r})\}$ basis. (The same Ψ can have different components if in two different bases.)

Expression for the scalar product in terms of the components

$$\varphi(\vec{r}) = \sum_i b_i u_i \quad , \quad \Psi(r) = \sum_j c_j u_j$$

Consider

$$(\varphi, \Psi) = \sum_{i.j} b_i^* c_j (u_i, u_j) = \sum_{i,j} b_i^* c_j \delta_{ij} = \sum_i b_i^* c_i$$

In particular, $(\Psi, \Psi) = \sum_i |c_i|^2$ (analogous to $\vec{v} \cdot \vec{w} = \sum_{ij}^{3} v_i w_i$ for vectors in R^3).

Closure relation

Expresses the fact that $\{u^i\}$ constitute a basis:

$$\Psi = \sum_i c_i u_i = \sum_i (u_i, \Psi) u_i = \sum_i \left[\int d^3r' u_1^*(\vec{r'})\, \Psi(\vec{r'})\right] u_i(r)$$

$$= \int d^3r'\, \Psi(\vec{r'}) \left[\sum_i u_i(\vec{r'}) u_i^*(\vec{r'})\right]$$

Thus, $\sum_i u_i(\vec{r'}) u_i^*(r^{*\prime}) = \delta\left(\vec{r} - \vec{r'}\right)$.

Introduction of "bases" not belonging to $\mathcal{F}$ (or L^2 at all)

Plane waves (in 1D)

Recall the advantage of the Fourier transform in analysis:

$\Psi(x) = \frac{1}{\sqrt{2\pi\hbar}} \int_{-\infty}^{\infty} dp\, \overline{\Psi}(p) e^{ipx/\hbar} = \int_{-\infty}^{\infty} dp\, \overline{\Psi}(p) v_p(x)$

$\overline{\Psi}(p) = \frac{1}{\sqrt{2\pi\hbar}} \int_{-\infty}^{\infty} dp\, \Psi(x) e^{-ipx/\hbar} = (v_p, \Psi) = \int_{-\infty}^{\infty} dx v_p^*(x)\, \Psi(x)$

So, consider the function $v_p(x) = \frac{1}{\sqrt{2\pi\hbar}} e^{ipx/\hbar}$. Now, $|v_p(x)|^2 = \frac{1}{\sqrt{2\pi\hbar}}$, which diverges upon integration over x, thus $v_p(x) \notin \mathcal{F}$.

Parseval's Relation

Parseval's relation follows from the above with: $(\Psi, \Psi) = \int_{-\infty}^{\infty} dp\, |\overline{\Psi}(p)|^2$ since we have the closure relation $\int_{-\infty}^{\infty} dp\, V_p(x) V_p^*(x') = \delta(x - x')$. The generalization to 3D then directly follows. Thus, $v_{\vec{p}}(\vec{r})$ can be considered to constitute a "continuous basis":

$$i \leftrightarrow \vec{p}$$

$$\sum_i \leftrightarrow \int d^3p$$

$$\delta_{ij} \leftrightarrow \delta(\vec{p} - \vec{p}')$$

Delta function basis

Consider the set $\{\varepsilon_{\vec{r}_0}(\vec{r})\}$ where $\varepsilon_{\vec{r}_0}(\vec{r}) = \delta(\vec{r} - \vec{r}_0)$, where $\varepsilon_{\vec{r}_0}(\vec{r}) \notin \mathcal{F}$. We have:

$$\Psi(\vec{r}) = \int d^3r_0\, \Psi(\vec{r}_0) \delta(\vec{r} - \vec{r}_0) = \int d^3r_0\, \Psi(\overline{r_0}) \varepsilon_{\vec{r}_0}(\vec{r})$$

and

$$\Psi(\vec{r_0}) = \int d^3r\delta(\vec{r}_0 - \vec{r})\,\Psi(\vec{r}) = (\varepsilon_{\vec{r}_0}, \Psi) = \int d^3r\varepsilon_{\vec{r}_0}{}^*(\vec{r})\,\Psi(\vec{r})$$

where, $(\varphi, \Psi) = \int d^3r_0\varphi^*(\vec{r}_0)\,\Psi(\vec{r})$ gives back the definition of scalar product orthogonalization (and closure condition also trivial).

Note: A physical state must always correspond to a square-integrable wave function. The continuous basis $v_p(x)$ and $\varepsilon_{\vec{r}_0}$ are not square-integrable and are only used as intermediaries in calculations.

Consider a general continuous "orthogonal" basis $\{w_\alpha(\vec{r})\}$:

$$(w_\alpha, w_{\alpha'}) = \int d^3r\, w_\alpha^*(\vec{r})\, w_{\alpha'}(\vec{r}) = \delta(\alpha - \alpha')$$

and

$$\int d\alpha\, w_\alpha(\vec{r})\, w_\alpha^*(\vec{r}') = \delta(\vec{r} - \vec{r}').$$

Note: for $\alpha = \alpha', (w_\alpha, w_\alpha')\ diverges,$ thus a continuous basis has $w_\alpha(\vec{r}) \notin \mathcal{F}.$

Mixed (discrete and continuous) Basis

$$\text{Orthonormality}\begin{Bmatrix} (u_i, u_j) = \delta_{ij} \\ (\omega_\alpha, \omega_{\alpha'}) = \delta(\alpha - \alpha') \\ (u_i, \omega_\alpha) = 0 \end{Bmatrix}$$

$$\text{Closure } \textstyle\sum_i u_i(\vec{r})u_i^*(\vec{r}') + \int d\alpha\, w_\alpha(\vec{r})w_\alpha^*(\vec{r}') = \delta(\vec{r} - \vec{r}')$$

Properties of linear operators

The Trace

$TrA = \sum_i < u_i|A|\, u_i >$ which is invariant, independent of chosen basis. If A is an observable, choose a diagonalizing basis, then $TrA = \sum_n g_n a_n$ where the $\{a_n\}$ are the eigenvalues and $\{g_n\}$ are the associated degrees of degeneracy.

$TrAB = TrBA$ and $Tr\ ABC = Tr\ BCA = Tr\ CAB$ (under cyclic permutations).

Commutator relations

$$[A, BC] = [A, B]C + B[A, C]$$
$$[A, [B, C]] + [B, [C, A]] + [C, [A, B]] = 0$$
$$[A, B]^\dagger = [B^\dagger, A^\dagger]$$

Thus

$$\text{If } [A, C] = [B, C] = 0 \text{ and } C = [A, B], \text{then } [A, B^n] = nCB^{n-1}$$

Thus

$$[A, F(B)] = [A, B]F'(B) \quad and \quad [B, G(A)] = [B, A]G'(A)$$

And it follows

$$[Q, P] = i\hbar \rightarrow [Q, F(P)] = i\hbar F'(P)$$

Restriction of an operator to a subspace

$$P_q = \sum_{i=1}^{q} |\varphi_i><\varphi_i|$$

is a projection onto a subspace, for which we have the restriction of an operator to that subspace given by:

$$\hat{A}_q = P_q A P_q$$

Functions of operators

A^n is defined as n successive applications of A. The inverse, A^{-1}, is defined by $A^{-1}A = 1$ (if it exists). Any function expressible by a power series $F(Z) = \sum_{n=0}^{\infty} f_n Z^n$ allows us to similarly define a corresponding operator $F(A)$. Consider, for example, $F(A) = e^A$, where

$$e^A = \sum_{n=0}^{\infty} \frac{A^n}{n!} = 1 + A + {A^2}/{2} + \cdots$$

If F(Z) is real function then the f_n are real, and if A is Hermitian, then $F(A)$ is Hermitian. Thus, if $A|\varphi_2> = a|\varphi a> then\ F(A)|\varphi_a> = F(a)|\varphi_a>$.

Note that if A is in its diagonalized basis, then $F(A)$ is the operator which is represented in the same basis by the diagonal matrix whose elements are $F(a_i)$. Take for example:

$$\sigma_Z = \begin{pmatrix} 1 & 0 \\ 0 & -1 \end{pmatrix} \quad \rightarrow \quad e^{\sigma_Z} = \begin{pmatrix} e & 0 \\ 0 & 1/e \end{pmatrix}.$$

Note that $e^A e^B \neq e^B e^A \neq e^{(A+B)}$, $unless\ [A, B] = 0$. Consider the following important example -- the potential operator:

$$V(X)|x> = V(x)|x>$$

So,

$$< x|V(X)|\Psi> = V(x)\,\Psi(x)$$

as used in Schrodinger equation.

Commutators involving functions of operators

Consider two observables P and Q satisfying $[Q, P] = i\hbar$ (canonical commutation relations for 1 degree of freedom). Using just $[Q, P] = i\hbar$ we have:

$$[Q, P^2] = [Q, P]P + P[Q, P] = 2i\hbar P,$$

which appears to act as a derivative. Let's now show that this generalizes to the relation:

$$[Q, P^n] = i\hbar n P^{n-1}.$$

Let's assume the relation is true and verify iteratively:

$$[Q, P^{n+1}] = [Q, P]P^n + P[Q, P^n] = i\hbar p^n + i\hbar p p^{n-1} = i\hbar(n+1)P^n$$

by recurrence from initial condition $[Q, P^2\ ] = i\hbar P^{n+1}$ it is thus shown. Thus, in general

$$[Q, F(P)] = i\hbar F'(P).$$

Analogously:

$$[P, G(Q)] = -i\hbar G'(Q).$$

Note that $[Q, \Phi(Q, P)]$ is complicated due to ordering problems.

Note: If $[A, C] = [B, C] = 0 \quad and\ C = [A, B]$ then we have

$[A, B^2] = [A, B]B + B[A, B] = CB + BC = 2CB$

Which generalizes to

$[A, B^n] = nCB^{n-1}$

Thus

$[A, F(B)] = CF'(B) = [A, B]F'(B)$

$[B, G(A)] = -CG'(A) = [B, A]G'(A)$

Thus, the essential communication properties of canonical variables follows not from $[Q, P] = constant$ but from the general property:

$$\big[Q, [Q, P]\big] = 0 \quad and \quad \big[P, [Q, P]\big] = 0.$$

So, $[Q, P]$ could even remain as an operator as long as it commutes with Q, P.

Differentiation of an operator

The above detail on essential commutation properties is revisited to obtain Glauber's Formula where the derivative is also used. The derivative, if it exists, will have the usual form:

$$\frac{dA}{dt} = \lim_{\Delta t \to 0} \frac{A(t + \Delta t) - A(t)}{\Delta t},$$

for the matrix elements of A(t) in an arbitrary basis of t-independent vectors $|u_j>$:

$$< u_i|A|u_j > = \mathrm{A}_{ij}$$

Then,

$$\left(\frac{dA}{dt}\right)_{ij} = < u_i \left|\frac{dA}{dt}\right| u_j >.$$

Consider

$$e^{At} = \sum_{n=0}^{\infty} \frac{(At)}{n!} \rightarrow \frac{d(e^{At})}{dt} = Ae^{At}$$

Similarly

$$\frac{d}{dt}(e^{At}e^{Bt}) = Ae^{At}e^{Bt} + e^{At}Be^{Bt}.$$

Note that $\frac{d}{dt}e^{A(t)}$ is generally not equal to $\frac{dA}{dt}e^{A(t)}$. In order for this equality to hold $A(t)$ and $d(A(t))/dt$ must commute.

Now to derive Glauber's formula:

$$e^A e^B = e^{A+B} e^{\frac{1}{2}[A+B]}$$

where both A and B are assumed to commute with their commutators. Consider $F(t) = e^{At}e^{Bt}$:

$$\frac{dF}{dt} = Ae^{At}e^{Bt} + e^{At}e^{At}Be^{Bt} = (A + e^{At}Be^{-At})F(t)$$

since

$$[e^{At}, B] = -[B, A]te^{At} = [A, B]te^{At}$$
$$\rightarrow e^{At}B = Be^{At} + t[A, B]e^{At}$$

we can write:

$$\frac{dF}{dt} = (A + [Be^{At} + t[A, B]e^{At}]e^{-At})F(t) = (A + B + t[A, B])F(t)$$

Thus,

$$\ln F(t) = (A + B)t + \frac{1}{2}[A, B]t^2 \ \ + const$$
$$F(t) = \exp\left\{(A + B)t + \frac{1}{2}t^2[A, B]\right\}$$

$Let\ t = 1$:

$$e^A e^B = e^{(A+B)} e^{\frac{1}{2}[A,B]}.$$

Schwarz inequality and other relations

Schwarz inequality can be derived using $< \varphi|\varphi > \geq 0$ with:

$$|\varphi> = |\varphi_1> + \lambda|\varphi_2>, \text{and choose } \lambda = \frac{-<\varphi_2|\varphi_1>}{<\varphi_2|\varphi_2>},$$

Then:

$$|<\varphi_1|\varphi_2>|^2 \leq <\varphi_1|\varphi_1><\varphi_2|\varphi_2>$$

Thus,

Schwartz inequality : $|<u|v>| \leq \sqrt{<u|u>}\sqrt{<v|v>}$

In 3-space $|\vec{A}\cdot\vec{B}| \leq |\vec{A}||\vec{B}|$

$|(\underline{u},\underline{v})| \leq \|\underline{u}\|^{1/2}\|\underline{v}\|^{1/2} = iff\ \underline{u} = \lambda\underline{v}$ Schwartz inequality

$\underline{\omega} = \underline{u} + \lambda\underline{v}$

$\|\omega\|^2 \geq 0 \rightarrow (\underline{u} + \lambda\underline{v}, \underline{u} + \lambda\underline{v}) \geq 0$

0 only if $\underline{u} = -\lambda\underline{v}$

$0 \leq (\underline{u},\underline{u}) + (\underline{v},\underline{v})|\lambda|^2 + \lambda(\underline{u},\underline{v}) + \lambda^*(\underline{v},\underline{u})$

$0 \leq u^2 + v^2\lambda^2 + [\lambda(\underline{u},\underline{v})] + c.c$

Choose $\lambda = -\frac{(\underline{v},\underline{u})}{\|\underline{u}\|^2} \rightarrow$

$0 \leq u^2 + v^2\left(\frac{(\underline{u},\underline{v})^2}{|v|^2}\right) - \left(\frac{(\underline{u},\underline{v})^2}{|v|^2}\right) - \left(\frac{|\underline{u},\underline{v}|^2}{|v|^2}\right) - (\underline{v},\underline{u})$

$0 \leq u^2 + \frac{|\underline{u},\underline{v}|^2}{|v|^2} \rightarrow u^2v^2 \geq |(\underline{u},\underline{v})|^2$

$|(\underline{u},\underline{v})| \leq \|\underline{u}\|^{1/2}\|\underline{v}\|^{1/2}$

Thus

$|(\underline{u}+\underline{v})| \leq \|\underline{u}\| + \|\underline{v}\|$

$|(\underline{u}+\underline{v})| = (\underline{u}+\underline{v},\underline{u}+\underline{v}) = \|\underline{u}\|^2 + \|\underline{v}\|^2 + (\underline{u}+\underline{v}) + (\underline{v}+\underline{u})$

$\leq \|\underline{u}\|^2 + \|\underline{v}\|^2 + 2|\|\underline{u}\|\|\underline{v}\||$

$\leq (\|\underline{u}\| + \|\underline{v}\|)^2$

Triangle inequality : $\sqrt{<u+v|u+v>} \leq \sqrt{<u|u>} + \sqrt{<v|v>}$

Translation operator :

$$\Omega(a)\varphi(x) = \varphi(x+a)$$

$$= \sum_{n=0}^{\infty}\frac{a^n}{n!}\frac{d^n}{dx^n}\varphi(x) = \sum_{n=0}^{\infty}\frac{1}{n!}\left(\frac{iap}{\hbar}\right)^n\varphi(x) = e^{\frac{i}{\hbar}ap}\varphi(x)$$

where

$$\Omega(a) = \exp(iap/\hbar)$$

Also recall some operator relations:

$[AB, C] = A[B, C] + [A, C]B$

If $\big[[A, B], A\big] = 0$ $then$ $[A^m, B] = mA^{m-1}[A, B]$

$[A, B^n] = \sum_{k=0}^{n-1} B^k [A, B] B^{n-k-1}$

$e^L A e^{-L} = A + [L, A] + \frac{1}{2!}\big[L, [L, A]\big] + \frac{1}{3!}\big[L, [L, A]\big] + \cdots$

$e^A e^B = e^{A+B} e^{-\frac{1}{2}[A,B]}$ if $\big[[A, B], A\big] = \big[[A, B], B\big] = 0$

Appendix C. Killing Vectors for Kasner (100)

Starting with

$$0 = \mathfrak{z}^{\mu}{}_{,\rho} g_{\sigma\sigma} + \mathfrak{z}^{\nu}{}_{,\sigma} g_{\rho\rho} + \mathfrak{z}^{0} \frac{\partial g_{\rho\sigma}}{\partial x^{0}}$$

1. $\rho = \sigma = 0$: $g\mathfrak{z}^0 + \mathfrak{z}^0{}_{,0} = 0$
2. $\rho = \sigma = 1$: $g\mathfrak{z}^0 + \mathfrak{z}^1{}_{,1} = 0$
3. $\rho = 0, \sigma = 1$: $\mathfrak{z}^1{}_{,0} - \mathfrak{z}^0{}_{,1} = 0$
4. $\rho = \sigma = 2 : \mathfrak{z}^2{}_{,2} = 0$
5. $\rho = \sigma = 3 : \mathfrak{z}^3{}_{,3} = 0$
6. $\rho = 2,\ \sigma = 3 : \mathfrak{z}^2{}_{,3} + \mathfrak{z}^3{}_{,2} = 0$
7. $\rho = 0, \sigma = 2 : \mathfrak{z}^2{}_{,0} - \mathfrak{z}^0{}_{,2} e^{2gt} = 0$
8. $\rho = 0, \sigma = 3 : \mathfrak{z}^3{}_{,0} - \mathfrak{z}^0{}_{,3} e^{2gt} = 0$
9. $\rho = 1, \sigma = 2$: $\mathfrak{z}^2{}_{,1} + \mathfrak{z}^1{}_{,2} e^{2gt} = 0$
10. $\rho = 1, \sigma = 3$: $\mathfrak{z}^3{}_{,1} + \mathfrak{z}^1{}_{,3} e^{2gt} = 0$

Consider $\mathfrak{z}^0$ a constant as in Minkowski, the first equation then implies $\mathfrak{z}^0 = 0$. Next consider
$\mathfrak{z}^0 = f(\vec{x})h(t)$, the first equation then gives:

$$gf(\vec{x})h(t) + f(\vec{x})\dot{h}(t) = 0 \quad \rightarrow \quad gh(t) + \dot{h}(t) = 0$$
$$\rightarrow \quad h(t) \sim e^{-gt}$$

thus, if $\mathfrak{z}^0 \neq 0$, then it must be $\sim e^{-gt}$. Moving on to solving $\mathfrak{z}^1$, consider $\mathfrak{z}^1$ equal to a constant, this would then be consistent with the $\mathfrak{z}^0 = 0$ solution for $\mathfrak{z}^0$. Thus, $\mathfrak{z}^1$ equal to a constant is a solution for $\mathfrak{z}^1$ (and that constant need not be zero).
Now consider $\mathfrak{z}^1 = h(t)$, a function of time only. By equations 2 and 3 we have $h(t)$ is a constant, thus the previous result, and there is not a time-only dependent solution for $\mathfrak{z}^1$.

Now consider $\mathfrak{z}^1 = f(x)$, a function of space only. By equations 2 and 3 we have $f(x)$ is a constant, thus the previous result, and there is not a space-only dependent solution for $\mathfrak{z}^1$.

Now consider $\mathfrak{z}^1 = h(t)f(x)$. We now have two separate relations whose solutions provide: $h(t) \sim e^{-gt}$ and $f(x) \sim A\sinh(gx) + B\cosh(gx)$, and consistency with the Killing Vector relations then give the two solutions:

$$\mathfrak{z}^0 = e^{-gt}\sinh(gx), \qquad \mathfrak{z}^1 = -e^{-gt}\cosh(gx)$$
$$\mathfrak{z}^0 = e^{-gt}\cosh(gx), \qquad \mathfrak{z}^1 = -e^{-gt}\sinh(gx)$$

Consider $\mathfrak{z}^2$ equal to a constant (the rest zero), this is consistent with all the relations as-is, thus is a solution as is. Similarly for $\mathfrak{z}^3$ equal to a constant.

Consider $\mathfrak{z}^2 = f_2(x_3)$ and $\mathfrak{z}^3 = h_3(x_2)$, then equation 6 requires:

$$\frac{\partial}{\partial x_3} f_2 + \frac{\partial}{\partial x_2} h_3 = 0$$

and there are no other equations for this case, thus the solution (providing two independent Killing Vectors) is simply:

$$\mathfrak{z}^2 = x_3 \,, \qquad \mathfrak{z}^3 = -x_2 \,.$$

Now consider $\mathfrak{z}^0 = f(\vec{x})e^{-gt}$ (since it must be $\sim e^{-gt}$ to be nontrivial and $f(x)$ has already been studied). To be consistent with the other equations we arrive at two more independent Killing Vectors:

$$\mathfrak{z}^0 = x_2 e^{-gt} \sinh(gx_1)\,, \qquad \mathfrak{z}^1 = x_2(-e^{-gt}\cosh(gx_1)),$$

$$\mathfrak{z}^2 = \frac{1}{g} e^{gt} \sinh(gx_1)$$

$$\mathfrak{z}^0 = x_1 e^{-gt} \cosh(gx_1)\,, \qquad \mathfrak{z}^1 = x_2(-e^{-gh}\sinh(gx_1))\,,$$

$$\mathfrak{z}^2 = \frac{1}{g} e^{gt} \cosh(gx_1)$$

Replacing $x_2 \leftrightarrow x_3$ in the above two equations yields the last two independent Killing vectors.

The Killing vectors derived for K(100) from the Killing vectors equation using the K(100) metric are precisely the same Killing vectors that can be derived by simply applying the coordinate transformation, relating Minkowski coordinates to K(100) coordinates to the Minkowski Killing vectors. The time-like Killing vector in K(100) space (in K(100) coordinates) is then:

$$\partial_\beta = e^{-gt}\{\cosh(gx)\partial_t - \sinh(gx)\partial_x\}$$

In Minkowski there are 10 killing vector fields that correspond to the inhomogeneous Lorentz group, or Poincare group, for the isometrics of the spacetime. In the quantum theory the generators of the isometries became operations. We ask that the transformation of a state vector by a given isometry be unitary so that probability is conserved:

$$u|\Psi> = e^{iA}|\Psi>$$

Working with scalar fields

$$\varphi(x') = u(\Lambda)\varphi(x)u^{-1}(\Lambda)$$
$$x'^{\mu} = \Lambda^{\mu}{}_{u}x^{u} + a^{\mu}$$

For displacement :

$$x'^{\mu} = x^{\mu} + a^{\mu}$$

Thus

$$\varphi(x + a) = u(a)\varphi(x)u^{-1}(a)$$

And using $u = e^{i\eta} \cong (1 + i\eta)$ we get:

$$\varphi(x) + a_{\mu}\partial^{\mu}\varphi(x) \cong (1 + i\eta)\varphi(x)(1 - i\eta) \cong \varphi(x) + i[\eta, \varphi(x)]$$

Or

$$ia_{\mu}[p^{\mu}, \varphi(x)] = i[\eta, \varphi(x)] \rightarrow \quad \eta = a_{\mu}p^{\mu},$$

Thus, p^{μ} is the generator of translations.

For the Lorentz transformation: $x'^{\mu} = \Lambda^{\mu}{}_{v}x^{u} = (\delta^{\mu}{}_{v} + \alpha^{\mu}{}_{v})x^{v}$. Thus

$$g_{\mu v}\Lambda^{\mu}{}_{\alpha}\Lambda^{v}{}_{\beta} = g_{\alpha\beta} \quad \rightarrow \quad g_{\mu\beta}\alpha^{\mu}_{\sigma} + g_{\sigma v}\alpha^{v}_{\beta} = 0 \quad \rightarrow \quad \alpha_{\beta\sigma} + \alpha_{\sigma\beta} = 0$$

The α's are, thus, antisymmetric. Continuing with fitting to a unitary form with the infinitesimal:

$$\varphi(x^{\mu} + \alpha^{\mu}{}_{v}x^{v}) = \varphi(x) + i[\eta, (x)] \rightarrow \quad i[\eta, (x)]$$
$$= \frac{1}{2}\alpha^{\mu v}(x_{v}\partial_{\mu} - x_{\mu}\partial_{v})\varphi(x)$$

Now write $\eta = \frac{1}{2}\alpha^{\mu v}M_{\mu v}$ and find that $M_{\mu v}$ are the generators of rotations in the standard form:

$$[M_{\mu v}, \varphi(x)] = -\frac{1}{i}(x_{\mu}\partial_{v} - x_{v}\partial_{\mu})\varphi(x) \equiv L_{\mu v}\varphi(x),$$

Where$\mu, v = 1,2,3$ and L_{23}, L_{31}, L_{12} are the angular momentum operators.

We specify a unique vacuum for inertial observers in Minkowski spacetime when we demand that the vacuum be invariant under the 10 parameter isometry group of the Poincare group:

$$u(\Lambda)|0> \; = \; |0>$$

for unique vacuum;

$$u(\Lambda) = e^{ia_{\mu}p^{\mu}}$$

for translations $\hat{p}^{\mu}|0> = 0$, and

$$u(\Lambda) = e^{i\frac{1}{2}\alpha^{\mu v}M_{\mu v}}$$

for rotations $\widehat{M}_{\mu v}|0> = 0$. Thus, we specify the vacuum by asking that the operators, which correspond to the generators of the isometry, annihilate the vacuum.

Appendix D. Emanator Theory Synopsis

D.1 Introduction

In quantum physics unitary propagation is a standard part of the description. Efforts to move to algebras to describe such propagation leads to formulations based on the normed division algebras (real, complex, quaternion, and octonion). In an effort to achieve maximal information propagation we relax the unitarity condition and show that multiplication (right) on a unit norm trigintaduonion base by a unit norm chiral trigintaduonions emanator results in a new unit norm product [91]. A path is comprised of repeated (right) multiplications. Each step of the 'emanation' arrived at is a multiplication by a chiral trigintaduonion. Use of methods from noise budget analysis, a constructive perturbation analysis, as well as analysis relating to maximal perturbation according to the Kato Rellich theorem, show that the chiral trigintaduonion with maximal perturbation has magnitude α, precisely the fine structure constant. A relation between α and π results. Suppose repeated achiral emanation steps can be described as an iterative mapping, with unit-norm constraint resulting in a quadratic relation on components, we then expect the Feigenbaum universal bifurcation parameter, C_{∞}, to appear according to the number of independent dimensions in a chiral trigintaduonion emanation step and the precise form of the "emanator" construction. The number of effective dimensions is shown to be 29 plus a little more, and a relation between α, π and C_{∞}results that is in agreement with the choice of emanator examined in computational studies shown here. The computational studies with the emanator have also been explored via "random walks" in the trigintaduonion space during emanation and to explore noise additivity effects. Component-level evolution is seen to behave like a random walk, with random walk asymptotics (established computationally). This helps to establish that the Emanation process is Martingale, since random walk processes are Martingale.

Just from the propagation structure on one path we already see core emergent structure that results in a universal emanation with structural parameters 10,22,78,137 and perturbation maximum α=~1/137. The central notion in the universal emanation hypothesis is that there should be *maximal information flow*, where this is accomplished by finding the highest theoretical dimensionality of unit-norm 'propagation', here called an emanation, which turns out to be 10, then add the maximal perturbation that still allows unit-norm propagation, where that perturbation is into the space the 10D motion is embedded in, here a 32 dimensional (trigintaduonion algebra) space.

The existing Standard Model, and reasonable extensions for the massive neutrino, cannot explain the parameters of the model. Why they are 19, or so, in number, and why the local gauge structure exists with the odd-looking product form: $U(1) \otimes SU(2)_L \otimes SU(3)$.

At the heart of the quantum formulation underlying the standard model is the fundamental theoretical element known as the quantum propagator (corresponds to a complex unitary matrix). Most notably, if the wavefunction has unit norm at the outset, after unitary propagation it remains unit norm.

Efforts to generalize the quantum formulation by considering hypercomplex but non-unitary matrices have been stymied for a variety of reasons [93]. Given the existing close agreement of experimental observation and quantum predictions, it is hard to work within the unitary propagator-type theoretical framework and arrive at anything other than an equivalent quantum formulation. Instead of working within the existing theory, in Emanator Theory the objective is to obtain a generalized theory that can project the quantized Lagrangian and Standard Model (or closely related extended Standard Model). To seek a larger theory that projects the existing theory is an excellent way to break free of the Godel-Incompleteness Trap that occurs when working within the existing standard physics and standard model. The drawback, of course, is it's likely to be a far-fetched idea, whatever it is, so it will face a higher bar to be seen as even interesting, not to mention valid. In what follows a synopsis of emanator theory will be given, as well as a few of the latest results, that will hopefully be convincing on both accounts.

If trying to generalize to a theory that can project to a renormalizable quantum field theory (with the Standard Model parameters), there is the issue of why it should project the particular formulation described. Of all the possible projections, how do you argue why it should be the unitary propagator-type physics seen? Here the aforementioned difficulty in generalizing the existing propagator theory (from the inside out) is a strength as it explains why the projection should be the standard physics and standard model as seen. That being the case, there is still the matter of finding the missing structure elsewhere (the Path Integral Action-

Lagrangian formulation itself, for example, and the Standard Model parameters) and then projecting that as the renormalizable quantum field theory with Standard Model parameters as seen experimentally.

D.1.1 Dirac used Lorentz Invariance

If trying to guess the mathematical basis of a generalized quantum theory, that would project the existing Lorentz Invariant quantum field theory, then Dirac provides a powerful lesson. Recall that Dirac's guess and derivation for the relativistic wave equation was purely based on seeking a representation that was Lorentz invariant. In doing this, Dirac discovered the Dirac Equation that describes spin-1/2 fermionic matter (all the fundamental particles).

In Dirac's approach 4-vectors, V^a, were used to describe the spinors (spin ½ fermions), where the length of the 4-vector is constant under Lorentz Transformation L:

$$L\left\{\frac{1}{2}\eta_{ab}V^aV^b\right\} = \frac{1}{2}\eta_{ab}V^aV^b.$$

In Penrose's book on spinors [94] we see how to write a 4-vector as a 2x2 Hermitian matrix, $\psi(V^a)$, where the length of the four vector is equal to the determinant of the equivalent 2x2 matrix:

$$det[\psi(V^a)] = \frac{1}{2}\eta_{ab}V^aV^b.$$

Suppose the length of the 4-Vector is 1 (it's a spinor probability amplitude), then the associated 2x2 Hermitian matrix will have determinant equal to 1, thus $SL(2,\mathbb{C})$, and we have a direct representation of the Lorentz Group.

Let's now consider a similar process involving transformational invariance but instead of encoding the Lorentz transform in the form of matrix transformation invariance let's use elements of the Cayley algebras instead. Specifically, consider the following transformation:

$$q' = aqa_c^*, \quad where \quad aa_c = 1,$$

where a is a (unitary) complex bi-quaternion: $H(\mathbb{C}) \times H(\mathbb{C})$, and $q = (ct, ix, iy, iz)$ (note this notation is for $q = (ReH_1, iImH_1, iReH_2, iImH_2)$). The $q' = (ct', ix', iy', iz')$ that results will correspond to a proper orthochronous Lorentz transform [95-97].

This is a remarkable result, but is there better (higher dimensionality/complexity)? Consider that a complex bi-quaternion is isomorphic to a complex octonion which is isomorphic to an 'achiral' sedenion, thus a theory for achiral sedenion emanation is indicated from this result as far back as 1917. The 'halting condition' on the generalization in 1917 seems to be that octonions are the highest-order of the division algebras that have inverses defined (which is necessary to have $aa_c = 1$ be defined). But they have already extended past octonions since these are complex octonions ≅ sedenions, so how are they guaranteed to have $aa_c = 1$ be defined? This is possible for the *chiral* sedenions, as shown in [91], if restricted to be unit norm. So now we have our answer based on the results from [91] – we can go one complexation order higher:

$$Q' = AQA_c^*, \ where \ \ AA_c = 1,$$

where A is a (unitary) bi-complex bi-quaternion: $H(\mathbb{C} \times \mathbb{C}) \times H(\mathbb{C} \times \mathbb{C})$, which is isomorphic to a unit norm quaternionic bi-quaternion, $H(\mathbb{H}) \times H(\mathbb{H})$, and $Q = (ct, ix, iy, iz)$.
where $Q = (ReH_1, iImH_1, iReH_2, iImH_2)$ as before, except $H_1 = H \times \mathbb{H}$ not $H_1 = H \times \mathbb{C}$
The $Q' = (ct', ix', iy', iz')$ that results will again correspond to a proper orthochronous Lorentz transform. Again, how do we know the critical operation $AA_c = 1$ can always be satisfied? Previously we saw that a complex bi-quaternion was equivalent to a 'chiral' sedenion (in the sense described in [91]), thus a bi-complex bi-quaternion is isomorphic to a complex chiral sedenion, which is isomorphic to a 'doubly chiral' trigintaduonion. This is precisely the construct examined in [91], so a generalization of the 1917 result to unitary quaternionic bi-quaternions appears possible. As shown in [91], however, there is no higher order construct. This latter form (on doubly-chiral trigintaduonions) establishes the Emanator as Lorentz Invariant.

Thus, in terms of Cayley algebras, we can generalize to encoding the Lorentz transformation into split-Cayley algebras as long as the Cayley transform has an inverse, and doing this for the highest order Cayley algebra possible. The Cayley algebras with inverses, the division algebras, consist of the first four Cayley Algebras: reals (1 parameter), complex numbers (2 parameter), quaternions (4 parameters), and octonions (8 parameters). Another aspect of the division algebras is that they have norm, and this then means that their unitary evolution can be described as unit norm propagation. Suppose we focus on this latter feature, unit norm propagation alone. Can we extend to even higher order Cayley Algebras in the sense of starting with unit norm and for a subset of that higher order algebra still effect a unit norm, invertible, propagation? In other words for the next higher algebras, sedenions (16 parameter) and trigintaduonion (32 parameter), etc., does there exist a subspace allowing unit norm propagation? This is equivalent to considering the maximal Cayley subalgebra order for which unit norms exist. This is the actual starting point of the Maximal Information Emanation Hypothesis

D.1.2 The Maximum Information Emanation (MIE) Hypothesis and Prior Results

Emanator theory stems from the Maximum Information Emanation (MIE) Hypothesis. The definition of information is context dependent, so how the MIE hypothesis will manifest depends on circumstance. We start with the fundamental notion of the quantum propagator, for which mathematical 'propagators' satisfy unitarity. We seek to extend this foundational element so start by asking what is the highest Cayley algebra that can remain unitary – where the answer comes down to the highest order division algebra, which are the octonions. What if we change the desired property from unitary to unit-norm preserving? Then we can extend 2 more dimensions beyond the 8D octonion algebra to a chiral subspace of the trigintaduonions that is 10D [91].

The 10D chiral trigintaduonions are then identified as the maximal information 'carriers' or emanators, operationally like the quantum propagators in a larger theory, where evolution of the system will be shown to result from sums on paths of emanators (similar to quantum evolution in terms of a path integral on propagators 'steps'). Having posited this maximal construct we see the classic signs of "asking the right question" since we get a variety of clear results:

(1) the chiral emanator is manifestly Lorentz Invariant as indicated above, and the emanation step or propagation is simply multiplication by a unit-norm 'emanator' according to the Cayley algebra multiplication rules.

(2) chiral emanation involves a 10D element in a 32D space (trigintaduonions) for which maximum perturbation is determined computationally (still permitting unit-norm transmission) to be ~1/137, e.g., 'alpha' – we therefore have the mysterious alpha by a computational definition.

(3) The emanation process involves multiplication on the current unit norm trigintaduonion base element by a unit norm chiral trigintaduonion emanator to arrive at a new, unit norm, trigintaduonion base (and then it

repeats infinitely). When the trigintaduonion multiplication steps are expanded to octonionic level, we find that there are 137 independent tri-octonionic terms that occur in the emanation process. Due to complex noise contributions, the effective number is 137*, which is slightly greater than 137. Thus, theoretically, the maximum perturbation that is allowed is 1/137* which happens to be exactly 'alpha'. We, thus, have a theoretical derivation for alpha, referred to as the $\{\alpha, \pi\}$ relation:

$$\alpha^{-1} = \frac{137}{\cos\beta}\cos\theta\frac{\theta}{\sin\theta},$$

where

$$\beta = \frac{\pi}{137} \; and \; \theta = \frac{\pi}{137x29}$$

and

$$\alpha^{-1} = 137.03599978669910\,,$$

which agrees with 2002 experimental observation:

$$\alpha^{-1} = 137.03599976(50).$$

(4) Chiral trigintaduonion (32D) emanation with perturbation does not have effective dimension 32D due to chiral and other constraints -- noise budget analysis or (equivalently) Kato Rellich operator analysis, both indicate effective dimension slightly greater than 29 referred to as "29*". It is hypothesized that the maximal level of information flow dimensionally should relate to the Universal fractal constant C_∞ according to:

$$\alpha^{-1} = (C_\infty)^\gamma, \;\; \gamma \equiv 29^*,$$

where $\gamma$is estimated by [7]:

$$\gamma \cong \frac{1}{2}\left(29 + \left(\frac{4\pi}{72}\right)\left[1 + \left(\frac{\pi}{137 \times 29}\right)\left\{\frac{\pi}{72} + \frac{3}{72}\right\}\right]\right)$$

thus

$$\alpha^{-1} \cong 137.035999206\,...$$

in agreement with the exact alpha with nine digit accuracy.

In the process of estimating 29* we must explore the definition of the emanation process in the sense of is it one step then normalization, or a chain of steps (analogous to hands greater than one in size being dealt) then normalization, with infinite repeat. The answer appears to be two steps (since this suffices to 'flood' or max-out the noise channels, so no further steps needed) and working within this construct we obtain the $\gamma \cong$ 29* estimate mentioned above. It's not an exact match to 16 decimal places and known measurement, but at a 9-decimal place match its pretty good, certainly indicating that there may well be a relation $\alpha^{-1} =$

$(C_{\infty})^{29^*}$ as hypothesized (sometimes called the $\{\alpha, \pi, C_{\infty}\}$ relation). And, if there is such a relation, it would indication that not only is evolution at maximum perturbation in the quantum sense, but it is also evolution at the edge-of-chaos in the thermal/statistical mechanical sense.

(5) the chiral emanator indicates 'motion' in a 10D subspace of 32D, suggesting 22 constants of the motion. This is a naïve analysis, but it turns out to be true upon deeper analysis within the emanator formalism as there are 22 types of emanation that result in no change to the base trigintaduonion describing the system. This, in turn, shows that the emanator theory will have 22 fixed parameters.

(6) By using the split form of the trigintaduonions, we not only have manifest Lorentz invariance, we also have an exact algebraic split to a space that is simply the direct product of 29* real dimensions (not a local approximation to such). This is important because the fundamental existence of a complex structure means that we trivially have the extension $\mathbb{R}^{29^*} \rightarrow \mathbb{C}^{29^*}$. Suppose we have point-like singular elements in $\mathbb{C}^{29^*}$, such will occur due to zero-divisors (zd's) in the 32D trigintaduonion space. To achieve a maximal domain of analyticity (an application of MIE), we must remove the zd-singularities. In doing so we obtain point-like matter in the theory and a small-h constant that enters the sum on emanator paths just as Planck's constant in the sum on propagator paths in the quantum formulation – suggesting that these small-h numbers are related.

(7) Three derivations of alpha are thus obtained: (i) Computational: $\{\alpha\}$ based on the maximal perturbation for which chiral emanation retains the unit-norm property; (ii) Theoretical: $\{\alpha, \pi\}$ based on the maximal noise transmission on a chiral emanation path; and (iii) Approximate: $\{\alpha, \pi, C_{\infty}\}$ based on the maximal noise transmission on an achiral emanation path (where maximal emanation is at "the edge of chaos" which is defined according to Feigenbaum Universality [89]).

(8) At component level in the emanation product, using 100's of millions of computational steps, we see an excellent asymptotic fit to random walk behavior. Since a random walk is a Martingale process, this strongly suggests that the achiral emanation process is Martingale (between normalization steps involved with zero crossings in their values). In turn, the projected quantum process (standard theory) would retain the imprint of that Martingale process.

(9) The achiral emanator, a sum of achiral emanation paths, can be shown to have the mathematical form $\sum \exp(i\mathbb{H} \times \mathbb{O}) \rightarrow \mathbb{C} \times \mathbb{H} \times \mathbb{O}$, which can be shown to give the gauge theory of the standard model: $U(1) \times SU(2)_L \times SU(3)$. Thus, there is no grand unified theory in terms of gauges that is fundamental (although a GUT may approximately occur at early times cosmologically, at high temperature, where there is conformal flatness). The 'ugly' product gauge that is observed is precisely what is predicted by emanator theory.

(10) Universal thermality is indicated by application of the MIE hypothesis to the choice of whether 'effective' achiral emanation is associative or not, at least at the sedenion-level of propagation. In effect, the mechanism to extend unit norm (chiral) propagation on a 10D subspace of the 32D trigintaduonions can be repeated for other mathematical properties from the lower-level Cayley algebras. Thus, the mechanism identified for extending 'nice' properties of lower-order Cayley algebras to higher order (no more than two orders higher to be precise), can be used for other than existence of a norm. By the chiral extension mechanism, associativity can be extended to the chiral octonions and chiral sedenions. This provides the basis for the universal thermality relation that appears to be ubiquitous (e.g., analytic time). It also provides the basis for an associative matching of terms for cancellation, e.g., renormalization. Thus provides the mechanism to correct the projected quantum field theory such that it has the renormalization counter-terms needed.

(11) In the emanation process there is a clear separation between spinorial elements and manifold elements. Manifold elements include geometry and thermality, have no alpha-perturbation effects, and appear to be part of the 'apparatus' from the perspective of the quantum theory.

D.2 Synopsis of Methods and Prior Results

D.2.1 The Cayley Algebras

The list representation for hypercomplex numbers will make things clearer in what follows so will be introduced here for the first seven Cayley algebras:

Reals: $X_0 \rightarrow (X_0)$.
Complex: $(X_0 + X_1\, \boldsymbol{i}) \rightarrow (X_0, X_1)$ with one imaginary number.

Quaternions: $(X_0 + X_1 \boldsymbol{i} + X_2 \boldsymbol{j} + X_3 \boldsymbol{k}) \rightarrow (X_0, X_1, X_2, X_3)$ with three imaginary numbers.
Octonions: $(X_0, \ldots, X_7)$ with seven imaginary numbers.
Sedenions: $(X_0, \ldots, X_{15})$ with fifteen imaginary numbers.
Trigintaduonions (Bi-Sedenions): $(X_0, \ldots, X_{31})$ with 31 imaginary numbers.
Bi-Trigintaduonions: $(X_0, \ldots, X_{63})$ with 63 types of imaginary number.

Consider how the familiar complex numbers can be generated from two real numbers with the introduction of a single imaginary number '$\boldsymbol{i}$', $\{X_0, X_1\} \rightarrow (X_0 + X_1 \boldsymbol{i})$. This construction process can be iterated, using two complex numbers, $\{Z_0, Z_1\}$, and a new imaginary number '$\boldsymbol{j}$':

$$(Z_0 + Z_1 \boldsymbol{j}) = (A+B\boldsymbol{i}) + (C+D\boldsymbol{i})\boldsymbol{j} = A+B\boldsymbol{i} + C\boldsymbol{j} + D\boldsymbol{ij} = A+B\boldsymbol{i} + C\boldsymbol{j} + D\boldsymbol{k},$$

where we have introduced a third imaginary number '$\boldsymbol{k}$' where '$\boldsymbol{ij=k}$'. In list notation this appears as the simple rule ((A,B),(C,D)) = (A,B,C,D). This iterative construction process can be repeated, generating algebras doubling in dimensionality at each iteration, to generate the 1,2,4,8,16, 32, and 64 dimensional algebras listed above. The process continues indefinitely to higher orders beyond that, doubling in dimension at each iteration, but we will see that the main algebras of interest for physics are those with dimension 1,2,4,and 8, and sub-spaces of those with dimension 16 and 32 dimensional algebras.

Addition of hypercomplex numbers is done component-wise, so is straightforward. For hypercomplex multiplication, list notation makes the freedom for group splittings more apparent, where any hypercomplex product ZxQ to be expressed as (U,V)x(R,S) by splitting Z=(U,V) and Q=(R,S). This is important because the product rule, generalized by Cayley, uses the splitting capability. The Cayley algebra multiplication rule is:

$$\mathbf{(A,B)(C,D) = ([AC-D^*B],[BC^*+DA])},$$

where conjugation of a hypercomplex number flips the signs of all of its imaginary components:

$$(A,B)^* = \mathrm{Conj}(A,B) = (A^*,-B)$$

The specification of new algebras, with addition and multiplication rules as indicated by the constructive process above, is known as the Cayley-Dickson construction, and this gives rise to what is referred to as the Cayley algebras in what follows.

If a Split Cayley algebra is used, then the multiplication rule has a single sign difference:

$$(A,B)(C,D) = ([AC+D^*B],[BC^*+DA]).$$

D.2.2 Unit-norm propagation

For a physical system, a unit norm object can be used to represent a system, and by repeated transformation to other unit norm objects, it thereby evolves. Mathematical objects that can effect this 'transformation' simply by the rule of multiplication would be objects like division algebras, ideals, and what I'll simply call projections or emanations. In the universal propagator we have a unit norm trigintaduonion (32D) and perform a right multiplication with a chiral (10D) unit norm 'alpha-step' (defined by a max perturbation α into the 29 free dimensions, given by 32 minus one for each chiral choice, and one for the unit normalization overall). Consider multiplication of a given (starting) trigintaduonion from the right with a chiral bi-sedenion as a 'projection' through the (chiral) step indicated. The repeated application and repeated 'chiral steps' thereby arriving at a path describing a chiral propagation. The resulting universal propagation consists of a 32D unit norm trigintaduonion with propagation via right multiplication using a unit-norm, chiral bi-sedenion, with max-α perturbation.

We thereby arrive at a 'Universe Propagator' that takes on the physics parameters desired (notably the fine-structure constant) and imprints them onto the evolution as seen from the 'internal reference frame' where we reference an object in the 4D spacetime with Standard Model gauge field, and where the standard Lagrangian emerges as the necessary 'propagate-able' structure (where Hilbert space must be complex, not real, quaternionic or octonionic, etc. [93]). From maximum information flow with the constructs, and the required emergent complex Hilbert space (thus complex path integral, thus standard quantum operator formalism) we arrive back at the familiar results with justification of their core mathematical representations (e.g., complex Hilbert space), and now with justification of all parameters, all from the emanation hypothesis.

Emanator Theory uses unit-norm propagation

Unit-norm right product propagation is trivial for the division algebras since norm(XY) = norm(X) $\times$ norm(Y). From this it is apparent that we have an automorphism group given by the norm itself (since an automorphism if A(XY)=A(X)A(Y)), and in the case of the octonions this automorphism group is G2 [98]. It can be shown that SU(3) is in G2 [98]. Let's now consider the situation with a higher-order Cayley algebra, the Sedenions, 'S'. We obviously don't have $norm(S_1S_2) = norm(S_1) \times norm(S_2)$ in general, as this would then allow S to join the ranks of the division algebras, and it is proven that such don't exist above the Octonions [99]. Can we still have a propagation structure? Is it possible to have a 'base' sedenion for which $norm(S_{base})=1$, and to have a right propagator (product) sedenion also $norm(S_{right})=1$, such that $norm(S_{base}\ x\ S_{right}) =1$? The answer is yes (see [91]), when the sedenion has the (chiral) form of an octonion crossed with a real octonion: $S_{chiral} = (O,O_{real})$ or $S_{chiral} = (O_{real},O)$. Can we continue this to arrive at a propagation structure on the Trigintaduonions? Again the answer is yes, with the chiral form generalizing off the chiral Sedenion as might be expected: $T_{chiral} = (S_{chiral}, S_{real})$ or (S_{real}, S_{chiral}) [91]. It is proven that this extension process will go no further [91]. What happens is that due to the chiral form we are still able to re-express all T products (or S) as collections of terms involving tri-octonionic products (which have nice properties as described in [91]), and this can no longer occur above the (chiral) trigintaduonion level.

Thus, we have achiral emanation, and to get achiral there must be a way to sum over all chiral to get an achiral result to arrive at the full emanator process. These details are described next.

D.2.3 Chiral T-emanation has 78 generators of change and 137 independent octonion terms

We begin with constructing the theoretical expression for a general element of the trigintaduonion algebra after two chiral trigintaduonion multiplicative propagation steps. A simple analysis of the number of terms in this expression, when reduced to three-element algebraic 'braid-level', results in a count on algebraic braids of 137, plus a little extra (e.g. some lagniappe for the best 'cooking') of a contribution towards a 138th braid when the "noise analysis" is done [91,7].

Consider a general Norm=1 (32D) Trigintaduonion (Bi-Sedenion): (A,B), where A and B are sedenions (16D). Then have (A,B) = ((a,b), (c,d)), where {a,b,c,d} are octonions. Slightly different than a propagator, we have an 'emanator' with the following notation and properties, where the emanator describes a 10D multiplicative step. The emanator is a chiral bi-sedenion: a trigintaduonion whose first sedenion half is itself a chiral bi-octonion, and the second sedenion half is a pure real (as is the second octonion half): $(\tilde{A},\beta)$, $\tilde{A} = (\tilde{a},\alpha)$, where the norm is 1, α is a real octonion, and β is a real sedenion. Thus:

Emanator: $(\tilde{A},\beta) = ((\tilde{a},\alpha), \beta)$.
Note: $\tilde{A}^* = (\tilde{a}^*,-\alpha)$.

Let's set up a description of the Universal 'Emanation' along a 'chiral path' resulting from a few emanation steps. To begin, suppose we have already arrived at, or received, a unit norm trigintaduonion (32D) state 'T', and suppose our emanations are the result of right multiplication with a chiral trigintaduonion (bi-sedenion) 'step', and suppose we consider one such path after just a few steps. Here's the notation to begin:

$\mathbf{T} = (A,B)$, a unit norm trigintaduonion.
$\boldsymbol{\tau} = (\tilde{A},\beta) = ((\tilde{a},\alpha), \beta)$, the 'emanator' above (so named to distinguish from a 'propagator').

Universal Emanation from T on single path with three steps: $((\mathbf{T} \bullet \boldsymbol{\tau}_1) \bullet \boldsymbol{\tau}_2) \bullet \boldsymbol{\tau}_3)$ …

Consider the first emanation step:
$\mathbf{T} \bullet \boldsymbol{\tau}_1 = (A,B) \bullet (\tilde{A},\beta) = ([A\bullet\tilde{A}-\beta^*\bullet B] , [B\bullet\tilde{A}^*+\beta\bullet A])$. (Standard Cayley algebra multiplication rules.)
$A\bullet\tilde{A} = (a,b) \bullet (\tilde{a},\alpha) = ([a\bullet\tilde{a}-\alpha^*\bullet b] , [b\bullet\tilde{a}^*+\alpha\bullet a])$
$B\bullet\tilde{A}^* = (c,d) \bullet (\tilde{a}^*,-\alpha) = ([c\bullet\tilde{a}^*+\alpha^*\bullet d] , [d\bullet\tilde{a}-\alpha\bullet c])$
Thus,
$\mathbf{T} \bullet \boldsymbol{\tau}_1 = (A,B) \bullet (\tilde{A},\beta) = ([(a\bullet\tilde{a}-\alpha^*\bullet b-\beta c) , (b\bullet\tilde{a}^*+\alpha\bullet a-\beta d)] , [(c\bullet\tilde{a}^*+\alpha^*\bullet d+\beta a) , (d\bullet\tilde{a}-\alpha\bullet c+\beta b)])$.

At the lowest octonion level, that covers the pure real trigintaduonion, we have:

$(a\bullet\tilde{a}-\alpha^*\bullet b-\beta c) \rightarrow 8x8 + 8 + 8 - 2 = 64+14 = 78$ independent octonion terms (78 independent generators of motion). The -2 comes from the unit norm constraints on T and τ.

Now consider the second propagation step:
$(\mathbf{T} \bullet \boldsymbol{\tau}_1) \bullet \boldsymbol{\tau}_2 = ([(a\bullet\tilde{a}-\alpha^*\bullet b-\beta c) , (b\bullet\tilde{a}^*+\alpha\bullet a-\beta d)] , [(c\bullet\tilde{a}^*+\alpha^*\bullet d+\beta a) , (d\bullet\tilde{a}-\alpha\bullet c+\beta b)]) \bullet (\tilde{A},\beta)$,

where $\boldsymbol{\tau}_2 = (\tilde{A}',\beta') = ((\tilde{a}',\alpha'), \beta')$.

Let $(\mathbf{T} \bullet \boldsymbol{\tau}_1) \bullet \boldsymbol{\tau}_2 = ([Z_{11},Z_{12}], [Z_{21}, Z_{22}])$.
$Z_{11} = (a\bullet\tilde{a}-b\alpha-c\beta)\bullet\tilde{a}' - (b\bullet\tilde{a}^*+\alpha a-\beta d)\,\alpha' - (c\bullet\tilde{a}^*+d\alpha+a\beta)\beta'$.

In Z_{11} we can replace the octonions with their unit component forms:

$$a = a_1e_1 + a_2e_2 + \ldots + a_8e_8 ,$$

where $\{e_1, e_2, \ldots, e_8\}$ are the unit octonions (one real, seven imaginary), while 'α'$=\alpha e_9$ and 'β'$=\beta e_{17}$, originally, but in expressions, are reduced to just their real part. All expressions, thus, involve 10 components: $\{e_1, e_2, \ldots, e_8, e_9, e_{17}\}$, and as the equations for Z_{11} shows, grouped in factors of three (three-element octonionic 'braids'). We don't have associativity but we do have alternativity and the braid rules on three-element octonionic products that allows their regrouping. Applying these rules to have only ordered $e_i\bullet e_j\bullet e_k$ products in a simplified expression, we will then have $10 \times 9 \times 8/3! = 120$ independent terms when the products involve different components. We have 8 independent terms when the first product are on the same component (equals 1), have 8 independent terms when the second product involves the same component, and have 1 independent term when the three-way product equals 1 (further details on this and the properties of the exponentiation map on hypercomplex numbers is given in the next section. There are, thus, 137 independent terms in Z_{11}, where each term has norm less than unity (since each octonionic component has norm less than one and the norm of a product of octonions is the product of their norms). The terms involving products with the same component, or with the components three-way product equal unity, correspond to the 'telescoping terms' in what follows.

When $\mathbf{T}=((\mathbf{a},\mathbf{b}),(\mathbf{c},\mathbf{d})) \rightarrow ((\mathbf{T} \bullet \boldsymbol{\tau}_1) \bullet \boldsymbol{\tau}_2)=((Z_{11},Z_{12}),(Z_{21}, Z_{22}))$. we have $a \rightarrow Z_{11}$ and the terms involving 'a' in Z_{11} are referred to as 'telescoping' due to their simple math properties with further emanation steps. In particular, the terms involving 'a' are:

$Z_{11}[\text{a terms}]= a\bullet\tilde{a}\bullet\tilde{a}' - a\alpha\alpha' - a\beta\beta'$.

We can see that the original 'a' information is passed along three (telescoping) channels, one involving repeated full octonionic factors $\tilde{a}$, one involving repeated real-octonion α factors, and one involving repeated real-octonion β factors:

(1) a → $(a \bullet \tilde{a}) \bullet \tilde{a}'$, if this product is continued indefinitely, then we have ***the random product of a collection of octonions***, all of which have norm less than one (although their norms can be quite close to one). If their norms were perfectly equal to one, then the addition of their random 'phases' would tend to cancel to zero, giving only a real octonionic component (same argument for phase cancelation on S1 as on S7 or S15). What results is a 'mostly' real octonion, having some imaginary part. A more precise, and lengthy, derivation is given in the next section.

(2) a → $a\alpha\alpha'$, if this product is continued indefinitely, 'telescoped' with repeated α products, we see that the original 8 independent terms arising from 'a' are passed forward with an overall real octonion product, giving rise to 8 independent terms.

(3) a → $a\beta\beta'$, as with (2), we have 8 independent terms.

From the above, we see an alternative accounting of the extra 17 independent terms to go with the 120 for a total of 137 independent terms in the propagation of the octonionic sectors of the universal emanation. A benefit of the telescoping analysis is it clarifies how in (1) an imaginary component may arise, and in perturbation expansions it will then be natural to refer to an overall imaginary component.

There are 137 terms in the dually chiral 'emanation', each with norm bounded by unity, with total bi-sedenion norm equal to unity. In the analysis that led to the computational discovery of α [6], an imaginary (non 10D) component was added of growing magnitude until unit-norm propagation failed. In essence, a maximum perturbation, from propagation strictly in the 10D subspace of the 32D trigintaduonions, was sought.

We identify maximal perturbation by doing an independent term analysis, and by adding a maximum perturbation term that implicitly identifies a definition of maximum antiphase. From this definition of maximum antiphase, there results the parameter π.

D.2.3.1 Exponential Map Properties when using hypercomplex numbers

For what follows, it helps to recall some important properties of the exponential, particularly its well-defined properties with hypercomplex numbers [7]. Important map relations:

(1) exponential map on Im(T) gives unit norm object: $\exp(\text{Im}(T)\theta) = \cos\theta + \text{Im}(T)\sin\theta$.

(2) exponential map on iT gives $C \times T$:
$\exp(iT) = \exp(i\text{Re}(T)) \times \exp(i\text{Im}(T)) = (\cos\theta + i\text{Re}(T)\sin\theta) \times (\cos\varphi + i\text{Im}(T)\sin\varphi) = C \times T$

Use (1) to focus on fluctuations in imaginary parameters free of normalization concerns.

Use (2) to get complex structure $C \times$ (object). Note that exponentiation into phase terms is precisely what occurs in the path integral propagator formalism, and will occur here as well for the emanator formalism, thus the "$C \times$" complex factor. When drawn upon in the emanator formalism, this method of achieving additional "$C \times$" complex structure will be forced by the zero-divisor handling (that will give rise to point-like matter with very small phase coupling, thus a highly oscillatory integral, and ties over to foundational aspects of the path integral formalism).

D.2.3.2 Alternate 137th count using Exponential Map

The derivation below follows [7], but with a more succinct accounting of the independent terms.

Consider a general norm=1 bisedenion in list notation: (A,B), where A and B are sedenions. Consider a propagator bisedenion (C,β), $C = (c,\alpha)$, where c is an octonion and α is shorthand for the real octonion $(\alpha,0,0,0,0,0,0,0)$, where α is a real number, and β is shorthand for the real sedenion $(\beta,0,0,0,0,0,0,0,0,0,0,0,0,0,0,0)$, where β is a real number. Using A=(a,b), B=(u,v), and the multiplication rule from Section 2, we have:

$(A,B)(C,\beta) = ([AC-\beta^*B], [BC^*+\beta A])$, where

$AC = (a,b)(c,\alpha) = ([ac-\alpha^*b],[bc^*+\alpha a])$; $BC^*=(u,v)(c^*,-\alpha) = ([uc^*+\alpha^*v],[vc^*-\alpha u])$.

Thus, we have:

$(A,B)(C,\beta) = ([(ac-\alpha^*b\ ,\ bc^*+\alpha a)-\beta^*(u,v)]\ ,\ [(uc^*+\alpha^*v\ ,\ vc-\alpha u)+\beta(a,b)])$,

so,

$(A,B)(C,\beta) = ([ac-\alpha^*b-\beta^*u\ ,\ bc^*+\alpha a-\beta^*v]\ ,\ [uc^*+\alpha^*v+\beta a\ ,\ vc-\alpha u+\beta b])$.

Now consider another propagator bisedenion (C',β'), $C' = (c',\alpha')$, and form the product corresponding to the next multiplicative step:

$((A,B)(C,\beta)) (C',\beta') = ([(ac)c' - \alpha^*bc' - \beta^*uc' - \alpha'^*(bc^*+\alpha a-\beta^*v) , \ldots] , [\ldots , \ldots])$, where only the first expression at octonionic level ($T=(O_1,O_2,O_3,O_4)$) is shown:

$$O_1 = (ac)c' - \alpha^*bc' - \beta^*uc' - \alpha'^*(bc^*+\alpha a-\beta^*v).$$

At octonionic-level there are 10x9x8/3x2=120 independent terms for 8 octonionic components (labeled a, b, c) plus a separate octonion component (α) and one sedenion component (β), e.g., have 10 choose 3. Also have telescoping terms with repeated real octonion factors, such as with the $a\alpha\alpha'^*$ term (think $a\alpha(\alpha'^*)^n$), which gives an additional 8 independent terms. Also have telescoping terms with alternating real octonion factors and real sedenion factors, such as with the $v\beta^*\alpha'^*$ term (think $v(\beta^*\alpha'^*)^n$), which gives another 8 independent terms. There is one other 'telescoping' term due to repeated octonion right products seen in $(ac)c'$ (now think $((ac)c')c'\ldots..c')$). The change in this term corresponds to an element of the automorphism group on octonions, G2, and as such provides one last independent term, for a total of 137 independent terms at octonion level.

All of the octonion products involve octonions with norms at most unity, and by the normed division algebra rules on octonions, their norm is simply the norm of the individual octonions multiplied together, all of which are bounded by unity, thus their product is bounded by unity. The overall bound for the expression, each individual term being bounded by unity, is therefore simply the counting on the independent terms.

The maximum magnitude of each component of the octonion in the product term is given with a 'channel multiplier' of 137. Also, in seeking the maximum information propagation we require that the real chiral component never cross zero (e.g., stay in its connected $\{\alpha,\beta\}$ quadrant), thus the strictest condition on evaluating evolution might be intuited to be when the imaginary components combine to have real component contribution that is antiphase, e.g., the total imaginary angle is π. The choice of antiphase will used in what follows and will be justified when "C ×" allows the antiphase to be understood in the context of the Universal Mandelbrot set [7] position on the negative real axis that gives the maximal magnitude of displacement from the origin: C_∞. We limit the maximum perturbation allowable by the antiphase worst case. At octonionic-level there is thus the channel multiplier: $137 + i\pi$.

D.2.3.3 The {α, π} relation using Exponential Map

The maximum perturbation, referred to as maximum noise in what follows, is first evaluated for a chiral emanation where we take a norm=1 T_{base}=(A,B) and take the right product with T_{chiral} in the form T_{chiral}=(C,β), with product (A,B)(C,β) proven to be unit norm above [91]. In the prior section we saw that there are 137 independent octonion terms at the octonion sub-level of the new unit norm trigintaduonion that results, which leads to 137 independent terms at component level. In order to use the map rules mentioned in the previous section, it is necessary to move from the trigintaduonion, T, space, to the C × T space. This is done in later sections anyway where we consider sums on exp(iT). The exponential function (map) provides a well-defined 'lift' of a hypercomplex (Cayley) algebra from T to C × T. The exponential map also provides a very useful maneuver when working with unit-norm hypercomplex numbers via the generalized deMoivre theorem exp(Im(T))=cos(θ)+Im(T)sin(θ), with the real part recoverable from cos(θ). More details on this follow later but for now, in evaluating the maximum noise allowed we have three structures to adopt: (1) the noise is generalized to be complex (as will be the case for the components themselves once the T→C × T structure is adopted). (2) At component-level, the noise (for maximum noise) is equipartitioned in both real and imaginary parts. (3) Total imaginary noise magnitude is π for maximal antiphase (to be justified later).

(I) Chiral emanation noise: have 137 terms with max unit norm each, for the real part, and for the imaginary part have a "phase angle" β such that 137β=π (here referred to as a phase angle in the sense that the exp(Im(T)) map is being used). The noise magnitude at octonionic-component level is then given by the right triangle with real part = 137 and angle β=π/137, thus maximum chiral emanation noise magnitude is:

$$H = 137/\cos(\pi/137)$$

(II) Achiral emanation noise: now have 29 "free" components, each with 137 independent terms. For maximum achiral emanation we thus have 137 x 29 independent terms that are built from the aforementioned chiral emanation terms (to make achiral). If we equipartition as before, with noise magnitude Hc, we have a "noise triangle" with magnitude (hypotenuse) Hc and with angle $\theta = \pi/137x29$. The imaginary part is then (Hc)sin(θ). As regards the H magnitude separated form (separating out the 'H' factor for now), we have for the imaginary part sin(θ)c. As before, we take maximal noise transmission when all the imaginary parts add to maximal antiphase. Given the equipartitioning assumption, we then simply have the factor 137x29:

$$\sin(\theta)\, c\ (137\text{x}29) = \pi \rightarrow c = \theta / \sin\theta.$$

The maximum real noise perturbation that the system can have is then α, where:

$$\alpha^{-1} = \frac{137}{\cos\beta}\cos\theta\frac{\theta}{\sin\theta}, \qquad where\ \beta = \frac{\pi}{137}\ and\ \theta = \frac{\pi}{137x29}$$

$$\alpha^{-1} = 137.03599978669910\,,$$

where the evaluation was done at WolframAlpha to high precision [100] (e.g., higher precision than that reported in earlier work). This matches the experimentally observed value to all 11 decimal places currently known. As of 2002, the measured value of α is:

$$\alpha^{-1} = 137.03599976(50).$$

Note that in quantum field theory the parameters are renormalized at a particular energy scale. Thus choice of energy scale impacts the value of α (as a coupling constant in the classical theory or a perturbation expansion factor in the quantum theory). At 0K we have the extreme low-energy end of the renormalization group (with the largest α value). We are at the 2.7K CMBR, so we have the max α to very high precision. (In studies at high energy scale at LEP, at the energy scale of the Z-boson (91GeV), we get the renormalized value to be: $\alpha^{-1}[M_Z] \cong 127.5$. Note that 91GeV is way above the energy scale of the familiar Hagedorn temperature at ~pion mass=150MeV or 1.7x10^12 K) [101], where hadronic matter 'evaporates' into quark matter.)

D.2.4 Trigintaduonion Emanation: achirality from chirality

There are four chiralities, and for a given chirality (with unit norm) there are 29 dimensions of freedom (10D + 19D of chiraly consistent perturbation). When analytic extension is taken to give maximal information flow, the effective dimension for each of the four chiralities is 29* (detailed in [102]). This clear decomposition into 29* independent effective dimensions is then revealed in the {α,π,C_∞} relation in [102]. The Mandelbrot Set is one of many that encounter the universal constant C_∞. The Mandelbrot set also describes a boundary with 2D fractal dimension at its "edge of chaos" [102]. If driven to similar optimality in approaching a zero-value (a zero-divider issue), we see a two-value zero-crossing specification effectively like a double zero. The parameterization of the zeros of the Emanator at chiral zero-divisor points will thus be as double-zeros.

For what follows we use the simple description of the emanator:

$$T_{chiral}^{(k)} = \begin{cases} ((0,\alpha),\beta) \\ ((\alpha,0),\beta) \\ (\beta,(0,\alpha)) \\ (\beta,(\alpha,0)) \end{cases}.$$

$$\text{Emanation}(\mathbf{T}) = \frac{1}{N} \sum_{k\in\{4x72\}^n} \mathbf{T} \bullet \boldsymbol{T}_{chiral}^{(k)} = \frac{1}{N} \sum_{K\in 4\ chiralities} \mathbf{T} \bullet \overline{\boldsymbol{T}}_{chiral}^{(K)}$$

If working with non-split T's, then we restrict to emanations that are perturbations of unity:

$$T_{chiral}^{(k)} = \begin{cases} ((0,\alpha),\beta) \\ ((\alpha,0),\beta) \\ (\beta,(0,\alpha)) \\ (\beta,(\alpha,0)) \end{cases}, where\ T_{chiral}^{(k)} = \mathbf{1} + i\boldsymbol{\delta}.$$

From unit norm we have $\alpha^2 = 1 - 0^2 - \beta^2$, with ± sign choice on α, similarly for β.

If working with split T's (bi-sedenions, etc.), then we have manifest Lorentz Invariance (shown in Chapter 7). So, it is often convenient to work with split-T' since this is manifest from the outset.

Issues with zero-divisors

Suppose we add the rule that emanation may not proceed when a particular chirality is zeroed-out, in other words:

$$\mathbf{T} \bullet \overline{\boldsymbol{T}}_{chiral}^{(K)} \neq \mathbf{0}.$$

For 'normal' numbers this goes without saying, since for real numbers if we have $r_1 \times r_2 = r_3$ then $r_3 \neq 0\ if\ neither\ r_1 = 0\ or\ r_2 = 0$. This holds true for the Real, Complex, Quaternion, and Octonion numbers. This does not hold true for Sedenions or higher. For sedenions the dimensionality of the zero-divisor event is mostly constrained, while for trigintaduonions it is significant. If such zeros were eliminated from the emanator description by using analytic extension component-wise (on 29* effective components) we see how a description devoid of matter (pure static field with no source or sink) might acquire matter by way of extending to a maximal domain of analyticity be removing zero-divisor events (a Wick transformation from real dimensionless action to pure imaginary action that is dimensionless but consisting of a dimensionful ratio). For further discussion along these lines, see Book 7 [7].

D.2.4.1 Achiral T-emanation has 29* effective dimensions

Let's estimate of the effective dimension of information transmission in an achiral T-emanation process. There are 4 chiralities, so to get an achiral emanator candidate, minimally need a "4-card deck" to emanate in the four chiralities, with emanator equal to normalized sum. The actual deck appears to require a normalized sum over sub-chiralities, as will be explicitly enumerated in what follows. Here are the four chiralities with real fluctuation noise shown:

$$\begin{gathered}
(\,(\,(\mathrm{O}[0] \pm \delta, \ldots\,), \alpha \pm \delta), \beta \pm \delta) \\
(\,(\,\alpha \pm \delta, (\mathrm{O}[0] \pm \delta, \ldots\,)), \beta \pm \delta) \\
(\,\beta \pm \delta, (\,(\mathrm{O}[0] \pm \delta, \ldots\,), \alpha \pm \delta)\,) \\
(\,\beta \pm \delta, (\,\alpha \pm \delta, (\mathrm{O}[0] \pm \delta, \ldots\,)\,)\,)
\end{gathered}$$

where α is a real octonion and β is a real sedenion, and Tem is an equal weight sum of the action of each of the sub-chiral propagations on the base T, with the fluctuations indicated each done separately. We have the constraints $\alpha \neq 0$, $\beta \neq 0$, and common octonion O not pure real.

Each of the δ's is an independent fluctuation corresponding to its own sub-chiral emanation, but no subscripting on δ's is used or shown. There are thus 9x2x4=72 independent *imaginary* noise fluctuations to consider in the exp(Im(Tem)) evaluation (that automatically provides unit-norm). The real noise fluctuations in the real (first) component are, thus, not counted. If our definition for Tem entails only one card being dealt, then the sum over those possibilities is the sum

$$\mathbf{T} \bullet \mathrm{T_{em}} \equiv \mathrm{Emanation}(\mathbf{T}) = \frac{1}{72} \sum_{k \in \{72\}} \mathbf{T} \bullet \boldsymbol{T}^{(k)}_{chiral}$$

For one-card, or a one-step, emanation, with real components and real noise, this makes sense from the counting shown, and it's what we use going forward. Using this will allow an entirely separate method for evaluating α (here at the one-card hand approximation). This will be done by determining the effective dimension 29*>29 of maximal information propagation (or maximal noise fluctuation). Before moving on, however, let's examine what happens when we allow complex noise fluctuations as this will trivially be allowed when we consider C × T via exp(iT) in later discussion anyway.

D.2.4.2 Effective Deck size is 72, which is consistent with the $\{\alpha,\pi,C_\infty\}$ relation

Maximum information transmission involves a complex extension to the T components and their noise fluctuations, but in doing this it must retain emanation structures such as the octonionic triple that occurs in previous expressions (starting with the proof of the T_{chiral} solution itself), which leads to the counting that gives 137 independent terms, etc. Thus, the maximal complex extension on the noise is that it remain real in the octonion components:

$$((\,(O[0] \pm \delta, \ldots), \alpha \pm i\delta), \beta \pm i\delta)$$
$$((\,\alpha \pm i\delta, (O[0] \pm \delta, \ldots)), \beta \pm i\delta)$$
$$(\,\beta \pm i\delta, (\,(O[0] \pm \delta, \ldots), \alpha \pm i\delta)\,)$$
$$(\,\beta \pm i\delta, (\,\alpha \pm i\delta, (O[0] \pm \delta, \ldots)\,)\,)$$

The first chiral T component is where new imaginary terms might arise (the others are already counted since in imaginary components). We see there are six more, so the deck is now78. In application, as we will see, those added six are precluded due to constraints such that the effective deck size (impacting the sums and numbers of independent terms in the emanator definition) remains 72.

All noise terms will be treated additively, including terms in different imaginary components as well as imaginary noise terms in the real component. The criterion for max noise (in-phase constructive interference) gives the extreme of linear additivity. (Not like Gaussian statistical noise that adds in quadrature.) Also note that the discussion in terms of "noise transmission" and "information transmission" will be used almost interchangeably, whenever one description or the other best suits the analysis it will be used. Note that with this kind of noise analysis we can effectively shift around T noise terms associatively. Also note that application of the Kato-Rellich theorem [103,104] is related to the noise budget analysis done here focusing on first order terms.

There are 137 independent tri-octonionic terms in each of 29 free components indicated by a particular chirality (within the 32 components of a general trigintaduonion). This is a nontrivial result since (T_{chiral} • T_{chiral}) is no longer T_{chiral} type (but still T_{norm1} type), so direct expansions are needed to identify the number of independent terms and this is briefly described below, with more detail in [7].

Obtaining an achiral emanation from a collection of chiral emanations requires that all chiralities be summed over (there are four) as well as sub-chiralities (there are 72). Noise analysis requires collecting of first-order terms. Analysis of noise transmission indicates 29* dimensions, where:

$$29^* \cong 29 + \left(\frac{4\pi}{72}\right)\left[1 + \left(\frac{\pi}{137 \cdot 29}\right)\left(\left(\frac{\pi}{72}\right) + \left(\frac{3}{72}\right)\right)\right]$$

The above result was obtained in [102] to describe the 72-card chiral 'deck' of chiral emanation products for a "single-step" emanation. In the new Results to follow this is reviewed and elaborated further.

D.2.4.3 'Edge of chaos' maximal perturbation hypothesis [102]
Consider the 'edge of chaos' maximal perturbation in each of the 29* dimensions to be at position C_∞, which is on the negative real axis, i.e., at π rotation to have -1 factor, ***thus at maximal antiphase***. This results in the relation for maximal perturbation at maximal antiphase (maximum reference angle with sign chosen positive by convention) has a lower bound on α given by:

$$\alpha_{\square}^{-1} = \left(\sqrt{C_\infty}\right)^{29^*}.$$

where

$$C_\infty = 1.4011551890920506004 \ldots$$

This ties $1/\alpha$ to the second Feigenbaum constant C_∞ in the context of the Mandelbrot set. It is well known that the Feigenbaum constants are universal, and part of a description of a universal transition to chaos regime. The Mandelbrot set is also universal [105], and maximal in that its fractal boundary has maximal fractal dimension of 2 [105], a detail that will be important in the meromorphic matter description given later.

For C_∞, most references only provide $C_\infty = 1.401155189 \ldots$, and a higher precision tabulation is not readily found, so use is made of the relation

$$C_n = a_n(a_n - 2)/4,$$

together with the tabulation on a_∞ [106]:

$$a_\infty = 3.5699456718709449018 \ldots$$

The resulting C_∞ is:

$$C_\infty = 1.4011551890920506004 \ldots$$

The resulting $\alpha_{\square}^{-1}$ is:

$$\alpha_{\square}^{-1} = 137.0359993370198263 \ldots$$

D.3 Implication of MIE

D.3.1 MIE requires a complex Hilbert Space

As mentioned previously, according to [93], a complex Hilbert space is selected by the quantum deFinetti theorem, since it is required for information propagation (and thereby consistent with the maximum information propagation concept in its selection). Because it's a complex Hilbert space, this explains why the path integral operates in a complex space, even though the underlying universal algebraic construct from which it is emergent is hypercomplex to the level of the trigintaduonions.

From [93], a simple derivation shows why the quantum deFinetti Theorem requires amplitudes to be complex. Suppose f(n) is the number of real parameters to specify an n-dimensional mixed state. For real amplitudes f(n)=n(n+1)/2, for complex amplitudes f(n)=n^2, and for quaternionic f(n) = n(2n-1). For propagation, etc., need f(n1n2)=f(n1)f(n2), which only works for complex amplitudes.

D.3.2 Time is analytic and Matter is meromorphic

A variety of efforts have been made to find a definition of time that is somehow implicit to the main QFT and GR formalisms, whether it be a choice of vacuum for QFT in curved spacetime (and even if the spacetime is not curved [107]) which is indirectly a choice of time. Or seeking an internal time-reference in a full-GR quantum minisuperspace analysis of dust-shell collapse [108]. Or in seeking a notion of time in full general relativistic (GR) models, in the equilibrium sense, with an assumption of euclideanizability [109,110]. For the latter, the self-consistent stable solutions that were indicated showed the general utility of the euclideanizability hypothesis on emanation/propagation solutions in general (that is especially relevant, or interpretable, when the system is in equilibrium). In none of these efforts, however, was there success in identifying some internal notion of time, time, it seems, is an added construction, and this is consistent with the results shown here, where we find that time is likely analytic and an emergent construct.

In Book 7 [7], we see matter as meromorphic residue precipitation, in amounts of one quantum given by a precursor to Planck's constant h*. The meromorphic residue winding number is also notable in that it gives an integer that stays constant in the meromorphic region. This raises the possibility that elementary particle attributes might encode by way of different winding numbers, with reference to their different winding numbers at residues, but this will not be discussed further here.

D.3.3 Emergent Evolution and Emergent Universal Learning

We see that the definition of the emanator process is not known, but that consistency arguments (such as achirality constructed from a collection of chiral emanations) lead to a certain set of forms. And that consistency with the $\{\alpha, \pi\ C_{\infty}\}$ relation imposes further constraints on the form of the emanator. What is hypothesized is that the emanator is selected for maximal information transmission, thus emergent itself under that criterion. Let's now consider the maximal information transmission idea from the receiving end, e.g., maximal information receiving, or learning, in this context. If we turn to the information geometry analysis of learning in neural nets [111-115] (which uses differential geometry) we obtain a fundamental origin for statistical entropy (Shannon Entropy), and we identify optimal learning processes, based on expectation/maximization, that involves two-steps (as the name suggests) that may be done according to two fundamentally different conventions, e.g. the optimal learning involves four types of step, or is doubly chiral, consistent with the emanator 4-chiral processes described [7]. The potential applications of these results to Trigintaduonion encoded neuromanifolds is beyond the scope of this paper.

D.3.4 Objective Reduction, Zero-Divisors, and possible origins of Planck 's constant

A new mechanism for objective reduction [92,116] is also indicated by the way π enters the theory as a maximum anti-phase amount comprising part of the maximal perturbation propagation. Consider in the context where there is a 'classical' trigintaduonion path in a congruence of paths (a flow-line description). On the classical path in the congruences, we have α calculated using a $+\pi$ maximal anti-phase, but this could also occur with $-\pi$ maximal anti-phase as well, thus we could have a $\pm\pi$ phase toggle when a zero divisor is encountered in the 32D propagation (given the perturbations extending outside the 10D somewhat into the entire 32D). The zero-divisor discontinuity requires the field to reformulate a new 'consistency' with the 32D algebraic propagation (and 64D and higher, as well), which could have the result that since the prior π phase had the discontinuity, then it must toggle to the other, negative, phase, e.g., objective reduction may occur as a zero-divisor phase-toggle event.

D.3.5 Where's the geometry?

The geometry side of emanation theory does not result from the action of the repeated emanator product directly, but from the accumulated product in the T base that results. Geometry is, in effect, emergent (projected) on the T_{base} 'space' of the T_{em} product action. Geometry appears as a manifold construct in both space-time curvature (where it is locally given with the standard model action) and as an intrinsic entropic property, via neuromanifold 'geodesic' motion being equivalent to, and possibly the origin of, the minimization of the relative entropy (and maximization of entropy, the 2nd Law) [88]. Setting aside thermodynamic issues in this discussion, this puts the Lagrangian formulation with standard model terms, and Hilbert action for GR, into better perspective. The representation of the geometry via the Hilbert action for GR suffices with maximal extension in whatever causally connected domain of interest. So, we've got the existing QFT in CST space-time formulations in the black hole exterior, for example. We may have the resolution at the black hole horizon causal boundary via String Theory on the surface (using Ads/CFT relation and related holographic hypothesis [118,119]).

We describe repeated chiral product action on the trigintaduonion spinor space. The emanation process, consisting of a chain of chiral trigintaduonion products, leads to a Lagrangian variational formalism *with the standard model.* The origins of the parameters of the model are beginning to be understood as well. Apparently state information memory/inertia is carried via the manifold curvature response to the matter density, where 'G' is the linkage for the balance on this 'learning' process. Presumably the G learning rate is set for optimal learning, e.g., maximal information flow, and as such its value may eventually be clarified theoretically.

D.4 Results

D.4.1 Two-card Hand consistent with 137 and Neuromanifold Two-step Learning

MIE may guide our choice of 'hand' size dealt. Consider that one chiral emanation step (multiplication) off a base trigintaduonion will only have 78 independent tri-octonionic noise terms introduced. With a second step we ‘flood the channels’ and get the full assortment of 137 tri-octonionic noise terms (as shown in the derivation that obtained the 137 terms in [102]). Thus, a minimum of a two card hand must be considered. Having achieved a maximal perturbation scenario, what need for further cards in the hand, i.e., or further multiplicative factors in the pre-achiral normalization step?

Previously, when speaking of “Single-step achiral” we presumed initial conditions for the noise that effectively performed an average two-card ‘hand’ emanation analysis. If not in such an approximate form the ***{α, π}*** relation is recovered. The approximate form, however, expresses a new hypothesized relationship: $\alpha_{\square}^{-1} = \left(\sqrt{C_{\infty}}\right)^{29^*}$, that is itself verified to many decimal places.

Two-card-hand and two-step emanation jibes with two-step neuromanifold ‘learning via ‘em’ and indicating relative entropy as optimal in this process (similar to choice of Euclidean distance in Riemann spatial change, here we have relative entropy as local entropic measure (with fixed reference, this then gives Boltzmann’s entropy.

D.4.2 The Chiral-Extension Cayley-Family relation

Recall that the unit-norm propagation was extended from octonion-level to trigintaduonion-level by introduction of chiral trigintaduonion emanators. The mathematical construct at the trigintaduonion-level that results can be represented in terms of tri-octonion products, thus all of the octonion properties are thereby inherited with such a construct since it is represented in terms of octonions, most notably a norm (here unit norm is of interest physically, for repetitive ‘propagation’). The same double-chiral extension can be done at any Cayley level, and we thereby have a Cayley-Family extension property, via use of chiral Cayley emanation only, that is general.

Associativity Extension most likely Candidate for Renormalization and Thermality reification

Application of the Chiral-Extension Cayley-Family relation is first considered for associativity. In the Cayley Algebras, the highest algebra that retains the property of associativity is the quaternions. If we consider extending this associativity to higher -order algebras in the context of chiral propagation we can do this up to algebras at the level of the octonions and the sedenions. Consider that ***unit-norm*** trigintaduonion emanation will reduce to sedenion propagation in a general sense, and it is in this space that we can now add another sedenion element that is associative.

D.4.2.1 Sedenion Associative Element and the Axiom of (Renormalization) Choice

The description of the renormalization process involves counter-terms. The emanation/selection of the standard gauge field group, and other elements (described in Section X), seems consistent with the Standard Model (and certain extended versions), but makes no mention of counterterms. Rather than add counter-terms as an external regularization element to the theory, here we can consider them as a more complete manifestation of the existing Emanator theory (again the MIE hypothesis justifies this). So, let's suppose there really is chiral trigintaduonion emanation, we can now see where a side-version of the theory presents that is sufficient to provide a (reified) version of the renormalization terms needed to renormalize the theory. Note that the application of the renormalization 'cancellation' is a fundamental application of the Axiom of Choice. More accurately, the universe that is hypothesized to be MIE, implicitly selects for the existence of the Axiom of Choice (and well-ordering foundation of mathematics) in its choice of renormalization cancellation.

D.4.2.2 Octonion Associative Element and Complex Periodic Time

If extension of associativity on a sub-class of chiral-trigintaduonion emanations was possible, via the aforementioned effective sedenion associativity on chiral propagation, what of the indicated octonion-associativity extension? At the level of the octonion we can see a more direct manifestation of Lorentz invariance, including terms that are spatial or temporal (see generalized Lorentz invariance representations in Book 4 [4]), so suppose it is here that the theory creates ties between the real time quantum dynamics on a manifold and the complex-time statistical dynamics on a neuromanifold. These ties are shown to exist during equilibrium analyses in a number of applications (ss Book 6 [6]), and in thermal quantum field theory applications (see Book 5 [5]), but here we speak to this being a more fundamental attribute, revealed in the indicated circumstances, yes, but also present in general (albeit perhaps not in a useful form).In essence, we use our second wish/demand (we only get two) of the associative-extension genie that it explain the recurrence of the appearance of a fundamental thermality by way of time having a complex periodicity proportional to inverse system temperature.

'Contact' Associativity from Higher-order Cayley Algebras (64 and 128 orders for example) was suggested previously, involving use of limit processes and infinitesimal processes on the non-propagatable orders (even chiraly) above 32-order trigintaduonion, to be at zero displacement from the 32-algebra (thus 'contact'), but still imparting possible nontrivial contributions, such that the above renormalizability and complex-time relations might be obtained. Here we find a better route to this result via the chiral-extension property.

D.4.2.3 Octonion Commutative Element and Analytic Time
If we are considering operations reduced to octonion-level, there are also chiral extensions on the commutative complex numbers to arrive as a subset of commutative octonion-level operations in the unit-norm chiral emanation process. This allows further modification to the octonions as regards the internal time parameter analyticity, not just its imaginary-time periodicity. Together these chiral extensions may provide for the analytic time that is observed in the many successful applications of such a notion (required in path integral descriptions of bound states, for example).

The multiplication properties of the Cayley algebras are: Commutativity, Associativity, Alternativity, and Power Associativity. As you go to higher order algebras you lose properties. The Complex numbers are the last to have commutativity, the quaternions the last to have associativity, and the octonions the last to have alternativity. All the algebras have power associativity. Thus, the chiral extensions described are now exhausted since the interesting cases to extend, from complex, quaternionic, and octonionic, have now been considered.

D.4.3 The Extended Standard Model predicted by Emanator Theory
According to the Standard model of Particle Physics, the Universe consists of point-like spin-1/2 fermions that reside in two families: the Leptons and the Quarks. These particles exist in a geometry with a gauge field, where they interact through that gauge field according to local gauge invariance and indirectly through the geometry according to general relativity. Even more indirectly, the gravitational interaction via mass has mass itself determined through exchange of spin-0 Higgs particles.

The local gauge invariance of the Standard model is:
$$U(1)_\gamma \otimes SU(2)_L \otimes SU(3)_C.$$

Let's consider each part separately:

$U(1)_\gamma \rightarrow$ gives rise to the electromagnetic force between charged
         particles.

$SU(2)_L \rightarrow$ gives rise to the force between left-handed particles.

$SU(3)_C \rightarrow$ gives rise to the strong force that operates on particles
         (hadrons) that carry color charge (three types).

In Book 5 [5] we see that each gauge group listed above derives from the general local gauge invariant group form:

$$SU(N)_L \otimes SU(N)_R.$$

For $N = 1$, we have the trivial case (no left or right): $U(1)$.
For $N = 2$, we have the asymmetric case of only left: $SU(2)_L$.
For $N = 3$, we have the symmetric case of left-right, or simply a fixed ratio: $SU(3)$.

In Book 5 we see that there is mixing between the $U(1)$ and $SU(2)$ parts to give the actual electroweak theory observed:

$$U(1) \otimes SU(2) \rightarrow U(1)_\gamma \otimes SU(2)_L.$$

The mixing introduces universal (globally gauge invariant) 'mixing angles' that number among the fundamental parameters of the theory (along with particle masses and a few other constants).

In Book 7 [7] we see that Emanator Theory predicts the form of the local gauge invariance as a general form, to be exactly that observed: $U(1)_\gamma \otimes SU(2)_L \otimes SU(3)_C$. Emanator theory also predicts a theory governed by 22 parameters, not 19 as currently listed for the Standard Model.

As strange as the form $U(1)_\gamma \otimes SU(2)_L \otimes SU(3)_C$ might seem, it is so far only applied to the interaction element of te theory. Let's now consider an interaction with what? The answer resides in the representations of the indicated groups (the left-handed particles interacting according to $SU(2)_L$ will be seen as interacting visa the weak force, etc.). So, saying there is a local gauge invariance is to say that there will be particles according to the irreducible, independent, representations of these groups. Thus, for any chosen representation there will be a collection of particles predicted. This is precisely what is observed experimentally. Thus, we not only know the groups $SU(2)_L$ and $SU(3)_C$, say, we also know their specific representations to obtain particle numbers and groupings observed experimentally.

So, the Standard Model is actually indicting the Model pair of Local Gauge Field and Representation $\mathcal{R}$:

$$\{U(1)_Y \otimes SU(2)_L \otimes SU(3)_C;\ \mathcal{R}\},$$

and it (the Standard Model) then says "times three", where the representation has there generations or copies:

$$\{U(1)_Y \otimes SU(2)_L \otimes SU(3)_C;\ 3 \times \mathcal{R}\}.$$

For the fundamental Leptons and Quarks this is shown in tabular form with the symbols for the particles as:

	electroweak left-handed Leptons	electro-only right-handed Leptons	Electro-only Quarks
1st Generation	$[e^-{}_L,\ \nu_{e,L}(m=0)]$	$[e^-{}_R, ---]$	d, u
2nd Generation	$[\mu^-{}_L,\ \nu_{\mu,L}(m=0)]$	$[\mu^-{}_R, ---]$	s, c
3rd Generation	$[\tau^-{}_L,\ \nu_{\tau,L}(m=0)]$	$[\tau^-{}_R, ---]$	b, t

Note that there are no right-handed neutrinos in the Standard Model and the left-handed neutrinos are massless. We already have ample evidence that the left-handed neutrinos are massive, so we know the Standard model will extend in this regard, which leaves the supposedly non-existent right-handed neutrinos. There probably are right-handed neutrino but if "electro-only" and having no charge, what results is a particle completely decoupled from the other matter, except gravitationally, precisely what describes dark matter. More on this will follow once we go from the 19 parameter model to the hypothesized 22 parameter model, as this suggests limits on extensions to the theory, as does cosmological data, which suggests a possible 4th neutrino and no more scenario consistent with the cosmological evolution, unless the other neutrinos are very massive (all probably the case).

With the Standard Model quantum field theory structure, with scatter results according to the indicated local gauge field and particle representation, we find additional conservation rules. One of these rules describes the conservation of Lepton number at each generation, in other words:

$$L_e = N(e^-) + N(\nu_e) - N(e^+) + N(\bar{\nu}_e)$$

is a constant, and the same for constants for the muon and tau generations. There is also conservation of baryon number and the familiar conservation of charge.

The Standard Model is a renormalizable/renormalized quantum field theory, and while this poses no difficulty with the $SU(2)$ elements alone or $SU(3)$ alone, here we have a product gauge and, what's more, it has handedness with use of $SU(2)_L$ not $SU(2)$. Not surprisingly, anomalous terms arise in he renormalization effort and for the renormalization to work, additional constraints are imposed on the structure of the theory. Most notable is that at a generational level across both leptons and baryons there should be a charge sum of zero. This is observed and was used to predict missing quarks (charm) in the early discovery process.

The structure identified by the standard model and the high decimal number agreement on key results from quantum field theory together provide a theory that can't be displaced directly by anything better, and only minimally augmented (with addition of right-handed neutrino), so trying to find a unified theory is impossible from 'within' the theory, and Godel's incompleteness theorem also suggests the hopelessness of such a task. This, then leaves no choice but to look for an encompassing theory that projects the quantum theory with the Standard Model structure in its entirety. And this is what is attempted with emanator theory (Book 7 [7]).

The Standard Model can't explain the three generations of matter, the origin of the local gauge group with its odd product form, the dimensionless constant alpha, and certainly can't explain the extension to massive(light) neutrinos, or the possible extension to massive (dark) right-handed neutrinos. In Emanator Theory (Book 7 [7]) we have:
(1) The local gauge $U(1) \otimes SU(2) \otimes SU(3)$ is projected by the theory. Why it should have three generations and why it should by asymmetric with use of $SU(2)_L$ is due to those choices providing the maximal packing of the Light matter sector (for maximal complexity information flow, an aspect of the MIE hypothesis), within he constraint of a 22-parameter emanation process.
(2) The 22 parameter, constants of the motion (emanation), result helps to constrain the particle representations such as to stay at 22-parameters, yet have maximally complex interaction.
(3) The MIE Hypothesis indicates the given generational structure given the constraint of working with the indicated product algebra. It also suggests any 'fine-tuning' on gravitational constant G might be dominated by the dark matter sector and its contribution to the evolution of the universe. Such fine-tuning would be most powerful if it indicated a fractal scale invariance property (see fractal G discussion in [59]).

Before considering the 22-parameter Emanator theory prediction of an extended Standard Model, let's first recount the 19-parameter Standard Model, which I'll separate into four groups:

> (I) 9 Yukawa coupling constants (masses) for the charged fermions
> (II) 5 constants for Weinberg Angle and the CKM matrix (with three mixing angles and CP-violating phase)
> (III) 3 Constants for electromagnetic coupling (alpha), for strong interaction (g3), and strong CP-violating phase ($\theta_3 \approx 0$).
> (IV) 2 Higgs parameters: Mass and Vacuum Expectation

If we allow for the left-handed neutrinos to have mass, then we get 3 more masses and another 4 constants for the PMNS matrix (three mixing angles and a CP-violating phase):

> (V) Extended model: 7 more constants → We, thus, have 26 parameters.

Let's update out table with this extended version of the theory:

	electroweak left-handed Leptons	electro-only right-handed Leptons	Electro-only Quarks
1st Generation	$[e^-{}_L, \nu_{e,L}(m \neq 0)]$	$[e^-{}_R, ---]$	d, u
2nd Generation	$[\mu^-{}_L, \nu_{\mu,L}(m \neq 0)]$	$[\mu^-{}_R, ---]$	s, c
3rd Generation	$[\tau^-{}_L, \nu_{\tau,L}(m \neq 0)]$	$[\tau^-{}_R, ---]$	b, t

There is now a problem. In order to maintain renormalizability, if there are left-handed neutrinos with mass there must be right handed neutrinos with mass [60]. So, just how sure are we about neutrinos having mass? There is strong evidence not only for neutrino mass, but for neutrino family dynamics (here seen as oscillation between neutrino mass states). We'll see more evidence of generational family kinematics/dynamics in a later section. Neutrino family dynamics was first seen in measurements indicating that neutrinos spontaneously changed flavor (the Solar neutrino experiments [61]), which not only indicates they have mass, but allows us to determine the mass differences.

Clearly the Extended Standard Model needs to be extended further, to allow for the massive right-handed neutrino that is hypothesized to have no weak interaction like its right-handed electron cousin (and no electric interaction since no electric charge and no strong interaction since no color charge, thus 'dark' matter). Such a right-handed neutrinos (with no charge) can act as their own antiparticle (which is consistent with formation of "Bright" supermassive Black Holes in the early universe as seen with Webb). Furthermore, in addition to the Dirac mass relation to the left-handed neutrino it also has a Majorana mass term not tied to the Higgs mechanism [62] What results is that instead of mass ${e^-}_L$ equal to mass ${e^-}_R$ here we have

$$m_{\nu_{e,L}} \propto \frac{1}{m_{\nu_{e,R}}},$$

which is known as the see-saw mechanism [63] Thus the very low-mass left-handed neutrinos indicate very large mass right-handed neutrinos. Large mass neutrinos are an excellent candidate for cold dark matter, precisely what is needed to complete the Cosmological Standard Model.

Given the renormalization constraints and the observation of neutrino mass we arrive at the following updated Model:

	electroweak left-handed Leptons	electro-only right-handed Leptons	Electro-Quarks
1st Generation	$[{e^-}_L, \nu_{e,L}(m \ll m_e)]$	$[{e^-}_R, \nu_{e,R}(m \gg m_e)]$	d, u
2nd Generation	$[{\mu^-}_L, \nu_{\mu,L}(m \ll m_e)]$	$[{\mu^-}_R, \nu_{\mu,R}(m \gg m_e)]$	s, c
3rd Generation	$[{\tau^-}_L, \nu_{\tau,L}(m \ll m_e)]$	$[{\tau^-}_R, \nu_{\tau,R}(m \gg m_e)]$	b, t

which is described by the 26 parameter theory indicated for massive left-handed neutrinos (if we assume that the right handed neutrino masses can be determined from the left-handed neutrino masses).

The standard counting to arrive at the 19 parameters (or, now, 26) includes $\theta_{QCD} \cong 0$ as a parameter of theory and its nearness to exactly zero is often referred to as the Strong CP problem. Going forward this is removed as a concern. The 'fine-tuning' that selects $\theta_{QCD} = 0$ is considered to be the same as that which selects the generational number, to arrive at 22 parameters while respecting the local gauge field constraint. Thus t is part of the MIE optimal selection.

Also included in the 19 parameter count is the U(1) gauge coupling g_1 and the SU(2) gauge coupling g_1, and these coupling are related to alpha by:

$$\alpha = \frac{1}{4\pi}\frac{{g_1}^2{g_2}^2}{{g_1}^2 + {g_2}^2}$$

Now, the emanation's projection of the quantizable Lagrangian theory involves various constants (22) in the resulting quantization. In that process, however, alpha already exists, in fact, from the derivation of alpha, we see that it establishes maximal perturbation starting at the level of the emanation process itself. Thus, alpha doesn't number among the 22, and the relation between coupling constants involving alpha means only one of those coupling constants is independently specifiable, so only one should be counted. At tis juncture we've reduced the count on 26 parameters to 24.

Notice in the prior counting, to arrive at the count of 26, we extended the count to be the same with dark neutrinos included since the dark neutrino mass was assumed to have some fixed (inverse) relation to the light neutrino mass. Let's not make this assumption, but retain the same mixing angle matrix, to now have 3 more masses in the theory (adding to the 24 count we now have 27).

In order to reduce the 27 parameter count to 22 we need to remove an 'overcount' of 5. Such an overcount would occur if there was a family-relationship generationally, for all masses. Consider such a mass relationship for the electron-muon-tau family, it is already known to exist and is known as the Koide relation. The Koide relation [64] was first observed for the three massive leptons currently known:

$$\frac{m_e + m_\mu + m_\tau}{\left(\sqrt{m_e} + \sqrt{m_\mu} + \sqrt{m_\tau}\right)^2} = \frac{2}{3}$$

To a lesser extent this relation is satisfied for the quarks as well, particularly for the three most massive, where the value is 0.6695. The problem with a simple application to the quark masses is that they are dependent on energy scale. A theoretical explanation for the Koide relation describes how this relation might exist for the masses of a given generation (or family group) [65]. As mentioned with neutrino oscillations, their oscillations involve the existence of such a family relationship in the quantum field theory implementation, so the hypothesis that it exists for each column in the table is not such a stretch. There are five columns of (independent) masses (the right handed electrons not independent of the left handed electrons removes them from the count). Thus, the five groups of three represent five groups with only two independent mass parameters, and the overall parameter count is thereby reduced by 5 from 27 to 22, as needed to be in agreement with Emanator theory. Thus Emanator theory, with the MIE hypothesis, would select for family group dynamics as observed with the neutrinos since it is what allows the maximal particle set within the 22-parameter constraint.

D.5 Conclusion

Maximal information propagation as an emergent construct appears to require two forms of propagation, an early hypercomplex 'emanation' that involves a chiral 10D propagation in a 32D trigintaduonion space, and standard propagation with complex propagators (consistent with the quantum deFinetti relation) operating inside the geometry and gauge field that is projected. From the 'emanation' stage we see the maximum dimensionality and fractal limits provide the fundamental constants that then imprint upon the emergent geometry and gauge field, including giving rise to the constants α. The origin of α has been a long-standing mystery. So much so that the central role of α in modern physics is literally engraved in stone, the tombstones of Sommerfeld (which displays $e^2/\hbar c$, which is α) and Schwinger (which displays $\alpha/2\pi$) for example. Its origin has eluded physics for over a century, and appears to reside in the algebra of trigintaduonions.

Emanator Theory results from a hypothesized maximal information propagation and this means maximal analyticity, maximal domain, etc. As a process, Emanator theory is also hypothesized to operate up to "the edge of chaos" to permit maximal perturbation (noise) domain. When taken with the results showing that Emanator theory is Martingale, thus has well-defined limits, we then must wonder if there are well-defined multi-scale (fractal) limits. In other words, is there a relation that would tie the micro scale constant to fundamental constants (as they are counted in the 22) with the cosmological scale 'constants' that have settled out, at macro scale, in the current evolution of the Universe? In this context, the Gravitational constant G is hypothesized to be a multiscale fractal coupling parameter.

In seeking a deeper theory we build on the sum-on-paths with propagator formulation to arrive at a sum-on-emanations with emanator formulation.

Propagation in a complex Hilbert space, however, in a standard QM/QFT formulation, requires the propagator function to be a complex number (not real or quaternionic, etc., [93]). This prohibits what would otherwise be an obvious generalization to hypercomplex algebras. In order to achieve this generalization, we have to introduce a new layer to the theory, one with universal emanation involving hypercomplex algebras (trigintaduonions) that is hypothesized to project to the familiar complex Hilbert space propagation with associated fixed elements (e.g., the emanator formalism projects out the observed constants and group structure of the standard model). The 'projection' is an induced mathematical construct, like having SU(3) on products of octonions, but here it we be the standard model U(1)xSU(2)xSU(3) on products of emanator trigintaduonions. Thus, a unified variational formulation is posed, one that arrives at alpha as a natural structural element, among other things, uniquely specified by the condition of maximal information emanation.

A 'deeper' phase of universal evolution is described by a theory of emanations, where mathematically invariant emergent structures appear:

$$\boldsymbol{emanation \rightarrow propagation \rightarrow trajectory}$$

At the emanation to propagation emergence, one of the emergent constructs is the familiar path integral based on standard (unitary) propagators in a complex Hilbert space.

We have α, 10,22,78,137 as parameters resulting from analysis on a single path maximal information flow construct, where the number 22 corresponds to the number of emergent parameters in the description of the propagating construct (exact derivation of the 22-parameter in [7]). In addition, the time choice is emergent via a multi-path construct, along with the propagator construct, and is coupled in both time step and imaginary time increment. The formulation is inherently embedded in a higher dimensional complex space, thus all of the QFT complex analysis analyticity methods are valid as the assumptions made are now part of the maximal information flow emergent construct.

References

[1] Winters-Hilt, S. Classical Mechanics and Chaos. 2023. (Physics Series: “Physics from Maximal Information Emanation” Book 1.)

[2] Winters-Hilt, S. The Dynamics of Fields, Fluids, and Gauges. 2023. (Physics Series: “Physics from Maximal Information Emanation” Book 2.)

[3] Winters-Hilt, S. The Dynamics of Manifolds. 2023. (Physics Series: “Physics from Maximal Information Emanation” Book 3.)

[4] Winters-Hilt, S. Quantum Mechanics, Path Integrals, and Algebraic Reality. 2024. (Physics Series: “Physics from Maximal Information Emanation” Book 4.)

[5] Winters-Hilt, S. Quantum Field Theory and the Standard Model. 2024. (Physics Series: “Physics from Maximal Information Emanation” Book 5.)

[6] Winters-Hilt, S. Thermal & Statistical Mechanics, and Black Hole Thermodynamics. 2024. (Physics Series: “Physics from Maximal Information Emanation” Book 6.)

[7] Winters-Hilt, S. Emanation, Emergence, and Eucatastrophe. 2023. (Physics Series: “Physics from Maximal Information Emanation” Book 7.)

[8] Hawking, S. W. (1974-03-01). "Black hole explosions?". Nature. 248 (5443): 30–31.

[9] Birrell, N.D. and Davies, P.C.W. (1982) Quantum Fields in Curved Space. Cambridge Monographs on Mathematical Physics. Cambridge University Press, Cambridge.

[10] Unruh, W. G, Phys. Rev. D 14, 870 (1976).

[11] Dirac, P.A.M. The quantum theory of the emission and absorption of radiation *Proc. R. Soc. Lond. A* **114**243–265. 1927.

[12] Ramond, P. Field Theory: A Modern Primer. Benjamin/Cummings, 1981.

[13] Mandle, F. and G. Shaw. Quantum Field Theory. Wiley. 1984.

[14] Ryder, Lewis H. (1985). *Quantum Field Theory*. Cambridge University Press.

[15] Winters-Hilt S. Topics in Quantum Gravity and Quantum field Theory in Curved Spacetime. UWM PhD Dissertation, 1997.

[16] Cheng, Ta-Pei and Ling-Fong Li. Gauge Theory of Elementary Particles. Clarendon Press. 1983.

[17] Collins, John. Renormalization. Cambridge University Press, 1984.

[18] Lopes, J. L. Gauge Field Theories: An Introduction. Pergamon Press. 1981.

[19] Aitchison, I.J.R. and A.J.G. Hey. Gauge Theories in Particle Physics. Adam Hilger 1982.

[20] Quigg, Chris. Gauge Theories of the Strong, Weak, and Electromagnetic Interactions. Princeton University Press, 2013.

[21] Itzykson, C. and J-B Zuber. Quantum Field theory. McGraw Hill 1980.

[22] De Wit, B. and J. Smith. Field theory in Particle Physics. North Holland 1986.

[23] Umezawa, Hiroomi. Advanced Field Theory: Micro, Macro, and Thermal Physics. AIP 1993.

[24] Fulling, S. A. Aspects of Quantum Filed Theory in Curved Space-time. Cambridge University Press 1989.

[25] Gupta, S. (1950), "Theory of Longitudinal Photons in Quantum Electrodynamics", Proc. Phys. Soc., **63A** (7): 681–691.

[26] Bleuler, K. (1950), "Eine neue Methode zur Behandlung der longitudinalen und skalaren Photonen", Helv. Phys. Acta (in German), **23** (5): 567–586, doi:10.5169/seals-112124.

[27] Feynman, R.P. and Hibbs, A.R. Quantum Mechanics and Path-Integral. McGraw-Hill, New York. (1965).

[28] Dirac, Paul A. M. (1933). "The Lagrangian in Quantum Mechanics". Physikalische Zeitschrift der Sowjetunion. 3: 64–72.

[29] Feynman, R.P. Space-Time Approach to Non-Relativistic Quantum Mechanics. Rev. Mod. Phys. 20, 367 – Published 1 April 1948.

[30] Laplace, P S (1774), "Mémoires de Mathématique et de Physique, Tome Sixième" [Memoir on the probability of causes of events.], Statistical Science, 1 (3): 366–367.

[31] Duru, İ. H.; Kleinert, Hagen (1979). "Solution of the path integral for the H-atom". *Physics Letters*. **84B** (2): 185–188. (1979).

[32] Albeverio, Sergio and Sonia Mazzucchi (2011) Path integral: mathematical aspects. Scholarpedia, 6(1):8832.

[33] Kumano-go, N and Fujiwara, D (2008). Feynman path integrals and semiclassical approximation. RIMS Kokyuroku Bessatsu B5: 241--263.

[34] Kleinert, H. Path Integrals in Quantum Mechanics, Statistics and Polymer Physics. World Scientific, Singapore. (1995).

[35] Nelson, E (1964). Feynman integrals and the Schrödinger equation. J. Math. and Phys. 5: 332-343.

[36] Klauder, J R (2000). Beyond conventional quantization. Cambridge University Press, Cambridge

[37] Hida, T; Kuo, H H; Potthoff, J and Streit, L (1993). White noise. An infinite-dimensional calculus. Mathematics and its Applications, 253. Kluwer Academic Publishers Group, Dordrecht.

[38] Itô, K (1961). Wiener integral and Feynman integral. Proc. Fourth Berkeley Symposium on Mathematical Statistics and Probability. California Univ. Press, Berkeley 2: 227-238.

[39] Albeverio, S; Høegh-Krohn, R and Mazzucchi, S (2008). Mathematical theory of Feynman path integrals. An Introduction. 2nd and enlarged edition. Lecture Notes in Mathematics 523. Springer-Verlag, Berlin.

[40] Elworthy, D and Truman, A (1984). Feynman maps, Cameron-Martin formulae and anharmonic oscillators. Ann. Inst. H. Poincaré Phys. Théor. 41(2): 115--142.

[41] Albeverio, S and Mazzucchi, S (2005). Generalized Fresnel integrals. Bull. Sci. Math. 129(1): 1--23.

[42] Feynman, R.P. (1950) Mathematical Formulation of the Quantum Theory of Electromagnetic Interaction. Physical Review, 80, 440-457.

[43] Feynman, R.P. Space-Time Approach to Quantum Electrodynamics. Phys. Rev 76, pg. 769. (1949).

[44] Gell-Mann, M.; Low, F. E. (1954). "Quantum Electrodynamics at Small Distances" Physical Review. 95 (5): 1300–1312.

[45] Stueckelberg, E.C.G.; Petermann,A. (1953). "La renormalisation des constants dans la théorie de quanta". Helv. Phys.Acta (in French). 26: 499–520.

[46] Kadanoff, Leo P. (1966). "Scaling laws for Ising models near " (https://doi.org/10.1103%2F PhysicsPhysiqueFizika.2.263). Physics Physique Fizika. 2 (6): 263.

[47] Wilson, K.G. (1975). "The renormalization group: Critical phenomena and the Kondo problem". Rev. Mod. Phys. 47 (4): 773.

[48] Fritzsch, Harald (2002). "Fundamental Constants at High Energy". Fortschritte der Physik. 50 (5–7): 518–524.

[49] Faddeev, L. D.; Popov, V. (1967). "Feynman diagrams for the Yang-Mills field". *Phys. Lett. B.* **25** (1): 29.

[50] Coleman, S. and E. Weinberg. Radiative Corrections as the Origin of Spontaneous Symmetry Breaking. Phys. Rev. D **7**, 1888. 1973.

[51] Yang & Mills, Phys. Rev. 96, 191 (1954).

[52] Utiyama, Phys. Rev. 100, 1597 (1956).

[53] Becchi, Rouet, and Stora. Phys. Letters 52B,344 (1974).

[54] Becchi, Rouet, and Stora. Ann. of Phys.,98,287 (1976).

[55] Frampton, Paul H. Gauge Field Theories. Wiley. 2008.

[56] Adler, S. L. (1969). "Axial-Vector Vertex in Spinor Electrodynamics". Physical Review. 177 (5): 2426–2438.

[57] Bell, J. S.; Jackiw, R. (1969). "A PCAC puzzle: π 0→γγ in the σ-model". Il Nuovo Cimento A. 60 (1): 47–61.

[58] Gell-Mann, M.. "The Eightfold Way: A theory of strong interaction symmetry". Synchrotron Laboratory. Pasadena, CA: California Institute of Technology. TID-12608; CTSL-20. (1961).

[59] Winters-Hilt S. Emanator Theory is shown to be an optimal Martingale process at the fractal edge of chaos, where the Gravitational constant G is hypothesized to be a multiscale fractal coupling parameter. Advanced Studies in Theoretical Physics, 2023.

[60] Peskin, M.E., Schroeder, D.V. (1995). An Introduction to Quantum Field Theory.

[61] Haxton, W.C.; Hamish Robertson, R.G.; Serenelli, Aldo M. (18 August 2013). "Solar Neutrinos: Status and Prospects". *Annual Review of Astronomy and Astrophysics*. **51** (1): 21–61.

[62] Sonjanovic, G. 2011. Probing the origin of neutrino mass: from GUT to LHC.

[63] Grossman, Y. 2003. TAST 2002 lectures on neutrinos.

[64] Koide, Y., Nuovo Cim. A 70 (1982) 411 [Erratum-ibid. A 73 (1983) 327].

[65] Sumino, Y. (2009). "Family Gauge Symmetry as an Origin of Koide's Mass Formula and Charged Lepton Spectrum". Journal of High Energy Physics. 2009 (5): 75. arXiv:0812.2103.

[66] Planck, Max (1901), "Ueber das Gesetz der Energieverteilung im Normalspectrum", *Ann. Phys.*, **309** (3): 553–63,

[67] Aharonov, Y; Bohm, D (1959). "Significance of electromagnetic potentials in quantum theory". *Physical Review.* **115** (3): 485–491.

[68] Gibbons, G.W. and M.J. Perry. Black Holes and Thermal Green's Functions. Proc. R. Soc.

[69] A. S. Wightman, "Fields as Operator-valued Distributions in Relativistic Quantum Theory," *Arkiv f. Fysik, Kungl. Svenska Vetenskapsak.* **28**, 129–189 (1964).

[70] Osterwalder, Konrad; Schrader, Robert (1975). "Axioms for Euclidean Green's functions II". *Communications in Mathematical Physics.* **42** (3). Springer Science and Business Media LLC: 281–305.

[71] Bogoliubov. 1958: Sov. Phys. JETP 7,51.

[72] Bogoliubov, N.N.; Shirkov, D.V. (1959). The Theory of Quantized Fields. New York, NY: Interscience.

[73] L. Parker, PhD Thesis, Harvard University (1966); L. Parker, Phys. Rev. 183, 1057 (1969).

[74] S. A. Fulling, Phys. Rev. D 7, 2850 (1973).

[75] P. C. W. Davies, J. Phys. A: Gen. Phys. 8, 609 (1975).

[76] A. Ashtekar and A. Magnon, Proc. R. Soc. London A346, 375 (1975); S. A. Fulling, Gen. Rel. and Gravit. 10, 807 (1979); U. H.

Gerlach, Phys. Rev. D 38, 514, 522 (1988); extensive references to be found in Ref 9.

[77] E. Kasner, Am. J. Math 43, 217 (1921).

[78] C. W. Misner, K. S. Thorne, and J. A. Wheeler, Gravitation, New York, W. H. Freeman and Co. 1973; Ch. 30.

[79] A. Ashtekar and A. Magnon, Proc. Roy. Soc. London A346, 375 (1975); A. Ashtekar and A. Magnon, Gen. Rel. and Gravit. 12, 205 (1980); T. Dray, J. Renn, and D. Salisbury, Lett. in Math. Phys. 7, 143 (1983); A.-H. Najmi and A. C. Ottewill, Phys. Rev. D 30, 1733 (1984); A.-H. Najmi and A. C. Ottewill, Gen. Rel. and Gravit. 17, 573 (1985); P. Chmielowski, report given at 1992 Midwest Relativity Conference, Univ. of Illinois at Champaign-Urbana.

[80] W. Rindler, Am. J. Phys. 34, 1174 (1966).

[81] G. T. Moore, J. Math. Phys. (NY) 9, 2679 (1970); S. A. Fulling and P. C. W. Davies, Proc. Roy. Soc. London A348, 393 (1976).

[82] Ch. Charach and L. Parker, Phys. Rev. D 24, 3023 (1981).

[83] T. Padmanabhan, Phys. Rev. Lett. 64, 2471 (1990); see also C. W. Misner, Phys. Rev. D 8, 3271 (1973); B. K. Berger, Phys. Rev. D 12, 368 (1975).

[84] K. Freese, C. T. Hill, and M. Mueller, Nucl. Phys. B255, 693 (1985).

[85] S. Kamefuchi and H. Umezawa, Nuovo Cimento 31, 492 (1964); L. Parker, Phys. Rev. 183, 1057 (1969); H. Umezawa, H. Matsumoto, and M. Tachiki, Thermo Field Dynamics and Condensed States, North Holland 1982.

[86] Parker, L. Quantized Fields and Particle Creation in Expanding Universes. I. Phys. Rev. **183**, 1057. (1969).

[87] Bunch, T.S. and L. Parker. Feynman propagator in curved spacetime: A momentum-space representation. Phys. Rev. D **20**, 2499 (1979).

[88] Winters-Hilt, S. Informatics and Machine Learning. Wiley Publishing. 9781119716747, Sept. 2021.

[89] Feigenbaum, M. J. (1976) "Universality in complex discrete dynamics", Los Alamos Theoretical Division Annual Report 1975-1976 .

[90] Balmer, J. J. "Notiz über die Spectrallinien des Wasserstoffs" [Note on the spectral lines of hydrogen]. Annalen der Physik und Chemie. 3rd series (in German). 25: 80–87. (1885).

[91] Winters-Hilt, S. Feynman-Cayley Path Integrals select Chiral Bi-Sedenions with 10-dimensional space-time propagation. Advanced Studies in Theoretical Physics, Vol. 9, 2015, no. 14, 667 – 683. dx.doi.org/10.12988/astp.2015.5881.

[92] Penrose, Roger (May 1996). "On Gravity's role in Quantum State Reduction". *General Relativity and Gravitation*. **28** (5): 581–600.

[93] Caves, C.M., C.A., Fuchs, R. Schack. Unknown quantum states: The Quantum de Finetti Representation. J. Math. Phys. 43, 4537 (2002).

[94] Penrose, R., W. Rindler (1984) Volume 1: Two-Spinor Calculus and Relativistic Fields, Cambridge University Press, United Kingdom.

[95] Cailler, C. 1917. Archs. Sci. Phys. Nat. ser. 4, 44 p. 237.

[96] Girard, P.R.. The Quaternion group and modern physics. Eur. J. Phys. 5 (1984): 25-32.

[97] Synge, J.L. Quaternions, Lorentz Transformations and the Conway-Dirac-Eddington Matrices.

[98] Gunaydin, M. and F. Gursey. Quark structure and the octonions. J. Math. Phys., 14, 1973.

[99] Hurwitz, A. (1923), "Über die Komposition der quadratischen Formen", Math. Ann., 88 (1–2): 1–25.

[100] https://www.wolframalpha.com

[101] Gaździcki, Marek; Gorenstein, Mark I. (2016), Rafelski, Johann (ed.), "Hagedorn's Hadron
Mass Spectrum and the Onset of Deconfinement", Melting Hadrons, Boiling Quarks – From Hagedorn Temperature to Ultra-Relativistic Heavy-Ion Collisions at CERN, Springer International Publishing, pp. 87–92.

[102] Winters-Hilt, S. Fiat Numero: Trigintaduonion Emanation Theory and its Relation to the Fine-Structure Constant α, the Feigenbaum Constant $C\infty$, and π. Advanced Studies in Theoretical Physics Vol. 15, 2021, no. 2, 71 - 98 HIKARI Ltd, www.m-hikari.com https://doi.org/10.12988/astp.2021.91517.

[103] Kato T. Fundamental Properties of Hamiltonian Operators of Schrodinger Type, Transactions of the American Mathematical Society, 1951, Pg. 195-211.

[104] Kato, T. Perturbation theory for linear operators. Springer 1980.

[105] McMullen, Curtis T. 2000. The Mandelbrot set is universal. In The Mandelbrot Set, Theme and Variations, ed. T. Lei, 1–18. Cambridge U.K.: Cambridge Univ. Press. Revised 2007.

[106] Briggs, K. A precise calculation of the Feigenbaum constants. Mathematics of Computation, Vol. 57, Num.195, July 1991, pages 435-439.

[107] Winters-Hilt S, I. H. Redmount, and L. Parker, "Physical distinction among alternative vacuum states in flat spacetime geometries," Phys. Rev. D 60, 124017 (1999).

[108] Friedman, J.L., J. Louko, and S. Winters-Hilt. Reduced phase space formalism for spherically symmetric geometry with a massive dust shell. Phys Rev. D Vol. 56, Num 12 (1997).

[109] J. Louko and S. N. Winters-Hilt, Phys. Rev. D 54, 2647 (1996). (grqc/ 9602003)

[110] J. Louko, J. Z. Simon, and S. N. Winters-Hilt, Phys. Rev. D, 55, 3525 (1997).

[111] Nielsen, Frank (2022). "The Many Faces of Information Geometry". Notices of the AMS. 69 (1). American Mathematical Society: 36-45.

[112] Nielsen, Frank (2018). "An Elementary Introduction to Information Geometry". Entropy. 22 (10).

[113] Amari S; Dualistic Geometry of the Manifold of Higher-Order Neurons. Neural Networks, Vol. 4(4), 1991:443-451.

[114] Amari S: Information Geometry of the EM and em Algorithms for Neural Networks. Neural Networks, Vol. 8(9), 1995:1379-1408.

[115] Amari S and Nagaoka H: Methods of Information Geometry. 2000. Translations of Mathematical Monographs Vol. 191.

[116] Diósi, L. (1989). "Models for universal reduction of macroscopic quantum fluctuations". Physical Review A. 40 (3): 1165–1174.

[117] Penrose, Roger (1996). "On Gravity's role in Quantum State Reduction". General Relativity and Gravitation. 28 (5): 581–600.

[118] Maldacena, J.. The Large N limit of superconformal field theories and supergravity.
Advances in Theoretical and Mathematical Physics. 2 (4): 231–252.

[119] Witten, Edward (1998). "Anti-de Sitter space and holography". Advances in Theoretical and Mathematical Physics. 2 (2): 253–291.

[120] Jackson, J.D. Classical Electrodynamics, 2nd Edition. Wiley 1975.

[121] Lorentz, Hendrik Antoon (1899), "Simplified Theory of Electrical and Optical Phenomena in Moving Systems" , Proceedings of the Royal Netherlands Academy of Arts and Sciences, 1: 427–442.

[122] D'Alembert, Jean Le Rond (1743). Traité de dynamique.

[123] Laplace, P S (1774), "Mémoires de Mathématique et de Physique, Tome Sixième" [Memoir on the probability of causes of events.], Statistical Science, 1 (3): 366–367.

[124] Winters-Hilt, S. Theory of Trigintaduonion Emanation and Origins of α and π. Researchgate 05/24/20.

[125] Winters-Hilt, S. Chiral Trigintaduonion Emanation Leads to the Standard Model of Particle Physics and to Quantum Matter. Advanced Studies in Theoretical Physics, Vol. 16, 2022, no. 3, 83-113.

[126] Winters-Hilt, S. Meromorphic precipitation of quantum matter with dimensionful action. May 2021. DOI:10.13140/RG.2.2.32294.24640.

[127] Winters-Hilt, S. Emanator Theory using split octonions is Manifestly Lorentz Invariant and reveals why the fundamental constant $\hbar$ should be so small. Advanced Studies in Theoretical Physics, 2023.

[128] Landau, Lev D.; Lifshitz, Evgeny M. (1969). Mechanics. Vol. 1 (2nd ed.). Pergamon Press.

[129] Goldstein, Herbert (1980). Classical Mechanics (2nd ed.). Addison-Wesley.

[130] Fetter, A.L and J.D Walecka, Theoretical Mechanics of Particles and Continua, Dover (2003).

[131] Percival, I.C. and D. Richards. Introduction to Dynamics. (1983) Cambridge University Press.

[132] Arnold, V.I. Ordinary Differential Equations. MIT Press. (1978).

[133] Arnold, Vladimir I. (1989). Mathematical Methods of Classical Mechanics (2nd ed.). New York: Springer.

[134] Woodhouse, N.M.J. Introduction to Analytical Dynamics. Springer, 2nd Edition. 2009.

[135] Bender, C.M. and S.A. Orszag. Advanced Mathematical Methods for Scientists and Engineers: Asymptotic Methods and Perturbation Theory. Springer. 1999.

[136] Robert L. Devaney. An Introduction to Chaotic Dynamical Systems. Addison -Wesley.

[137] Landau, Lev D.; Lifshitz, Evgeny M. (1971). The Classical Theory of Fields. Vol. 2 (3rd ed.). Pergamon Press.

[138] Penrose, Roger (1965), "Gravitational collapse and space-time singularities", Phys. Rev. Lett., 14 (3): 57.

[139] Hawking, Stephen & Ellis, G. F. R. (1973). The Large Scale Structure of Space-Time. Cambridge: Cambridge University Press.

[140] Peebles, P. J. E. (1980). Large-Scale Structure of the Universe. Princeton University Press.

[141] B. Abi et al. Measurement of the Positive Muon Anomalous Magnetic Moment to 0.46 ppm. Phys. Rev. Lett. 126, 141801 (2021).

[142] Einstein, A. "On a heuristic point of view concerning the production and transformation of light" . Ann. Phys., Lpz 17 132-148.

[143] Heisenberg, Werner. "Über quantentheoretische Umdeutung kinematischer und mechanischer Beziehungen". Zeitschrift für Physik (in German). 33 (1): 879–893. ("Quantum theoretical re-interpretation of kinematic and mechanical relations"). (1925).

[144] Schrödinger, E. (1926). "An Undulatory Theory of the Mechanics of Atoms and Molecules". Physical Review. 28 (6): 1049–1070.

[145] Born, Max; J. Robert Oppenheimer. "Zur Quantentheorie der Molekeln" [On the Quantum Theory of Molecules]. Annalen der Physik (in German). 389 (20): 457–484. (1927).

[146] Dirac, P. A. M. "The Quantum Theory of the Electron". Proceedings of the Royal Society A: Mathematical, Physical and Engineering Sciences. 117 (778): 610–624. (1928).

[147] Dirac, P. A. M. . The Principles of Quantum Mechanics. Oxford: Clarendon Press. (1930).

[148] Feynman, Richard P. (1942). The Principle of Least Action in Quantum Mechanics (PhD). Princeton University.

[149] Erdeyli, A. Asymptotic Expansions. 1956 Dover.

[150] Erdeyli, A. Asymptotic Expansions of differential equations with turning points. Review of the Literature. Technical Report 1, Contract Nonr-220(11). Reference no. NR 043-121. Department of Mathematics, California Institute of Technology, 1953.

[151] Carrier, G.F, M. Crook and C.E. Pearson. Functions of a complex variable. 1983 Hod Books.

[152] Sommerfeld, A., Atombau und Spektrallinien. Friedrich Vieweg und Sohn, Braunschweig, 1919.

[153] Tolkien, J.R.R. (1990). The Monsters and the Critics and Other Essays. London: HarperCollins Publishers.

Index

A

Abelian, 195, 202, 204, 209, 232
abelian, 135–137, 139, 185, 187–188, 190, 193, 195, 197, 206
absorption, 5, 17, 27, 39, 43–45, 235, 366
accelerating, 223–224, 235, 259
acceleration, 3, 223–224, 235, 255, 258–259
Achiral, 345, 348
achiral, 314, 329, 331–332, 335–336, 339, 345, 348, 350, 354
Action, 23, 56, 62–63, 65, 90, 103, 121, 174–175, 190, 193, 195, 225–226, 271, 330, 376
action, 14, 23, 52, 55, 57, 63, 120, 122, 134, 138, 169–170, 191, 205, 308, 348, 353, 375
Actions, 63
actions, 191
Adiabatic, 237, 248, 271, 302–305
adiabatic, 237, 240–241, 248, 250, 302–306
adiabatically, 314
adjoint, 26, 168, 186
adjoints, 164
AdS, 291–292, 314
Ads, 290, 353
affine, 243
Aharonov, 224, 370
Alembert, 374
Algebra, 132
algebra, 1, 27, 132, 161, 167, 218, 257, 271, 313, 330, 332–333, 337–340, 344–345, 354, 356, 359, 363
Alpha, 158, 311
alpha, 1–2, 60, 158, 160, 218, 220, 241, 311–315, 333–336, 338, 359–360, 362, 364
amplitude, 34–35, 45, 55, 61–62, 67, 141, 233–234, 300, 331
amplitudes, 15, 35, 62, 262–263, 265–266, 351
Analytic, 224, 356
analytic, 52–54, 109, 160, 223–224, 226–228, 232–233, 251–252, 336, 346–347, 351, 356
Analyticity, 264
analyticity, 2, 55, 66, 223, 226–227, 251, 335, 348, 356, 363–364
annihilation, 12, 15, 18, 27, 29, 162, 229, 258, 261, 263, 266, 299
Anomalies, 202
anomalies, 188–189, 202, 210
Anomalous, 202, 376
anomalous, 158, 202, 204, 208, 217, 359
Anomaly, 204
anomaly, 189, 204, 209–210, 306
anticommutation, 38–39, 132
anticommutations, 38
anticommutator, 47, 132
anticommutators, 230
anticommute, 232
anticommuting, 197
antimatter, 230
Antiparticle, 50
antiparticle, 219, 361
antiparticles, 38
antiphase, 313–314, 342, 344–346, 350

antiquark, 214
antisqueezed, 267
antisymmetric, 134, 189, 196, 208, 263, 266, 327
antisymmetry, 209
area, 101, 284–285
areas, 6–7, 23, 51, 211, 315
Associative, 355
associative, 336, 355
associatively, 349
Associativity, 354, 356
associativity, 336, 341, 354–356
asymmetric, 158, 215, 218, 357, 359
asymmetry, 268–269
Asymptotic, 375, 377
asymptotic, 3, 34, 66, 162, 211, 237, 240, 295, 297, 301, 313, 335
asymptotically, 28, 131–132, 183, 211, 235–236, 294, 305
asymptotics, 240, 329
atom, 17, 52, 224, 367
atomic, 17, 21, 311–312
Atoms, 376
atoms, 17
automorphism, 338–339, 344
Axial, 369
axial, 190, 192, 203–204, 206, 210

B

backreaction, 235
Balmer, 312, 372
baryon, 217, 358
baryons, 217, 359
Basis, 13, 319
basis, 13–14, 30, 52, 55, 89, 135, 164, 212, 252, 304, 317–321, 331, 336
Bath, 230
bath, 224, 230–231, 234–235
Becchi, 195, 369
Bernoulli, 283
Bessel, 237, 240
bicontinuous, 243
bifurcation, 259, 329
bijective, 243
binding, 21
bioctonion, 339
Birrell, 365
bisedenion, 339, 343
BiSedenions, 372
blackbody, 5, 302
Bogoliubov, 3, 236, 238, 241, 250–252, 254–255, 261, 267–268, 297–298, 300, 370
Boguliubov, 264
Bohm, 224, 370
Boiling, 373
Boltzmann, 224, 230, 354
Born, 376
Bose, 3, 40, 132, 228–229, 299
BoseEinstein, 299
Boson, 45, 174, 180–181
boson, 38, 43, 174, 181–182, 194, 300, 346
bosonic, 161
Bosons, 229
bosons, 203, 274
bound, 1, 44, 52, 100, 224, 344, 350, 356
boundaries, 2, 6, 241, 259
Boundary, 243
boundary, 2–3, 6, 8, 23, 30, 53, 61, 90, 252, 275, 278, 282, 284, 287, 289–292, 294, 303, 308, 311, 346, 350, 353
bounded, 251–252, 342, 344
box, 6, 8, 264, 295
Bra, 271
brace, 300
Bracket, 189, 274
bracket, 274

brackets, 188
braid, 339, 341
braidlevel, 339
braids, 339, 341
break, 5, 120, 166, 169, 174, 183, 274, 330
breakdown, 165, 302
Breaking, 163, 166, 369
breaking, 8, 164, 166–169, 172–174, 179–182, 185, 203, 308
breaks, 121, 167, 169, 179, 203, 241
Bremsstrahlung, 149
Briggs, 373

C

Canonical, 23, 62, 132, 229, 261, 271
canonical, 1, 6, 26–29, 41–42, 47, 52, 63, 69, 84, 104, 135, 228, 262, 271, 274, 294, 296, 321
Cauchy, 238, 242, 244–246, 251–252, 290–292
causal, 2, 223, 226, 353
Causality, 305
causality, 28–29, 163, 169, 211, 223–224, 232
causally, 353
Caves, 372
Cayley, 313, 331–333, 336–340, 345, 354, 356, 372
CayleyDickson, 337
Chakraborty, 305
Chaos, 365
chaos, 311, 335, 346, 350, 363, 369
Chaotic, 375
chaotic, 311
Chapman, 57, 193
Charge, 189
charge, 7, 20, 25, 47–48, 117–118, 141, 155, 161, 188, 204, 210, 215–217, 219, 312, 357–359, 361
Charged, 285, 370
charged, 5, 7, 215, 218, 312, 357, 360
charges, 17–20, 151, 161, 204–205
charm, 210, 217, 359
Chiral, 202, 204, 213, 334, 339, 345, 354, 372, 375
chiral, 188–189, 202, 204, 311, 313–314, 329, 332–336, 338–340, 342, 344–345, 347–350, 352–356, 363
chiralities, 210, 346, 348, 350
chirality, 169, 205, 346–347, 349
chiraly, 346, 356
Chromodynamics, 202, 210–211
Coleman, 75–76, 92, 180, 182, 369
collapse, 351, 376
Collins, 366
collision, 44
Collisions, 373
color, 203, 211, 213, 215, 219, 357, 361
colors, 210–211
Commutation, 30, 42
commutation, 27–30, 41–43, 47, 62, 104, 229, 261–262, 276, 280, 291, 296, 300, 321
commutations, 38
Commutative, 356
commutative, 356
Commutativity, 356
commutativity, 35, 356
Commutator, 317, 319
commutator, 28, 163, 230, 232, 274, 297, 299, 317
Commutators, 320
commutators, 188, 206, 296, 322

commute, 167, 232, 272, 322
commutes, 162, 321
commuting, 262
comoving, 236, 241, 302, 304
compact, 136, 164, 169, 232, 271, 287, 291
compactifying, 286, 288
compaction, 170–171
complete, 3, 5, 9, 29, 34, 37–38, 41, 53, 137, 153, 169, 187, 203, 219, 224, 245, 261–262, 280, 355, 361
completeness, 245, 250
Complex, 25, 48, 180, 336, 347, 355–356
complex, 2, 25–26, 33, 45–47, 63, 100, 107, 161, 164, 167, 179, 185, 195, 218, 223, 226, 251, 257, 285, 313, 329–332, 334–335, 337–338, 343, 345, 349, 351, 355–356, 359, 363–364, 372, 377
complexation, 332
Complexification, 229
complexification, 225
Complexified, 225
complexified, 224, 226, 311
complexity, 46, 63, 218, 331, 359
Compton, 5
Computation, 373
computation, 152
Computational, 335
computational, 329, 333, 335, 342
computationally, 329, 333
Compute, 95, 112, 128
compute, 25, 122, 148, 164, 168, 187, 280, 294
computed, 36, 124
Condensed, 371
condensed, 297
Conformal, 286, 292, 298
conformal, 236, 260, 286–289, 292–293, 298, 306–308, 336
Conformally, 298
conformally, 291–292, 298–299, 308
congruence, 223, 256, 258–260, 352
congruences, 256–258, 352
conjugate, 26, 28, 189, 245, 262–263, 273, 275
conjugates, 262, 266
conjugation, 141, 155, 203, 337
connect, 63, 75, 78, 178
connected, 70, 88, 92–93, 95, 108, 119, 121, 169, 258, 344, 353
conservation, 34, 217, 231, 234, 292, 358
conserve, 272
conserved, 25–26, 29, 43, 47–48, 139, 161, 188, 204, 213, 249–250, 261, 277, 279, 290, 296–297, 326
constrain, 218, 359
constrained, 100, 299, 347
constrains, 224
constraint, 82, 189–190, 200, 204, 218, 220–221, 299, 329, 359, 361, 363
constraints, 104, 188, 190, 217, 219, 334, 340, 348–349, 352, 359, 361
Contact, 356
contact, 356
Continua, 375
continuation, 54, 105, 228, 233
Cosmological, 219, 297–298, 361
cosmological, 217, 224, 302, 358, 363
cosmologically, 336
cosmology, 63
Coulomb, 7, 19–20, 150, 190, 193, 232

coulomb, 43, 151, 213
Covariant, 30, 42, 271, 290
covariant, 25, 40–41, 139, 242, 271, 290–291
covariantly, 212
Creation, 371
creation, 12, 15, 18, 27, 29, 39, 43–45, 162, 229, 241, 258, 261, 263, 266, 292, 299, 307–308
curvature, 2, 261–262, 292, 353
Curved, 286, 308, 365–366
curved, 2–3, 6, 23, 51–52, 224, 236, 257, 271, 279, 289–290, 292, 296, 298, 300, 308, 351, 371
cyclic, 177, 319
cylindrical, 284

D

Dark, 2
dark, 169, 217–220, 358–359, 361–362
Daubechies, 55
Davies, 257, 365, 370–371
DeBroglie, 9
Deck, 349
deck, 348–350
Deconfinement, 373
decoupled, 9, 217, 358
deFinetti, 351, 363
deform, 245
deformation, 228, 244, 290
deformations, 245
Degeneracy, 12
degeneracy, 12, 319
degenerate, 10, 12–13, 119, 162, 173, 238, 256–258, 260
degree, 15, 55, 158, 184, 199, 321
degrees, 24, 40, 47, 226, 265, 319
Density, 86, 99
density, 7, 23–24, 37, 41, 43–44, 46, 91, 99, 229–231, 260, 262, 279, 284, 289, 302, 353
deSitter, 243, 291
Detector, 234
detector, 233–235, 237, 302, 304
detectors, 233, 236, 309
determinant, 134–135, 188, 192, 194, 331
20
diagonalize, 213
diagonalized, 167, 213, 320
diagrammatic, 88, 93, 125
diagrams, 1, 5, 69–70, 73–74, 78–79, 82–83, 86, 88–89, 93–95, 97, 100, 106–107, 109, 111, 114, 119–120, 125, 140–141, 152, 156–159, 175, 271, 286, 369
dielectric, 144
diff, 30
Diffeomorphism, 243
diffeomorphism, 243
Differential, 375
differential, 55, 59–60, 92, 127, 243, 252, 262, 297, 352, 377
Dimension, 213
dimension, 2, 23, 67, 104, 184, 213, 311, 334, 337, 346, 348, 350
Diósi, 374
Dirac, 1, 5–6, 26, 37–40, 47–48, 52–53, 100, 134, 139, 143, 151–152, 155, 188, 202, 204, 206, 219, 229, 271, 281, 331, 361, 366–367, 376
Distributions, 370
Divergence, 279, 290
divergence, 24, 82–83, 101, 113, 140–141, 149, 157–158, 177, 184, 199, 201, 209, 231, 277
Divergences, 305

divergences, 82, 95–96, 101, 260, 303, 305–306
divergent, 74, 82–83, 95, 100, 102, 113, 141, 148–149, 156, 158, 199–201, 207, 209, 259, 268–269, 306
diverges, 265, 305, 318
Dualistic, 374
duality, 169
dually, 342
Dynamical, 375
dynamical, 26, 40, 212, 223, 236, 265, 272–273, 299, 301
Dynamics, 365, 371, 375
dynamics, 20, 219, 221, 226, 239, 262, 274, 299, 355, 360, 363, 372
Dyson, 44

E

Earth, 224
Eddington, 372
Eigenfunctions, 10
eigenfunctions, 256–257, 263
eigenstate, 48, 61, 183
Eigenstates, 13
eigenstates, 12–13, 30, 34, 36, 64, 174, 276
eigenvalue, 9, 11–13, 166, 257–258
Eigenvalues, 10
eigenvalues, 10–13, 135–136, 166, 319
eigenvector, 11–12, 166, 174
Eigenvectors, 13
eigenvectors, 11
Eightfold, 211, 369
Einstein, 5, 40, 228–229, 238, 262, 289, 291–292, 299, 376
Electric, 18
electric, 7, 16–18, 25, 150, 219, 312, 361
Electrical, 374
electrically, 5
Electro, 216, 219, 358, 360
electro, 216, 219–220, 358, 360–361
Electrodynamics, 210, 367–369, 374
electrodynamics, 40, 52, 139, 143, 179, 204
Electrohanded, 220, 361
Electromagnetic, 40, 284, 366, 368
electromagnetic, 5–6, 18–20, 41–43, 117, 135, 139, 215, 218, 284, 312, 357, 360, 370
Electromagnetism, 7, 135
electromagnetism, 5–7, 26, 60, 118, 135, 180, 185–188, 204, 209, 224, 311–312
Electron, 376
electron, 1, 5–6, 21, 44–45, 83, 118, 141, 148, 151–152, 155, 158–159, 210, 219, 221, 312, 361–362
electrons, 21, 39, 47, 147, 221, 363
electroproduction, 211
electrostatics, 151
electroweak, 51–52, 209, 215–216, 219–220, 357–358, 360–361
Element, 355–356
element, 37, 103, 191, 202, 216, 261, 290, 330, 333, 339, 341, 344, 355, 357, 364
elements, 5, 14, 57, 64, 83, 88, 125, 186, 189, 217, 272, 320–321, 331, 335–336, 355, 359, 364
Emanation, 311, 313, 329, 333, 340, 346–348, 365, 373, 375
emanation, 218, 220, 313–314, 329, 332–336, 338–342, 345, 347–351, 353–356, 359, 362–364
emanations, 314, 338, 340, 347, 350, 352, 355, 363–364

Emanator, 2, 53, 100, 169, 216, 218, 221, 312–313, 329–330, 332–333, 338, 340, 347, 355–357, 359–360, 363, 369, 375
emanator, 53, 60, 63, 100, 217, 314, 329–330, 333, 335–336, 339–340, 343, 347–349, 352–353, 359, 363–364
emanators, 333, 354
embed, 223
embedded, 248, 330, 364
Emergence, 161, 365
emergence, 364
Emergent, 352
emergent, 329, 338, 351–353, 363–364
emerges, 338
emerging, 289
emission, 5, 17, 19, 235, 366
emitted, 17–18, 235
encode, 352
encoded, 352
encodes, 226
encoding, 331–332
Ensemble, 229
ensemble, 228
entire, 34, 63, 87, 118, 152, 229, 244, 259–260, 302, 309, 352
entropic, 353–354
Entropy, 230, 352, 374
entropy, 230–232, 352–354
enumerate, 199
enumerated, 348
enveloping, 224
environment, 229
equilibrium, 223, 229, 234, 351, 355
equipartition, 314, 345
equipartitioned, 313, 345
equipartitioning, 314, 346
Eucatastrophe, 365
Euclidean, 208, 223, 233, 354, 370
euclideanizability, 351
Euclideanization, 3, 54, 149, 229–230, 232
Euclideanize, 177
Euclideanized, 105, 149, 225–228
Euler, 21, 24, 52, 71, 82, 187, 275, 279, 283
Evolution, 352
evolution, 29, 60, 63, 217–218, 225, 230, 256, 259–260, 262, 266, 272–273, 282, 329, 332–333, 335, 338, 344, 358–359, 363–364
expansion, 3, 6, 9, 38, 44–45, 74, 86, 118, 120–121, 134, 165, 170, 174–175, 177–178, 208, 236, 263, 281, 292, 295, 298, 302–305, 312, 346
Expansions, 377
expansions, 2, 42, 266, 304, 342, 350
Expectation, 218, 360
expectation, 53, 70, 72, 84, 164, 169, 228, 259–260, 268, 273, 303–304, 352
exponent, 68, 134, 193, 238
Exponential, 342–343, 345
exponential, 92, 171, 241, 313, 342–343, 345
exponentiated, 188, 193
extensibility, 260
Extension, 354
extension, 2–3, 51, 218, 224, 226–227, 243, 251–252, 335–336, 339, 346–347, 349, 353–356, 359
extensions, 3, 51, 217, 330, 356, 358
extensive, 260, 371
exterior, 100, 353
External, 110, 150–151

external, 78–80, 82–83, 93–94, 106, 110–111, 147, 150–151, 158, 169, 175, 184, 188, 194, 198–199, 235, 292, 355
extremum, 122, 311

F

Feigenbaum, 311, 329, 335, 350, 372–373
Fermi, 3, 40–41, 132, 211, 229
Fermion, 40, 47, 49, 139
fermion, 38, 45, 155, 157, 183–184, 188, 194
Fermionic, 132, 139
fermionic, 49, 140, 161, 183, 188, 194, 331
Fermions, 230
fermions, 47, 215, 218, 232, 331, 356, 360
Feynman, 1–2, 5–6, 40, 42, 51–56, 59–60, 62–67, 69, 72, 78, 82–83, 87, 90, 93–94, 98, 104, 109, 113–114, 135, 137, 139–140, 142, 148, 158–160, 174–176, 182, 190, 194, 208, 223–224, 226–228, 303, 367–369, 371–372, 376
Flavor, 213
flavor, 203, 212, 219, 360
flavors, 212
fluctuation, 17, 348–349
fluctuations, 343, 348–349, 374
Fluids, 365
flux, 21, 224, 241
Fock, 247, 250, 256–261, 265, 268–269, 300
Fockspace, 260
foliation, 238–239, 247, 258, 262–263
folium, 258
Force, 284
force, 213, 215–216, 235, 252, 284, 357
Frampton, 369
frequencies, 292, 303
Frequency, 301
frequency, 5, 18, 144, 237, 240–241, 249, 251, 256–259, 264, 297–298, 301–305
Fresnel, 55, 368
Friedman, 373
Friedrich, 377
FRW, 236, 243, 247, 298–299, 302–304
Fuchs, 372
Fujiwara, 367
Furry, 141, 155

G

gamma, 145, 205
gases, 312
Gauge, 168, 182, 190, 211, 216, 358, 366, 369–370
gauge, 1, 6–8, 19–20, 23, 25–26, 41, 51, 83, 100, 120, 135–142, 149, 152–153, 163, 169, 171–172, 174–175, 180–183, 185–197, 199–202, 204–205, 209, 211–213, 215–218, 220, 224, 232, 262, 285, 330, 336, 338, 355–359, 361–363
Gauges, 365
gauges, 182–183, 336
Gauss, 188, 190, 277, 279, 290
Gaussian, 15, 51, 56, 58, 90, 121, 134–135, 139, 194, 252, 349
generations, 169, 204, 216–218, 358–359
generative, 60, 63
Generator, 89
generator, 52, 68, 84, 119, 134, 152–153, 167, 169, 174, 188, 190, 210, 327

generators, 166, 186–188, 190, 205, 210, 326–327, 339–340
geodesic, 243, 353
geodesically, 34
geodesics, 243
geodetics, 34
geometric, 125, 289, 311
geometries, 256, 373
Geometry, 353, 374
geometry, 3, 215, 260, 286, 289, 302, 314–315, 336, 352–353, 356, 363, 373
Gerlach, 371
ghost, 188, 194–195, 198–199, 201
ghosts, 188, 192, 201
Gibbons, 370
gluon, 198–199, 201, 212–213
gluons, 203, 210, 212–213
Godel, 217, 330, 359
Goldstone, 164, 172–174, 180–182
Gordon, 1, 6, 24–30, 35, 40–41, 46, 59, 66, 90, 150–151, 155, 227, 239, 248–249, 252, 261, 263, 279
Grassman, 3, 132–134, 140, 194
grassman, 133
Grassmanian, 188
Grassmannian, 188, 194, 196–197, 201
Gravitation, 371–372, 374
gravitation, 236, 279, 314
graviton, 315
Gravitons, 299
gravitons, 299
Gravity, 366, 372, 374
gravity, 224, 314–315
Green, 51–54, 61, 63–67, 79, 82, 88–89, 95, 119–120, 169, 172–173, 182, 223, 226–228, 234–235, 291, 303, 308, 370
green, 115
Greens, 125, 128, 170
greens, 121
Gribov, 193
Gunaydin, 372
Gupta, 43, 367
Gursey, 372
GUT, 209–210, 313, 336, 369

H

Hadron, 160, 373
hadron, 211
hadronic, 346
Hadrons, 373
hadrons, 159–160, 204, 211, 215, 357
Hagedorn, 346, 373
Hamilton, 63
Hamiltonian, 10, 13, 19–20, 27, 53–54, 64, 118, 135, 162, 164, 188–189, 225, 229–230, 262–263, 266–267, 271–272, 274, 276, 296, 373
Hamiltonians, 55, 107, 263
Hamish, 369
Hand, 353
hand, 35, 76, 164, 197, 201–202, 236, 289, 291, 296, 302, 348, 353–354
handed, 215–221, 357–361, 363
handedness, 217, 359
Hankel, 240
Harmonic, 9
harmonic, 1, 9, 12, 15, 27–28, 55, 65–66, 84, 89, 225, 229, 263, 266, 281
Hawking, 2–3, 257, 365, 376
Heat, 230
heat, 54, 230, 232
heated, 312
Heavy, 373

heavy, 211
Heisenberg, 1, 15, 29, 52–53, 64, 224, 236, 261, 273–274, 277, 290, 376
Herbert, 375
Hermite, 15
Hermitian, 14, 107, 167–168, 211–213, 245, 262–263, 266, 271, 273, 279, 320, 331
heuristic, 311, 376
Hibbs, 55, 367
Hida, 55, 367
hidden, 312
Higgs, 179–180, 215, 218–219, 356, 360–361
Hilbert, 33–34, 56, 232, 249, 259, 271, 290–291, 304, 338, 351, 353, 364
holographic, 353
holography, 374
homogeneity, 236
homogeneous, 236, 290
Hooft, 100, 182–183
horizon, 2, 243, 259, 291, 353
horizons, 259–260, 314
Hurwitz, 372
Huygen, 60
Hydrogen, 52, 312
hydrogen, 224, 312, 372
hydrogenic, 312
Hyperbolic, 242
hyperbolic, 59, 238, 290–291
hyperbolicity, 291
hypercomplex, 313, 330, 336–337, 341–342, 345, 351, 363–364
hyperplane, 252
hypersurface, 244–245, 247, 261–263, 266, 290–291
hypersurfaces, 250, 258, 262, 278
hypotenuse, 314, 345

I

idempotent, 133
imaginary, 54–55, 64–66, 105, 107, 173, 223–228, 313–314, 336–337, 341–346, 348–349, 356, 364
imbedded, 242, 248
Imbeddings, 248
immersed, 234
implosion, 289
inertia, 353
inertial, 223, 233–235, 237, 256–259, 264, 291, 308–309, 327
Infinite, 55
infinite, 1, 6, 27, 51, 55, 63, 87, 99, 102, 123, 135–136, 144, 163, 169, 188, 191, 194, 237, 263, 265, 281, 287, 334, 367
Infinitesimal, 278
infinitesimal, 25, 37, 99, 153, 186, 188, 190–192, 278, 292–293, 327, 356
infinitesimals, 161, 171, 293
infinity, 6, 8, 27, 35, 62, 74, 105, 117, 188, 190, 238, 243–244, 260, 287, 290–292
inflationary, 301
Information, 1, 333, 365, 374
information, 2, 18, 33, 36, 118, 218, 231, 291, 303, 314, 329, 333–334, 338, 341, 344, 346, 348–349, 351–353, 359, 363–364
infrared, 101
ingoing, 33, 244–245, 247, 250
inhomogeneous, 326
Invariance, 211, 213, 331, 347
invariance, 7–8, 25–26, 29, 41, 47–48, 77, 100, 109, 133, 135–137, 139–141, 149, 153, 161–163, 172, 174, 179–180, 185, 188, 190–191, 197, 199–203, 215–216, 218,

234, 278, 285, 292, 298, 308, 331, 335, 355–357, 359
invariances, 164
Invariant, 298, 331–333, 375
invariant, 25, 28, 41, 127–128, 136–137, 139–140, 142, 157, 161, 166–168, 175, 187, 191, 197, 201, 203, 211–212, 215, 234, 236, 242, 245, 285, 287, 298–299, 308–309, 319, 327, 331, 357, 364
invariants, 5, 128, 169
inverse, 53, 118, 172, 213, 220, 224, 263, 266, 320, 332, 355, 362
inverses, 332
inversion, 46, 133
invert, 160, 170
inverted, 118
invertible, 332
Inverting, 151
irreducible, 79–80, 89, 111, 128, 185, 216, 357
Ising, 368
isolate, 136, 149
isolated, 102, 211
isometries, 243, 326
Isometry, 243
isometry, 260, 279, 304, 326–327
Isomorphic, 247
isomorphic, 34, 331–332
Isomorphism, 34
isotropic, 235, 299
isotropically, 236, 294

J

Jackiw, 204, 369
Jackson, 374
Jacobian, 172

K

kabbalistic, 312
Kac, 54
Kadanoff, 368
Kamefuchi, 371
Kasner, 238, 242, 244–245, 247–248, 256–258, 260, 325, 371
kasner, 238
Kato, 329, 334, 349, 373
Ket, 271–273
ket, 12
kets, 12–13
Killing, 223, 242–243, 256–260, 263, 289, 325–326
killing, 243, 289, 307, 326
kinematic, 183, 376
kinematical, 272
kinematically, 2
kinematics, 219, 360
kinetic, 26, 140, 161, 175, 186–187, 211–212
Kinoshita, 160
Kirckoff, 82, 184
Klein, 1, 6, 24–30, 35, 40–41, 46, 59, 66, 90, 155, 227, 239, 248–249, 252, 261, 263, 279
Kleinert, 367
Koide, 221, 362, 370
Kokyuroku, 367
Kolmogorov, 57, 193
Krohn, 368
Kronecker, 281
Kruskal, 287, 289
Kumano, 367
Kungl, 370
Kuo, 367

L

Lagendre, 276
lagniappe, 339
Lagrange, 21, 24, 52, 187, 275, 277, 279
Lagrangian, 20, 23–26, 28–29, 37, 41, 44, 46–47, 56, 59, 61–62, 67,

74, 86, 88, 91, 95, 99, 106–107, 113, 119–120, 122, 125, 135–136, 139–142, 152, 161, 164, 173–175, 180–183, 185, 187–189, 191, 195, 197, 199–200, 202–204, 212–213, 220, 233, 260–261, 271, 275, 278–279, 285, 289, 330–331, 338, 353, 362, 367
Lagrangians, 56, 119
Lamb, 151
Landau, 139, 182–183, 375–376
Laplace, 52, 55, 367, 374
learn, 223
Learning, 352–353, 372
learning, 352–354
Lepton, 217, 358, 370
lepton, 140, 210
Leptons, 215–216, 219–220, 356, 358, 360–361
leptons, 152, 159–160, 217, 221, 359, 362
Lie, 1, 161, 169, 171, 185, 257–258
Light, 218, 359
light, 156, 169, 218, 220, 224, 239, 243, 247–248, 258, 260, 292, 312, 359, 362, 376
linear, 17, 54, 148, 165, 209, 290, 315, 319, 349, 373
Local, 216, 302, 358
local, 5, 25–26, 63, 118, 191, 211, 215–218, 220, 224, 231–232, 241, 243, 330, 335, 354, 356–359, 361
locally, 241, 309, 353
logarithmic, 141, 158, 199
logarithmically, 156, 207
Longitudinal, 367
longitudinal, 19, 38–39, 42, 139, 144, 154, 197–198, 201–202
Loop, 79
loop, 74, 79, 81, 83, 93, 100, 104, 121, 140, 142, 147, 150, 156, 160, 174–178, 184, 188, 194–195, 207, 209
Loops, 82
loops, 79, 82, 100, 120, 124, 155, 157, 175, 188, 194, 198
Lorentz, 20, 28–29, 37, 41, 100, 138, 189–190, 192–194, 197, 205, 211, 326–327, 331–333, 335, 347, 355, 372, 374–375
Lorentzian, 223, 226
LSZ, 29, 31, 34, 45, 48–49, 86, 99, 170

M

Maclaurin, 283
macroscopic, 118, 374
Magnetic, 376
magnetic, 1, 7, 18, 20, 151, 158
magnitude, 160, 313–314, 329, 342, 344–345
Majorana, 219, 361
Maldacena, 374
Mandelbrot, 344, 346, 350, 373
Mandelstam, 105–106
Mandl, 6, 23, 51
Manifold, 336, 374
manifold, 260, 286–287, 289–290, 311, 314, 336, 353, 355
Manifolds, 365
manifolds, 290
Map, 342–343, 345
map, 243, 313, 341–343, 345
mapped, 291
mapping, 243, 292, 329
maps, 238, 368
Martingale, 329, 335, 363, 369
Mass, 218, 360, 370, 373
mass, 2, 21, 41, 74, 87–89, 99, 106, 114, 117, 125, 127–128, 141,

144, 146–147, 149–150, 155, 165, 174–175, 177–179, 181, 183–184, 188, 194, 203, 213, 215, 218–221, 235–236, 260, 298, 302, 312, 346, 356, 360–363, 369
masses, 2, 131–132, 211, 215, 218, 220–221, 276, 285, 357, 360–363
massive, 181–182, 216–221, 244, 246, 260, 330, 358–359, 361–362, 373
massless, 177–180, 182, 186, 203, 205, 216, 244, 246, 253, 262, 282, 284, 295, 298–299, 358
Matrices, 372
matrices, 37, 142, 167–168, 212–213, 330
Matrix, 14
matrix, 33, 35, 37, 39, 44, 122, 128, 137, 155, 165, 167–168, 174, 181, 183, 188, 210–211, 218, 220, 230–231, 254, 272, 285, 317, 320–321, 330–331, 360, 362
Matsumoto, 371
Matter, 2, 351, 375
matter, 3, 5–6, 19–20, 88, 141, 169, 182, 188, 212, 217–219, 269, 297, 312, 314–315, 330–331, 335, 343, 346–348, 350, 352–353, 358–359, 361, 375
Maximal, 1, 333, 363, 365
maximal, 218, 220–221, 224, 311–314, 329, 333–335, 342, 344–346, 348–350, 352–354, 359, 362–364
maximally, 53, 218, 223, 256, 289, 359
maximization, 352–353
Maximum, 333, 349
maximum, 2, 313–314, 329, 333–335, 338, 342, 344–346, 350–352, 363
Maxwell, 7, 60
measurable, 2, 99, 273
measure, 34, 53, 56, 102–103, 133, 135–137, 140, 169, 171–172, 177, 180, 190–191, 195, 197, 313, 354
Measurement, 376
measurement, 150, 308, 311, 314, 334
measurements, 2, 219, 360
Mechanical, 56
mechanical, 2, 52, 60, 63, 223, 335, 376
Mechanics, 63, 223, 271, 312, 365, 367, 375–376
mechanics, 1–2, 9, 51–53, 60–62, 67, 89, 271, 274, 276, 281, 311–312
medium, 144
memory, 353
Meromorphic, 375
meromorphic, 350–352
mesh, 241
metric, 223, 238–239, 242–243, 247–248, 255, 279, 286–288, 294–295, 326
microcasuality, 40
MIE, 218, 220–221, 333, 335–336, 351, 353, 355, 359, 361, 363
Milne, 246–250
minimal, 87, 107, 123, 125, 127, 149, 173, 185, 262, 292, 302, 306
minimally, 217, 295, 308, 348, 359
minimization, 353
minimizes, 165
minimum, 179, 269, 354
minisuperspace, 351
Minkowski, 3, 208, 223, 233–234, 237–238, 241–252, 255–260, 262–269, 279, 284, 286, 288–289, 295–297, 302, 305, 308–309, 325–327
mirror, 252

mirrors, 259
Misner, 371
Mode, 249
mode, 9, 229, 235–237, 251, 255, 260–261, 263, 265, 281, 295, 300–302, 304–305, 309
Molecules, 376
molecules, 229, 312
Moment, 376
moment, 1, 158, 233, 252
momenta, 21, 37, 100, 111, 128, 131–132, 175, 184, 189, 209, 300
Momentum, 77, 109, 113
momentum, 21, 26–28, 33–34, 37, 45, 48, 55, 63, 66–67, 77, 80, 104, 109, 114, 117, 128, 153–154, 172, 177, 179, 199, 211, 227, 262, 272, 275, 279, 292, 296, 327, 371
monopole, 233
Moore, 371
multiscale, 363, 369
Multispecies, 161
multivalued, 47
Muon, 160, 376
muon, 2, 159, 217, 221, 358, 362
Myrberg, 311

N

Nagaoka, 374
Najmi, 371
Nambu, 120
natural, 224, 236, 241, 284, 342, 364
Nature, 365
nature, 28, 53, 60, 82, 140, 169, 203, 303, 309
neighborhood, 133
neighbouring, 24, 45
Neumann, 230
Neural, 374
neural, 352
Neuromanifold, 353
neuromanifold, 311, 353–355
neuromanifolds, 352
Neurons, 374
neutral, 204, 279, 282, 292
Neutrino, 219, 360
neutrino, 169, 210, 216–217, 219–221, 330, 358–362, 369
Neutrinos, 369
neutrinos, 2, 216–221, 299, 358–363, 370
Noether, 25, 29, 47, 161–162, 204
Noise, 350
noise, 55, 313–314, 329, 334–335, 339, 345–346, 348–350, 353–354, 363, 367
NonAbelian, 232
nonabelian, 137
nonanalyticity, 252
noncompact, 291
noncovariant, 60
nondegenerate, 12–13
nonlocal, 247
nonrenormalizable, 83, 141, 182
nonsingularity, 260
Norm, 339
norm, 34, 163, 232, 242, 273, 313, 329–330, 332–333, 335–336, 338–348, 354, 356
normalization, 8–9, 12, 29–30, 62, 74, 88, 112, 114, 121, 133, 137, 187, 263, 334–335, 338, 343, 354
normalize, 38
normalized, 14, 252, 348

O

observable, 14, 43, 273, 277, 311–312, 319
observables, 13, 29, 262–263, 272, 321
Observer, 233, 235, 241

observer, 2–3, 223–224, 241, 243, 247, 258–259, 291, 309
observers, 236, 247, 256–260, 309, 327
observes, 265
octet, 214
Octonion, 347, 355–356
octonion, 313, 329, 331, 333, 339–345, 348–349, 354–356
octonionassociativity, 355
octonionic, 313, 334, 338–339, 341–342, 344–345, 349, 353–354, 356
Octonions, 337, 339
octonions, 332–333, 336, 338–339, 341–342, 344, 354, 356, 364, 372, 375
Operator, 317, 370
operator, 12–13, 18, 27, 33–35, 39, 46, 53–54, 60, 74, 143, 162, 200–202, 206, 225, 228–230, 233, 244, 260, 262–266, 271, 273, 276, 282, 297, 299, 309, 320–321, 323–324, 334, 338
Operators, 373
operators, 9–10, 14–15, 18, 27, 29, 38–39, 44–47, 53, 63–64, 162, 181, 213, 229, 258, 261, 263, 266, 271–272, 274, 290, 317, 319–320, 327, 373
Oppenheimer, 376
Optical, 374
optical, 18
optimal, 220, 314, 352–354, 361, 369
optimality, 346
orbit, 247
orbital, 37
ordered, 53, 64, 71, 95, 132, 162, 303, 308, 341
Ordering, 27
ordering, 27, 71–72, 74, 83, 91, 272, 281, 304, 308, 321
orthochronous, 331–332
orthogonal, 14, 42, 186, 263, 265, 317, 319
orthogonalization, 319
orthonormal, 46, 245, 261, 304, 317
Orthonormality, 319
orthonormality, 38, 280
Orthonormalization, 14
oscillating, 30, 305
oscillation, 219, 360
oscillations, 34, 221, 362
Oscillator, 9
oscillator, 1, 9, 12, 15, 27–28, 55, 65–66, 84, 89, 225, 229, 263, 281
oscillators, 9, 263, 266, 368
oscillatory, 52, 55, 259, 343
Osterwalder, 232–233, 370
outgoing, 33, 244–245, 247
outlined, 107

P

Padmanabhan, 371
Palatini, 23
Papastamatiou, 6, 23, 51
parameterization, 347
parameters, 105–106, 127, 158, 175, 192, 200, 215–216, 218, 220–221, 306, 312, 329–332, 335, 338, 343, 346, 351, 353, 357, 359–364
parametric, 104
parametrizations, 223
parametrized, 225
parity, 10, 15, 213
Parker, 303, 370–371, 373
Parseval, 55, 318
Particle, 49–50, 215, 233, 292, 297–298, 308, 356, 366, 371, 375

particle, 2, 5–6, 32, 34, 38, 49, 56, 59, 61–63, 79, 88–89, 99, 111, 125, 128, 147, 158, 181, 188, 194, 215–218, 221, 233, 235–237, 241, 243, 257–258, 260, 268–269, 272, 282, 292, 297–298, 300, 302, 304, 307–309, 352, 357–359, 363
Particles, 279, 309, 366, 375
particles, 2, 5, 7, 35, 37–39, 41, 44–45, 48, 50, 189, 194, 215–216, 223, 229, 231, 234–237, 241, 244, 257, 260, 292, 296–297, 300–302, 308, 331, 356–358
Partition, 223
partition, 2, 52–53, 60, 63, 73, 118, 225, 229
Parton, 211
Path, 1, 51, 53, 55, 60, 67, 89, 132, 135, 190, 194, 330, 365, 367, 372
path, 1, 37, 51–53, 55–57, 59–60, 62, 64, 67–68, 72, 74, 84–85, 90, 103, 135–137, 152–153, 175, 188, 190–191, 193, 224, 226, 230, 232, 329, 333, 335, 338, 340, 343, 351–352, 356, 364, 367–368
PathIntegral, 367
paths, 51–52, 54–57, 62–63, 225, 333, 335–336, 352, 363
Pauli, 37
Penrose, 238, 271, 286–287, 315, 331, 372, 374, 376
Periodic, 355
periodic, 6, 8, 37, 53, 224–225, 282, 284, 286
periodicity, 224, 228, 291, 355–356
Perturbation, 312–313, 373, 375
perturbation, 1, 6, 18, 20, 74–76, 82–83, 87–88, 99, 118, 120, 170, 174–175, 203, 220, 233, 274, 299, 311–314, 329–330, 333–336, 338, 342, 344–346, 350, 352, 354, 362–363
perturbations, 347, 352
perturbative, 6–7, 52, 274, 302
phase, 13–14, 25–26, 52, 55, 57, 64–65, 135, 137, 180, 218, 263, 276, 313, 342–343, 345, 349, 352–353, 360, 364, 373
phases, 57, 342
photoelectric, 5
Photon, 43
photon, 17–18, 20–22, 42, 83, 138, 140–141, 144, 148–149, 155, 179, 181–182
Photons, 367
photons, 5, 21–22, 139, 188, 299
Pion, 27, 45
pion, 45, 279, 346
pions, 45, 204
Poincare, 234, 244, 309, 326–327
Poincaré, 368
point, 19, 23, 52–53, 55, 60–61, 63–64, 68, 70, 74, 83, 86, 88–89, 93, 97, 107–108, 111, 118, 123, 125, 128, 134, 137, 177, 181, 198–199, 215, 230, 233, 245, 263, 333, 335, 343, 356, 376
points, 64, 93–94, 258–259, 347, 377
Poisson, 20, 188–189, 274
polar, 180–182
polarizable, 144
polarization, 19, 21, 42, 143, 150, 284, 312
polarizations, 42
pole, 33, 66–67, 88–89, 102, 112, 125, 128, 144–146
Poles, 112
poles, 33, 66, 102, 128, 183
Polymer, 367
polynomial, 15, 55, 124

Popov, 135–136, 188, 190–192, 194, 369
potential, 7, 9–10, 20, 40–41, 43, 53, 78, 87, 114, 121, 151, 157–158, 169, 173–177, 179–180, 213, 224, 229, 320, 352
potentials, 41, 60, 76, 370
Potthoff, 367
Power, 356
power, 86, 93, 106, 113, 123, 158, 175, 182, 191, 195, 230–231, 278, 305, 320, 356
precipitation, 352, 375
projection, 53, 143, 169, 205, 220, 314–315, 320, 330, 338, 362, 364
projections, 330, 338
projects, 217, 314, 330, 359, 364
propagatable, 356
propagating, 364
Propagation, 364
propagation, 8, 42, 140, 169, 193, 311, 329–330, 332–333, 336, 338–340, 342, 344, 349, 351–352, 354–355, 363–364, 372
propagations, 348
Propagator, 60, 223, 225–226, 338
propagator, 2, 33, 40, 43, 51, 53, 58, 60–67, 69, 80, 83, 112, 123–125, 138–139, 147, 153–155, 172, 178, 180, 182–183, 194, 198–199, 201, 223–228, 267, 330, 333, 335, 338–340, 343, 363–364, 371
propagators, 64, 93, 104, 140, 177, 183, 194–195, 333, 363–364
proton, 211
pseudoscalar, 203
pseudoscalars, 205

Q

QCD, 189, 210, 213, 312–313
QED, 3, 5, 26, 44, 60, 83, 117–118, 139–141, 147, 149, 152, 161, 198–199, 213, 311–312
QFT, 46, 161, 249, 271, 291–292, 351, 353, 364
QM, 364
quantizable, 220, 362
Quantization, 1, 5–7, 23, 26–28, 37–38, 51, 67, 188, 190, 271, 294, 296
quantization, 1, 6–7, 9, 24, 26, 28–29, 38, 41, 46–47, 51–52, 132, 188, 195, 210, 220, 224, 252, 261, 271, 274, 290–292, 294, 362, 367
quantize, 15, 26, 38, 40, 42, 180, 188
Quantized, 370–371
quantized, 1, 5, 9, 12, 15, 33, 42, 188, 256–260, 299, 330
Quantum, 2, 5, 56, 60, 63, 139, 202, 210–211, 223, 248–249, 271, 312, 365–370, 372, 374–376
quantum, 1–3, 5–6, 9, 23, 27–28, 47, 51–53, 55–56, 59–63, 65, 67, 89, 117–118, 152, 175–176, 204, 210, 212, 217, 221, 223–224, 226, 229, 232–233, 235–236, 247, 249, 257, 262, 268, 271, 274, 276, 281, 290–291, 302, 304, 308, 311–312, 314–315, 326, 329–331, 333, 335–336, 338, 346, 351–352, 355, 358–359, 362–363, 366, 370, 372, 374–375
Quark, 372
quark, 203, 210–212, 214, 221, 346, 362
quarkonia, 211
Quarks, 210, 215–216, 219–220, 356, 358, 360–361, 373
quarks, 152, 203–205, 210–211, 213, 217, 221, 359, 362

Quaternion, 347, 372
quaternion, 329, 331–332
quaternionic, 332, 338, 351, 356, 364
Quaternions, 337, 372
quaternions, 332, 354, 356
Quigg, 366

R

Radiation, 3
radiation, 2, 5, 7–9, 15, 17, 149, 234–235, 259, 291, 302, 366
Radiative, 17, 159–160, 369
radiative, 139, 147, 159, 179, 213
Ramond, 6, 23, 51, 366
random, 329, 335, 342
range, 6, 23, 51, 101–102, 235, 268
recurrence, 12–13, 321, 355
Recursion, 13
reduce, 53, 120, 221, 272, 354, 362
Reduced, 373
reduced, 53, 144, 220–221, 339, 341, 356, 362–363
reducible, 79, 88
Reduction, 33–34, 162, 352, 372, 374
reduction, 7, 36–37, 48–49, 69, 79, 165, 314–315, 352–353, 374
Regge, 211
region, 18, 236, 238, 243–244, 251, 258, 282, 289, 352
regions, 227, 255, 282, 302, 305
regular, 47, 199, 251, 256–257, 259–260
regularity, 257
Regularization, 114, 271, 306
regularization, 2, 83, 100, 102–103, 105, 113, 149, 179, 188, 199, 201–202, 240, 303–305, 355
Regularize, 114
regularized, 106
regularizer, 60
reification, 354
reified, 355
Relativistic, 367, 370, 372–373
relativistic, 1, 8, 21, 62, 331, 351
relativistically, 6
Relativity, 280, 371–372, 374
relativity, 5, 215, 223, 311, 356
Rellich, 329, 334, 349
renormalisation, 368
Renormalizability, 182, 211
renormalizability, 83, 182–185, 188, 198, 201, 219, 314, 356, 360
Renormalizable, 83, 182, 197
renormalizable, 83, 95–96, 105, 117, 124, 141, 182–185, 188, 202, 212, 217, 313, 330–331, 359
Renormalization, 1, 86, 99, 117–119, 128, 141, 354–355, 366
renormalization, 1–2, 6, 23, 51–53, 66, 74, 82, 87–89, 95, 99, 106–107, 113–114, 117–119, 122–128, 141, 144, 146, 149–150, 177–179, 191, 195, 199–200, 213, 217, 219, 276, 284, 308, 312–313, 336, 346, 355, 359, 361, 368
renormalizations, 105
renormalize, 88, 107, 178, 355
renormalized, 88, 95, 99, 107, 113, 117, 141, 144, 150, 152, 198–200, 260, 346, 359
reparameterization, 148
Representation, 77, 216, 358, 372
representation, 1, 10, 12, 14, 26, 39, 51, 63–64, 161, 166–168, 170, 185–186, 194, 208, 210, 216–217, 226–227, 232, 238, 258–262, 272, 276, 281, 285, 317, 331, 336, 353, 357–358, 371

representations, 1, 100, 216, 218, 256–259, 338, 355, 357, 359
representative, 161
representatives, 186, 210
represented, 174, 180, 188, 256, 259–260, 264, 272, 285, 287, 320, 354
rescale, 213
rescaled, 295
rescaling, 118
reservoir, 230–231
resonances, 211
Rindler, 3, 223, 247, 251–253, 255–260, 264–268, 371–372
rotating, 105, 232
rotation, 81, 98, 112, 114, 136, 226–228, 232–233, 243–244, 285, 350
rotational, 5
rotations, 167, 174, 205, 236, 242, 327
Rydberg, 312
Ryder, 6, 23, 51, 366

S

Salam, 202, 209–210
Scalar, 285, 296
scalar, 8, 26, 28, 42, 45–47, 66–67, 90, 100, 104, 147, 161, 175, 179, 181–189, 194, 205, 233, 239, 245–247, 252, 260–262, 279–280, 282, 284–285, 290, 292, 298–299, 304, 306, 317, 319, 327
Scale, 376
scale, 30, 117–118, 125, 179, 213, 218, 221, 236, 295, 297–298, 302, 312–313, 346, 359, 362–363
scales, 118–119
Scaling, 368
scaling, 118, 211
scatter, 217, 358
scattered, 22, 33, 148
Scattering, 21, 27, 44–45
scattering, 21–22, 28, 35, 44–45, 52, 93, 99, 156, 158, 211
Schrodinger, 1, 52, 54, 60–63, 67, 104, 113, 224–225, 260, 262, 265, 272–274, 276, 281, 320, 373
Schrödinger, 60, 261–262, 367, 376
Schrôdinger, 261–262
Schroeder, 369
Schwarzschild, 287
Schwinger, 159, 303, 363
Sedenion, 339, 355
sedenion, 332, 336, 338–340, 342–344, 348, 354–355
Sedenions, 337, 339, 347
sedenions, 332, 336, 339, 343, 347, 354
Semiclassical, 120
semiclassical, 52, 120–121, 367
semigroup, 6, 52, 118
semigroups, 54
Shannon, 230, 352
signature, 223
singlet, 211, 214
singlets, 211
singular, 259, 335
singularities, 260, 269, 335, 376
singularity, 243, 260, 289
Solar, 219, 360, 369
Sommerfeld, 363, 377
spatial, 24, 57, 226, 236, 238, 243–244, 256, 258, 263, 265, 278, 290–291, 354–355
Spatially, 294
spatially, 236, 294, 298–299
spectra, 271, 312
spectral, 5, 46, 312, 372
Spectrum, 370, 373

spectrum, 5, 10–11, 27, 29, 33–34, 181, 211, 223–224, 255, 301
speed, 224, 312
speeds, 258
sphere, 101
spherical, 177
spherically, 373
Spin, 299
spin, 2–3, 37–40, 45, 55, 63, 139, 183, 215, 232, 299, 331, 356
Spinor, 47, 369, 372
spinor, 5, 38, 47, 100, 139, 232, 274, 331, 353
spinorial, 6, 336
spinors, 204, 331
Split, 338
split, 88, 169, 179, 207, 230, 254, 332, 335, 347, 375
splits, 120
Splitting, 193
splitting, 20, 88, 257, 259, 337
splittings, 88, 337
Spontaneous, 163, 369
spontaneous, 19, 164–165, 174, 179–181, 185
spontaneously, 163, 166, 183, 203, 219, 360
States, 256, 371
states, 12–14, 17–19, 33–34, 42, 44–45, 52, 151, 162, 219, 224, 231, 244, 256–260, 262, 265, 268–269, 271–272, 284, 302–303, 356, 360, 372–373
Static, 291–292
static, 62, 150–151, 158, 213, 235–236, 302, 305, 348
statics, 226
stationary, 14, 23, 52, 55, 68, 134
Strong, 220, 361, 366
strong, 52, 54, 210–211, 213, 215, 218–219, 313, 357, 360–361, 369
Sumino, 370
superconformal, 374
supergravity, 374
supermassive, 219, 361
superposition, 62, 241, 302
Superpositions, 257
superpositions, 264
surface, 24, 35, 101, 239, 244–246, 251, 261, 290–291, 293, 353
surfaces, 238–239, 247, 249, 252, 291
symmetric, 65, 143, 157–158, 196, 208, 215, 256, 263, 266, 279, 357, 373
Symmetries, 202, 289
symmetries, 29, 140, 161, 177, 213, 289, 304
symmetrized, 280
Symmetry, 163, 166, 213, 307, 369–370
symmetry, 46, 120, 132, 139, 142, 163–164, 166, 169, 171–174, 177, 179–183, 185, 203–205, 211, 213, 236, 242, 369

T

Tachiki, 371
tadpole, 95, 101, 113
Takahashi, 154
Taylor, 92, 165, 177, 195, 197
telescoped, 342
telescoping, 341–342, 344
Temperature, 373
temperature, 53, 169, 224–226, 228, 234, 268, 302, 336, 346, 355
Thermal, 223, 228, 365–366, 370
thermal, 2, 63, 223–224, 228–231, 234–235, 255, 268, 291, 335, 355
Thermality, 354
thermality, 2, 223–224, 336, 355
thermally, 235

Thermo, 371
thermodynamic, 3, 60, 353
thermodynamically, 3
Thermodynamics, 230–232, 365
thermodynamics, 2, 63, 223–224, 229, 231
topological, 184, 198, 212
trajectories, 211, 234, 258
trajectory, 233–234, 311
Transform, 6, 119, 182, 227, 250, 276
transform, 32, 55, 140, 166, 181, 186, 191, 202, 247, 251–252, 308, 318, 331–332
Transition, 234
transition, 1, 17, 19, 233–235, 350
Translation, 109, 323
translation, 77, 137, 234, 243, 256, 259, 278
Translations, 374
translations, 236, 242, 247, 278–279, 327
transmission, 2, 314, 333, 335, 346, 348–350, 352
transmittable, 313
transmutation, 179
transverse, 8, 18–19, 42, 139, 198, 201
Trigintaduonion, 339, 346, 352, 373, 375
trigintaduonion, 311, 313, 329–330, 332–335, 338–340, 345, 350, 352–356, 363
Trigintaduonions, 337, 339
trigintaduonions, 329, 332–333, 335–336, 342, 347, 351, 363–364
trioctonionic, 334
Trotter, 53–54

U

Umezawa, 366, 371
uncertainty, 10, 236
unitarity, 34, 107, 211, 329, 333
Unitary, 182–183, 272
unitary, 34, 107, 162, 171, 182, 186, 213, 225, 230, 238, 265, 272, 326–327, 329–333, 364
Universal, 334, 336, 340, 344, 352
universal, 215, 224, 291, 309, 311, 313, 329, 336, 338, 342, 346, 350–351, 357, 364, 373–374
Universality, 311, 335, 372
universality, 118, 313
Universe, 169, 215, 291–292, 338, 356, 363, 376
universe, 218–219, 236, 292, 297–299, 301–302, 304, 355, 359, 361
Universes, 371
universes, 238, 299
Unruh, 3, 224, 251, 257, 365
Utiyama, 185, 369

V

variation, 23, 25, 90, 117–118, 136, 275, 293
variational, 63, 275, 353, 364
Variations, 373
variations, 18, 23, 278
Vector, 9, 223–224, 242, 272, 325, 331, 369
vector, 7–9, 20, 26–27, 40, 44, 167, 180–182, 185–187, 189, 210, 242–243, 256–259, 263, 272–274, 277, 289–290, 304, 306–307, 326, 331
Vectors, 242, 325–326
vectors, 10, 13–14, 42, 174, 242–243, 285, 289, 304, 318, 321, 326, 331
velocities, 260

Vertex, 369
vertex, 94, 147, 150, 183–184, 199, 201–202, 211, 213
vertices, 78, 82, 94, 96, 120, 175, 183, 198–199, 209, 213

W

Walecka, 375
Ward, 121, 142, 149, 152–155, 172–173, 188, 195, 197, 200, 202, 204–206, 208–209
Wave, 14
wave, 5–7, 9, 13, 27, 30, 34, 36, 41, 114, 144, 240, 247, 249–250, 262–263, 265–268, 272, 299, 317, 319, 331
wavecollapse, 315
wavefunction, 54, 106, 141, 146, 178, 199, 230, 244–245, 330
wavefunctions, 17, 36, 263
wavelength, 17
wavelengths, 18
wavepacket, 252
wavepackets, 252
Weinberg, 180, 182, 202, 209–210, 218, 360, 369
Wick, 2, 44, 53, 71, 74, 81, 98, 105, 112, 114, 226–228, 232–233, 348
Wiener, 55, 368
Wightman, 163, 232–235, 370

Y

Yang, 185, 187–190, 193, 195, 197–200, 202–203, 369
Yukawa, 218, 360

Z

Zuber, 366